DISCARDED

Complex Analysis and Algebraic Geometry

Complex Analysis and Algebraic Geometry

A Volume in Memory of
Michael Schneider

Editors

Thomas Peternell
Frank-Olaf Schreyer

Walter de Gruyter · Berlin · New York 2000

Editors
Thomas Peternell, Frank-Olaf Schreyer
Fakultät für Mathematik und Physik
Universität Bayreuth
95440 Bayreuth, Germany

1991 Mathematics Subject Classification:
14 E25, H50, J25, J28, J30, J40, J45, J80, K10, N30, P25, P99;   32 G13, J25, J27, M05, Q15, Q25, Q30, Q55, Q57;   53 D20.

Keywords:
tangent and cotangent bundle, real algebraic manifold, abelian variety, Peterson-Weil metric, hermitian, metric, symplectic geometry, Kähler manifold, projective manifold, fundamental group, Calabi-Yau threefold, K3 surface, Godeaux surface, Seiberg-Witten invariant, nodal curve.

∞ Printed on acid-free paper which falls within the guidelines of the ANSI to ensure permanence and durability.

Library of Congress – Cataloging-in-Publication Data

Complex analysis and algebraic geometry: a volume in memory of Michael Schneider / editors, Thomas Peternell, Frank-Olaf Schreyer.
    p.   cm.
One contribution in French.
Includes bibliographical references.
ISBN 3110162040 (alk. paper)
1. Functions of complex variables. 2. Mathematical analysis. 3. Geometry, Algebraic.   I. Peternell, Th. (Thomas), 1954–   II. Schreyer, Frank-Olaf
QA331.7.C66   1999
515'.9–dc21
                                              99-052677

Die Deutsche Bibliothek – Cataloging-in-Publication Data

Complex analysis and algebraic geometry : a volume in memory of Michael Schneider / ed. Thomas Peternell ; Frank-Olaf Schreyer. – Berlin ; New York : de Gruyter, 2000
ISBN 3-11-016204-0

© Copyright 2000 by Walter de Gruyter GmbH & Co. KG, D-10785 Berlin
All rights reserved, including those of translation into foreign languages. No part of this book may be reproduced or transmitted in any form or by any means, electronic or mechanical, including photocopy, recording or any information storage and retrieval system, without permission in writing from the publisher.
Printed in Germany.
Typeset using the author's T_EX files: I. Zimmermann, Freiburg. Printing: WB-Druck GmbH & Co., Rieden/Allgäu. Binding: Lüderitz & Bauer, Berlin. Cover design: Thomas Bonnie, Hamburg.

*Am 29. August 1997 verstarb unser Freund Michael Schneider.
Dieser Band und jeder Beitrag sind seinem Andenken gewidmet.*

Michael Schneider 18. 5. 1942 – 29. 8. 1997

Table of Contents

Lucian Bădescu, Mauro C. Beltrametti and Paltin Ionescu
Almost-lines and quasi-lines on projective manifolds 1

Daniel Barlet et Jón Magnusson
Transfert de métrique 29

Wolf P. Barth
On the classification of K3 surfaces with nine cusps 41

Arnaud Beauville
Complex manifolds with split tangent bundle 61

Mauro C. Beltrametti, Alan Howard, Michael Schneider and Andrew J. Sommese
Projections from subvarieties 71

Indranil Biswas and Georg Schumacher
Generalized Petersson–Weil metric on the Douady space
of embedded manifolds 109

Fabrizio Catanese and Roberto Pignatelli
On simply connected Godeaux surfaces 117

Ciro Ciliberto, Angelo Felice Lopez, and Rick Miranda
On the Wahl map of plane nodal curves 155

Lawrence Ein, Bo Ilic and Robert Lazarsfeld
A remark on projective embeddings of varieties with non-negative
cotangent bundles 165

David Garber and Mina Teicher
The fundamental group's structure of the complement of some
configurations of real line arrangements 173

Peter Heinzner and Alan Huckleberry
Kählerian structures on symplectic reductions 225

Klaus Hulek
Nef divisors on moduli spaces of abelian varieties 255

Klaus Hulek and Kristian Ranestad
Abelian surfaces with two plane cubic curve fibrations
and Calabi–Yau threefolds 275

János Kollár
Real algebraic threefolds IV. Del Pezzo fibrations 317

Christian Okonek and Andrei Teleman
Seiberg–Witten invariants for 4-manifolds with $b_+ = 0$ 347

Jeroen G. Spandaw
A geometric proof of Ax' theorem 359

Sheng-Li Tan and Eckart Viehweg
A note on the Cayley–Bacharach property for vector bundles 361

The scientific work of Michael Schneider *by Thomas Peternell* 375

Michael Schneider – an alpine vita *by Ulf Persson* 389

Lectures of the Symposium 397

List of Authors and Participants 399

Almost-lines and quasi-lines on projective manifolds

Lucian Bădescu, Mauro C. Beltrametti and Paltin Ionescu

1. Introduction

Let X be a fixed complex projective manifold of dimension $n \geq 2$, and let Y be a smooth closed (connected) subvariety of X. For every integer $m \geq 0$ denote by $X(m)$ the m-th infinitesimal neighbourhood of Y in X, i.e. $X(m)$ is the algebraic scheme over $\mathbb{C}$ whose underlying topological space coincides with the underlying topological space of Y, and whose structural sheaf $\mathcal{O}_{X(m)}$ is by definition $\mathcal{O}_X / I_Y^{m+1}$, where $I_Y \subset \mathcal{O}_X$ denotes the ideal sheaf of Y in X. Of course, $Y = X(0)$. For every integer $m \geq 0$ we may consider the natural restriction maps

$$\alpha_m : \operatorname{Pic}(X) \to \operatorname{Pic}(X(m))$$

One may ask under which conditions the restriction map $\alpha_1 : \operatorname{Pic}(X) \to \operatorname{Pic}(X(1))$ is surjective.

There are many examples of submanifolds of $\mathbb{P}^n$ of dimension ≥ 2 for which the map α_1 is surjective. These include the following:

(i) If Y is a smooth complete intersection of dimension ≥ 2 in $X := \mathbb{P}^n$, or

(ii) If Y is a smooth subvariety of dimension $> \frac{n}{2}$ in $\mathbb{P}^n$ ($n \geq 3$).

Indeed, using the results of [E-G-P-S] (see also [Br]), it turns out that the surjectivity of the map α_1 is equivalent to verifying that $\operatorname{Pic}^\tau(Y) = 0$ (i.e. every numerically trivial line bundle on Y is trivial), and that $\mathcal{O}_Y(1) := \mathcal{O}_{\mathbb{P}^n}(1)|Y$ is not divisible in $\operatorname{Pic}(Y)$. But these two last properties follow from Lefschetz theorems in case (i), and from Barth-Larsen theorems in case (ii) (see [Ha2]).

Motivated by some work by Griffiths and Harris [G-H] and d'Almeida [dA], we shall consider the case when Y is a smooth projective curve. Indeed, in [G-H], pp. 698–699, the authors reinterpreted in modern language the classical condition of Reiss concerning the existence of a plane curve of degree d in $\mathbb{P}^2$ cutting out d different points of a fixed line ℓ with prescribed tangents and second-order conditions. This involves considering the Picard group of the second infinitesimal neighbourhood of the curve ℓ in $\mathbb{P}^2$. Later on, in connection with this problem, d'Almeida studied in

[dA] conditions under which the map

$$\alpha_1 : \mathrm{Pic}(X) \to \mathrm{Pic}(X(1))$$

is surjective in the case when Y is a curve with positive selfintersection in a smooth projective surface X, and showed that this happens essentially when Y is a line in $\mathbb{P}^2$.

In Section 2 we study the surjectivity of the map $\alpha_1 : \mathrm{Pic}(X) \to \mathrm{Pic}(X(1))$ when Y is a smooth connected curve in a projective manifold X of arbitrary dimension $n \geq 2$, whose normal bundle $N_{Y|X}$ is ample. The main result here is Theorem (2.1), which tells us that α_1 is surjective if and only if the following three conditions hold: (i) $Y \cong \mathbb{P}^1$, (ii) $N_{Y|X} \cong (n-1)\mathcal{O}_{\mathbb{P}^1}(1)$, and (iii) the map $\alpha_0 : \mathrm{Pic}(X) \to \mathrm{Pic}(Y)$ is surjective. As a corollary of this result we recover d'Almeida's theorem quoted above (see Corollary (2.3)). One may ask to what extent d'Almeida's result can be generalized to dimensions $n \geq 3$. We show by an example that this is not the case for any $n \geq 3$ (see Example (2.7)). However, under some additional assumptions, we prove that $(X, Y) \cong (\mathbb{P}^n, \text{line})$ (see Corollary (2.4) and Theorem (2.6)).

In Section 3 we undertake a systematic study of those projective manifolds containing curves Y satisfying the conditions (i) and (ii) above (which we call *quasi-lines*), or curves Y satisfying all three conditions (i), (ii), and (iii) (which we call *almost-lines*). To our surprise, there are large classes of 3-folds containing quasi-lines but no almost-lines (e.g. some Fano 3-folds of index 2 with Picard group $\mathbb{Z}$, see Theorem (3.2)).

We prove that every projective manifold containing a quasi-line (resp. an almost-line) is filled up by such curves passing through a general point of it (see Theorems (3.5) and (3.7)); in particular, they are very special examples of rationally connected manifolds in the sense of Campana [C] and Kollár–Miyaoka–Mori [K-M-M]. We also show that these classes of manifolds (which we call *quasi-linearly connected*, resp. *almost-linearly connected*) are stable under small deformations, but, unlike the rationally connected manifolds, are not stable under arbitrary deformations.

In the fourth section we introduce and study a special class of rational manifolds, which we call *strongly rational*. A projective n-fold X is *strongly rational* if X contains a Zariski open subset which is isomorphic to a Zariski open subset V of $\mathbb{P}^n$ such that $\mathrm{codim}_{\mathbb{P}^n}(\mathbb{P}^n \setminus V) \geq 2$. For example, every projective n-fold which dominates birationally $\mathbb{P}^n$ is strongly rational. We also give very simple examples of rational manifolds which are not strongly rational. Hironaka's resolution results show however that any rational manifold is dominated by a strongly rational one. By using the geometry of quasi-lines as investigated in the previous section we are able to characterize strongly rational manifolds geometrically as follows. A projective n-fold X is strongly rational if and only if X contains a quasi-line Y such that there is a divisor D on X with the following two properties: $(D \cdot Y) = 1$ (hence Y is an almost-line) and $\dim |D| \geq n$ (see Theorem (4.4)). As a first application of Theorem (4.4), we prove a "birational version" of Kobayashi–Ochiai's theorem (see Corollary (4.6)), which improves Corollary (2.4). It is easy to see that a projective surface S is strongly rational if and only if S dominates the projective plane. One can ask whether the same holds true in higher dimensions. Another application of Theorem (4.4) is Example (4.7), which shows that in any dimension $n \geq 3$ this is not the case.

In the last section we discuss some open questions.

After the paper was made we become aware of [O] and [L] where 3-dimensional quasi-linearly connected manifolds ("twistor spaces" in the terminology of [O] and [L]) of low degree with respect to a given polarization where studied from a different point of view (see also [Hi] where twistor spaces were introduced for the first time).

Throughout this paper we shall work with varieties or schemes defined over the field $\mathbb{C}$ of complex numbers. The terminology and notations used are standard.

Acknowledgements. A first version of this paper was written in July 1997, when the first two named authors were visiting the Max-Planck-Institut für Mathematik, Bonn. They thank this institution for financial support and excellent conditions to work. The first and third named authors acknowledge financial support from the National Swiss Funds for Scientific Research during their visit at the Department of Mathematics of the University of Geneva, where they enjoyed hospitality and stimulating atmosphere in the final stage of the paper.

We thank the referee for pointing out that the previous formulation of Lemma (4.5) was incorrect.

2. The Picard group of the first infinitesimal neighbourhood of a curve in an n-fold

In the notations of the introduction, the first result is the following.

Theorem (2.1). *Let X be a projective manifold of dimension $n \geq 2$, and let Y be a smooth curve in X. Assume that the normal bundle $N_{Y|X}$ of Y in X is ample. Then the map $\alpha_1 : \mathrm{Pic}(X) \to \mathrm{Pic}(X(1))$ (from the introduction) is surjective if and only if $Y \cong \mathbb{P}^1$, $N_{Y|X} \cong (n-1)\mathcal{O}_{\mathbb{P}^1}(1)$, and the map $\alpha_0 : \mathrm{Pic}(X) \to \mathrm{Pic}(X(0)) = \mathrm{Pic}(Y)$ is surjective.*

Proof. Assume first $Y \cong \mathbb{P}^1$, $N_{Y|X} \cong (n-1)\mathcal{O}_{\mathbb{P}^1}(1)$ and that the map α_0 is surjective. Considering the truncated exponential sequence

$$0 \to N^*_{Y|X} \to \mathcal{O}^*_{X(1)} \to \mathcal{O}^*_Y \to 1,$$

we get the exact sequence of cohomology

$$H^1(Y, N^*_{Y|X}) \to \mathrm{Pic}(X(1)) \to \mathrm{Pic}(Y) \to H^2(Y, N^*_{Y|X}).$$

The last space is zero because Y is a curve, and the first space is also zero by our hypotheses. Therefore the restriction map $\beta : \mathrm{Pic}(X(1)) \to \mathrm{Pic}(Y)$ is an isomorphism. Since $\alpha_0 = \beta \circ \alpha_1$, it follows that α_1 is surjective.

Conversely, assume that the map α_1 is surjective. If Z is a projective scheme over $\mathbb{C}$, let us denote by $\underline{\operatorname{Pic}}^0(Z)$ the Picard scheme of Z (see [FGA] for the general theory of the Picard scheme). Then the surjectivity of α_1 implies that the natural map of Picard schemes $\underline{\operatorname{Pic}}^0(X) \to \underline{\operatorname{Pic}}^0(X(1))$ is also surjective. But $\underline{\operatorname{Pic}}^0(X)$ is an abelian variety because X is a smooth projective variety over $\mathbb{C}$. Therefore $\underline{\operatorname{Pic}}^0(X(1))$ is also an abelian variety.

On the other hand, the above truncated exponential sequence yields the exact sequence of cohomology

$$H^0(\mathcal{O}^*_{X(1)}) \to H^0(\mathcal{O}^*_Y) \to H^1(Y, N^*_{Y|X})$$
$$\to \operatorname{Pic}(X(1)) \to \operatorname{Pic}(Y) \to H^2(Y, N^*_{Y|X}) = 0.$$

Since $H^0(\mathcal{O}^*_Y)$ consists only of non-zero constants, the first map is surjective, and therefore we get the exact sequence

$$0 \to H^1(Y, N^*_{Y|X}) \to \operatorname{Pic}(X(1)) \to \operatorname{Pic}(Y) \to 0.$$

This yields the exact sequence of algebraic groups

$$0 \to H^1(Y, N^*_{Y|X}) \to \underline{\operatorname{Pic}}^0(X(1)) \to \underline{\operatorname{Pic}}^0(Y) \to 0,$$

where $H^1(Y, N^*_{Y|X})$ is regarded as a product of $h^1(Y, N^*_{Y|X})$ copies of the additive group $\mathbb{G}_a$. Since we just proved that $\underline{\operatorname{Pic}}^0(X(1))$ is an abelian variety and since $H^1(Y, N^*_{Y|X})$ is a linear algebraic group, this implies that

(1) $$H^1(Y, N^*_{Y|X}) = 0.$$

Applying duality on Y, (1) becomes

(2) $$H^0(Y, N_{Y|X} \otimes \omega_Y) = 0,$$

and by Riemann–Roch we get

(3) $$h^0(Y, N_{Y|X} \otimes \omega_Y) \geq \deg(N_{Y|X} \otimes \omega_Y) + (n-1)(1-g)$$
$$= \deg(N_{Y|X}) + (g-1)(n-1),$$

where g is the genus of Y. Since by hypothesis $N_{Y|X}$ is ample, $\deg(N_{Y|X}) > 0$, whence the right hand side is > 0 when $g \geq 1$. Therefore by (2) and (3) we get that $g = 0$, i.e. $Y \cong \mathbb{P}^1$.

Then by a theorem of Grothendieck [Gr], there are integers a_i, $i = 1, \ldots, n-1$, such that

$$N_{Y|X} \cong \mathcal{O}(a_1) \oplus \cdots \oplus \mathcal{O}(a_{n-1}),$$

where $\mathcal{O}(a) := \mathcal{O}_{\mathbb{P}^1}(a)$, for each $a \in \mathbb{Z}$. Since by hypothesis $N_{Y|X}$ is ample, $a_i > 0$, for each $i = 1, \ldots, n-1$.

Recalling (1) and the explicit cohomology of the projective line we get $a_1 = \cdots = a_{n-1} = 1$, which finishes the proof of the theorem. □

Example (2.2). The ampleness assumption for the normal bundle $N_{Y|X}$ in Theorem (2.1) is essential, as the following example shows.

Fix an arbitrary smooth projective curve C of genus $g \geq 0$, and let L be a line bundle on C of degree $d \leq 2 - 2g$. Set $E := \mathcal{O}_C \oplus L$ and $X := \mathbb{P}(E)$, and let $\pi : X \to C$ denote the canonical projection. Let $i : Y \hookrightarrow X$ be the section of π corresponding to the canonical map $E = \mathcal{O}_C \oplus L \to L$. Then the normal bundle of Y in X is isomorphic to L (and in particular, $Y^2 = d \leq 2 - 2g$). Since $\pi \circ i = \mathrm{id}_C$, we get $i^* \circ \pi^* = \mathrm{id}$, and in particular, the restriction map $i^* : \mathrm{Pic}(X) \to \mathrm{Pic}(Y)$ is surjective.

On the other hand, consider the exact sequence of cohomology

$$H^1(Y, N^*_{Y|X}) = H^1(Y, L^{-1}) \to \mathrm{Pic}(X(1)) \to \mathrm{Pic}(Y) \to H^2(Y, L^{-1}) = 0.$$

By duality $H^1(Y, L^{-1}) \cong H^0(Y, L \otimes \omega_Y)^*$, and if $L \not\cong \omega_Y^{-1}$, then this last space is zero under the above assumptions. Thus, if $d \leq 2 - 2g$ and $L \not\cong \omega_Y^{-1}$, the map $\mathrm{Pic}(X(1)) \to \mathrm{Pic}(Y)$ is an isomorphism.

Putting everything together we get the following

Fact. Under the above assumptions, the map $\alpha_1 : \mathrm{Pic}(X) \to \mathrm{Pic}(X(1))$ is surjective if $d \leq 2 - 2g$ and $L \not\cong \omega_Y^{-1}$.

In conclusion, under the absence of the ampleness assumption of the normal bundle, there are examples of curves of arbitrary genus Y on a smooth projective surface X such that the map $\alpha_1 : \mathrm{Pic}(X) \to \mathrm{Pic}(X(1))$ is surjective.

The first consequence of Theorem (2.1) is the following result, which is a slight improvement of the main result of d'Almeida's work [dA].

Corollary (2.3) (d'Almeida [dA]). *Let X be a smooth projective surface containing a closed smooth connected curve Y such that $Y^2 > 0$. Assume that the map $\alpha_1 : \mathrm{Pic}(X) \to \mathrm{Pic}(X(1))$ is surjective. Then there is a birational morphism $f : X \to \mathbb{P}^2$ such that f is an isomorphism in a Zariski open neighbourhood of Y in X and $f(Y)$ is a line in $\mathbb{P}^2$.*

Proof. By Theorem (2.1) the surjectivity of the map α_1 implies $Y \cong \mathbb{P}^1$ and $Y^2 = 1$. Then it is well known (and easy to see) that the linear system $|Y|$ is base point free and yields a birational morphism $f : X \to \mathbb{P}^2$ with the desired properties. $\square$

As another consequence of Theorem (2.1) and of the general adjunction theory (see e.g. [B-S]) we get the following result.

Corollary (2.4). *Let X be a smooth projective variety of dimension $n \geq 2$ such that $\mathrm{Pic}(X) \cong \mathbb{Z}$, and let Y be a smooth connected curve in X with ample normal bundle such that the restriction map $\alpha_1 : \mathrm{Pic}(X) \to \mathrm{Pic}(X(1))$ is surjective. Then $X \cong \mathbb{P}^n$ and Y is a line in $\mathbb{P}^n$.*

Proof. By Theorem (2.1), the surjectivity of the map α_1 implies that $Y \cong \mathbb{P}^1$, $N_{Y|X} \cong (n-1)\mathcal{O}(1)$ and that the map $\alpha_0 : \text{Pic}(X) \to \text{Pic}(Y)$ is surjective. In particular, there is a line bundle $L \in \text{Pic}(X)$ such that $L|Y \cong \mathcal{O}(1)$ (i.e. $(L \cdot Y) = 1$). Since $\text{Pic}(X) \cong \mathbb{Z}$ and the restriction $L|Y$ is ample, we infer that L is ample on X. From $N_{Y|X} \cong (n-1)\mathcal{O}(1)$ we get $\det(N_{Y|X}) \cong \mathcal{O}(n-1)$. Then the adjunction formula yields in our situation $-(K_X \cdot Y) = n+1$ (where K_X denotes the canonical class on X). This shows in particular that the line bundle $K_X + (n+1)L$ is not ample (because its restriction to Y is trivial). Then applying a general result from the adjunction theory (see e.g. [B-S], (7.2.1)), we infer that $X \cong \mathbb{P}^n$ and $L \cong \mathcal{O}_{\mathbb{P}^n}(1)$ and, since $(L \cdot Y) = 1$, Y is a line.

Alternatively, an even more direct argument is to notice that, since $\text{Pic}(X) \cong \mathbb{Z}$ and $(K_X + (n+1)L)|Y \cong \mathcal{O}_Y$, it follows that $K_X + (n+1)L$ is numerically trivial, and one can conclude by the Kobayashi–Ochiai characterization of the projective space (see e.g. [B-S]). □

Remark (2.5). 1) The proof of Corollary (2.4) yields in fact the following stronger statement. Let $Y \cong \mathbb{P}^1$ be a curve on the projective n-fold X (without the hypothesis that $\text{Pic}(X) \cong \mathbb{Z}$) such that $N_{Y|X} \cong (n-1)\mathcal{O}(1)$. Assume moreover that there exists an *ample* line bundle $L \in \text{Pic}(X)$ such that $(L \cdot Y) = 1$ (in particular, the map α_0 is surjective, and so, by Theorem (2.1), the map α_1 is also surjective). Then $X \cong \mathbb{P}^n$ and Y is a line.

2) Corollary (2.4) applied to $X = \mathbb{P}^n$ yields the statement that a smooth connected curve Y in $\mathbb{P}^n$ has the map α_1 surjective if and only if Y is a line in $\mathbb{P}^n$. However, this fact can be easily proved also directly.

Theorem (2.6). *Let X be a projective manifold of dimension $n \geq 2$. Let E be an ample vector bundle over X of rank $n-1$, and let $s \in H^0(X, E)$ be a section such that its zero locus $Y := Z(s)$ is a smooth curve. Assume that the map $\alpha_1 : \text{Pic}(X) \to \text{Pic}(X(1))$ is surjective. Then $(X, E) \cong (\mathbb{P}^n, (n-1)\mathcal{O}_{\mathbb{P}^n}(1))$.*

Proof. Since in this situation $N_{Y|X} \cong E|Y$, we infer that $N_{Y|X}$ is ample (because E is ample). Then by Theorem (2.1) the surjectivity of the map α_1 implies that $Y \cong \mathbb{P}^1$ and $N_{Y|X} \cong (n-1)\mathcal{O}_{\mathbb{P}^1}(1)$. Thus by the adjunction formula we conclude that $K_X + \det(E)$ is not numerically effective.

Note that $H^1(X, \mathcal{O}_X) = 0$ (see also Remark (3.8)). Indeed, otherwise, consider the (non trivial) Albanese morphism $f : X \to \text{Alb}(X)$. Since Y is a rational curve and $\text{Alb}(X)$ is a positive dimensional abelian variety, $f(Y)$ is a point of $\text{Alb}(X)$. But this contradicts the ampleness of the normal bundle $N_{Y|X}$.

Therefore we can apply [Y-Z], Theorem 3, to deduce that (X, E) is isomorphic either to $(\mathbb{P}^n, (n-1)\mathcal{O}_{\mathbb{P}^n}(1))$, or to one of the following:
(i) $(\mathbb{P}^n, \mathcal{O}_{\mathbb{P}^n}(2) \oplus (n-2)\mathcal{O}_{\mathbb{P}^n}(1))$, or
(ii) $(\mathcal{Q}, (n-1)\mathcal{O}_\mathcal{Q}(1))$, where $\mathcal{Q}$ is a hyperquadric in $\mathbb{P}^{n+1}$.

(iii) $(\mathbb{P}(F), \bigoplus_{j=1}^{n-1}(\mathcal{O}_{\mathbb{P}(F)}(1) \otimes \pi^*(\mathcal{O}_{\mathbb{P}^1}(b_j))))$, where F is a vector bundle of rank n on $\mathbb{P}^1$, $\pi : \mathbb{P}(F) \to \mathbb{P}^1$ is the canonical projection, and $b_j \in \mathbb{Z}$ for each $j = 1, \ldots, n-1$.

In case (i) we have $N_{Y|X} \cong \mathcal{O}_{\mathbb{P}^1}(2) \oplus (n-2)\mathcal{O}_{\mathbb{P}^1}(1)$, while in case (ii), $N_{Y|X} \cong (n-1)\mathcal{O}_{\mathbb{P}^1}(2)$. Therefore, by what we said above, cases (i) and (ii) are ruled out.

Note that in case $(\mathbb{P}^n, (n-1)\mathcal{O}_{\mathbb{P}^n}(1))$ the map α_1 is surjective.

It remains to rule out the case (iii). By a theorem of Grothendieck, F is of the form $F = \mathcal{O}_{\mathbb{P}^1}(a_1) \oplus \mathcal{O}_{\mathbb{P}^1}(a_2) \oplus \cdots \oplus \mathcal{O}_{\mathbb{P}^1}(a_n)$, with $a_1 \leq a_2 \leq \cdots \leq a_n$. By tensoring F with $\mathcal{O}_{\mathbb{P}^1}(-a_1)$, we may also assume that $a_1 = 0$. Set $L_j := \mathcal{O}_{\mathbb{P}(F)}(1) \otimes \pi^*(\mathcal{O}_{\mathbb{P}^1}(b_j))$, for each $j = 1, \ldots, n-1$.

Note that, since $H^0(X, E) = \bigoplus_{j=1}^{n-1} H^0(X, L_j)$, $Y = Z(s_1, \ldots, s_{n-1})$, with $s_j \in H^0(X, L_j)$, is the transversal intersection of members $A_j \in |L_j|$, $j = 1, \ldots, n-1$. Using this observation we easily infer that Y is a section of π.

The normal sequence of Y in $X = \mathbb{P}(F)$,

$$0 \to T_Y \cong \mathcal{O}_{\mathbb{P}^1}(2) \to T_X|Y \to N_{Y|X} \to 0$$

gives

$$\deg(T_X|Y) = \deg(N_{Y|X}) + 2 = (n-1) + 2 = n+1.$$

Moreover, the Euler sequence of $X = \mathbb{P}(F)$ restricted to Y yields

$$0 \to \mathcal{O}_Y \to [\pi^*(F^*) \otimes \mathcal{O}_{\mathbb{P}(F)}(1)]|Y \to T_{\mathbb{P}(F)|\mathbb{P}^1}|Y \to 0,$$

where $T_{\mathbb{P}(F)|\mathbb{P}^1}$ is the tangent bundle relative to π. Set $d := a_2 + \cdots + a_n = \deg(F)$. Since the middle term is isomorphic to $\mathcal{O}_{\mathbb{P}^1}(d) \oplus \mathcal{O}_{\mathbb{P}^1}(d - a_2) \oplus \cdots \oplus \mathcal{O}_{\mathbb{P}^1}(d - a_n)$ we get

$$\deg(T_{\mathbb{P}(F)|\mathbb{P}^1}|Y) = nd - (a_2 + \cdots + a_n).$$

Here we used the well known formula $(\mathcal{O}_{\mathbb{P}(F)}(1)^n) = d$. We also have the canonical exact sequence

$$0 \to T_{\mathbb{P}(F)|\mathbb{P}^1}|Y \to T_{\mathbb{P}(F)}|Y \to \pi^*(T_{\mathbb{P}^1})|Y \cong \mathcal{O}_{\mathbb{P}^1}(2) \to 0,$$

which gives

$$\deg(T_{\mathbb{P}(F)}|Y) = \deg(T_{\mathbb{P}(F)|\mathbb{P}^1}|Y) + 2 = nd - (a_2 + \cdots + a_n) + 2.$$

The comparison of the two formulae yields $(n-1)(a_2 + \cdots + a_n) = n - 1$, or else, $d = a_2 + \cdots + a_n = 1$ $(n \geq 2)$.

On the other hand, first we have

$$(L_j \cdot Y) = (\mathcal{O}_{\mathbb{P}(F)}(1) \otimes \pi^*(\mathcal{O}_{\mathbb{P}^1}(b_j))) \cdot Y = d + b_j,$$

and second, since $E|Y \cong N_{Y|X} \cong (n-1)\mathcal{O}_{\mathbb{P}^1}(1)$, $L_j|Y \cong \mathcal{O}_{\mathbb{P}^1}(1)$, which implies $(L_j \cdot Y) = 1$. Therefore $d + b_j = 1$, and since $d = 1$, we get $b_j = 0$, for each $j = 1, \ldots, n-1$. So, $E \cong \bigoplus_{j=1}^{n-1} \mathcal{O}_{\mathbb{P}(F)}(1)$, which implies that F is ample. But this is absurd since $a_1 = 0$. □

Corollary (2.4) and Theorem (2.6) might suggest that Corollary (2.3) extends to arbitrary dimensions $n \geq 3$. However this is not the case even if $n = 3$ because of the following example.

Example (2.7). Start with $X' := \mathbb{P}^n$, for each $n \geq 3$, and consider the group G (of order $n + 1$) generated by the automorphism $\sigma : \mathbb{P}^n \to \mathbb{P}^n$ given by

$$\sigma([x_0, x_1, \ldots, x_n]) = [x_0, \zeta x_1, \zeta^2 x_2, \ldots, \zeta^n x_n],$$

where $\zeta \in \mathbb{C}$ is a (fixed) primitive root of unity of order $n + 1$. Let U' be the open subset of $X' := \mathbb{P}^n$ in which G acts freely. Consider the line L of X' given by the equations:

$$x_0 = x_1, \quad x_2 = x_3, \quad x_4 = \cdots = x_n = 0.$$

It is easily checked that the $n + 1$ lines ξL (for each $\xi \in G$) are mutually disjoint and $\xi L \subset U'$, for all $\xi \in G$. In particular, the ramification locus $X' \setminus U'$ of the action of G on X' is of codimension ≥ 2 in X'. Let $f : X' \to Z := X'/G$ be the canonical morphism onto the quotient, and set $U := f(U')$. Thus $U' = f^{-1}(U)$ and the restriction $f' := f|U' : U' \to U$ is an étale morphism. If we set $Y := f(L)$, it follows that $Y \subset U$ and the restriction $f|L : L = \mathbb{P}^1 \to Y$ is an isomorphism. Since f' is étale, by a general result (see e.g. [Gi2], Theorem (4.2)) the morphism f yields an isomorphism $\widehat{X}'_{/L} \cong \widehat{Z}_{/Y}$ between the formal completions. This implies in particular that f yields an isomorphism $X'(1) \cong Z(1)$, where $X'(1) = U'(1)$ is the first infinitesimal neighbourhood of L in X' and $Z(1) = U(1)$ is the first infinitesimal neighbourhood of Y in Z. Moreover, $N_{Y|Z} \cong N_{L|X'} = (n-1)\mathcal{O}_{\mathbb{P}^1}(1)$.

On the other hand, $\mathcal{O}_{\mathbb{P}^n}(1)$ is a G-sheaf in the sense of [Mu], p. 69 (since G acts linearly on $\mathbb{P}^n$), and since f' is étale, we can apply Proposition 2 of [Mu], p. 70 to deduce that there exists an $\mathcal{L} \in \text{Pic}(U)$ such that $f'^*(\mathcal{L}) \cong \mathcal{O}_{U'}(1) := \mathcal{O}_{\mathbb{P}^n}(1)|U'$. Thus the map $f'^* : \text{Pic}(U) \to \text{Pic}(U') = \mathbb{Z}[\mathcal{O}_{U'}(1)]$ is surjective, whence the map $\alpha_1 : \text{Pic}(U) \to \text{Pic}(U(1))$ is also surjective (because $U(1) \cong U'(1)$ and L is a line of $\mathbb{P}^n$ contained in U'). Let $\varphi : X \to Z$ be a desingularization of Z such that $\varphi^{-1}(U) \to U$ is an isomorphism (recall that U is contained in the smooth locus of Z). Then the embedding $Y \hookrightarrow Z$ lifts to an embedding $Y \hookrightarrow X$ such that $X(1) \cong Z(1) = U(1)$, where $X(1)$ is the first infinitesimal neighbourhood of Y in X. Since X is smooth, the restriction map $\text{Pic}(X) \to \text{Pic}(U)$ is surjective, whence the map

$$\alpha_1 : \text{Pic}(X) \to \text{Pic}(X(1))$$

is also surjective.

Before stating the result, let us recall some definitions. Two closed embeddings $Y \hookrightarrow X$ and $Y' \hookrightarrow X'$ are said to be *Zariski equivalent* if there is a Zariski open neighbourhood U of Y in X, a Zariski open neighbourhood U' of Y' in X', and an isomorphism $f : U \to U'$ such that $f(Y) = Y'$. In particular, if the embeddings $Y \hookrightarrow X$ and $Y' \hookrightarrow X'$ are Zariski equivalent, the varieties Y and Y' are isomorphic.

Let Y be a closed subvariety of an (irreducible) algebraic variety X. Denote by $K(X)$ the field of rational functions on X, and by $K(\widehat{X}_{/Y})$ the ring of formal-rational functions of X along Y (see [H-M], or also [Ha1]). Then there is a canonical map of $\mathbb{C}$-algebras

$$\varphi : K(X) \to K(\widehat{X}_{/Y}).$$

According to [H-M], we say that Y is $G3$ in X if the map φ is an isomorphism; we also say that Y is $G2$ in X if $K(\widehat{X}_{/Y})$ is a field and φ makes $K(\widehat{X}_{/Y})$ a finite field extension of $K(X)$.

With these definitions we have the following result (in the setting of (2.7)).

Proposition (2.7.1). (i) *The embedding* $Y \hookrightarrow X$ *of (2.7) has the property that the map* $\alpha_1 : \mathrm{Pic}(X) \to \mathrm{Pic}(X(1))$ *is surjective, Y is $G2$ (but not $G3$) in X, and X is simply connected.*

(ii) *The embedding* $Y \hookrightarrow X$ *is not Zariski equivalent with the embedding of a line in* $\mathbb{P}^n$.

(iii) *For any desingularization* $X \to Z$ *which does not change* U, $\mathrm{Pic}(X)$ *is a finitely generated free abelian group of rank* ≥ 2.

Proof. (i) Since the normal bundle $N_{Y|X} \cong (n-1)\mathcal{O}_{\mathbb{P}^1}(1)$ is ample, we can apply a result of Hartshorne (see [Ha1], Theorem (6.7), p. 438) to deduce that Y is $G2$ in X. Moreover, Hartshorne's arguments from case $n = 3$ (see [Ha1], p. 440) apply in our situation to show that Y is not $G3$ in X. Since X contains the curve $Y \cong \mathbb{P}^1$ with $N_{Y|X} \cong (n-1)\mathcal{O}_{\mathbb{P}^1}(1)$, X is simply connected by results of Campana and Kollár–Miyaoka–Mori (see [C] and [K-M-M], (2.5)). The rest of (i) follows from the discussion above.

For part (ii) we give two arguments. The first one uses the fact Y is ($G2$ but) not $G3$ in X. By a result of Hironaka (see [H-M], or also [Ha1]) any line of $\mathbb{P}^n$ is $G3$ in $\mathbb{P}^n$ (for all $n \geq 2$). This yields (ii).

The second argument for (ii) is more elementary and uses only the observation that the open set U is not simply-connected (because it has the connected étale covering $f' : U' \to U$ of degree $n+1$). On the other hand, any Zariski open neighbourhood of a line in $\mathbb{P}^n$ is simply-connected.

(iii) By (i), X is simply connected, whence $\mathrm{Pic}(X)$ is a finitely generated free abelian group. By Corollary (2.4) and (ii), the rank of $\mathrm{Pic}(X)$ cannot be 1. □

Note (2.7.2). The statement about $G2$ and $G3$ in Proposition (2.7.1) belongs to Hartshorne (see [Ha1], p. 440). In fact, he used the above quotient $Z = \mathbb{P}^n/G$ (in case $n = 3$) to produce examples of $G2$ curves which are not $G3$ (loc. cit.).

3. The geometry of quasi-lines and almost-lines

Let us begin this section by introducing the following definitions.

Definition (3.1). Let X be a complex projective manifold of dimension $n \geq 2$. Let Y be a smooth connected curve in X. The curve Y is called a *quasi-line* if $Y \cong \mathbb{P}^1$ and $N_{Y|X} \cong (n-1)\mathcal{O}_{\mathbb{P}^1}(1)$. The curve Y is called an *almost-line* if Y is a quasi-line and the restriction map $\alpha_0 : \mathrm{Pic}(X) \to \mathrm{Pic}(Y)$ is surjective.

Theorem (2.1) gives a good motivation to study geometrical properties of a variety X containing a quasi-line (or an almost-line).

First we note that a quasi-line on a smooth projective surface X is automatically an almost-line. In fact, a quasi-line Y on a smooth projective surface is nothing but a curve Y such that $Y \cong \mathbb{P}^1$ and $Y^2 = 1$.

In Example (2.7) we already gave non-trivial examples of n-folds X (for every $n \geq 3$) containing almost-lines Y. Let us give explicit examples of quasi-lines lying on polarized 3-folds with Picard group $\mathbb{Z}$, which are no longer almost-lines. This will show in particular that in Corollary (2.4), unlike in the surface case, one cannot replace the hypothesis that α_1 is surjective by the weaker assumption that Y is a quasi-line in X.

If Y is a quasi-line in X, $\deg(N_{Y|X}) = n - 1$, so, by adjunction formula, $-(K_X \cdot Y) = n + 1$. Assume now that X is a Fano n-fold of index r, i.e. r is the largest positive integer such that $-K_X = rH$, for some ample $H \in \mathrm{Pic}(X)$. Then the above formula becomes $r \deg_H(Y) = n + 1$, where $\deg_H(Y) = (H \cdot Y)$. If $\deg_H(Y) = 1$, then the Kobayashi–Ochiai theorem (see e.g. [B-S]) shows that $X \cong \mathbb{P}^n$ and Y is a line. This formula also shows that X cannot be isomorphic to a hyperquadric ($r = n$ in this case).

In dimension 3, the formula above suggests that conics in Fano 3-folds of index 2 are natural candidates for quasi-lines. The following result is not new. We refer to [O] for a complete classification of threefolds X containing a quasi-line Y of degree $(Y \cdot H) \leq 2$ with respect to a given polarization H on X, and to [L] for the case $(Y \cdot H) \leq 4$. In particular, from [O], §4, one knows exactly all Fano 3-folds of index 2 which contain quasi-lines. However the proof we give here is quite different from the arguments used in [O].

Theorem (3.2). *Let X be a smooth Fano 3-fold of index 2 such that $\mathrm{Pic}(X) = \mathbb{Z}[\mathcal{O}_X(H)]$, with H very ample. Then X contains a quasi-line Y, with $\deg_H(Y) = 2$ (in particular, $\mathrm{Coker}(\mathrm{Pic}(X) \to \mathrm{Pic}(Y)) \cong \mathbb{Z}/2\mathbb{Z}$). Moreover, X does not contain any almost-line.*

Proof. By Fano–Iskovskih classification (see [Is]), X is one of the following:
 (i) a cubic hypersurface in $\mathbb{P}^4$, or
 (ii) a complete intersection of two hyperquadrics in $\mathbb{P}^5$, or

(iii) a section of the Plücker embedding of the Grassmann variety $\mathbb{G}(2,5)$ (of lines in $\mathbb{P}^4$) by three general hyperplanes of $\mathbb{P}^9$.

Let $Z \in |H|$ be a general hyperplane section of X. Then Z is a Del Pezzo surface of degree d ($d = 3$ in case (i), $d = 4$ in case (ii), and $d = 5$ in case (iii)). By the classification of Del Pezzo surfaces, Z is isomorphic to the blowing up of $\mathbb{P}^2$ in $9-d$ points $P_1, \ldots, P_{9-d}$ in general position ($9-d \geq 4$). Let C be the proper transform of a smooth conic in $\mathbb{P}^2$ passing through $P_1, \ldots, P_4$.

It follows easily that C is a smooth conic on Z such that $C^2 = 0$. Thus $N_{C|Z} \cong \mathcal{O}_C$. Now consider the exact sequence of normal bundles

(4) $$0 \to N_{C|Z} \cong \mathcal{O}_{\mathbb{P}^1} \to N_{C|X} \to N_{Z|X}|C \cong \mathcal{O}_{\mathbb{P}^1}(2) \to 0.$$

In particular, $\deg(N_{C|X}) = 2$, and so, by the theorem of Grothendieck [Gr], we get

(5) $$N_{C|X} \cong \mathcal{O}_{\mathbb{P}^1}(a) \oplus \mathcal{O}_{\mathbb{P}^1}(2-a), \quad a \geq 0.$$

Assume that sequence (4) does not split. Since $\mathcal{O}_{\mathbb{P}^1}(2)$ is ample, a general result due to Gieseker [Gi1] (or by an easy direct argument) shows that $N_{C|X}$ is ample, and so by (5) we get

(6) $$N_{C|X} \cong \mathcal{O}_{\mathbb{P}^1}(1) \oplus \mathcal{O}_{\mathbb{P}^1}(1).$$

In other words C is a quasi-line with $\deg_H(C) = 2$, and in this case we are done.

We can therefore assume that sequence (4) splits, i.e.

(7) $$N_{C|X} \cong \mathcal{O}_{\mathbb{P}^1} \oplus \mathcal{O}_{\mathbb{P}^1}(2).$$

Consider the Hilbert scheme $\mathcal{P}$ parametrizing all curves in X having the same Hilbert polynomial as C. Pick a general point $x \in C$ and consider the closed subscheme $\mathcal{P}_x$ of $\mathcal{P}$ parametrizing all curves of $\mathcal{P}$ which pass through the point x. Let us denote by $[C]$ the closed point of $\mathcal{P}_x$ corresponding to the conic C. From general deformation theory (see e.g. [K], Theorem (1.7), p. 95), the tangent space $T_{\mathcal{P}_x,[C]}$ of $\mathcal{P}_x$ at $[C]$ is isomorphic to $H^0(C, N_{C|X} \otimes \mathcal{I}_x)$, where $\mathcal{I}_x$ is the ideal sheaf of x in C. Moreover if

(8) $$H^1(C, N_{C|X} \otimes \mathcal{I}_x) = 0,$$

then $\mathcal{P}_x$ is smooth at the point $[C]$ (loc. cit).

In our case, by (7) we get $N_{C|X} \otimes \mathcal{I}_x \cong \mathcal{O}_{\mathbb{P}^1}(-1) \oplus \mathcal{O}_{\mathbb{P}^1}(1)$, so that (8) holds true. We thus conclude that $[C]$ is a smooth point of $\mathcal{P}_x$ and $\dim_{[C]}(\mathcal{P}_x) = \dim(T_{\mathcal{P}_x,[C]}) = 2$.

In particular, $[C]$ belongs to a unique irreducible component $\mathcal{P}' := \mathcal{P}_{x,[C]}$ of $\mathcal{P}_x$ containing $[C]$. If $\mathcal{T} \subset \mathcal{P} \times X$ is the universal family of the Hilbert scheme $\mathcal{P}$, denote by $\mathcal{T}' := \mathcal{T}|\mathcal{P}' \times X$. Denote also by $p : \mathcal{T}' \to \mathcal{P}'$ the restriction to $\mathcal{T}'$ of the first projection of $\mathcal{P}' \times X$, and by $q : \mathcal{T}' \to X$ the restriction of the second projection of $\mathcal{P}' \times X$. Finally, set $S := q(\mathcal{T}')$. Thus S is the locus of all curves in X parametrized by $\mathcal{P}'$ (all of them are conics in X).

Since $\dim(\mathcal{P}') = 2$, we have $\dim(\mathcal{T}') = 3$. Assume that the map q is surjective, i.e.

(9) $$\dim(S) = 3.$$

Then, by using [K], Corollary (3.10.1), p. 117, we conclude that C can be deformed to a smooth rational curve C' such that $T_X|C'$ is ample (more precisely, C' is the image via q of a general fiber of p). Since $N_{C'|X}$ is a quotient of $T_X|C'$, $N_{C'|X}$ is also ample. (Note that there is a perfect analogue of Corollary (3.10.1) of [K] in the context of Hilbert schemes, which allows one to get directly the ampleness of $N_{C'|X}$.) Now, by construction, C' is also a conic, whence $\deg(N_{C'|X}) = 2$. Using (5) for C' and the ampleness of $N_{C'|X}$ we get $N_{C'|X} \cong 2\mathcal{O}_{\mathbb{P}^1}(1)$, i.e. C' is a quasi-line in X (so we are done if (9) holds).

It remains therefore to prove (9). Assume the contrary. Then

(10) $$\dim(S) = 2.$$

Observe that $q^{-1}(x) \cong \mathcal{P}'$ via the morphism p, so $\dim(q^{-1}(x)) = 2$. Furthermore, for a general point $y \in S$, the theorem of dimension of fibers applied to q yields $\dim(q^{-1}(y)) = 1$. In other words, there is a 2-dimensional family of (smooth) conics passing through x and contained in S. Moreover, there is a 1-dimensional family of conics passing through x and a fixed general point y, and which are contained in S.

Assume that S is not a plane. Then we claim that the surface S spans a $\mathbb{P}^3$. To see this, consider the projection π from the point x. Then the image $S' := \pi(S)$ is a surface which satisfies the condition that for each general point $y' \in S'$ there is a 1-dimensional family of lines through y'. This implies that S' is a plane. This yields the fact that S spans a linear space L of dimension 3. Observe also that S is not a cone with vertex x because otherwise every conic passing through x would be degenerate.

Now consider the linear system $\mathcal{L}_x \cong (\mathbb{P}^2)^*$ of all planes in L passing through x. Since S is not a cone, the linear system cut out on S by $\mathcal{L}_x$ is not composed with a pencil. Then by Bertini, the intersection of S with a general plane $P \in \mathcal{L}_x$ is an irreducible reduced curve Γ_P.

Consider also the algebraic variety $\mathcal{C}_x$ (which is isomorphic to $\mathcal{P}'$) consisting of all conics in S passing through x. Then we have the natural rational map $\varphi_x : \mathcal{C}_x \dashrightarrow \mathcal{L}_x$ which sends a conic $D \in \mathcal{C}_x$ (which is not a double line) into the plane generated by D. Since two conics $D, D' \in \mathcal{C}_x$ have the same image P if and only if they are contained in $P \cap S$, the general fibre of φ_x is finite. This, together with the equalities $\dim(\mathcal{C}_x) = \dim(\mathcal{L}_x) = 2$, implies that every general plane $P \in \mathcal{L}_x$ is of the form $P = \varphi(C_P)$, with C_P a conic in S. It follows that $\Gamma_P = C_P$, i.e. the intersection of S with a general plane P of L through x is a conic. This last fact clearly implies that S is a quadric surface in L. So, assuming that S is a surface, we have proved it is either a plane, or a quadric.

On the other hand, $\mathrm{Pic}(X) = \mathbb{Z}[\mathcal{O}_X(H)]$. Therefore there is a positive integer b such that $\mathcal{O}_X(S) \cong \mathcal{O}_X(bH)$. It follows that $\deg(S)$ is a multiple of $\deg_H(X) = (H^3) = d \geq 3$, a contradiction.

The last statement follows from the fact that $\text{Pic}(X) \cong \mathbb{Z}$ and from Theorem (2.1) and Corollary (2.4). $\square$

Our next objective is to study the geometry of the projective manifolds containing quasi-lines or almost-lines. Let Y be a quasi-line on a projective n-fold X, and let $[Y]$ be a point of the Hilbert scheme $\text{Hilb}(X)$ which corresponds to the curve Y. Since $H^1(Y, N_{Y|X}) = H^1(\mathbb{P}^1, (n-1)\mathcal{O}_{\mathbb{P}^1}(1))$ we conclude that $[Y]$ is a smooth point of $\text{Hilb}(X)$ (see [FGA], éxposé 221, 1960/61; see also [K], p. 95). Therefore $[Y]$ belongs to a unique irreducible component, $\mathcal{H}$, of $\text{Hilb}(X)$.

Consider the universal flat family in $\text{Hilb}(X) \times X$

$$\begin{array}{ccccc} \mathcal{F} & \xrightarrow{i} & \mathcal{H} \times X & \xrightarrow{p_2} & X \\ {\scriptstyle p}\downarrow & & \downarrow{\scriptstyle p_1} & & \\ \mathcal{H} & \xrightarrow{\text{id}} & \mathcal{H} & & \end{array}$$

Denote $q := p_2 \circ i : \mathcal{F} \to X$. Note that for each $t \in \mathcal{H}$, $Y_t := q(p^{-1}(t))$ is the curve on X given by t and the restriction map

$$q|p^{-1}(t) : p^{-1}(t) \to Y_t$$

is an isomorphism.

For any given points $x, x' \in X$, set $\mathcal{H}_x := p(q^{-1}(x))$ and $\mathcal{H}_{x,x'} := \mathcal{H}_x \cap \mathcal{H}_{x'}$. Then $\mathcal{H}_x$ (respectively $\mathcal{H}_{x,x'}$) is the closed subscheme of $\mathcal{H}$ consisting of those points $t \in \mathcal{H}$ such that the curve Y_t contains x (respectively x, x').

Lemma (3.3). *Let X be a complex projective manifold of dimension $n \geq 2$. Let $Y_0 := q(p^{-1}(t_0))$ be a quasi-line (resp. an almost-line) in X, $t_0 \in \mathcal{H}$. Then there exists an open neighbourhood $\mathcal{U}$ of t_0 in $\mathcal{H}$ such that for each closed point $t \in \mathcal{U}$, the curve $Y_t := q(p^{-1}(t))$ is a quasi-line (resp. an almost-line) in X.*

Proof. Since the morphism $p : \mathcal{F} \to \mathcal{H}$ is flat and $p^{-1}(t_0) \cong \mathbb{P}^1$, there exists an open neighbourhood $\mathcal{U}$ of t_0 in $\mathcal{H}$ such that $Y_t \cong \mathbb{P}^1$ for each $t \in \mathcal{U}$. By the base-change theorems (see e.g. [Ha3], III, §9) the functions $f(t) := \dim H^0(Y_t, N_{Y_t|X})$, $g(t) := \dim H^1(Y_t, N_{Y_t|X})$ are upper semicontinuous. Thus, shrinking $\mathcal{U}$ if necessary, $g(t) = 0$ for each $t \in \mathcal{U}$. Hence $f(t) = f(t) - g(t) = \chi(Y_t, N_{Y_t|X})$ is constant in $\mathcal{U}$ by the base-change theorems again. Moreover, since ampleness is an open condition, shrinking $\mathcal{U}$ once again, we can assume that, for each $t \in \mathcal{U}$,

$$N_{Y_t|X} \cong \mathcal{O}(a_1) \oplus \cdots \oplus \mathcal{O}(a_{n-1}),$$

where $a_i = a_i(t)$ are positive integers depending on t, $i = 1, \ldots, n-1$. Since

$$\dim H^0(Y_t, N_{Y_t|X}) = \sum_{i=1}^{n-1}(a_i + 1) = n - 1 + \sum_{i=1}^{n-1} a_i$$

and
$$\dim H^0(Y_0, N_{Y_0|X}) = \dim H^0(\mathbb{P}^1, (n-1)\mathcal{O}_{\mathbb{P}^1}(1)) = 2n - 2,$$

we conclude that $\sum_{i=1}^{n-1} a_i = n - 1$ (because the function f is constant in $\mathcal{U}$), or else, $a_i = 1$ for each $i = 1, \ldots, n-1$ (because $a_i > 0$ for each $i = 1, \ldots, n$). This proves the lemma for quasi-lines.

Assume now that Y_0 is an almost-line in X. This amounts to the existence of an $L \in \text{Pic}(X)$ such that $(L \cdot Y_0) = 1$. Then $(q^*(L) \cdot p^{-1}(t)) = (L \cdot Y_t)$ for all $t \in \mathcal{H}$ (since $p^{-1}(t) \cong Y_t$ for each $t \in \mathcal{H}$). So, on the same open subset $\mathcal{U}$ of $\mathcal{H}$, Y_t is an almost-line. □

Fix now a quasi-line Y and two distinct points $x, x' \in Y$. From general deformation theory (see e.g. [K], Theorem (1.7), p. 95), we know that the tangent space to $\mathcal{H}_x$ at $[Y]$ is isomorphic to $H^0(Y, N_{Y|X} \otimes \mathcal{J}_x)$ and the tangent space to $\mathcal{H}_{x,x'}$ at $[Y]$ is isomorphic to $H^0(Y, N_{Y|X} \otimes \mathcal{J}_{x,x'})$, where $\mathcal{J}_x$, $\mathcal{J}_{x,x'}$ denote the ideal sheaves in Y of the subsets $\{x\}$, $\{x, x'\}$ with reduced structure. Note that, since $N_{Y|X} \cong (n-1)\mathcal{O}(1)$, one has

$$N_{Y|X} \otimes \mathcal{J}_x \cong N_{Y|X} \otimes \mathcal{O}_Y(-x) \cong N_{Y|X} \otimes \mathcal{O}_{\mathbb{P}^1}(-1) \cong (n-1)\mathcal{O}_{\mathbb{P}^1}.$$

Similarly we get
$$N_{Y|X} \otimes \mathcal{J}_{x,x'} \cong (n-1)\mathcal{O}_{\mathbb{P}^1}(-1).$$

We conclude that $H^1(Y, N_{Y|X} \otimes \mathcal{J}_x) = H^1(Y, N_{Y|X} \otimes \mathcal{J}_{x,x'}) = 0$. By loc.cit. we infer that $\mathcal{H}_x$ and $\mathcal{H}_{x,x'}$ are smooth at the point $[Y]$. Thus there exists a unique component, $\mathcal{H}_{x,[Y]}$, of $\mathcal{H}_x$ containing $[Y]$. From Lemma (3.3) we know that there exists an open subset, $\mathcal{U}_x \subset \mathcal{H}_x$, such that for each $t \in \mathcal{U}_x$, $q(p^{-1}(t))$ is a quasi-line containing x. Set $\mathcal{H}_x^\circ := \mathcal{H}_{x,[Y]} \cap \mathcal{U}_x$ and $\mathcal{H}_{x,x'}^\circ := \mathcal{H}_x^\circ \cap \mathcal{H}_{x,x'}$.

The following result shows that there are only finitely many quasi-lines passing through two distinct points, x, x', of a given quasi-line Y. In particular, $[Y]$ will be an isolated point of $\mathcal{H}_{x,x'}^\circ$. It will also follow that there are only finitely many quasi-lines passing through any two different points of X.

Lemma (3.4). *Let X be a complex projective manifold of dimension $n \geq 2$. Let Y be a quasi-line in X, and let $x, x' \in Y$. With the notation and assumptions as above we have:*

(i) $\dim_{[Y]}(\mathcal{H}_x^\circ) = n - 1$;

(ii) $\dim_{[Y]}(\mathcal{H}_{x,x'}^\circ) = 0$.

Proof. The tangent space to $\mathcal{H}_x^\circ$ at $[Y]$ is isomorphic to $H^0(Y, N_{Y|X} \otimes \mathcal{J}_x)$ and the tangent space to $\mathcal{H}_{x,x'}^\circ$ at $[Y]$ is isomorphic to $H^0(Y, N_{Y|X} \otimes \mathcal{J}_{x,x'})$. Moreover, $\mathcal{H}_x^\circ$ and $\mathcal{H}_{x,x'}^\circ$ are both smooth at $[Y]$. Thus we get

$$\dim_{[Y]} \mathcal{H}_x^\circ = \dim H^0(Y, N_{Y|X} \otimes \mathcal{J}_x) = \dim H^0(Y, (n-1)\mathcal{O}_{\mathbb{P}^1}) = n - 1,$$

$\dim_{[Y]} \mathcal{H}^\circ_{x,x'} = \dim H^0(Y, N_{Y|X} \otimes \mathcal{J}_{x,x'}) = \dim H^0(Y, (n-1)\mathcal{O}_{\mathbb{P}^1}(-1)) = 0.$ □

The following result shows in particular that the family of quasi-lines (resp. of almost-lines) in X fills up a dense open set of X.

Theorem (3.5). *Let X be a complex projective manifold of dimension $n \geq 2$. Let Y be a quasi-line (resp. an almost-line) in X, and let x be a fixed point on Y. With the notation and assumptions as above, we have:*

(i) *There exists a non-empty open subset U_x in X such that for each point $y \in U_x$ there exists a quasi-line (resp. an almost-line) passing through x and y;*

(ii) *The restriction map $q_x := q|p^{-1}(\mathcal{H}_x) : p^{-1}(\mathcal{H}_x) \to X$ is surjective;*

(iii) *For each point $y \in X$ there exists a connected curve with rational (but possibly singular) irreducible components passing through x and y.*

Proof. (i) Since $\mathcal{H}^\circ_x$ is an irreducible open set and the fiber of p over an arbitrary closed point of $\mathcal{H}^\circ_x$ is isomorphic to $\mathbb{P}^1$ (and hence all fibers of $p^{-1}(\mathcal{H}^\circ_x) \to \mathcal{H}^\circ_x$ are irreducible of dimension 1), we infer that $p^{-1}(\mathcal{H}^\circ_x)$ is irreducible and of dimension n (by Lemma (3.4)). We claim that the morphism $q^\circ_x := q|p^{-1}(\mathcal{H}^\circ_x) : p^{-1}(\mathcal{H}^\circ_x) \to X$ is dominant. Indeed, using the theorem on dimension of the fibers of a morphism between two irreducible schemes, it will be sufficient to show that there is a fiber of q°_x with an isolated point. To this extent, pick an arbitrary point $y \in Y$, $y \neq x$. Since there is a one-to-one correspondence between $(q^\circ_x)^{-1}(y)$ and all the curves parametrized by $\mathcal{H}^\circ_{x,y}$, and since $[Y] \in \mathcal{H}^\circ_{x,y}$, it follows from Lemma (3.4), (ii) that $[Y]$ corresponds to an isolated point of $(q^\circ_x)^{-1}(y)$.

Now, by the theorem of Chevalley, there exists a non-empty open subset U_x in X such that $U_x \subseteq q^\circ_x(p^{-1}(\mathcal{H}^\circ_x))$. This means that for every point $y \in U_x$ there is a quasi-line (resp. an almost-line, if Y is an almost-line) passing through x and y. This proves (i).

Part (ii) follows from (i) because $p^{-1}(\mathcal{H}_x)$ is a projective scheme.

Part (iii) is a consequence of the following well known lemma (applied to the proper flat morphism $p|p^{-1}(\mathcal{H}_x) : p^{-1}(\mathcal{H}_x) \to \mathcal{H}_x$). □

Lemma (3.5.1). *Let $f : U \to V$ be a proper flat morphism of algebraic schemes over $\mathbb{C}$ such that the general fiber of f is isomorphic to $\mathbb{P}^1$. Then every fiber of f is a connected curve with rational (but possibly singular) irreducible components.*

Remark (3.6). It is worth noting that, in the case of a surface X, given a quasi-line and two distinct points on it, there is a unique quasi-line passing through those points.

To see this, let C be a quasi-line and let x, x' be two distinct points on C. Then $C \cong \mathbb{P}^1$ and $C^2 = 1$. Thus the linear system $|C|$ is base point free and defines a birational morphism $f : X \to \mathbb{P}^2$, such that $f^*\mathcal{O}_{\mathbb{P}^2}(1) \cong \mathcal{O}_X(C)$. Therefore C is the pullback of a line, ℓ, of $\mathbb{P}^2$. Since C is irreducible we thus conclude that C does not meet the locus of X where f is not biregular. Then $C \cong \ell$ and, since there is a unique line in $\mathbb{P}^2$ passing through two distinct points, we are done.

Theorem (3.7). *Let X be a complex projective manifold of dimension $n \geq 2$. Then the following conditions are equivalent.*

(i) *X contains a quasi-line (resp. an almost-line) Y.*

(ii) *The locus of all quasi-lines (resp. almost-lines) passing through a general point of X is dense in X.*

(iii) *There exists a non-empty open subset V of $X \times X$ such that for every point $(x, y) \in V$, there is a quasi-line (resp. an almost-line) joining x and y.*

If furthermore $n \geq 3$, the above conditions which refer to quasi-lines, are also equivalent to any of the following ones:

(iv) *There is a morphism $f : \mathbb{P}^1 \to X$ such that $f^*(T_X) \cong \mathcal{O}_{\mathbb{P}^1}(2) \oplus (n-1)\mathcal{O}_{\mathbb{P}^1}(1)$.*

(v) *There is a morphism $f : \mathbb{P}^1 \to X$ such that $\deg(f^*(T_X)) = n+1$ and the locus of all deformations of f sending $0 \in \mathbb{P}^1$ to a given general point of X (see [K]) is dense in X.*

Proof. The equivalence (i)⇔(ii) is a direct consequence of Theorem (3.5). Clearly (iii)⇒(i).

Assume therefore that (i) holds. Consider the open dense subset $\mathcal{H}'$ of $\mathcal{H}$ which parametrizes the set of quasi-lines of X (which is the same as the subset of $\mathcal{H}$ which parametrizes the set of almost-lines, if Y is an almost-line), and define the incidence variety

$$\Gamma := \{([C], x, x') \in \mathcal{H}' \times X \times X \mid x, x' \in C\}.$$

Denote by $\rho : \Gamma \to X \times X$ the restriction to Γ of the projection onto $X \times X$. By Lemma (3.4), (ii), there is a point $(x, x') \in X \times X$, with $x \neq x'$, such that $\rho^{-1}(x, x')$ is finite. Thus, by the theorem of dimension of the fibers, the morphism ρ is dominant. Then, as above, there is a non-empty open subset V of $X \times X$ such that $V \subseteq \rho(\Gamma)$. This yields the required open subset of (iii), i.e. (i)⇒(iii).

Using the normal exact sequence of Y in X we easily get (i)⇒(iv). Noting that the condition (iv) can be restated as $\deg(f^*(T_X)) = n+1$ and $f^*(T_X)$ is ample, the equivalence (iv)⇔(v) follows from [M-P], Lemma (2.6), p. 101. Finally, (iv)⇒(i) follows from [K], Theorem (3.14), (iii) (with $B = \emptyset$), since $n \geq 3$. □

Remark (3.8). The projective manifolds containing a quasi-line are rationally connected in the sense of [K-M-M]. Theorem (3.7) can be considered, in the case of quasi-lines, a strengthened form of Theorem (2.1) in [K-M-M]. In particular, using Proposition (2.5) of [K-M-M], every projective manifold which contains a quasi-line is simply-connected and the following vanishings hold: $H^0(X, \otimes^m \Omega_X^1) = 0$, for all $m > 0$, and $H^i(X, \mathcal{O}_X) = 0$, for all $i > 0$.

Remark (3.8) suggests the following definition.

Definition (3.9). Let X be a projective manifold of dimension ≥ 2. We say that X is *quasi-linearly connected* (resp. *almost-linearly connected*) if X contains a quasi-line (resp. an almost-line). Theorem (3.7) gives various characterizations of such manifolds (which in particular motivate the terminology chosen).

It is well known that any smooth projective deformation of a rationally connected manifold is again rationally connected (see [M-P], p. 107, Proposition (2.13)). This statement is no longer true neither for quasi-linear, nor for almost-linear connectedness. For example the Hirzebruch surface $\mathbb{F}_1$ (the projective plane blown up at a point), which is obviously almost-linearly (and hence also quasi-linearly) connected, can degenerate into the Hirzebruch surface $\mathbb{F}_{2e+1}$, with $e \geq 1$, which is not quasi-linearly connected. However the following holds.

Proposition (3.10). (i) *Any small (smooth) projective deformation of a quasi-linearly connected manifold X is again quasi-linearly connected.*

(ii) *Any small (smooth) projective deformation of an almost-linearly connected manifold X is also almost-linearly connected.*

Proof. Let $f : \mathcal{X} \to T$ be a smooth projective morphism such that there is a point $t_0 \in T$ with the property that $f^{-1}(t_0) \cong X$. We may assume that T is a smooth curve. Let Y be a quasi-line (resp. an almost-line) in X.

(i) The proof of the openness of the deformations of rationally connected manifolds works in our situation as well (see the first part of the proof of Proposition (2.13) of [M-P], p. 107). In fact it becomes even simpler, working with the Hilbert scheme (instead of the Hom-scheme), and with the exact sequence

$$0 \to N_{Y|X} \to N_{Y|\mathcal{X}} \to N_{X|\mathcal{X}}|Y \cong \mathcal{O}_Y \to 0$$

(instead of the normal bundle sequence $0 \to T_X \to T_{\mathcal{X}}|X \to N_{X|\mathcal{X}} \cong \mathcal{O}_X \to 0$).

(ii) Assume now that Y is an almost-line, i.e. Y is a quasi-line and the map $\alpha_0 : \operatorname{Pic}(X) \to \operatorname{Pic}(Y)$ is surjective. In particular, there is an $L \in \operatorname{Pic}(X)$ such that $L|Y \cong \mathcal{O}_{\mathbb{P}^1}(1)$, i.e. $(L \cdot Y) = 1$. For every $t \in T$ set $X_t := f^{-1}(t)$. From Miyaoka's arguments (loc. cit.) we see that (shrinking T if necessary) there is a closed subscheme $\mathcal{Y}$ of $\mathcal{X}$ such that the restriction $g := f|\mathcal{Y} : \mathcal{Y} \to T$ is smooth, $Y_t := g^{-1}(t)$ is a quasi-line for each $t \in T$, and $Y_{t_0} = Y$.

Now, by Remark (3.8), $H^i(X, \mathcal{O}_X) = 0$ for $i = 1, 2$. Let $\mathcal{X}(m)$ ($m \geq 0$) be the m-th infinitesimal neighbourhood of $X = X_{t_0}$ in $\mathcal{X}$. Since the conormal bundle of X_{t_0} in $\mathcal{X}$ is trivial, the truncated exponential sequence yields the exact sequence of cohomology

$$0 = H^1(X, \mathcal{O}_X) \to \operatorname{Pic}(\mathcal{X}(m+1)) \to \operatorname{Pic}(\mathcal{X}(m)) \to H^2(X, \mathcal{O}_X) = 0,$$

i.e. the restriction maps $\mathrm{Pic}(\mathcal{X}(m+1)) \to \mathrm{Pic}(\mathcal{X}(m))$ are isomorphisms for all $m \geq 0$. It follows that the restriction map $\mathrm{Pic}(\widehat{\mathcal{X}}_{/X}) \to \mathrm{Pic}(X)$ is an isomorphism. Using Artin's approximation theory (see [A], Theorem (3.5)), we may replace T by an appropriate étale neighbourhood of t_0, such that (after base-change) there is an $\mathcal{L} \in \mathrm{Pic}(\mathcal{X})$ inducing L by restriction to X. As the intersection number is constant in a flat family, we get $(\mathcal{L}_t \cdot Y_t) = (L \cdot Y) = 1$. This proves that ($Y_t$ is a quasi-line in X_t and) the restriction map $\mathrm{Pic}(X_t) \to \mathrm{Pic}(Y_t)$ is surjective for t close to t_0, and we are done. □

Note that the property of containing a quasi-line is not birationally invariant. For instance, the quadric surface $\mathbb{F}_0 = \mathbb{P}^1 \times \mathbb{P}^1$ does not contain any quasi-line. However, we have the following proposition.

Proposition (3.11). *Let $\varphi : X \to X'$ be a birational morphism between smooth projective manifolds of dimension $n \geq 2$. Assume that X' contains a quasi-line (resp. an almost-line). Then X also contains a quasi-line (resp. an almost-line).*

Proof. Let E be the locus where φ is not biregular. Then the image $\varphi(E)$ is a closed subset of X' of dimension $\leq n - 2$. Fix a general point x on X'. Then there is a quasi-line through x which does not meet $\varphi(E)$. Indeed, by Lemma (3.4), (ii), there is only a finite number of quasi-lines through x and some point y of $\varphi(E)$. Thus the locus of quasi-lines through x which meet $\varphi(E)$ is of dimension $\leq \dim(\varphi(E)) + 1 \leq n - 1$. But from Theorem (3.5) we know that the quasi-lines through x fill up a dense open subset of X'. Thus we conclude that there exists a quasi-line, C, on X' such that $C \cap \varphi(E) = \emptyset$. Therefore $Y := \varphi^{-1}(C)$ is a quasi-line on X. The case of almost-lines follows immediately from the case of quasi-lines. □

Let us point out that the converse does not hold in (3.11): if φ is as above and X contains a quasi-line (resp. an almost-line), it does not follow that X' does. It is enough to consider some elementary transformation of the Hirzebruch surface $\mathbb{F}_1$.

4. Strongly rational varieties

In this section we study some geometrical properties of a special class of rational varieties in terms of properties of quasi-lines.

Throughout this section X always denotes a complex projective manifold of dimension $n \geq 2$.

Definition (4.1). We say that X is *strongly rational* if there exist non empty open sets $U \subseteq X$, $V \subseteq \mathbb{P}^n$ such that U is isomorphic to V and $\mathrm{codim}_{\mathbb{P}^n}(\mathbb{P}^n \setminus V) \geq 2$.

Remark (4.2). 1) Any (smooth) hyperquadric of dimension $n \geq 2$, which is certainly rational, is easily seen not to be strongly rational. Indeed, any strongly rational manifold contains an almost-line. But we have seen at the beginning of the previous section that a hyperquadric does not contain even quasi-lines.

2) If there is a birational morphism $\psi : X \to \mathbb{P}^n$, X is strongly rational. Indeed, if E is the locus of X where ψ is not biregular, then $\operatorname{codim}_{\mathbb{P}^n}(\psi(E)) \geq 2$, whence X contains an open subset isomorphic to $\mathbb{P}^n \setminus \psi(E)$. It is easy to see that any strongly rational surface is of this type, i.e. it dominates $\mathbb{P}^2$. This no longer holds if $n \geq 3$ as we shall see in Example (4.7) below.

Now, if X' is any (smooth) rational variety and $\beta : X' \dashrightarrow \mathbb{P}^n$ is a fixed birational map, it follows from Hironaka's results [H] that we can find a composition of blowing ups with smooth centers, $\varphi : X \to X'$, and a birational morphism $\psi : X \to \mathbb{P}^n$ such that $\psi = \beta \circ \varphi$. Thus we see that any rational variety is dominated by a strongly rational one.

We need the following general fact.

Lemma (4.3). *Let Y be a quasi-line on X. Then there exist only finitely many prime divisors H on X such that $(H \cdot Y) = 0$.*

Proof. Let H be a prime divisor on X such that $(H \cdot Y) = 0$. We claim that $H \cap Y = \emptyset$. Indeed, if $x \in H \cap Y$ is a point, by Theorem (3.5), (i), there exists a quasi-line Y' passing through x, not contained in H, and which is numerically equivalent to Y. In particular, $(H \cdot Y) = 0$ implies $(H \cdot Y') = 0$. Since Y' is not contained in H, this last equality implies $H \cap Y' = \emptyset$, contradicting the fact that $x \in H \cap Y'$.

This shows that H does not meet the locus of quasi-lines which are in the same family with Y, and hence H is contained in a proper closed subset of X. In particular, there are only finitely many prime divisors of X that are orthogonal to Y. □

We can now prove the main result of this section.

Theorem (4.4). *Let X be a projective manifold of dimension n. The following conditions are equivalent.*

(i) *X is strongly rational;*

(ii) *X contains a quasi-line Y such that the pair (X, Y) is Zariski equivalent to $(\mathbb{P}^n, \text{line})$;*

(iii) *X contains a quasi-line Y and there exists a divisor D on X such that $(D \cdot Y) = 1$ and $\dim |D| \geq n$.*

Proof. The implications (i)⇔(ii) and (ii)⇒(iii) are easy consequences of the definitions. To show (iii)⇒(i), we will use induction on n. We first prove this implication in the case $n = 2$. Consider the exact sequence

$$0 \to \mathcal{O}_X(D - Y) \to \mathcal{O}_X(D) \to \mathcal{O}_Y(D) \cong \mathcal{O}_{\mathbb{P}^1}(1) \to 0.$$

Assume $\dim H^0(X, \mathcal{O}_X(D)) \geq 4$. Then $\dim |D - Y| \geq 1$, so that $D - Y$ moves. Since $(D - Y) \cdot Y = 0$, this contradicts Lemma (4.3). Therefore we conclude that $\dim H^0(X, \mathcal{O}_X(D)) = \dim H^0(X, \mathcal{O}_X(Y)) = 3$ and $\dim |D - Y| = 0$. Therefore $E := D - Y$ is an effective divisor which must be the fixed locus of $|D|$. Thus the rational maps associated to $|D|$ and $|Y|$ coincide. But, as is well known, the linear system $|Y|$ has no base points and yields a birational morphism $X \to \mathbb{P}^2$ such that Y is the pull-back of a line.

Assume now $n \geq 3$. Write $D = E + M$, where E, M are the fixed part and the moving part of the linear system $|D|$ respectively. Since Y moves, we have $(E \cdot Y) \geq 0$, $(M \cdot Y) \geq 0$. Therefore, by Lemma (4.3), the assumption $(D \cdot Y) = 1$ yields $(M \cdot Y) = 1$ and $(E \cdot Y) = 0$. Since $\dim |M| = \dim |D| \geq n$, we conclude that M satisfies the same assumptions as D. Thus we can assume that $|D|$ is free from fixed components. From Hironaka's desingularization theory [H], there exists a morphism $\sigma : X' \to X$ which is a composition of blowing ups along smooth centers such that $\sigma^*(|D|) = E' + |D'|$, with $|D'|$ a base point free linear system on X', $E' \geq 0$ the fixed part of $|E' + D'|$, $\dim(\sigma(E')) \leq n - 2$. In particular, it follows that $(\sigma^{-1}(Y) \cdot D') = (Y \cdot D) = 1$. Therefore by the previous argument $|D'|$ is not composed with a pencil. Since $|D'|$ is base point free and not composed with a pencil, by Bertini it follows that a general member, Δ', of $|D'|$ is smooth and connected. Fix such a Δ' and set $\Delta := \sigma(\Delta')$.

Take two general points $x, y \in X'$. Since $\dim |D| \geq n \geq 2$ we may assume that $x, y \in \Delta'$. From Theorem (3.7), (iii) it follows that there exists a quasi-line Y on X passing through $\sigma(x), \sigma(y)$. If Z denotes the locus of points where σ^{-1} is not defined we have $\mathrm{codim}_X(Z) \geq 2$. The proof of Proposition (3.11) then shows that we can further assume that Y does not meet Z. So $Y' := \sigma^{-1}(Y)$ is a quasi-line on X' passing through x and y. Therefore $\{x, y\} \subset \Delta' \cap Y'$ and the assumption that $(\Delta \cdot Y) = 1$ implies $(\Delta' \cdot Y') = 1$. Since $\{x, y\} \subset \Delta' \cap Y'$ this last equality forces $Y' \subset \Delta'$.

Now we claim that Y' is a quasi-line on Δ'. To see this, consider the exact sequence

$$0 \to N_{Y'|\Delta'} \to N_{Y'|X'} \cong (n-1)\mathcal{O}_{\mathbb{P}^1}(1) \to N_{\Delta'|X'}|Y' \cong \mathcal{O}_{\mathbb{P}^1}(1) \to 0.$$

By tensoring with $\mathcal{O}_{\mathbb{P}^1}(-1)$ and dualizing we get the exact sequence

$$0 \to \mathcal{O}_{\mathbb{P}^1} \to (n-1)\mathcal{O}_{\mathbb{P}^1} \to N^*_{Y'|\Delta'}(1) \to 0$$

on $\mathbb{P}^1$. From this we see that $N^*_{Y'|\Delta'}(1)$ is globally generated and of degree zero. Hence it is trivial. This is equivalent to $N_{Y'|\Delta'} \cong (n-2)\mathcal{O}_{\mathbb{P}^1}(1)$, i.e. Y' is a quasi-line on Δ'.

Consider the exact sequence

(11) $$0 \to \mathcal{O}_{X'} \to \mathcal{O}_{X'}(\Delta') \to \mathcal{O}_{\Delta'}(\Delta') \to 0.$$

Since $\dim |\Delta'| \geq n$, it follows that $h^0(\Delta', \mathcal{O}_{\Delta'}(\Delta')) \geq n$. Now apply induction on Δ'.

Then the rational map given by $H^0(\Delta', \mathcal{O}_{\Delta'}(\Delta'))$ maps a neighbourhood of a fixed quasi-line, $Y' \subset \Delta'$, isomorphically to some open subset of $\mathbb{P}^{n-1}$, such that Y' corresponds to a line. Since $H^1(X', \mathcal{O}_{X'}) = 0$ (see Remark (3.8)) we see from the cohomology sequence associated to the exact sequence (11) that the rational map given by $H^0(X', \mathcal{O}_{X'}(\Delta'))$ maps a neighbourhood of $Y' \subset X'$ isomorphically to some open subset of $\mathbb{P}^n$, such that Y' corresponds to a line. We obtain the same conclusion for the pair (X, Y) via the birational morphism σ. This completes the proof of our theorem. □

Before passing to applications we need the following simple lemma:

Lemma (4.5). *Let X be a projective manifold of dimension $n \geq 1$, and let D be a divisor on X such that the linear system $|D|$ has no fixed components and yields a birational map $\varphi : X \dashrightarrow \mathbb{P}^n$ which is an isomorphism between an open subset U of X and an open subset V of $\mathbb{P}^n$ whose complement is of codimension ≥ 2 in $\mathbb{P}^n$. Then for every $m \geq 0$, φ induces isomorphisms $H^0(X, \mathcal{O}_X(mD)) \cong H^0(\mathbb{P}^n, \mathcal{O}_{\mathbb{P}^n}(m))$. In particular, if $|mD|$ is base point free for some $m > 0$, then $|D|$ is also base point free.*

Proof. Let U' be the open subset of X on which φ is defined. Since $|D|$ has no fixed components, $\mathrm{codim}_X(X \setminus U') \geq 2$, the restriction map $H^0(X, \mathcal{O}_X(mD)) \to H^0(U', \mathcal{O}_X(mD)|U')$ is an isomorphism. By hypothesis $\mathrm{codim}_{\mathbb{P}^n}(\mathbb{P}^n \setminus V) \geq 2$, whence the restriction map $\psi : H^0(\mathbb{P}^n, \mathcal{O}_{\mathbb{P}^n}(m)) \to H^0(V, \mathcal{O}_{\mathbb{P}^n}(m)|V)$ is also an isomorphism. Now the lemma follows easily from the commutative diagram

$$\begin{array}{ccc} H^0(\mathbb{P}^n, \mathcal{O}_{\mathbb{P}^n}(m)) & \xrightarrow{\varphi^*} & H^0(U', \mathcal{O}_X(mD)|U') \\ \psi \downarrow & & \downarrow \\ H^0(V, \mathcal{O}_{\mathbb{P}^n}(m)|V) & \longrightarrow & H^0(U, \mathcal{O}_X(mD)|U) \end{array}$$

in which the second vertical (restriction) map is injective and the bottom horizontal map is an isomorphism.

For the last assertion, observe that φ is a morphism, being the composition of the inverse of the isomorphism $v_m : \mathbb{P}^n \to v_m(\mathbb{P}^n)$ given by the m-th Veronese embedding of $\mathbb{P}^n$, and the morphism associated to $|mD|$. □

A first consequence of Theorem (4.4) is the following "birational version" of Kobayashi–Ochiai's result (see e.g. [B-S], (3.1.6)) which improves (2.5), 1).

Corollary (4.6). *Assume that X contains a quasi-line Y. Let H be a nef and big divisor on X such that $(H \cdot Y) = 1$. Let $n := \dim X$. Then we have:*

(i) *The pair (X, Y) is Zariski equivalent to $(\mathbb{P}^n, \text{line})$;*

(ii) *Assume moreover that there is no effective divisor orthogonal to Y. Then $(X, Y) \cong (\mathbb{P}^n, \text{line})$.*

Proof. (i) Let $d := H^n > 0$. Consider the degree-n Hilbert polynomial $p(t) := \chi(\mathcal{O}_X(tH))$. By duality and by the Kawamata–Viehweg vanishing theorem we have, for $i = 1, \ldots, n$,

$$p(-i) = \chi(\mathcal{O}_X(-iH)) = (-1)^n \chi(\mathcal{O}_X(K_X + iH)) = (-1)^n h^0(\mathcal{O}_X(K_X + iH)).$$

Since $(K_X \cdot Y) = -n-1$, we get $(K_X + iH) \cdot Y < 0$ and hence $h^0(\mathcal{O}_X(K_X + iH)) = 0$, $i = 1, \ldots, n$. Therefore

$$p(t) = \frac{d}{n!}(t+1)\ldots(t+n).$$

On the other hand,

$$p(-(n+2)) = (-1)^n d(n+1) = (-1)^n h^0(\mathcal{O}_X(K_X + (n+2)H)).$$

Thus $h^0(\mathcal{O}_X(K_X + (n+2)H)) = d(n+1) \geq n+1$, so that there exists an effective divisor $D \in |K_X + (n+2)H|$ such that $\dim |D| \geq n$. Note that $(D \cdot Y) = 1$. Then Theorem (4.4) applies to give the result.

(ii) With the same notation and arguing as in the proof of part (i), we have

$$p(-(n+1)) = (-1)^n d = (-1)^n h^0(\mathcal{O}_X(K_X + (n+1)H)) \neq 0.$$

Thus $K_X + (n+1)H$ is an effective divisor. Since $(K_X + (n+1)H) \cdot Y = 0$ we conclude that $K_X + (n+1)H$ is trivial and hence H is linearly equivalent to $D := K_X + (n+2)H$. Since H is nef and big, we can apply the Kawamata–Reid–Shokurov base point free theorem (see e.g. [B-S], (1.5.1)) to deduce that $|mD|$ is base point free for some $m \geq 1$. By Lemma (4.5) it follows that $|D|$ is base point free. So $|D|$ defines a birational morphism to $\mathbb{P}^n$ which has to be an isomorphism by Zariski's Main Theorem and our hypothesis. □

As a further application of Theorem (4.4), we give examples of strongly rational manifolds in any dimension $n \geq 3$ which do not dominate birationally $\mathbb{P}^n$.

Example (4.7). Let $n \geq 3$ and let $\sigma : Z \to \mathbb{P}^{n-1}$ be the blowing up of a point. Denote by E' the exceptional locus of σ and by H' the pullback of a hyperplane in $\mathbb{P}^{n-1}$. Moreover, let $L' := H' - E'$. Define X to be the projective bundle $\mathbb{P}(\mathcal{O}_Z \oplus \mathcal{O}_Z(L'))$ and let $\pi : X \to Z$ be the projection. Denote by S the section of π given by the natural surjection $\mathcal{O}_Z \oplus \mathcal{O}_Z(L') \to \mathcal{O}_Z(L')$. Set $E := \pi^*E'$, $H := \pi^*H'$ and $L := \pi^*L'$.

We first claim that there is no birational morphism from X to $\mathbb{P}^n$. To prove the claim it is enough to see that for any nef divisor, say A, on X, we have $A^n \neq 1$. To this end we shall compute intersection numbers with respect to the basis of $\text{Pic}(X)$ given by the classes of L, H and S. Note that the normal bundle of S in X is identified via π to $\mathcal{O}_Z(L')$. We get the following: $(L^i \cdot H^{n-i}) = (L^i \cdot S^{n-i}) = 0$ for any $0 \leq i \leq n$, and $(S^i \cdot L^j \cdot H^{n-i-j}) = 1$ if $i+j \leq n-1$ and $1 \leq i \leq n-1$, $0 \leq j \leq n-2$. Now write $A = aL + bH + cS$, where a, b, c are integers. Assuming A nef and noting that H and L are also nef, we get $(A \cdot S^{n-1}) = b \geq 0$, $(A \cdot H^{n-1}) = c \geq 0$. Assume moreover that $A^n = 1$. Computing A^n by the preceeding formulae we first see that A^n is divisible by b and c, so we get $b = c = 1$; thus, the expression for A^n becomes:

$$(12) \qquad A^n = \sum_{i=1}^{n-1} \sum_{j=0}^{i-1} \binom{n}{i}\binom{i}{j} a^j.$$

Next we get $(A^2 \cdot H^{n-2}) = 3 + 2a \geq 0$, giving $a \geq -1$ since a is an integer. Now (12) gives $A^n > 1$ if $a \geq 0$, while for $a = -1$ we get $A^n = 2$ if n is even and $A^n = 0$ if n is odd. In conclusion, we proved that $A^n \neq 1$ if A is nef, which shows that there is no birational morphism from X to $\mathbb{P}^n$.

Secondly, consider the linear system $|D|$, where $D := S + E$. It follows easily that $H^1(X, \mathcal{O}_X(E)) = 0$, $H^0(X, \mathcal{O}_X(E)) \cong \mathbb{C}$ and $H^0(X, \mathcal{O}_X(D)) \cong \mathbb{C}^{n+1}$. Let $Y \subset S$ be the curve corresponding via π to the pullback of a general line in $\mathbb{P}^{n-1}$ via σ. Clearly Y is a quasi-line on S; as we also have $(Y \cdot S) = (Y \cdot L) = 1$, it follows from the normal bundles sequence that Y is in fact a quasi-line of X. Moreover we have $(D \cdot Y) = 1$ and $\dim |D| = n$. So, it follows from Theorem (4.4) that X is strongly rational. □

As far as we know it is an open problem whether every small projective deformation of a rational manifold of dimension ≥ 3 is again rational. One can ask the same question for strongly rational manifolds. Some evidence towards a positive answer to this last question is the following simple result:

Proposition (4.8). *Any small projective deformation of a manifold X which dominates birationally $\mathbb{P}^n$ is again a projective manifold which dominates birationally $\mathbb{P}^n$.*

Proof. Let $f : \mathcal{X} \to T$ be a smooth projective morphism such that there is a point $t_0 \in T$ with the property that $f^{-1}(t_0) \cong X$. We may assume that T is a smooth curve. Since X dominates $\mathbb{P}^n$, there exists a divisor D on X such that $|D|$ is base point free and $D^n = 1$. As in the proof of (3.10), by using Artin's approximation theory, we see that, shrinking T if necessary, there exists a divisor $\mathcal{D}$ on $\mathcal{X}$, inducing D by restriction to X.

For every $t \in T$, set $X_t := f^{-1}(t)$ and $D_t := \mathcal{D}|X_t$. By standard arguments, the fact that the linear system $|D|$ is base point free implies that for any t close to t_0, $|D_t|$ is also base point free. We conclude the proof by noting that $D_t^n = D^n = 1$. □

Let us conclude this section by giving the reader some (not indispensable) motivation for our choice of candidates of manifolds containing non trivial quasi-lines in Theorem (3.2). The reason why this comment was not made earlier is that it makes use of Lemma (4.3) of this section. We remark that the adjunction-theoretic point of view, as in the following lines, was already systematically used in Langer's paper [L].

Let Y be a quasi-line on X and let

$$a := \min\{(Y \cdot H) \mid H \text{ ample divisor on } X\}.$$

Note that if $a = 1$, (2.5), 1) says that $(X, Y) \cong (\mathbb{P}^n, \text{line})$.

Lemma (4.9). *Let Y be a quasi-line on X and let H be an ample divisor on X such that $a = (Y \cdot H)$. Assume that $aK_X + (n+1)H$ is nef, $n = \dim X$. Then $aK_X + (n+1)H$ is trivial (hence, in particular, X is a Fano manifold).*

Proof. By the Kawamata–Reid–Shokurov base point free theorem we know that the complete linear system $|m(aK_X + (n+1)H)|$ is base point free for $m \gg 0$. Thus either $aK_X + (n+1)H$ is numerically trivial (and hence it is trivial since X is a Fano manifold), or $h^0(\mathcal{O}_X(m(aK_X + (n+1)H))) \geq 2$. Since $(aK_X + (n+1)H) \cdot Y = 0$ the latter possibility is ruled out by Lemma (4.3). □

Now the first interesting case to search for non trivial quasi-lines is $a = 2$, $n = 3$. If $2K_X + 4H$ is not nef, then we know (see e.g. [B-S]) that either $(X, H) \cong (\mathbb{P}^3, \mathcal{O}_{\mathbb{P}^3}(1))$, or (X, H) is a scroll over a curve C, or $(X, H) \cong (\mathcal{Q}, \mathcal{O}_\mathcal{Q}(1))$, where $\mathcal{Q}$ is a hyperquadric in $\mathbb{P}^4$. The first case is excluded since $a = 2$; the third case is excluded since, as already noted, a hyperquadric does not contain quasi-lines. In the second case, since we know that $H^1(X, \mathcal{O}_X) = 0$, we conclude that $C \cong \mathbb{P}^1$. If $2K_X + 4H$ is nef, X is a Fano threefold of index two by Lemma (4.9) above.

5. Concluding remarks

(5.1). Example (2.7) and Proposition (2.7.1) show that there are almost-lines Y in projective n-folds X which are not Zariski equivalent to a line in $\mathbb{P}^n$. It is then natural to find necessary and sufficient conditions for an almost-line Y in a projective n-fold to be Zariski equivalent to a line in $\mathbb{P}^n$. For instance, some conditions of this type are given in Theorem (4.4). We can further ask:

Question (5.2). Let Y be an almost-line in the projective n-fold X. Is it true that (X, Y) is Zariski equivalent to $(\mathbb{P}^n, \text{line})$ if and only if Y is G3 in X?

Note that the G3-condition is necessary because by [H-M] any line is G3 in $\mathbb{P}^n$. On the other hand any quasi-line Y in X is G2 by [Ha1], Theorem (6.7), because $N_{Y|X} \cong (n-1)\mathcal{O}_{\mathbb{P}^1}(1)$.

Question (5.2) would follow if the following would have an affirmative answer:

Question (5.3). Let Y be an almost-line in the projective n-fold X. Is it true that there is an isomorphism of formal schemes $\widehat{X}_{/Y} \cong \widehat{\mathbb{P}^n}_{/\text{line}}$?

(5.4). It is natural to look for examples and to try to classify quasi-lines lying on polarized or embedded n-folds X, at least for small values of their degree d. We refer to [O] and [L] for the case $n = 3$ and $d \leq 4$. An alternative question is to find examples of quasi-lines Y in X such that $\operatorname{Coker}(\alpha_0) \cong \mathbb{Z}/b\mathbb{Z}$ for $b \geq 3$.

Theorem (3.2) and Propositions (3.10) and (3.11) show that quasi-linearly connected manifolds form a quite large class of varieties. A possible attempt to understand them would be to study their minimal models in the following sense (suggested by Proposition (3.11)):

Definition (5.5). A pair (X, Y), with X a projective n-fold, and Y a quasi-line in X is said to be *relatively minimal* if there is no non-trivial birational morphism $f : X \to X'$, with X' a projective n-fold such that f is an isomorphism along Y and $f(Y)$ is a quasi-line in X'.

Corollary (4.6), (ii) suggests the following stronger definition of "minimality".

Definition (5.6). Let X be a projective n-fold, and Y a quasi-line in X. Let us introduce the *degeneration locus* of the pair (X, Y) as the closed subset

$$\mathcal{D}(X, Y) := \overline{X \setminus \{\text{locus of quasi-lines}\}}.$$

Then we say that the pair (X, Y) is *minimal* if $\operatorname{codim}_X \mathcal{D}(X, Y) \geq 2$. Note that this is the same as saying that there is no effective divisor orthogonal to Y.

There is clear evidence that a better understanding of the degeneration locus $\mathcal{D}(X, Y)$ would give more information about the geometry of manifolds containing quasi-lines.

Question (5.7). Let Y be a quasi-line in X. Is it true that if (X, Y) is minimal in the above sense, then X is a Fano n-fold?

Note that in [O], in the case of threefolds, a positive answer to the above question was conjectured under the stronger assumption that $\mathcal{D}(X, Y) = \emptyset$.

References

[dA] J. d'Almeida, Une caractérisation du plan projectif complexe, Enseign. Math. 41 (1995), 135–139.

[A] M. Artin, Algebraic approximation of structures over complete local rings, Inst. Hautes Études Sci. Publ. Math. 36 (1969), 23–58.

[B-S] M. C. Beltrametti, A. J. Sommese, The Adjunction Theory of Complex Projective Varieties, de Gruyter Exp. Math. 16, Walter de Gruyter, Berlin–New York 1995.

[Br] R. Braun, On the normal bundle of Cartier divisors on projective varieties, Arch. Math. 59 (1992), 403–411.

[C] F. Campana, Connexité rationnelle des variétés de Fano, Ann. Sci. École Norm. Sup. 25 (1992), 539–545.

[E-G-P-S] G. Ellingsrud, L. Gruson, C. Peskine, S. A. Strømme, On the normal bundle of curves on smooth projective surfaces, Invent. Math. 80 (1985), 181–184.

[Gi1] D. Gieseker, P-ample bundles and their Chern classes, Nagoya Math. J. 43 (1971), 91–116.

[Gi2] D. Gieseker, On two theorems of Griffiths about embeddings with ample normal bundles, Amer. J. Math. 99 (1977), 1137–1150.

[G-H] Ph. Griffiths, J. Harris, Principles of Algebraic Geometry, Wiley Interscience, New York 1978.

[Gr] A. Grothendieck, Sur la classification des fibrés vectoriels sur la sphère de Riemann, Amer. J. Math. 79 (1957), 121–138.

[FGA] A. Grothendieck, Fondements de la Géométrie Algébrique, Extraits du Séminaire Bourbaki, Paris 1957–1962.

[Ha1] R. Hartshorne, Cohomological dimension of algebraic varieties, Ann. of Math. 88 (1968), 403–450.

[Ha2] R. Hartshorne, Varieties of small codimension in projective space, Bull. Amer. Math. Soc. 80 (1974), 1017–1031.

[Ha3] R. Hartshorne, Algebraic Geometry, Springer Verlag, Berlin–Heidelberg–New York 1977.

[H] H. Hironaka, Resolution of singularities of an algebraic variety over a field of characteristic zero, Ann. of Math. 79 (1964), 109–326.

[H-M] H. Hironaka, H. Matsumura, Formal functions and formal embeddings, J. Math. Soc. Japan 20 (1986), 52–82.

[Hi] N. J. Hitchin, Kählerian twistor spaces, Proc. London Math. Soc. 43 (1981), 133–150.

[Is] V. A. Iskovskih, Fano 3-folds. I, Math. USSR Izvestija 11 (1977), 485–527.

[K] J. Kollár, Rational Curves on Algebraic Varieties, Ergeb. Math. Grenzgeb. (3) 32, 1996.

[K-M-M] J. Kollár, Y. Miyaoka, S. Mori, Rationally connected varieties, J. Algebraic Geom. 1 (1992), 429–448.

[L] A. Langer, On twistor curves of small degree, Quart. J. Math. Oxford 49 (1998), 43–58.

[M-P] Y. Miyaoka, T. Peternell, Geometry of Higher Dimensional Algebraic Varieties, DMV Seminar 26, Birkhäuser, Basel 1997.

[Mu] D. Mumford, Abelian Varieties, TATA Institute of Fundamental Research, Bombay, Oxford Univ. Press, 1970.

[O]   W. M. Oxbury, Twistor spaces and Fano threefolds, Quart. J. Math. Oxford 45 (1994), 343–366.

[Y-Z] Y.-G. Ye, Q. Zhang, On ample vector bundles whose adjunction bundles are not numerically effective, Duke Math. J. 60 (1990), 671–687.

Transfert de métrique

Daniel Barlet et Jón Magnusson

0. Introduction

Dans notre précédent travail (voir [B-M 2]) nous avons montré comment l'amplitude du fibré normal à une intersection complète locale Y de codimension $n+1$ d'une variété complexe lisse Z est transférée au fibré normal du diviseur d'incidence $\Sigma_Y \subset S$ défini par Y relativement à la famille analytique de n-cycles $(X_s)_{s \in S}$ de Z paramètrée par l'espace analytique réduit S, sous des hypothèses d'incidence convenables.

L'étape suivante est, bien sûr, de préciser une métrique sur le fibré normal à Y et d'essayer de la transporter de façon "naturelle" en une métrique du fibré normal à Σ_Y.

Un contrôle de la courbure de la métrique ainsi transportée en fonction de la courbure de la métrique initiale devrait alors permettre d'obtenir des résultats plus fins que la méthode précédente qui consiste à "transférer" des sections des puissances symétriques du fibré normal à Y dans Z en des sections des puissances symétriques du fibré normal à Σ_Y dans S (pour préciser cette assertion, le lecteur se reportera à la construction de l'intégration des classes de cohomologie à support Y du théorème 7 de [B-M 1] et à la proposition 1 de [B-M 2]).

L'objectif de cet article est d'expliquer comment on peut transporter une métrique hermitienne C^2 sur le fibré conormal à Y pour obtenir, de façon "naturelle" (c'est à dire avec des bonnes propriétés fonctorielles) une métrique C^0 singulière (donc éventuellement nulle en certains points) sur le fibré conormal du diviseur d'incidence de Y. De plus nous donnons une formule simple calculant la courbure (au sens des $(1, 1)$ courants) de la métrique ainsi obtenue: c'est la somme d'une image réciproque convenable de la courbure de la métrique initiale et du courant d'intégration sur les composantes de codimension 1 du lieu de dégénérescence des intersections avec Y le long de Σ_Y (avec des multiplicités convenables). Cette écriture correspond à la décomposition de Lebesgue (partie absolument continue et partie singulière) de cette mesure.

Ces points sont précisés dans le théorème et son corollaire.

1. Métriques hermitiennes singulières

Dans cet article on confondra systématiquement un fibré vectoriel avec le faisceau localement libre des sections holomorphes de ce fibré; nous utiliserons souvent le même symbôle pour les deux.

Soit L un fibré en droites sur un espace analytique X.

Définition. Une *métrique (hermitienne) singulière* sur L est une métrique $\mathcal{H}$ telle que pour toute trivialisation $\theta : L|\mathcal{U} \to \mathcal{U} \times \mathbb{C}$ on ait:

$$\mathcal{H}(v) = |\theta(v)|e^{-\varphi(x)}, \ x \in \mathcal{U} \text{ et } v \in L_x, \text{ où } \varphi \in L^1_{\text{loc}}(\mathcal{U}).$$

Pour une discussion détaillée des métriques singulières le lecteur pourra consulter par exemple [De].

Considérons un exemple simple mais fondamental de métrique singulière.

Exemple. Soit X un espace analytique réduit et soit D un diviseur de Cartier effectif sur X. Pour k dans $\mathbb{Z}$ on note $\mathcal{O}_X(kD)$ le fibré en droites sur X défini par

$$\mathcal{O}_X(kD) := \{f \in \mathcal{M}_X \mid \text{div}(f) + kD \geq 0\}.$$

On appelle métrique singulière canonique sur $\mathcal{O}_X(kD)$, notée $\mathcal{H}_k$, la métrique (hermitienne) singulière définie par $\mathcal{H}_k(f) = |f|$.

On constate que pour $k \leq 0$ la métrique $\mathcal{H}_k$ est C^0 alors que pour $k > 0$ la métrique $\mathcal{H}_k$ explose le long de D.

On a, bien sûr, $\mathcal{O}_X(kD) = \mathcal{O}_X(D)^{\otimes k}$ et $\mathcal{H}_k = \mathcal{H}_1^{\otimes k}$.

La courbure de $\mathcal{H}_k$ au sens de courants de type $(1, 1)$ est donnée par

$$\text{curv}(\mathcal{H}_k) = k[D],$$

où $[D]$ désigne le courant d'intégration sur D. Ceci se déduit de la formule de Schwartz-Lelong

$$\frac{i}{\pi}\partial\bar{\partial} \log|f| = [f = 0].$$

En particulier si on a un isomorphisme

$$\theta : L \xrightarrow{\sim} \mathcal{O}_X(-D) \otimes M,$$

où L est un fibré en droites sur X et M est une fibré en droites sur X muni d'une métrique hermitienne h de classe C^2, on en déduira sur L la métrique singulière

$$\xi := \theta^*(\mathcal{H}_{-1} \otimes h)$$

qui sera C^0 et de courbure

$$\text{curv}(\xi) = \text{curv}(h) - [D]$$

dont $\text{curv}(h)$ sera la partie absolument continue et $-[D]$ la partie singulière.

Proposition. *Soit Λ un fibré en droites sur un espace analytique réduit Σ. Soit $\pi : \Sigma^\sharp \to \Sigma$ une modification propre de Σ et soit*
$$\varphi : \pi^*(\Lambda) \longrightarrow E$$
un morphisme linéaire dans les fibres et de rang générique 1 sur $\Sigma^\sharp$ à valeurs dans un fibré vectoriel E sur $\Sigma^\sharp$ muni d'une métrique hermitienne h de classe C^2.

A ces données on associe une métrique singulière L^∞_{loc} sur Λ, notée $\xi(h, \varphi)$. Cette correspondance vérifie les propriétés fonctorielles suivantes:

1) Localisation: *si U est un ouvert de Σ on a*
$$\xi\bigl(h_{|\pi^{-1}_{(U)}}, \varphi_{|\pi^{-1}_{(U)}}\bigr) = \xi(h, \varphi)|_U.$$

2) Sous fibré: *si $\theta : E \hookrightarrow F$ est un morphisme injectif de fibrés vectoriels (θ de rang constant égal à $\mathrm{rg}(E)$) et si k est une métrique hermitienne C^2 sur F, on a*
$$\xi\bigl(\theta^*(k), \varphi\bigr) = \xi(k, \theta \circ \varphi).$$

3) Singularité: *si $\pi = \mathrm{id}_\Sigma$ et si $\varphi : \Lambda \to \mathcal{O}_\Sigma$ définit le diviseur de Cartier τ de Σ (effectif), on a*
$$\xi(h, \varphi) = \tilde{\varphi}^*\bigl(\mathcal{H}_{-1}(\tau) \otimes h\bigr)$$
où $\tilde{\varphi} : \Lambda \to \mathcal{O}_\Sigma(-\tau) \otimes \mathcal{O}_\Sigma$ est l'isomorphisme déduit de φ.[1]

4) Invariance par modification: *Si $\tilde{\Sigma} \xrightarrow{q} \Sigma^\sharp$ est une modification propre de $\Sigma^\sharp$ et si $q^*(\varphi) : (\pi \circ q)^*(\Lambda) \to q^*E$ est l'image réciproque de φ sur $\tilde{\Sigma}$, on a*
$$\xi(h, \varphi) = \xi\bigl(q^*(h), q^*(\varphi)\bigr).$$

De plus, les propriétés 1) à 4) caractérisent cette correspondance.

La preuve de la proposition s'appuiera sur le lemme suivant:

Lemme. *Soit $\Sigma^\sharp$ un espace analytique réduit et soit $\varphi : L \to E$ un morphisme de $\mathcal{O}_{\Sigma^\sharp}$-modules localement libres de rangs respectifs 1 et $n+1$. Soit $\tau^\sharp$ le "lieu de zéros" de φ et supposons $\tau^\sharp$ d'intérieur vide dans $\Sigma^\sharp$. Soit $q : \tilde{\Sigma} \to \Sigma^\sharp$ l'éclatement de $\tau^\sharp$ dans $\Sigma^\sharp$ et notons $\tilde{\tau}$ le diviseur de Cartier $q^*(\tau^\sharp)$. Notons $\Phi : \tilde{\Sigma} \to \mathbb{P}(E)$ l'application holomorphe associée à φ et notons T le fibré en droites tautologique sur $\mathbb{P}(E)$. Alors le morphisme*
$$q^*\varphi : q^*L \longrightarrow q^*E$$
se factorise en un isomorphisme "naturel"
$$\theta : q^*L \longrightarrow \mathcal{O}_{\tilde{\Sigma}}(-\tilde{\tau}) \otimes \Phi^*T$$
et l'inclusion canonique
$$\mathcal{O}_{\tilde{\Sigma}}(-\tilde{\tau}) \otimes \Phi^*T \longrightarrow \mathcal{O}_{\tilde{\Sigma}} \otimes q^*E \xrightarrow{\sim} q^*E.$$

---
[1] φ donne les isomorphismes suivants $\Lambda \simeq \mathrm{Im}\,\varphi = \mathcal{I}_\tau \simeq \mathcal{O}_\Sigma(-\tau)$.

Remarques. 1) Si on considère l'application holomorphe des espaces totaux associées à φ, alors l'idéal de définition de $\tau^\sharp$ est localement engendré par les "composantes" de $\tau^\sharp$.

2) Le sous-espace $\tau^\sharp$ est d'intérieur vide dans $\Sigma^\sharp$ si et seulement si le morphisme φ est injectif (comme morphisme de faisceaux de $\mathcal{O}_{\Sigma^\sharp}$-modules).

3) Précisons que l'inclusion canonique ci-dessus est donnée par

$$i \otimes j : \mathcal{O}_{\tilde{\Sigma}}(-\tilde{\tau}) \otimes \Phi^* T \longrightarrow \mathcal{O}_{\tilde{\Sigma}} \otimes q^* E \xrightarrow{\sim} q^* E.$$

où $i : \mathcal{O}_{\tilde{\Sigma}}(-\tilde{\tau}) \to \mathcal{O}_{\tilde{\Sigma}}$ et $j : \Phi^* T \to q^* E$ sont respectivement l'inclusion de l'idéal de $\tilde{\tau}$ dans $\mathcal{O}_{\tilde{\Sigma}}$ et l'image réciproque par Φ de l'inclusion de T dans $\text{pr}^* E$ où $\text{pr} : \mathbb{P}(E) \to \Sigma^\sharp$ est la projection.

Preuve du lemme. Ce résultat est "standard". Nous le détaillons pour être complet. L'assertion est locale sur $\Sigma^\sharp$ et nous pouvons supposer que :

1) $\Sigma^\sharp \subset \mathcal{U}$ est un sous ensemble analytique fermé (muni de sa structure réduite) de l'ouvert de Stein $\mathcal{U}$ de $\mathbb{C}^n$

2) $L = \mathcal{O}_{\Sigma^\sharp}$ et $E = \mathcal{O}_{\Sigma^\sharp}^{\oplus n+1}$

3) le morphisme φ est induit par un $\mathcal{O}_\mathcal{U}$-morphisme :

$$(\varphi_0 \ldots \varphi_n) : \mathcal{O}_\mathcal{U} \longrightarrow \mathcal{O}_\mathcal{U}^{n+1}$$

où $(\varphi_0 \ldots \varphi_n)$ sont des fonctions holomorphes sur $\mathcal{U}$.

On a alors $\tau^\sharp \equiv \Sigma^\sharp \cap V(\varphi_0 \ldots \varphi_n)$ c'est à dire que $\tau^\sharp$ est défini dans $\Sigma^\sharp$ par l'idéal de $\mathcal{O}_{\Sigma^\sharp}$ engendré par $(\varphi_0 \ldots \varphi_n)$.

Soit $\tilde{\Sigma}$ l'adhérence du graphe dans $\Sigma^\sharp \times \mathbb{P}_n(\mathbb{C})$ de l'application holomorphe

$$\emptyset = (\varphi_0 \ldots \varphi_n) : \Sigma^\sharp - \tau^\sharp \longrightarrow \mathbb{P}_n(\mathbb{C}).$$

Alors $\tilde{\Sigma}$ est la réunion des composantes irréductibles de

$$\Gamma = \left\{ (\sigma, \xi) \in \Sigma^\sharp \times \mathbb{P}_n(\mathbb{C}) \,/\, \text{rg}\left(\emptyset(\sigma), \xi\right) \leq 1 \right\}$$

qui dominent une composante irréductible de $\Sigma^\sharp$.

La première projection $q : \tilde{\Sigma} \to \Sigma^\sharp$ est une modification propre et la seconde projection $\Phi : \tilde{\Sigma} \to \mathbb{P}_n(\mathbb{C})$ incarne l'application holomorphe $\Phi : \tilde{\Sigma} \to \mathbb{P}(E)$ déduite de φ.

Soit $\tilde{\tau} = q^*(\tau^\sharp)$; c'est un diviseur de Cartier de $\tilde{\Sigma}$ car sur l'ouvert $\{\xi_i \neq 0\}$ de $\tilde{\Sigma}$ on a

$$\emptyset_j(\sigma) = \frac{\xi_j}{\xi_i} \emptyset_i(\sigma)$$

ce qui montre que l'idéal $q^*\varphi_0 \ldots q^*\varphi_n$ est principal engendré par $q^*\varphi_i$ sur cet ouvert.

Le morphisme φ induit alors un morphisme $\tilde{\varphi} : \mathcal{O}_{\tilde{\Sigma}} \to \mathcal{O}_{\tilde{\Sigma}}^{\oplus n+1}$ (correspondant à $\tilde{\varphi} : q^* L \to q^* E$) donné par

$$\tilde{\varphi}(1)_{\sigma,\xi} = \left((\varphi_0)_\sigma, \ldots, (\varphi_n)_\sigma\right)$$

et sur l'ouvert $\{\xi_i \neq 0\}$ de $\tilde{\Sigma}$ on aura

$$\tilde{\varphi}(1)_{\sigma,\xi} = (\varphi_i)_\sigma \left(\left(\frac{\xi_0}{\xi_i}\right)_{\sigma,\xi}, \ldots, \left(\frac{\xi_n}{\xi_i}\right)_{\sigma,\xi} \right).$$

Ceci met en évidence le fait que $\tilde{\varphi}$ induit un isomorphisme de $\mathcal{O}_{\tilde{\Sigma}}$ sur $\mathcal{O}_{\tilde{\Sigma}}(-\tilde{\tau}) \otimes \Phi^*(\mathcal{O}_{\mathbb{P}_n}(-1))$ qui correspond à

$$\theta : q^*L \to \mathcal{O}_{\tilde{\Sigma}}(-\tilde{\tau}) \otimes \Phi^*(T)).$$

Ceci achève la preuve du lemme. □

Preuve de la proposition. Posons $L = \pi^*(\Lambda)$ sur $\Sigma^\sharp$. En utilisant le lemme précédent on obtient le diagramme commutatif:

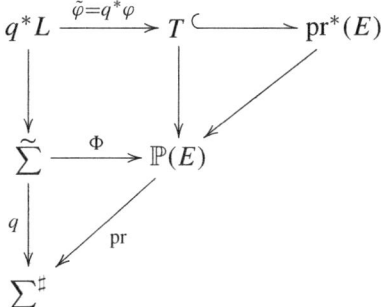

où $\tilde{\varphi}$ induit l'isomorphisme θ du lemme:

$$\theta : q^*L \xrightarrow{\sim} \mathcal{O}_{\tilde{\Sigma}}(-\tilde{\tau}) \otimes \Phi^*(T).$$

Munissons le fibré en droites T de la métrique C^2 induite par $\mathrm{pr}^*(h)$ et notons $q^*(h)$ la métrique correspondante sur $\Phi^*(T)$.

Comme $\pi \circ q : \tilde{\Sigma} \to \Sigma$ est une modification propre, on définit alors $\xi(h, \varphi)$ comme induite par $\theta^*(\mathcal{H}_{-1}(\tilde{\tau}) \otimes q^*(h))$ sur Λ.

Comme $\mathcal{H}_{-1}(\tilde{\tau}) \otimes q^*(h)$ est singulière C^0 on obtient bien ainsi une métrique singulière L^∞_{loc} sur Λ.

Les propriétés 1) à 4) de cette construction sont immédiates à vérifier.

Pour voir qu'elles caractérisent cette construction il suffit, grâce à 4) de remonter à $\tilde{\Sigma}$ et de constater que 2) et 3) imposent alors la suite des opérations puisque $\tilde{\varphi}$ se factorise par θ et $i \otimes j$ qui est une inclusion de rang 1 de fibrés vectoriels □

Remarque. Si $\pi : \Sigma^\sharp \to \Sigma$ est finie et si l'on a $\mathrm{rg}\, \varphi_x = 0$ dès que π n'est pas un isomorphisme local près de x, alors la métrique $\xi(h, \varphi)$ sera nécessairement C^0 sur Λ: en effet $\mathcal{H}_{-1}$ s'annule le long de $\tilde{\tau}$ et donc aussi $\theta^*(\mathcal{H}_{-1}(\tilde{\tau}) \otimes q^*(h))$.

Elle descend donc en une métrique continue sur Λ, nulle là où π n'est pas un isomorphisme local.

Ceci sera le cas dans l'application que nous donnerons de la proposition pour le théorème de transfert de métrique.

Complément (étude de la courbure). Sous les hypothèses de la proposition, la courbure au sens des $(1, 1)$ courants sur Σ de $\xi(h, \varphi)$ est donnée par

$$\operatorname{curv}\bigl(\xi(h, \varphi)\bigr) = (\pi \circ q)_*\bigl(\operatorname{curv} q^*(h)\bigr) - (\pi \circ q)_*[\tilde{\tau}]$$

où $[\tilde{\tau}]$ désigne le courant d'intégration sur $\tilde{\tau}$ et où $q^*(h)$ est la métrique induite par h sur $\Phi^*(T)$. Le premier terme est donc la partie absolument continue L^1_{loc} et le second la partie singulière de cette mesure.

En particulier $\operatorname{curv}(h) \leq 0$ (ou même seulement $\operatorname{curv}\bigl(\operatorname{pr}^*(h)|_T\bigr) \leq 0$) implique

$$\operatorname{curv}\bigl(\xi(h, \varphi)\bigr) \leq 0.$$

Par ailleurs, si $\tau_1 := (\pi \circ q)([\tilde{\tau}])$ est de codimension ≥ 2 le terme singulier de la courbure disparait (la métrique est singulière mais sa courbure est L^1_{loc}).

On remarquera aussi que $\Phi : \tilde{\Sigma} \to \mathbb{P}(E)$ est un plongement générique et donc que la stricte négativité de $\operatorname{curv}(\operatorname{pr}^*(h)|_T)$ sur $\Phi(\tilde{\Sigma})$ implique la stricte négativité du terme $\operatorname{curv}\bigl(q^*(h)\bigr)$ génériquement sur $\tilde{\Sigma}$ (donc de $\operatorname{curv}\bigl(\xi(h, \varphi)\bigr)$ génériquement sur Σ).

2. Le théorème

Commençons par une définition qui correspond à l'hypothèse d'incidence que nous ferons dans le théorème :

Définition. Soit $\mathcal{E} = (\pi, \tilde{X}, p)$ une n-équerre et soit $Y \subset Z$ un pôle pour Σ qui soit intersection complète locale relative dans Z. Notons Σ_Y le diviseur d'incidence de Y dans S et soit

$$\Sigma^\sharp = \bigl\{\, \pi^{-1}(\Sigma_Y) \cap p^{-1}(Y) \,\bigr\}_{\text{red}}.$$

On dira que l'incidence avec Y est *génériquement bonne* si :

1) $\Sigma^\sharp \xrightarrow{\pi} |\Sigma_Y|$ est de degré générique 1 (sur chaque composante irréductible de $|\Sigma_Y|$),

2) pour s générique dans $|\Sigma_Y|$ le cycle $X_s = p_*\bigl(\pi^{-1}(s)\bigr)^{(2)}$ est non tangent à Y en son point d'intersection[3].

Remarque. Dans $\mathbb{C}^N$ les couples de sous-espaces vectoriels (E_n, F_{N-n-1}) qui ne sont pas en somme directe est de codimension ≥ 2 dans le produit des grassmanniennes

---
(2) Ceci est bien défini près de Y car $p^{-1}(Y) \cap \pi^{-1}(s)$ est fini.
(3) C'est à dire que si $f_0 \ldots f_n$ engendrent localement $\mathcal{I}_Y$ près de $X_s \cap Y$, $f_*(X_s)$ est un diviseur lisse près de 0 dans $\mathbb{C}^{n+1}$.

correspondantes. On peut donc s'attendre, dans la situation "générique" à avoir bonne incidence sauf en codimension ≥ 2 sur $|\Sigma_Y|$.

Soit $\mathcal{E} = (\pi, \tilde{X}, p)$ une n-équerre de dimension pure sur un espace analytique réduit S avec $p : \tilde{Z} \to Z$ où $\tilde{Z}$ est un espace analytique réduit. Soit Y un pôle pour $\mathcal{E}$, localement intersection complète relative dans Z de codimension $n+1$. Notons Σ_Y ou $\Sigma(\mathcal{E}, Y)$ le diviseur d'incidence de Y dans S et supposons que l'incidence de $\mathcal{E}$ avec Y soit génériquement bonne. Soit h une métrique hermitienne C^2 sur le conormal $N^*_{Y|Z}$.

Théorème. *A la donnée de $\mathcal{E}, Y$ et h comme ci-dessus on associe une métrique singulière, notée $\xi(\mathcal{E}, Y, h)$, sur le fibré en droites $N^*_{\Sigma_Y|S} := \mathcal{I}_{\Sigma_Y}/\mathcal{I}^2_{\Sigma_Y}$ sur $\Sigma := |\Sigma_Y|$. Cette correspondance vérifie les propriétés suivantes.*

1) (Aspect local). *Pour W ouvert de $\tilde{Z}$ tel que $\mathcal{E}|W$ admette Y comme pôle posons*
$$\Sigma_W := \left\{ s \in \Sigma \mid \pi^{-1}(s) \cap p^{-1}(Y) \subset W \right\}_{\text{red}}.$$
Alors $\xi(\mathcal{E}|W, Y, h) = \xi(\mathcal{E}, Y, h)|_{\Sigma_W}$.

2) (Changement de base). *Soit $\lambda : T \to S$ un morphisme d'espaces analytiques réduits tel que l'incidence de $\lambda^*\mathcal{E}$ avec Y soit génériquement bonne. Alors on a*
$$\xi(\lambda^*\mathcal{E}, Y, h) = \lambda^*\xi(\mathcal{E}, Y, h).$$

3) (Morphisme plat). *Soit $q : Z \to Z_1$ un morphisme plat d'espaces analytiques et soit $Y_1 \subset Z_1$, une intersection complète locale relative dans Z_1, vérifiant $q^*Y_1 = Y$. Alors l'incidence de $\mathcal{E}_1 := (\pi, \tilde{X}, q \circ p)$ avec Y_1 est génériquement bonne et pour h_1 métrique C^2 sur $N^*_{Y_1|Z_1}$ on a*
$$\xi(\mathcal{E}_1, Y_1, h_1) = \xi(\mathcal{E}, Y, q^*h_1).$$

4) (Image directe). *Soit un diagramme commutatif*

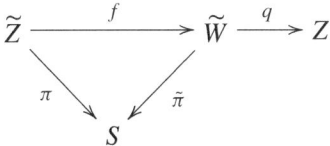

d'espaces analytiques avec $\tilde{W}$ réduit, f propre et surjectif, $\tilde{\pi}$ n-équidimensionnel et $p = q \circ f$. Alors
$$\xi(f_*\mathcal{E}, Y, h) = \xi(\mathcal{E}, Y, h)$$
où $f_*\mathcal{E} = (\tilde{\pi}, f_*\tilde{X}, q)$.

Démonstration du théorème. La démonstration du théorème consiste essentiellement à construire la métrique singulière $\xi(\mathcal{E}, Y, h)$ car une fois construite il sera assez évident que la correspondance vérifie les quatre propriétés ci-dessus. Cette construction sera donnée au paragraphe 3. Nous allons d'abord faire quelques remarques sur les énoncés et donner un corollaire important du théorème qui explique comment on peut, à l'aide du complément "étude de la courbure" de la proposition, expliciter la courbure de la métrique singulière $\xi(\mathcal{E}, Y, h)$.

Remarques. 1) Dans le n°2, pour que l'incidence de $\lambda^*\mathcal{E}$ et Y soit génériquement bonne il suffit que chaque composante irréductible de $\lambda^{-1}(\Sigma)$ domine une composante irréductible de Σ.

2) La formule $\xi(\lambda^*\mathcal{E}, Y, h) = \lambda^*\xi(\mathcal{E}, Y, h)$ dans le n°2 a un sens car l'on sait que sous ces hypothèses, $\Sigma(\lambda^*\mathcal{E}, Y) = \lambda^*\Sigma(\mathcal{E}, Y)$ et donc le fibré conormal au diviseur de Cartier $\Sigma(\lambda^*\mathcal{E}, Y)$ dans T est l'image réciproque du fibré conormal à $\Sigma(\mathcal{E}, Y)$ dans S.

3) On notera que sous les hypothèses du n°3 on a $\Sigma_Y = \Sigma_{Y_1}$ dans S et que l'égalité
$$N^*_{Y|Z} = q^*(N^*_{Y_1|Z_1})$$
montre que q^*h_1 est bien une métrique C^2 sur $N^*_{Y|Z}$.

Corollaire. *Dans la situation du théorème on a les implications suivantes:*[4]

1) *Si* $\mathrm{curv}(h) \leq 0$ *sur* $N^*_{Y|Z}$ *ou même seulement* $\mathrm{curv}(\mathrm{pr}^* h|_T) \leq 0$ *le long de* $\Phi(\tilde{\Sigma})$ *alors* $\mathrm{curv}(\xi(\mathcal{E}, Y, h)) \leq 0$ *au sens des courants de type* $(1, 1)$ *sur* Σ.

2) *Si l'incidence de $\mathcal{E}$ avec Y est bonne sauf peut-être en codimension ≥ 2 sur Σ, alors* $\mathrm{curv}(\xi(\mathcal{E}, Y, h))$ *est absolument continue sur* Σ.

3) *Sous l'hypothèse du 2) si on a* $\mathrm{curv}(h) \geq 0$ *sur* $N^*_{Y|Z}$ *ou même seulement* $\mathrm{curv}(\mathrm{pr}^* h|_T) \geq 0$ *le long de* $\Phi(\tilde{\Sigma})$, *alors on aura* $\mathrm{curv}(\xi(\mathcal{E}, Y, h)) \geq 0$ *sur* Σ.

3. Démonstration du théorème

Dans ce qui suit nous utiliserons systématiquement la terminologie introduite dans [B-M 1].

Notons respectivement $\mathcal{I}_Y$ et $\mathcal{I}_\Sigma$ les idéaux de définition de Y et Σ_Y dans Z et S.

Soit $\mathcal{J}$ l'idéal de définition de $p^*(Y)$ (donc $\mathcal{J}$ est l'image dans $\mathcal{O}_{\tilde{Z}}$ de $p^*(\mathcal{I}_Y)$). On a alors un $\mathcal{O}_{\tilde{Z}}$ morphisme canonique
$$\varphi_0 : \pi^*(\mathcal{I}_\Sigma) \longrightarrow \mathcal{J}$$
d'après [B-M 1], proposition 9.

---

(4) Voir la proposition pour la terminologie.

On en déduit un morphisme de $\mathcal{O}_{\tilde{Z}}$-modules

$$\varphi_1 : \mathcal{O}_{\tilde{Z}}/\mathcal{J} \underset{\mathcal{O}_{\tilde{Z}}}{\otimes} \pi^*\bigl(\mathcal{I}_\Sigma/\mathcal{I}_\Sigma^2\bigr) \longrightarrow \mathcal{J}/\mathcal{J}^2$$

puisque l'on a $\pi^*\mathcal{I}_\Sigma/\pi^*\mathcal{I}_\Sigma^2 \simeq \pi^*\bigl(\mathcal{I}_\Sigma/\mathcal{I}_\Sigma^2\bigr)$.

Montrons maintenant que la surjection naturelle

$$\eta_1 : p^*\bigl(\mathcal{I}_Y/\mathcal{I}_Y^2\bigr) \longrightarrow \mathcal{J}/\mathcal{J}^2$$

induit un isomorphisme

$$\eta_2 : p^*\bigl(\mathcal{I}_Y/\mathcal{I}_Y^2\bigr) \underset{\mathcal{O}_{\tilde{Z}}}{\otimes} \mathcal{O}_{\tilde{Z}}/\sqrt{\mathcal{J}} \longrightarrow \mathcal{J}/\mathcal{J}^2 \underset{\mathcal{O}_{\tilde{Z}}}{\otimes} \mathcal{O}_{\tilde{Z}}/\sqrt{\mathcal{J}}.$$

Comme $p^{-1}(Y)$ est de codimension $n+1$ dans $\tilde{Z}$ on a

$$\dim_{\mathbb{C}} \left((\mathcal{J}/\mathcal{J}^2)_{\tilde{z}} \underset{\mathcal{O}_{\tilde{Z},\tilde{z}}}{\otimes} \mathbb{C}_{\tilde{z}}\right) \geq n+1$$

pour chaque $\tilde{z} \in p^{-1}(Y)$.

D'autre part, $\mathcal{I}_Y/\mathcal{I}_Y^2$ est un $\mathcal{O}_Y$-module localement libre de rang $n+1$ d'après notre hypothèse sur Y; il s'en suit que

$$\dim_{\mathbb{C}} \left((\mathcal{J}/\mathcal{J}^2)_{\tilde{z}} \underset{\mathcal{O}_{\tilde{Z},\tilde{z}}}{\otimes} \mathbb{C}_{\tilde{z}}\right) = n+1$$

pour tout $\tilde{z} \in p^{-1}(Y)$. Par conséquent le morphisme η_1 induit un isomorphisme sur $p^{-1}(Y)_{\text{red}}$. Donc η_2 est bien un isomorphisme.

Soit maintenant $\tilde{z}_0 \in \pi^{-1}(\Sigma_Y) \cap p^{-1}(Y)$ et posons $s_0 = \pi(\tilde{z}_0)$. Quitte à remplacer $\tilde{Z}$ par un voisinage ouvert convenable de $p^{-1}(Y) \cap \pi^{-1}(s_0)$, Z par un voisinage ouvert convenable de $Y \cap p(\pi^{-1}(s_0))$ et S par un voisinage ouvert convenable de s_0, on peut supposer, d'après [B-M 1] lemme 8, que $(p_*\tilde{X}_s)_{s \in S}$ est une famille analytique de n-cycles de Z paramétrée par S, que Y est donné dans Z par des fonctions holomorphes $f_0, \ldots, f_n : Z \to \mathbb{C}$ qui induisent une application holomorphe plate (notée f) sur un voisinage ouvert de l'origine dans $\mathbb{C}^{n+1}$ ($Y = f^{-1}(0)$). On peut de plus supposer que ce voisinage ouvert de l'origine est de la forme $U \times D$ où U et D sont des polydisques de centre 0 relativement compacts dans $\mathbb{C}^n$ et $\mathbb{C}$ respectivement.

On peut également supposer que f est finie sur chaque cycle X_s de sorte que $\bigl(f_*(X_s)\bigr)_{s \in S}$ soit une famille analytique de diviseurs de $\mathbb{C}^{n+1}$ donnant des revêtements ramifiés sur U via la projection $U \times D \to U$ (c'est à dire que $U \times D$ est le "centre" d'une écaille adaptée à la famille $\bigl(f_*(X_s)\bigr)_{s \in S}$).

Notons par $(z, t_1, \ldots, t_n)$ les coordonnées sur $D \times U$ (dans cet ordre !) et posons $t = (t_1, \ldots, t_n)$. Soit $P(s,t)[z]$ le polynôme de Weierstrass en z que donne l'équation canonique du revêtement ramifié $f_*(X_s)$ dans l'écaille $U \times D$. Ecrivons

$$P(s,t)[z] = P(s,0)[0] + L(s,t,z) + F(s,t,z)$$

où $F(s, t, z) \in (t, z)^2$ et où $L(s, t, z)$ est linéaire en (t, z). Alors Σ est défini au voisinage de s_0 par l'équation $P(s, 0)[0] = 0$ qui est un générateur (local) de $\mathcal{I}_\Sigma$ (voir [B-M 1], théorème 7).

La fonction holomorphe sur $\tilde{Z}$
$$g : \tilde{z} \to P\big(\pi(\tilde{z}), 0\big)[0]$$
est donc générateur local de $\pi^* \mathcal{I}_\Sigma$ près de $\tilde{z}_0$.

Comme l'idéal $\mathcal{J}$ est engendré dans $\mathcal{O}_{\tilde{Z}, \tilde{z}_0}$ par les fonctions
$$\tilde{z} \to f_0(p(\tilde{z})), \ldots, f_n(p(\tilde{z}))$$
et que $\tilde{z} \to F\big(\pi(\tilde{z}), f_1(p(\tilde{z})), \ldots, f_n(p(\tilde{z})), f_0(p(\tilde{z}))\big)$ est dans $\mathcal{J}^2$, la fonction $-g$ et la fonction
$$h(\tilde{z}) = L\big(\pi(\tilde{z}), f_1(p(\tilde{z})), \ldots, f_n(p(\tilde{z})), f_0(p(\tilde{z}))\big)$$
auront la même image dans $\mathcal{J}/\mathcal{J}^2$.

On décrit donc le morphisme
$$\varphi_1 : \mathcal{O}_{\tilde{Z}}/\mathcal{J} \otimes \pi^*(\mathcal{I}_\Sigma/\mathcal{I}_\Sigma^2) \longrightarrow \mathcal{J}/\mathcal{J}^2$$
en disant que $\varphi_1(1 \otimes g) = -h$.

Remarquons bien que pour $s \in \Sigma$ donné, la forme linéaire $L(s, -, -)$ sur $\mathbb{C}^{n+1}$ est nulle dès que X_s coupe Y en au moins deux points distincts ou bien dès que X_s est tangent à Y en l'unique point d'intersection. Ceci décrit exactement l'ensemble τ_1 des points de Σ_Y au dessus desquels φ_1 n'est pas de rang 1.

Introduisons alors l'hypothèse de bonne incidence générique (voir paragraphe 2) et posons
$$\Sigma^\sharp := \big\{p^{-1}(Y) \cap \pi^{-1}(\Sigma_Y)\big\}_{\text{red}}$$

et
$$L := \pi^*(\mathcal{I}_\Sigma/\mathcal{I}_\Sigma^2)\big|_{\Sigma^\sharp}$$

$$E = p^*(\mathcal{I}_Y/\mathcal{I}_Y^2)\big|_{\Sigma^\sharp} \simeq \mathcal{J}/\mathcal{J}^2\big|_{\Sigma^\sharp}$$
grâce à l'isomorphisme η_2.

Le morphisme φ_1 donne alors un morphisme de $\mathcal{O}_{\Sigma^\sharp}$-modules localement libres de rang 1 et $n + 1$ respectivement
$$\varphi : L \longrightarrow E$$
qui est de rang générique 1.

On est alors ramené aux résultats de la proposition du paragraphe 1 qui donnent facilement le théorème et le corollaire. $\square$

Références

[B-M 1] D. Barlet, J. Magnusson, Intégration de classes de cohomologie méromorphes et diviseurs d'incidence, Ann. Sci. École Norm. Sup. (4) 31 (1998), 811–842.

[B-M 2] D. Barlet, J. Magnusson, Transfert de l'amplitude du fibré normal au diviseur d'incidence, J. Reine Angew. Math. 513 (1999), 71–95.

[De] J. P. Demailly, Singular hermitian metrics on positive line bundles, Lecture Notes in Math. 1507, Springer-Verlag, 1992.

On the classification of K3 surfaces with nine cusps

Wolf P. Barth *

Abstract. By a $K3$-surface with nine cusps I mean a compact complex surface with nine isolated double points A_2, but otherwise smooth, such that its minimal desingularisation is a $K3$-surface. In an earlier paper I showed that each such surface is a quotient of a complex torus by a cyclic group of order three. Here I try to classify these $K3$-surfaces, using the period map for complex tori. In particular I show:

A $K3$-surface with nine cusps carries polarizations only of degrees 0 or 2 modulo 6. This implies in particular that there is no quartic surface in projective three-space with nine cusps. (T. Urabe pointed out to me how to deduce this from Nikulin's Theorem 1.12.2 in [N2].)

In an appendix I give explicit equations of quartic surfaces in three-space with eight cusps.

1991 Mathematics Subject Classification: MSC (1991): 14J28, 14J15

Contents

1	Introduction	42
2	Topology	43
	2.1 The action on $H_1(A, \mathbb{Z})$	43
	2.2 The action on $H_2(A, \mathbb{Z}) = \bigwedge^2 H_1(A, \mathbb{Z})$	46
	2.3 The maps q_* and q^*	47
3	Moduli	49
	3.1 The period on A	49
	3.2 Polarizations on X	52
	3.3 Polarizations on A	52
	3.4 The classification problem	53
4	Appendix: Quartic surfaces with eight cusps	55

*Supported by the HCM project AGE (contract number ERBCH RXCT 940557)

1. Introduction

In [B] it was shown that each compact complex surface with nine cusps (A_2-type double points), but no further singularities, such that its minimal desingularization is a $K3$-surface, arises as a 3 : 1 quotient of a complex torus. It was claimed there, that there are nonalgebraic surfaces of this kind, and it was suggested that the examples of [BL] are the only algebraic surfaces of this type: double covers of the plane, branched over the dual sextic to a smooth cubic curve.

The aim of this note is to prove the first claim, and to show that there are lots of other examples than those in [BL]. In fact, the author was informed by P. Vanhaecke of his joint note [BV] with J. Bertin, where sextic surfaces in $\mathbb{P}_4$ are constructed, complete intersections of a quartic and a cubic hypersurface, which have nine cusps. They are 3 : 1 quotients of Jacobians of genus-2 curves with an automorphism of order three.

It is not an accident, that the surfaces of [BL] carry a polarization of degree two, while the surfaces of [BV] have a polarization of degree six. It is shown below, that only polarizations of degrees 0 or 2 modulo 6 appear on algebraic $K3$-surfaces with nine cusps. This has the following consequence:

There is no quartic surface in $\mathbb{P}_3$ with nine cusps, and no further singularities (or with other isolated rational double points only).

(T. Urabe pointed out to me that this is also a consequence of Nikulin's theorem 1.12.2 in [N2]. J. Stevens told me that it follows from the Arnold–Varchenko spectral bound. It seems however not to have been mentioned explicitly before.)

This leads to the following obvious question: Are there quartic surfaces in $\mathbb{P}_3$ with eight cusps? The existenc follows indeed from Urabe's partial classification of quartic surfaces with simple singularities [U, thm. 0.2]. But it seems quite hard to write down their equations directly. Fortunately [BV] describe their sextic surfaces in $\mathbb{P}_4$ with nine cusps quite explicitly. So it is fairly easy to project them from one of their cusps into $\mathbb{P}_3$. There probably is some general reason for the fact, that none of the projections degenerates. They all have eight cusps and no further singularities. However I check this by simple, but tedious direct computation (see the appendix).

I use the period (= holomorphic 2-form) of the covering two-dimensional torus to classify, at least to some extent, complex two-dimensional tori admitting a cyclic symmetry group of order three such that the quotient is a $K3$-surface with nine cusps. The period domain Ω is an open dense subset in a smooth quadric of dimension two, in particular it is connected. The moduli space for pairs (A, t) with A a complex torus of dimension two and t an automorphism of order three as above, is a quotient of Ω by an infinite arithmetic group. The general surface A of this type is not algebraic.

Fixing a polarization α on A reduces the period domain to a curve $\Omega_\alpha \subset \Omega$. Funnily enough this curve consists of two disjoint copies of the upper half-plane, interchanged by conjugation of the complex structure on A. Here I mean by a polarization a fixed divisor class $\alpha \in H^2(A, \mathbb{Z})$ with $\alpha^2 > 0$. Of course there are infinitely many different classes α, which are equivalent under the group $SL(4, \mathbb{Z})$. To induce a polarization on the quotient $X = A/t$, the class α has at least to be t-invariant. And the group identifying isomorphic triplets A, t, α is the subgroup $G \subset SL(4, \mathbb{Z})$ of elements commuting with t. Unfortunately it seems quite difficult to classify G-orbits on the set of classes α of fixed square α^2. At least it is a problem which I cannot solve.

Finally I give examples of algebraic tori A with automorphism t which are simple, i.e., not isogenous to a product of elliptic curves.

Just as the note [B] was essentially parallel to the first pages of Nikulin's paper [N1], which treats the natural involution $a \mapsto -a$, the basic method here is parallel to Remark 2 in section 1 of [N1]. Using the discriminant of its quadratic form we identify the orthogonal complement in the $K3$-lattice of the sub-lattice I spanned by the 18 classes of the rational curves resolving the nine cusps.

2. Topology

2.1. The action on $H_1(A, \mathbb{Z})$

In this section let $A = \mathbb{C}^2/\Gamma$ be a complex torus of dimension two. Let $t : A \to A$ be an automorphism of order three having the origin as an isolated fixed point. Assume also that $t^*(\omega) = \omega$, where ω is the holomorphic 2-form induced by the constant form $dz_1 \wedge dz_2$ on $\mathbb{C}^2$. (This assumption assures the existence of a nontrivial holomorphic 2-form on the quotient surface.)

t induces a linear automorphism $\tilde{t} : \mathbb{C}^2 \to \mathbb{C}^2$ on the universal covering, with determinant $= 1$. Since the origin is an isolated fixed point, it has no eigenvalue $= 1$. So it must have the two eigenvalues $\rho = e^{2\pi i/3}, \rho^2$.

In this section I want to identify the action of t on the homology $H_1(A, \mathbb{Z})$, or what is the same, the action of $\tilde{t}$ on the lattice $\Gamma \simeq \mathbb{Z}^4 \subset \mathbb{C}^2$.

As there is no t-invariant real line in $\mathbb{C}^2$, nor a t-invariant real subvector space of dimension three, there is no invariant sub-lattice in $H_1(A, \mathbb{Z})$ of rank one or three. But the t-orbit of each period $\gamma \in \Gamma$ spans an invariant sub-lattice of rank ≤ 3. This implies that each orbit $0 \neq \gamma, t(\gamma), t^2(\gamma)$ spans a sub-lattice of rank two.

Consider some primitve vector $\alpha \in \Gamma$ and denote the primitive vector $t(\alpha)$ by β. Then

$$t(\beta) = p \cdot \alpha + q \cdot \beta \quad \text{with} \quad p, q \in \mathbb{Q}$$

and

$$\begin{pmatrix} 0 & p \\ 1 & q \end{pmatrix}^3 = \begin{pmatrix} pq & p(p+q^2) \\ p+q^2 & q(2p+q^2) \end{pmatrix} = \begin{pmatrix} 1 & 0 \\ 0 & 1 \end{pmatrix}.$$

Here $pq = 1$ and $-p = q^2 = 1/p^2$ imply $p = q = -1$. So the action of t on the invariant sub-lattice generated by α and β is given by the matrix

$$\begin{pmatrix} 0 & -1 \\ 1 & -1 \end{pmatrix}.$$

Proposition 1. *There is a $\mathbb{Z}$-basis of the lattice $H_1(A, \mathbb{Z})$*

$$\alpha_1, \beta_1, \alpha_2, \beta_2,$$

in which the action of t is

$$t(\alpha_1) = \beta_1, \ t(\beta_1) = -\alpha_1 - \beta_1, \quad t(\alpha_2) = \beta_2, \ t(\beta_2) = -\alpha_2 - \beta_2.$$

Proof. Consider $H_1(A, \mathbb{Z})$ as a sub-lattice of the real vector space $H_1(A, \mathbb{R}) \simeq \mathbb{R}^4$ and choose on this vector space some t-invariant inner product $(-, -)$. Let $\alpha_1 \in H_1(A, \mathbb{Z})$ be some lattice vector of smallest length $\| \alpha_1 \| = \sqrt{(\alpha_1, \alpha_1)} \neq 0$ and put $\beta_1 := t(\alpha_1)$. On the plane spanned by α_1 and β_1 the automorphism t is an isometry of order three. This implies

$$(\alpha_1, \beta_1) = -\frac{1}{2} \| \alpha_1 \|^2.$$

If the vectors α_1 and β_1 would not span a primitive sub-lattice of $H_1(A, \mathbb{Z})$, there would be some nonzero lattice vector $\gamma = u\alpha_1 + v\beta_1 \in H_1(A, \mathbb{Z})$ with $0 \leq u, v < 1$. But such a vector would have squared length

$$(u^2 + v^2) \| \alpha_1 \|^2 + 2uv \cdot (\alpha_1, \beta_1) = (u^2 + v^2 - uv) \| \alpha_1 \|^2.$$

Either $u^2 \leq uv$ or $v^2 \leq uv$, hence

$$u^2 + v^2 - uv \leq \max\{u^2, v^2\} < 1.$$

Such a vector $\gamma \neq 0$ of length $< \| \alpha_1 \|$ cannot exist. So α_1 and β_1 indeed span a primitive sub-lattice in $H_1(A, \mathbb{Z})$.

Now choose a lattice vector $\alpha_2 \in H_1(A, \mathbb{Z})$ of smallest distance $\neq 0$ from the subvector space generated by α_1 and β_1 in $H_1(A, \mathbb{R})$ with $\beta_2 = t(\alpha_2)$. Exactly the same argument shows that the residues $\bar{\alpha}_2$ and $\bar{\beta}_2$ form a $\mathbb{Z}$-basis of the quotient lattice $H_1(A, \mathbb{Z})/(\mathbb{Z} \cdot \alpha_1 + \mathbb{Z} \cdot \beta_1)$. So $\alpha_1, \beta_1, \alpha_2, \beta_2$ form a $\mathbb{Z}$-basis of $H_1(A, \mathbb{Z})$. □

With respect to a $\mathbb{Z}$-basis as in the proposition the action of t is given by the matrix

$$T := \begin{pmatrix} 0 & -1 & 0 & 0 \\ 1 & -1 & 0 & 0 \\ 0 & 0 & 0 & -1 \\ 0 & 0 & 1 & -1 \end{pmatrix}.$$

As a sub-lattice of $\mathbb{C}^2$ the lattice Γ carries a natural orientation.

Proposition 2. *Each basis α_1, $t(\alpha_1)$, α_2, $t(\alpha_2)$ as above is negatively oriented with respect to the natural orientation on Γ.*

Proof. As a complex linear map, t has the two eigenvalues ω and ω^2. Let $c_1, c_2 \in \mathbb{C}^2$ be a complex basis of eigenvectors, so $t(c_1) = \omega c_1$ and $t(c_2) = \omega^2 c_2$. Clearly c_1 and ωc_1 represent the natural orientation of the complex line containing these two vectors, while c_2 and $\omega^2 c_2$ represent the opposite of the natural orientation of their line. This shows:

There is some real basis $\alpha_1 = c_1, t(\alpha_1), \alpha_2 = c_2, t(\alpha_2)$ of $\mathbb{C}^2$ representing the opposite of the natural orientation. The assertion follows, if we prove the next

Lemma 1. *All $\mathbb{R}$-bases of the form α_1, $t(\alpha_1)$, α_2, $t(\alpha_2)$ of $\mathbb{C}^2$ represent the same orientation.*

Proof. Let us change from the basis $\alpha_1, \beta_1 = t(\alpha_1), \alpha_2, \beta_2 = t(\alpha_2)$ to another $\mathbb{R}$-basis $\alpha_1', \beta_1' = t(\alpha_1'), \alpha_2', \beta_2' = t(\alpha_2')$ of $\mathbb{C}^2$.

a) If α_1' and β_1' span the same real plane as α_1 and β_1, and if α_2' and β_2' span the same real plane as α_2 and β_2, then the pairs α_i' and β_i' are obtained from α_i and β_i, $i = 1, 2$ in their respective planes by orientation preserving rotations. So $\alpha_1', \beta_1', \alpha_2', \beta_2'$ represent the same orientation as $\alpha_1, \beta_1, \alpha_2, \beta_2$.

b) If $\mathbb{R}\alpha_1' + \mathbb{R}\beta_1' = \mathbb{R}\alpha_1 + \mathbb{R}\beta_1$, but $\mathbb{R}\alpha_2' + \mathbb{R}\beta_2' \neq \mathbb{R}\alpha_2 + \mathbb{R}\beta_2$, then, by a), $\alpha_1', \beta_1', \alpha_2', \beta_2'$ has the same orientation as $\alpha_1, \beta_1, \alpha_2', \beta_2'$. Write

$$\alpha_2' = a_1 \alpha_1 + a_2 \beta_1 + a_3 \alpha_2 + a_4 \beta_2.$$

Then

$$\beta_2' = -a_2 \alpha_1 + (a_1 - a_2)\beta_1 - a_4 \alpha_2 + (a_3 - a_4)\beta_2$$

and

$$\det(\alpha_1, \beta_1, \alpha_2', \beta_2') = \det \begin{pmatrix} 1 & 0 & a_1 & -a_2 \\ 0 & 1 & a_2 & a_1 - a_2 \\ 0 & 0 & a_3 & -a_4 \\ 0 & 0 & a_4 & a_3 - a_4 \end{pmatrix} = a_3^2 + a_4^2 - a_3 a_4 \geq 0.$$

Again the orientation of $\alpha_1', \beta_1', \alpha_2', \beta_2'$ is the same as the one of $\alpha_1, \beta_1, \alpha_2, \beta_2$.

c) If $\mathbb{R}\alpha_i' + \mathbb{R}\beta_i' \neq \mathbb{R}\alpha_j + \mathbb{R}\beta_j$ for $i, j = 1, 2$, then by b) the orientation does not change, if we pass from $\alpha_1, \beta_1, \alpha_2, \beta_2$ to $\alpha_1, \beta_1, \alpha_2', \beta_2'$, and if we pass from $\alpha_1, \beta_2, \alpha_2', \beta_2'$ to $\alpha_1', \beta_1', \alpha_2', \beta_2'$. □

2.2. The action on $H_2(A, \mathbb{Z}) = \bigwedge^2 H_1(A, \mathbb{Z})$

On the exterior products of the basis vectors from the last section the automorphism t acts as follows:

$$\alpha_1 \wedge \beta_1 \mapsto \beta_1 \wedge (-\alpha_1 - \beta_1) = \alpha_1 \wedge \beta_1$$
$$\alpha_2 \wedge \beta_2 \mapsto \alpha_2 \wedge \beta_2$$
$$\alpha_1 \wedge \alpha_2 \mapsto \beta_1 \wedge \beta_2$$
$$\alpha_1 \wedge \beta_2 \mapsto -\beta_1 \wedge \alpha_2 - \beta_1 \wedge \beta_2$$
$$\beta_1 \wedge \alpha_2 \mapsto -\alpha_1 \wedge \beta_2 - \beta_1 \wedge \beta_2$$
$$\beta_1 \wedge \beta_2 \mapsto \alpha_1 \wedge \alpha_2 + \alpha_1 \wedge \beta_2 + \beta_1 \wedge \alpha_2 + \beta_1 \wedge \beta_2.$$

Using this table one finds the following t-invariant classes in $H_2(A, \mathbb{Z})$:

$$\gamma_1 := -\alpha_1 \wedge \beta_1,$$
$$\gamma_2 := \alpha_2 \wedge \beta_2,$$
$$\gamma_3 := \alpha_1 \wedge \beta_2 - \beta_1 \wedge \alpha_2,$$
$$\gamma_4 := \alpha_1 \wedge \alpha_2 + \alpha_1 \wedge \beta_2 + \beta_1 \wedge \beta_2.$$

The wedge product induces on $H_2(A, \mathbb{Z})$ an integral, unimodular quadratic form

$$(\alpha, \alpha') := \frac{\alpha \wedge \alpha'}{\alpha_1 \wedge \alpha_2 \wedge \beta_1 \wedge \beta_2}$$

with discriminant -1. (Recall from the last section that $\alpha_1 \wedge \alpha_2 \wedge \beta_1 \wedge \beta_2$ represents the natural orientation.) The matrix (γ_i, γ_j) is

$$\begin{pmatrix} 0 & 1 & 0 & 0 \\ 1 & 0 & 0 & 0 \\ 0 & 0 & 2 & 1 \\ 0 & 0 & 1 & 2 \end{pmatrix}$$

with determinant -3. This shows that the invariant classes $\gamma_1, \ldots, \gamma_4$ span a primitive sub-lattice $L_A \subset H_2(A, \mathbb{Z})$ of rank four.

Proposition 3. *The lattice L_A is the sublattice $H_2(A, \mathbb{Z})^{inv} \subset H_2(A, \mathbb{Z})$ of all t-invariant classes.*

Proof. As the sub-lattice $L_A \subset H_2(A, \mathbb{Z})$ is primitive, it suffices to show that the invariant subspace $H_2(A, \mathbb{R})^{inv} \subset H_2(A, \mathbb{R})$ has dimension at most four. But from the table above, exhibiting the action of t on the exterior products of the α_i and β_j one reads off that the action of t on $H_2(A, \mathbb{Z})$ has trace $= 3$. So $H_2(A, \mathbb{R})^{inv}$ is a proper subspace of $H_2(A, \mathbb{R})$. Its dimension cannot be five, since then there would be a t-invariant one-dimensional complement, and all of $H_2(A, \mathbb{R})$ would be t-invariant. So its dimension is at most four. □

The dual lattice $\check{L}_A \subset H_2(A, \mathbb{R})$ consists of all classes $\gamma \in \mathbb{R} \cdot L_A \subset H_2(A, \mathbb{R})$ with

$$(\alpha, \gamma) \in \mathbb{Z} \text{ for all } \alpha \in L_A.$$

Proposition 4. *The dual lattice $\check{L}_A$ has a $\mathbb{Z}$-basis consisting of*

$$\gamma_1, \ \gamma_2, \ \gamma_3, \ \frac{1}{3}(\gamma_3 + \gamma_4).$$

Proof. Let

$$\gamma = c_1 \gamma_1 + \cdots + c_4 \gamma_4 \in \mathbb{R} \cdot L_A, \quad c_i \in \mathbb{R}.$$

Then

$$(\gamma, \gamma_1) = c_2, \qquad (\gamma, \gamma_2) = c_1,$$
$$(\gamma, \gamma_3) = 2c_3 + c_4, \qquad (\gamma, \gamma_4) = c_3 + 2c_4.$$

So $\gamma \in \check{L}_A$ if and only if $c_1, c_2 \in \mathbb{Z}$ and if

$$2c_3 + c_4 \in \mathbb{Z}, \quad c_3 + 2c_4 \in \mathbb{Z}.$$

But the latter is equivalent with

$$c_3 - c_4 \in \mathbb{Z}, \quad 3(c_3 + c_4) \in \mathbb{Z}. \qquad \square$$

In the basis $\gamma_1, \gamma_2, \gamma_3, (\gamma_3 + \gamma_4)/3$ the quadratic form has the matrix

$$\begin{pmatrix} 0 & 1 & 0 & 0 \\ 1 & 0 & 0 & 0 \\ 0 & 0 & 2 & 1 \\ 0 & 0 & 1 & 2/3 \end{pmatrix}$$

and the discriminant $-1/3$.

2.3. The maps q_* and q^*

As in [B], let $q : A \to A/t = X$ be the quotient map. Let $\tilde{A}$ be the blow-up of A in the nine fixed points of t and $\tilde{X}$ the minimal desingularization of X. One has $H_2(\tilde{A}, \mathbb{Z}) = \mathbb{Z}^9 \perp H_2(A, \mathbb{Z})$ with $\mathbb{Z}^9$ spanned by the classes of the nine exceptional curves.

As in [B] denote by $I \subset H_2(\tilde{X}, \mathbb{Z})$ the lattice spanned by the eighteen (-1)-curves resolving the nine cusps on X. It is a lattice of rank 18 and discriminant 3^9. In [B] it was shown that it is contained in a primitive sublattice $\bar{I} \subset H_2(\tilde{X}, \mathbb{Z})$ with the quotient $\bar{I}/I$ being a group of order 3^3. This implies that the discriminant $d(\bar{I})$ of the quadratic form on $\bar{I}$ is $3^9/(3^3)^2 = 3^3$.

The orthogonal complement $L_X := I^\perp = \tilde{I}^\perp \subset H_2(\tilde{X}, \mathbb{Z})$ is a lattice of rank $22 - 18 = 4$. Since $H_2(\tilde{X}, \mathbb{Z})$ is unimodular with discriminant -1 the discriminant of L_X is

$$d(L_X) = -d(\tilde{I}) = -3^3.$$

The induced map $\tilde{q}_*$ clearly maps $H_2(A, \mathbb{Z})$ into L_X.

We identify

$$H_2(\tilde{A}, \mathbb{Z}) = H^2(\tilde{A}, \mathbb{Z}), \quad H_2(\tilde{X}, \mathbb{Z}) = H^2(\tilde{X}, \mathbb{Z})$$

via Poincaré-duality. Then we get a morphism

$$\tilde{q}^* : H_2(\tilde{X}, \mathbb{Z}) \to H_2(\tilde{A}, \mathbb{Z}).$$

It satisfies

$$(\tilde{q}^*\xi, \tilde{q}^*\xi') = 3 \cdot (\xi, \xi') \text{ for all } \xi, \xi' \in H_2(\tilde{X}, \mathbb{Z}),$$

and the projection formula

$$(\tilde{q}_*\alpha, \xi) = (\alpha, \tilde{q}^*\xi) \text{ for all } \alpha \in H_2(\tilde{A}, \mathbb{Z}), \xi \in H_2(\tilde{X}, \mathbb{Z}).$$

Proposition 5. *The map* $\tilde{q}_* : H_2(\tilde{A}, \mathbb{Z}) \to H_2(\tilde{X}, \mathbb{Z})$ *induces an isomorphism*

$$\check{L}_A(3) \to L_X$$

and the map $\tilde{q}^* : H_2(\tilde{X}, \mathbb{Z}) \to H_2(\tilde{A}, \mathbb{Z})$ *induces an isomorphism*

$$L_X(3) \to 3 \cdot \check{L}_A.$$

Proof. Since all classes $q^*(\xi)$, $\xi \in H_2(\tilde{X}, \mathbb{Z})$, are t-invariant, $\tilde{q}^*(L_X)$ is a sub-lattice of L_A. The endomorphism $\tilde{q}_* \tilde{q}^* : L_X \to L_X$ has the property

$$\tilde{q}_* \tilde{q}^*(\xi) = 3\xi$$

for all $\xi \in L_X$. This implies that the maps $\tilde{q}^* : L_X \to L_A$ and $\tilde{q}_* : L_A \to L_X$ are injective. So $q_*(L_A) \subset L_X$ is a sublattice of rank four, the rank of L_A, and $\tilde{q}^*(L_X)$ spans L_A over $\mathbb{Q}$.

For all $\xi, \xi' \in L_X$ we have

$$(\tilde{q}^*\xi, \tilde{q}^*\xi') = 3(\xi, \xi') \text{ and } (\tilde{q}_*\tilde{q}^*\xi, \tilde{q}_*\tilde{q}^*\xi') = (3\xi, 3\xi') = 3(\tilde{q}^*\xi, \tilde{q}^*\xi').$$

This implies

$$(\tilde{q}_*\alpha, \tilde{q}_*\alpha') = 3(\alpha, \alpha') \text{ for all } \alpha, \alpha' \in L_A.$$

If $\alpha \in L_A^\perp \subset H_2(A, \mathbb{Z}) \subset H_2(\tilde{A}, \mathbb{Z})$, then $q_*(\alpha) \in L_X$ with

$$(\xi, q_*(\alpha)) = (q^*(\xi), \alpha) = 0$$

for all $\xi \in L_X$. This implies $\tilde{q}_*(\alpha) = 0$.

And conversely, if $\tilde{q}_*(\alpha) = 0$, then $(\tilde{q}^*\xi, \alpha) = (\xi, \tilde{q}_*(\alpha)) = 0$ for all $\xi \in L_X$. So $\alpha \in L_A^\perp$. Hence $\tilde{q}_*$ defines an injective map

$$\tilde{q}_* : H_2(A, \mathbb{Z})/L_A^\perp \simeq \check{L}_A \to L_X.$$

The discriminant of the image lattice is

$$3^4 \cdot d(\check{L}_A) = -3^3 = d(L_X).$$

This shows $\tilde{q}_*(\check{L}_A) = L_X$. $\square$

As a corollary we obtain

Proposition 6. *The lattice L_X has an integral basis, in which its quadratic form has the matrix*

$$\begin{pmatrix} 0 & 3 & 0 & 0 \\ 3 & 0 & 0 & 0 \\ 0 & 0 & 6 & 3 \\ 0 & 0 & 3 & 2 \end{pmatrix}.$$

3. Moduli

3.1. The period on A

A complex structure on the real four-dimensional torus $A = H_1(A, \mathbb{R})/H_1(A, \mathbb{Z})$ is given by two complex coordinates z_1, z_2. These are $\mathbb{R}$-linear maps $z_k : H_1(A, \mathbb{R}) \to \mathbb{C}$ inducing an $\mathbb{R}$-isomorphism $(z_1, z_2) : H_1(A, \mathbb{R}) \to \mathbb{C}^2$. This complex structure determines a *period* $\omega = z_1 \wedge z_2 \in H^2(A, \mathbb{C})$. ω is uniquely determined by the complex structure up to multiplication by complex scalars. This period satisfies the following two *period relations*

(i) $\omega \wedge \omega = 0,$ (ii) $\omega \wedge \bar{\omega} > 0.$

Relation (i) is obvious. Relation (ii) means the following: Write $z_k = x_k + iy_k$ with $\mathbb{R}$-linear maps $x_k, y_k : H_1(A, \mathbb{R}) \to \mathbb{R}$. Then

$$\begin{aligned}
\omega \wedge \bar{\omega} &= (x_1 + iy_1) \wedge (x_2 + iy_2) \wedge (x_1 - iy_1) \wedge (x_2 - iy_2) \\
&= -(x_1 + iy_1) \wedge (x_1 - iy_1) \wedge (x_2 + iy_2) \wedge (x_2 - iy_2) \\
&= -[-2i \cdot (x_1 \wedge y_1) \wedge (-2i) \cdot (x_2 \wedge y_2)] \\
&= 4 \cdot x_1 \wedge y_1 \wedge x_2 \wedge y_2
\end{aligned}$$

is a positive multiple of the form $x_1 \wedge y_1 \wedge x_2 \wedge y_2$ defining the orientation by the complex structure.

There is also the converse:

Given a class $\omega \in H^2(A, \mathbb{C})$ satisfying the period relations (i) and (ii), there is a unique complex structure on A belonging to this form.

Proof. Since $\omega \wedge \omega = 0$ by (i), the form $\omega \in \Lambda^2 H^1(A, \mathbb{C})$ decomposes, say $\omega = z_1 \wedge z_2$ with $z_1, z_2 \in H^1(A, \mathbb{C}) = \mathrm{Hom}_\mathbb{R}(H_1(A, \mathbb{R}), \mathbb{C})$. These functions z_1 and z_2 are uniquely determined by ω up to complex linear combination. Write $z_k = x_k + i y_k$ as above. Then (ii) implies $x_1 \wedge y_1 \wedge x_2 \wedge y_2 \neq 0$ and the map $(z_1, z_2) : H_1(A, \mathbb{R}) \to \mathbb{C}^2$ is bijective, i.e., it defines a complex structure on A. □

Proposition 7. *If $t^*\omega = \omega$, then the map $t : A \to A$ is $\mathbb{C}$-linear.*

Proof. By assumption
$$t^*(z_1) \wedge t^*(z_2) = t^*(\omega) = \omega = z_1 \wedge z_2.$$
So $t^*(z_1)$ and $t^*(z_2)$ are complex linear combinations of z_1 and z_2. □

Proposition 8. *Isomorphism classes (A, t) of complex tori A with an order-three automorphism t as in 1.1 are classified by the period domain Ω/G where*
$$\Omega = \{\mathbb{C} \cdot \omega \in \mathbb{P}(L_A) : \omega \wedge \omega = 0, \omega \wedge \bar{\omega} > 0\},$$
G is the group of orientation-preserving $\mathbb{Z}$-isomorphisms $g : L_A \to L_A$ commuting with the t-action specified in 2.1.

Proof. Let (A_1, t_1) and (A_2, t_2) be two such complex tori with automorphisms. Fix isomorphisms $H_1(A_k, \mathbb{Z}) = \mathbb{Z}^4$ such that t_k acts as in 2.1. An isomorphism $\phi : A_1 \to A_2$ preserves the automorphism if
$$\phi \circ t_1 = t_2 \circ \phi.$$
It induces an isomorphism $\phi_* : \mathbb{Z}^4 \to \mathbb{Z}^4$ commuting with t. So $\phi_* \in G$. It sends Ω to Ω and ω_1 to ω_2. □

This group G can be described a little more explicitly: Indeed
$$g = \begin{pmatrix} A & B \\ C & D \end{pmatrix}$$
with A, B, C, D integral 2×2-matrices belongs to G, if it is invertible, of determinant 1, and satisfies
$$AT = TA, \ldots, DT = TD, \text{ with } T = \begin{pmatrix} 0 & -1 \\ 1 & -1 \end{pmatrix}.$$
And the set of 2×2-matrices commuting with T consists of the $\mathbb{Z}$-algebra generated by the unit matrix $\mathbb{1}$ and T.

Theorem 1. *The moduli space Ω/G is connected, of dimension two.*

Proof. It suffices to show that Ω is connected. The intersection form on L_A has signature $(3, 1)$. Let us choose real coordinates $x_1, \ldots, x_4$ on $L_A \otimes \mathbb{R}$ such that in these coordinates this form is
$$x_1^2 + x_2^2 - x_3 x_4.$$
Let $z_k = x_k + i \cdot y_k$ be the corresponding complex coordinates on $L_A \otimes \mathbb{C} = H^2(A, \mathbb{C})^{inv}$. If ω has the coordinates $(c_1, \ldots, c_4)$, then
$$\omega \wedge \omega = c_1^2 + c_2^2 - c_3 c_4 = 0.$$
Now, if $c_3 = 0$, then $c_1^2 + c_2^2 = 0$ too. The second period condition in this case is
$$\omega \wedge \bar\omega = |c_1|^2 + |c_2|^2 > 0,$$
satisfied unless $\omega = (0, 0, 0, 1)$.

If $c_3 \neq 0$ we may assume $c_3 = 1$. Hence $\omega = (z_1, z_2, 1, z_1^2 + z_2^2)$ and
$$\begin{aligned}\omega \wedge \bar\omega &= |c_1|^2 + |c_2|^2 - \Re(c_1^2 + c_2^2) \\ &= 2(\Im(c_1)^2 + \Im(c_2)^2) \\ &> 0\end{aligned}$$
unless $\Im(c_1) = \Im(c_2) = 0$. So $\Omega \cap \{c_3 \neq 0\}$ is just a copy of $\mathbb{C}^2 \setminus \mathbb{R}^2$, hence connected. And $\Omega \cap \{c_3 = 0\}$ lies in its boundary. $\square$

The condition $\omega \wedge \omega = 0$ defines a non-degenerate quadric in $\mathbb{P}_3 = \mathbb{P}(L_A \otimes \mathbb{C})$. It is not contained in any hyper-plane. So the open subset Ω of this quadric is not contained in a hyper-plane too.

Now let A (and X) be algebraic with $C \subset A$ the pull-back of some ample divisor on X. It determines a class $\gamma_C \in L_A$. Since
$$\gamma_C \wedge \omega \sim \int_C \omega = 0,$$
in this case the period ω lies in the hyper-plane $\gamma_C^\perp \subset L_A \otimes \mathbb{C}$. This proves

Theorem 2. *The general complex torus A with an order-three automorphism t as in 2.1 is not algebraic.*

There is also this converse: If a class $\gamma \in H^2(A, \mathbb{Z})$ has the property $\gamma \wedge \omega = 0$, then it is of type $(1, 1)$. Being integral it is the first chern class of a line bundle on A. For $\omega \in L_A \otimes \mathbb{C}$, in particular all classes γ in the two-dimensional lattice $L_A^\perp \subset H^2(A, \mathbb{Z})$ have this property. This proves

Proposition 9. *For all complex tori A with an automorphism t as above, the group*
$$\mathrm{Pic}(A)/\mathrm{Pic}^0(A)$$
has rank at least two.

But beware: The lattice $L_A^\perp$ is negative definite. So all classes $\gamma \in L_A^\perp$ belong to line bundles on A, but if A is not algebraic, these line bundles do not come from divisors.

3.2. Polarizations on X

Let $\xi_1, \ldots, \xi_4 \in L_X \subset H^2(\tilde{X}, \mathbb{Z})$ the basis from the proposition 6. Each class $n_1 \xi_1 + \cdots + n_4 \xi_4$, $n_i \in \mathbb{Z}$, has self-intersection

$$6n_1 n_2 + 6(n_3^2 + n_3 n_4) + 2n_4^2 \equiv 0 \text{ or } 2 \mod 6.$$

Theorem 3. *There is no quartic surface $X \subset \mathbb{P}_3$ with nine cusps and no other singularities.*

Proof. A general hyperplane section $C \subset X$ would define a divisor class $\xi = [C] \in J_X$ with self-intersection $4 \not\equiv 0$ or $2 \mod 6$. □

On the other hand, for each integer $d \equiv 0$ or $2 \mod 6$ there are $K3$-surfaces with nine cusps carrying a polarization of degree d: If $d \equiv 0 \mod 6$, then put $n_1 = 1$, $n_3 = n_4 = 0$ and $n_2 = d/6$. If $d = 6k + 2$, then put $n_1 = n_4 = 1$, $n_3 = 0$ and $n_2 = (d-2)/6$. In fact, [BL] explicitly describe all surfaces with polarization of degree two, and [BV] gave explicit examples of surfaces with polarizations of degree six.

3.3. Polarizations on A

Fix some class $\alpha \in L_A$ with $\alpha \wedge \alpha > 0$. It defines a hyper-plane $\alpha^\perp \subset L_A \otimes \mathbb{C}$ and a curve $\Omega_\alpha = \Omega \cap \alpha^\perp$. For all periods $\omega \in \Omega_\alpha$ the complex torus A defined by ω carries line-bundles of class α. Since $\alpha \wedge \alpha > 0$, these line-bundles are ample, and A is algebraic.

Proposition 10. *All these complex curves $\Omega_\alpha \subset \Omega$ consist of two connected components. In fact, they are a union of two copies of the complex upper half-plane.*

Proof. As $\alpha \wedge \alpha > 0$, the lattice $L_\alpha := L_A \cap \alpha^\perp$ has signature $(2, 1)$. So there are real coordinates x_1, x_2, x_3 of $L_\alpha \otimes \mathbb{R}$ in which the intersection form is $x_1^2 - x_2 x_3$. Let z_k be the corresponding complex coordinates on $L_\alpha \otimes \mathbb{C}$. Each $\omega \in \Omega_\alpha$ has coordinates

$$(c_1, c_2, c_3), \quad c_k \in \mathbb{C}$$

with

$$\omega \wedge \omega = c_1^2 - c_2 c_3 = 0.$$

If $c_3 = 0$, then $c_1 = 0$ implies $\omega \wedge \bar\omega = 0$, a contradiction. So $c_3 \ne 0$. We may assume $c_3 = 1$ and $\omega = (c_1, c_1^2, 1)$.
Then
$$\omega \wedge \bar\omega = |c_1|^2 - \Re(c_1^2) = 2\Im(c_1)^2 > 0$$
unless $\Im(c_1) = 0$. It follows that $\Omega_\alpha = \mathbb{C} \setminus \mathbb{R}$. □

3.4. The classification problem

Of course, it would be nice to have a classification for the set of abelian surfaces with an automorphism t of order three as in section 2.1, together with a t-invariant polarization. Polarizations of degree $2d$ on abelian surfaces (without t) are classified by their elementary divisors

$$d_1, d_2 \in \mathbb{N}, \quad d_1 | d_2, \quad 2 \cdot d_1 d_2 = 2d$$

in the sense that they belong to the same connected moduli space, if these elementary divisors coincide. However for fixed d_1, d_2, not all pairs (A, t) admitting a polarization of type (d_1, d_2) are topologically equivalent. There is an obvious reason: If two polarizations, i.e. primitive vectors α and $\alpha' \in L_A$ with $\alpha^2 = (\alpha')^2 > 0$ are topologically equivalent, they are conjugate under the arithmetic group G from 3.1. In particular they must be conjugate under the orthogonal group $O(L_A)$. And then $\alpha \in L_A \subset \check L_A$ is primitive, if and only if α' is primitive in $\check L_A$.

It is easy to see that the vector

$$\alpha := \gamma_1 + 3\gamma_2$$

of length $\alpha^2 = 6$ is primitive in $\check L_A$, while the vector

$$\alpha' = \gamma_3 + \gamma_4$$

of the same length 6 is not. So the polarizations α and α' are not topologically equivalent. (In fact, by 2.3, the polarization α' descends to the quotient $X = A/t$, while the t-invariant polarization α does not do this).

The real classification problem is this:

Problem 1. *Classify the G-orbits on the sets of primitive vectors $\alpha \in L_A$ of the same length $\alpha^2 > 0$ and primitive / not primitive in $\check L_A$.*

I expect the groups G and $O(L_A)$ to be more or less the same, but I have no idea, whether they act transitively or not on primitive vectors $\alpha \in L_A$ of the same length, primitive, resp. not primitive in $\check L_A$.

Each abelian surface (= algebraic torus) A with automorphism t has Picard number three. Let me give examples of simple abelian surfaces A with automorphism t.
To do this, I have to identify the orthogonal complement $L_A^\perp \subset H^2(A, \mathbb{Z})$:

Proposition 11. *The lattice $L_A^\perp$ admits a $\mathbb{Z}$-basis δ_1, δ_2 in which the quadratic form is given by the matrix*

$$\begin{pmatrix} -2 & 1 \\ 1 & -2 \end{pmatrix}.$$

Proof. Recall the basis $\alpha_1, \beta_1, \alpha_2, \beta_2$ from section 2.1. Put

$$\delta_1 := \alpha_1 \wedge \alpha_2 - \beta_1 \wedge \beta_2, \quad \delta_2 = \alpha_1 \wedge \beta_2 + \beta_1 \wedge \alpha_2 + \beta_1 \wedge \beta_2.$$

One easily checks that these classes are orthogonal to the basis $\gamma_1, \ldots, \gamma_4$ of L_A from section 1.2, and

$$\delta_1 \wedge \delta_1 = \delta_2 \wedge \delta_2 = -2, \quad \delta_1 \wedge \delta_2 = 1.$$

So δ_1 and δ_2 span a sublattice of $L_A^\perp$ in which the form has the matrix above. Its discriminant is 3. This sublattice therefore is primitive, and must coincide with $L_A^\perp$. $\square$

The essential observation is: This lattice does not represent $-12 \cdot n^2$ for any $n \in \mathbb{N}$.

Proof. Assume there are $k, l \in \mathbb{Z}$ with

$$(k\delta_1 + l\delta_2)^2 = -2(k^2 + l^2 - kl) = -12n^2.$$

We solve the quadratic equation

$$k^2 - kl + l^2 - 6n^2 = 0$$

for k to find the two solutions

$$k_{1,2} = \frac{1}{2}(l \pm \sqrt{24n^2 - 3l^2}).$$

This shows

$$24n^2 - 3l^2 = (2k_{1,2} + l)^2 = w^2$$

is a square. Obviously 3 divides w, so write $w = 3w'$ and

$$8n^2 - l^2 = 3(w')^2.$$

If neither n nor l are divisible by three, we find

$$8n^2 \equiv 2, \quad l^2 \equiv 1 \quad \text{modulo } 3,$$

impossible. So $n = 3n'$ and $l = 3l'$. But this leads to

$$24(n')^2 - 3(l')^2 = (w')^2,$$

the original equation. As we may repeat this argument infinitely often, this is a contradiction. $\square$

Proposition 12. *There are abelian surfaces with an automorphism t as above, and not carrying elliptic curves.*

Proof. Recall the basis $\gamma_1, \ldots, \gamma_4 \in L_A$ from section 2.2. Clearly

$$\gamma := -\gamma_1 + 6\gamma_2 + 2(\gamma_3 + \gamma_4)$$

is a primitive vector of length 12. Let A be a surface carrying a polarization with class γ, and with Neron–Severi group of rank three. Then this Neron–Severi group is spanned by γ, δ_1 and δ_2. If it carried elliptic curves, there would be a class $n\gamma + k\delta_1 + l\delta_2$ of length

$$12n^2 - 2(k^2 - kl + l^2) = 0,$$

in conflict with the observation above. □

4. Appendix: Quartic surfaces with eight cusps

In this section I compute the equations of those quartic surfaces in $\mathbb{P}_3(\mathbb{C})$, which are the projections of the sextic surfaces in $\mathbb{P}_4$ from [BV] out of one of their nine cusps. Then I show by direct computation, that these projected surfaces have precisely eight cusps, and no further singularities.

So recall the sextic surfaces from [BV]. On p.141 they are presented as a complete intersection of a hyperplane P, a quadric Q and a cubic $C \subset \mathbb{P}_5$ with

$$P: \quad x_1 + x_2 + x_3 + x_4 + x_6 + x_6 = 0,$$
$$Q: \quad (1+k)(x_1x_2 + x_1x_3 + x_2x_3) + (1-k)(x_4x_5 + x_4x_6 + x_5x_6) = 0,$$
$$C: \quad (1+k)^2 x_1 x_2 x_3 + (1-k)^2 x_4 x_5 x_6 = 0.$$

Here $k \neq 0$ is a complex parameter. The nine cusps of these surfaces are the points

$$(1:0:0-1:0:0) \quad \text{etc.}$$

with one coordinate x_1, x_2 or x_3 equal to 1 and one coordinate x_4, x_5 or x_6 equal to -1.

I fix the cusp $(1:0:0:-1:0:0)$ and project from it onto the 3-plane

$$x_1 = x_4 = -\frac{1}{2}(x_2 + x_3 + x_5 + x_6) =: \sigma.$$

The rays of projection are parametrized by

$$(\lambda + \mu s : \mu x_2 : \mu x_3 : -\lambda + \mu \sigma : \mu x_5 : \mu x_6) \quad \text{with} \quad (\lambda : \mu) \in \mathbb{P}_1.$$

On this ray the equation of Q restricts to

$$\lambda[(1+k)(x_2 + x_3) - (1-k)(x_5 + x_6)]$$
$$+ \mu[(1+k)(\sigma(x_2 + x_3) + x_2 x_3) + (1-k)(\sigma(x_5 + x_6) + x_5 x_6)] = 0$$

and the equation of C becomes
$$\lambda[(1+k)^2 x_2 x_3 - (1-k)^2 x_5 x_6] + \mu \cdot \sigma[(1+k)^2 x_2 x_3 + (1-k)^2 x_5 x_6] = 0.$$
Eliminating $(\lambda : \mu)$ from these two equations, and replacing the coordinates x_2, x_3, x_5, x_6 by x_0, x_1, x_2, x_3 leads to the equation
$$(1+k)^3 x_0^2 x_1^2 + 2k(1-k^2) x_0 x_1 x_2 x_3 - (1-k)^3 x_2^2 x_3^2$$
$$+ (1-k^2)(x_0 + x_1 + x_2 + x_3)[(1-k)x_2 x_3 (x_0 + x_1) - (1+k)x_0 x_1 (x_2 + x_3)] = 0$$
for the projected surface.

The eight nodes of the sextic surface, different from the center of projection, go onto the four coordinate vertices
$$(1:0:0:0), \ldots, (0:0:0:1)$$
and the four points
$$(1:0:-1:0),\ (1:0:0:-1),\ (0:1:-1:0),\ (0:1:0:-1)$$
on the coordinate lines in the plane $x_0 + x_1 + x_2 + x_3 = 0$.

The center of projection blows up to the pair of lines
$$(1+k)x_0 - (1-k)x_2 = (1+k)x_1 - (1-k)x_3 = 0$$
$$(1+k)x_0 - (1-k)x_3 = (1+k)x_1 - (1-k)x_2 = 0$$
on the projected surface.

The projected surfaces have the obvious $\mathbb{Z}_2 \times \mathbb{Z}_2$-symmetry
$$x_0 \leftrightarrow x_1 \quad \text{and} \quad x_2 \leftrightarrow x_3,$$
while the symmetry
$$(x_0, x_1) \leftrightarrow (x_2, x_3), \quad k \leftrightarrow (-k)$$
interchanges two surfaces in the family (if $k \neq 0$).

Proposition 13. *For $k \neq \pm 1$ the quartic surfaces have no other singularities than the eight ones specified.*

Proof. To have some (not very big) computational advantages, let me pass to the coordinates
$$y_0 = (1+k)x_0,\ y_1 = (1+k)x_1,\ y_2 = (1-k)x_2,\ y_3 = (1-k)x_3,$$
in which the equation (multiplied by $(1-k^2)$) takes the form
$$f(y) = (1-k) y_0^2 y_1^2 + 2k y_0 y_1 y_2 y_3 - (1+k) y_2^2 y_3^2$$
$$+ ((1-k)(y_0 + y_1) + (1+k)(y_2 + y_3))((y_0 + y_1) y_2 y_3 - (y_2 + y_3) y_0 y_1)$$
$$= 0.$$

We have to compute the derivative

$$\partial_0 f = 2(1-k)y_0 y_1^2 + 2k y_1 y_2 y_3 + (1-k)((y_0+y_1)y_2 y_3 - (y_2+y_3)y_0 y_1)$$
$$+ ((1-k)(y_0+y_1) + (1+k)(y_2+y_3))(y_2 y_3 - (y_2+y_3)y_1).$$

To take advantage of the symmetries, let me abbreviate

$$s := y_0 + y_1, \ t := y_2 + y_3, \ p := y_0 y_1, \ q := y_2 y_3.$$

Then

$$\partial_0 f = 2((1-k)p + kq)y_1 + (1-k)(sq - tp) + ((1-k)s + (1+k)t)(q - ty_1),$$

and by the symmetries

$$\partial_1 f = 2((1-k)p + kq)y_0 + (1-k)(sq - tp) + ((1-k)s + (1+k)t)(q - ty_0).$$

The difference of these two derivatives is

$$\partial_0 f - \partial_1 f = (y_1 - y_0) \cdot (2(1-k)p + 2kq - ((1-k)s + (1+k)t)t).$$

So, in a singularity, either

$$y_0 = y_1,$$

or

$$((1-k)s + (1+k)t)t = 2(1-k)p + 2kq.$$

Inserting the latter in the expression for $\partial_0 f$ leads to

$$\partial_0 f = 2(1-k)sq - (1-k)tp + (1+k)tq.$$

Using the symmetry $(x_0, x_1, k) \to (x_2, x_3, -k)$ we find: At a singular point

either $x_0 = x_1$ or $\quad -(1-k)t \cdot p + (2(1-k)s + (1+k)t) \cdot q = 0,$
$\qquad\qquad\qquad\qquad 2(1-k) \cdot p + 2k \cdot q = ((1-k)s + (1+k)t)t,$
either $x_2 = x_3$ or $\quad (2(1+k)t + (1-k)s) \cdot p - (1+k)s \cdot q = 0,$
$\qquad\qquad\qquad\qquad -2k \cdot p + 2(1+k) \cdot q = ((1+k)t + (1-k)s)s.$

The two homogeneous equations for p and q form a system with determinant

$$D = (1-k^2)st - (2(1-k)s + (1+k)t)(2(1+k)t + (1-k)s) = -2((1-k)s + (1+k)t)^2.$$

So, if they hold, either

$$\frac{s}{1+k} + \frac{t}{1-k} = x_0 + x_1 + x_2 + x_3 = 0,$$

or $p = q = 0$. In this case the two inhomogeneous equations for p and q show

$$((1-k)s + (1+k)t) \cdot t = ((1-k)s + (1+k)t) \cdot s = 0.$$

And this again implies $x_0 + x_1 + x_2 + x_3 = 0$.

So, if a surface for $k \neq \pm 1$ is singular at $(x_0 : x_1 : x_2 : x_3)$, then we are in one of the following cases:

Case I: $x_0 + x_1 + x_2 + x_3 = 0$. But then

$$\frac{\partial_0 f - \partial_1 f}{2(y_1 - y_0)} = (1-k)p + kq = 0, \quad \frac{\partial_2 f - \partial_3 f}{2(y_3 - y_2)} = -kp + (1+k)q = 0.$$

This system for p and q has determinant

$$(1 - k^2) + k^2 = 1,$$

showing $p = q = 0$. So x is one of the four points

$$(1:0:-1:0), \ (1:0:0:-1), \ (0:1:-1:0), \ (0:1:0:-1).$$

Case II: $x_0 = x_1$, hence $y_0 = y_1 =: y$, but $x_2 \neq x_3$. Then there still are the two equations

$$(2(1+k)t + 2(1-k)y) \cdot y^2 - 2(1+k)y \cdot q = 0,$$
$$-2k \cdot y^2 + 2(1+k) \cdot q = ((1+k)t + (1-k)2y) \cdot 2y$$

from the four equations above. Unless $y = 0$ this leads to the two quadratic equations

$$(1-k) \cdot y^2 + (1+k)t \cdot y - (1+k)q = 0,$$
$$(k-2) \cdot y^2 - (1+k)t \cdot y + (1+k)q = 0.$$

Adding both equations we see $y = 0$. And if $y = 0$, the second equation shows $q = 0$. The point in question is one of the four coordinate vertices.

Case III: $x_2 = x_3$, but $x_0 \neq x_1$. This leads to a coordinate vertex, just like case II.

Case IV: $x_0 = x_1$ and $x_2 = x_3$, hence $y_0 = y_1 =: y$ and $y_2 = y_3 =: z$. Then

$$\partial_0 f = 2(1-k)y^3 + 2kyz^2 + (1-k)(2yz^2 - 2y^2z) + 2((1-k)y + (1+k)z)(z^2 - 2yz) = 0,$$

and

$$\frac{1}{2}\partial_0 f = (1-k)y^3 - 3(1-k)y^2z - 3kyz^2 + (1+k)z^3 = 0,$$

$$\frac{1}{2}\partial_2 f = (1-k)y^3 + 3ky^2z - 3(1+k)yz^2 + (1+k)z^3 = 0.$$

The difference of both these equations is

$$3yz \cdot (z - y) = 0.$$

Now $yz = 0$ leads to the contradiction $y = z = 0$. And for $y = z$ both equations become $-y^3 = 0$, again a contradiction. □

The projected surface has no other singularities than eight images of the cusps on the sextic surface. In particular it is *normal*. And the blow-up of the sextic surface in the point $(1:0:0:-1:0:0)$ is a normal surface too. Therefore the projection induces an isomorphism between these two surfaces. This shows that the eight singularities on the projected surface are cusps too.

References

[B] W. Barth, K3-surfaces with nine cusps, Geom. Dedicata 72 (1998), 171–178.

[BL] C. Birkenhake, H. Lange, A family of abelian surfaces and curves of genus four. Manuscripta Math. 85 (1994), 393–407.

[BV] J. Bertin, P. Vanhaecke, The even master system and generalized Kummer surfaces. Math. Proc. Cambridge Philos. Soc. 116 (1994), 131–142.

[N1] V.V. Nikulin, On Kummer surfaces. Math. USSR Izv. 9 (2) (1975), 261–275.

[N2] V.V. Nikulin, Integral symmetric bilinear forms and some of their applications. Math. USSR. Izv. 14 (1) (1980), 103-167.

[U] T. Urabe, Elementary transformations of Dynkin graphs and singularities on quartic surfaces. Invent. Math. 87 (1987), 549–572.

Complex manifolds with split tangent bundle

Arnaud Beauville

Abstract. Let X be a compact Kähler manifold. We expect that any direct sum decomposition $T_X = \oplus_{i \in I} E_i$ of its tangent bundle comes from a splitting of the universal covering space of X as a product $\prod_{i \in I} U_i$, in such a way that the given decomposition $T_X = \oplus_{i \in I} E_i$ lifts to the canonical decomposition $T_{\prod U_i} = \oplus_i T_{U_i}$. We prove this assertion when X is a Kähler–Einstein manifold or a Kähler surface, and discuss a general conjecture.

1991 Mathematics Subject Classification: 32J15

Introduction

The theme of this note is to investigate when the tangent bundle of a compact complex manifold X splits as a direct sum of sub-bundles. This occurs typically when the universal covering space $\widetilde{X}$ of X splits as a product $\prod_{i \in I} U_i$ of manifolds on which the group $\pi_1(X)$ acts diagonally (that is, $\pi_1(X)$ acts on each U_i and its action on $\widetilde{X} = \prod U_i$ is the diagonal action $g.(u_i) = (gu_i)$): the vector bundles* T_{U_i} on $\widetilde{X}$ are stable under $\pi_1(X)$, hence the decomposition $T_{\widetilde{X}} = \oplus_i T_{U_i}$ descends to a direct sum decomposition of T_X. For *Kähler* manifolds, we ask whether the converse is true, namely whether any direct sum decomposition of the tangent bundle T_X gives rise to a splitting of the universal covering. We will show that this is indeed the case in three different situations:

a) X admits a Kähler–Einstein metric;
b) T_X is a direct sum of line bundles of negative degree;
c) X is a Kähler surface.

In case a) the properties of Hermite–Einstein metrics imply that the tangent bundle splits as a direct sum of *hermitian* sub-bundles; we then conclude with a holonomy argument (a slightly less precise statement appears already in [Y]). Case b) is a small improvement of a uniformization result of Simpson [S]. To treat case c) we use the classification of surfaces and some simple remarks about connections. The result in this case is actually an easy consequence of the paper [KO], where the authors classify surfaces with a holomorphic conformal structure – this turns out to be closely related to the question we are studying here. However we found simpler and more

---

*   Throughout the paper we will abuse notation and write T_{U_i} instead of $pr_i^* T_{U_i}$.

enlightening to give an independent proof rather than extracting from [KO] the pieces of information that we need.

In §2 we give examples which show that the Kähler assumption, as well as some integrability assumptions, are necessary, and we propose a general conjecture.

1. Kähler–Einstein manifolds

Theorem A. *Let X be a compact complex manifold admitting a Kähler–Einstein metric. Assume that the tangent bundle of X has a decomposition $T_X = \oplus_{i \in I} E_i$. Then the universal covering space of X is a product $\prod_{i \in I} U_i$ of complex manifolds, in such a way that the decomposition $T_X = \oplus_{i \in I} E_i$ lifts to the decomposition $T_{\prod U_i} = \oplus_{i \in I} T_{U_i}$; the group $\pi_1(X)$ acts diagonally on $\prod_{i \in I} U_i$.*

Proof. (1.1) A Kähler–Einstein metric on X is a *Hermite–Einstein* metric on the vector bundle T_X, that is a hermitian metric whose curvature endomorphism, contracted with the Kähler form ω, is scalar (a good reference for the properties of Hermite–Einstein metrics that we will use is [K]). By Theorem V.8.3 of [K], the hermitian bundle T_X is the direct sum of a family $(F_j)_{j \in J}$ of ω-stable, hermitian vector bundles having the same slope as T_X. These bundles are preserved by the Levi-Civita connection, hence the holonomy representation of X is the direct sum of a family of representations corresponding to the F_j's. By the De Rham theorem, the universal covering space of X splits as a product $\prod_{j \in J} U_j$, such that the decomposition $T_X = \oplus_{j \in J} F_j$ pulls back to the decomposition $T_{\prod U_j} = \oplus_{j \in J} T_{U_j}$.

(1.2) We observe that the fact that the group $\pi_1(X)$ preserves the decomposition $T_{\prod U_j} = \oplus_{j \in J} T_{U_j}$ implies that it acts diagonally on $\prod_{j \in J} U_j$. Let indeed γ be an automorphism of $\prod U_i$; for $j \in I$, put $\gamma_j = pr_j \circ \gamma$. The condition $\gamma^* T_{U_j} = T_{U_j}$ means that the partial derivatives of γ_j in the directions of U_k for $k \neq j$ vanish, hence $\gamma_j((u_i)_{i \in I})$ depends only on u_j, which gives our claim.

(1.3) The bundles F_j are indecomposable, and we can assume that each E_i is indecomposable. By the Krull–Remak–Schmidt theorem, we can identify J to I in such a way that F_i is isomorphic to E_i for every $i \in I$. We want to compare the decompositions $T_X = \oplus_{i \in I} E_i$ and $T_X = \oplus_{i \in I} F_i$.

Lemma 1.4. *If $\mathrm{Hom}(F_i, F_j) \neq 0$ for some distinct indices i, j in I, the bundles F_i and F_j are isomorphic and admit a holomorphic connection.*

In particular, all Chern classes of F_i vanish.

Proof. Since F_i and F_j are stable with the same slope, our hypothesis implies that F_i and F_j are isomorphic ([K], 7.11 and 7.12); this is equivalent to the existence of an isomorphism $\varphi : T_{U_i} \to T_{U_j}$ compatible with the actions of $\pi_1(X)$.

Recall that if $f : T \to S$ is a holomorphic map between two manifolds, and E a vector bundle on S, the bundle f^*E carries a canonical relative flat connection $\nabla_{T/S} : f^*E \to f^*E \otimes \Omega^1_{T/S}$, characterized by the property $\nabla_{T/S}(f^*s) = 0$ for every local holomorphic section s of E; if moreover f is equivariant with respect to a group Γ acting on T, S and E, the connection $\nabla_{T/S}$ is Γ-equivariant. Applying this to the projection $\prod_i U_i \to U_i$ we obtain for each $k \neq i$ a partial, $\pi_1(X)$-equivariant, connection $\nabla_k : T_{U_i} \to T_{U_i} \otimes \Omega^1_{U_k}$. Similarly we have for each $k \neq j$ a partial connection $\nabla'_k : T_{U_j} \to T_{U_j} \otimes \Omega^1_{U_k}$. Put $\nabla_i = (\varphi \otimes 1)^{-1} \circ \nabla'_i \circ \varphi$; then $\sum_{k \in I} \nabla_k$ is a connection on T_{U_i} which is $\pi_1(X)$-equivariant, and therefore descends to a connection on F_i. □

(1.5). Let $i \in I$. If F_i does not admit any holomorphic connection, it follows from the Lemma that the only sub-bundle of T_X isomorphic to F_i is F_i itself, hence $E_i = F_i$.

Now assume that F_i has a holomorphic connection. Since F_i has the same slope as T_X, this can only occur if $c_1(X) = 0$. According to the structure theorem for manifolds with $c_1 = 0$ ([B2], Theorem 1), the set I splits into two subsets J and K, such that U_i is isomorphic to a vector space for $i \in J$ and is compact for $i \in K$; the vector bundle F_i has trivial Chern classes if and only if $i \in J$. Put $F = \oplus_{j \in J} F_j$; according to Lemma 1.4 we have $E_j \subset F$ for $j \in J$. We saw already that $E_k = F_k$ for $k \in K$, hence $\oplus_{j \in J} E_j = F$.

Put $V = \prod_{j \in J} U_j$, $M = \prod_{k \in K} U_k$. There exists a complex torus A with universal covering V and a finite étale covering $\pi : A \times M \to X$ (*loc. cit.*). We have $\pi^*F = T_A$; the decomposition $F = \oplus_{j \in J} E_j$ pulls back to a decomposition of the trivial bundle T_A, which corresponds to a decomposition $V = \oplus_{j \in J} V_j$ of V into vector subspaces. The splitting $\tilde{X} - \prod_{j \in J} V_j \times \prod_{k \in K} U_k$ has the requested properties. □

2. Discussion of the conjecture

Let us first show that the Kähler assumption is necessary.

(2.1) Hopf manifolds. Let $T = \text{diag}(\alpha_1, \ldots, \alpha_n)$ be a diagonal matrix, with $n \geq 2$ and $0 < |\alpha_i| < 1$ for each i. The cyclic group $T^{\mathbf{Z}}$ generated by T acts freely and properly on $\mathbf{C}^n - \{0\}$; the quotient X is a compact complex manifold, called a Hopf manifold. For each non-zero complex number θ, denote by L_θ the flat line bundle associated to the character of $\pi_1(X) = T^{\mathbf{Z}}$ mapping T to θ; in other words, L_θ is the quotient of the trivial line bundle $(\mathbf{C}^n - \{0\}) \times \mathbf{C}$ by the action of the automorphism (T, θ). By construction we have $T_X = \oplus_{i=1}^n L_{\alpha_i}$, but the universal covering space

$\mathbf{C}^n - \{0\}$ of X is clearly not a product. Note that all direct sums $\oplus_{j\in J} L_{\alpha_j}$, for $J \subset [1, n]$, are integrable sub-bundles of T_X.

(2.2) Integrability conditions. Let X be a compact Kähler manifold. If a decomposition $T_X = \oplus_{i\in I} E_i$ is associated as above to a splitting $\widetilde{X} \cong \prod_{i\in I} U_i$ of the universal covering space of X, the vector bundles E_i and their direct sums $\oplus_{i\in J} E_i$, for every subset J of I, are integrable (that is, stable under the Lie bracket). It is easy to produce examples where the tangent bundle splits into non-integrable factors: take for instance $X = A \times \mathbf{P}^1$, where A is an abelian surface. Let (U, V) be a basis of $H^0(A, T_A)$, and S, T two vector fields on $\mathbf{P}^1$ which do not commute. The vector fields $U + S$ and $V + T$ span a (trivial) rank 2 sub-bundle of T_X, supplementary to $T_{\mathbf{P}^1}$, but not integrable.

In view of the above examples the natural conjecture is the following:

(2.3). *Let X be a compact Kähler manifold such that $T_X = \oplus_{i\in I} E_i$, each sub-bundle $\oplus_{i\in J} E_i$, for $J \subset I$, being integrable. Then the universal covering space of X is isomorphic to a product $\prod_{i\in I} U_i$, in such a way that the given decomposition $T_X = \oplus_{i\in I} E_i$ lifts to the canonical decomposition $T_{\prod U_i} = \oplus_i T_{U_i}$.*

In the case when all the E_i's are line bundles and X is projective, this conjecture has just been proved by S. Druel [D].

In the situations a), b), c) considered here it turns out that the integrability is automatic. One may ask whether this holds whenever the canonical bundle K_X is *nef*.

3. Simpson's uniformization result

The following lemma*, which is a variation on the Baum–Bott theorem [B-B], will allow us to slightly improve Simpson's result:

Lemma 3.1. *Let X be a complex manifold, and E a direct summand of T_X. The Atiyah class $\mathrm{at}(E) \in H^1(X, \Omega^1_X \otimes \mathcal{E}nd(E))$ comes from $H^1(X, E^* \otimes \mathcal{E}nd(E))$. In particular, any class in $H^r(X, \Omega^r_X)$ given by a polynomial in the Chern classes of E vanishes for $r > \mathrm{rk}(E)$.*

Proof. Write $T_X = E \oplus F$; let $p : T_X \to E$ be the corresponding projection. For sections U of E and V of F over some open subset of X, put $D_V U = p([V, U])$. This expression is $\mathcal{O}_X$-linear in V and satisfies the Leibniz rule $D_V(fU) = f D_V(U) + (Vf)U$, so that D is a F-connection on E [B-B]: if we denote by $\mathcal{D}^1(E)$

---

* F. Bogomolov reminded me that this lemma appears already in his IHES preprint *Kählerian varieties with trivial canonical class* (1981).

the sheaf of differential operators $\Delta : E \to E$, of degree ≤ 1, whose symbol $\sigma(\Delta)$ is scalar, this means that D defines an $\mathcal{O}_X$-linear map $F \to \mathcal{D}^1(E)$ such that $\sigma(D_V) = V$ for all local sections V of F. Thus the exact sequence

$$0 \to \mathcal{E}nd(E) \longrightarrow \mathcal{D}^1(E) \xrightarrow{\sigma} T_X \to 0$$

splits over the sub-bundle $F \subset T_X$; therefore its extension class at$(E) \in H^1(X, \Omega^1_X \otimes \mathcal{E}nd(E))$ vanishes in $H^1(X, F^* \otimes \mathcal{E}nd(E))$, hence comes from $H^1(X, E^* \otimes \mathcal{E}nd(E))$. The last assertion follows from the definition of the Chern classes in terms of the Atiyah class. □

We denote as usual by **H** the Poincaré upper half-space.

Theorem B. *Let X be a compact Kähler manifold, with Kähler class ω. Assume that the tangent bundle T_X is a direct sum of line bundles $L_1, \ldots, L_n$ with $\omega^{n-1}.c_1(L_i) < 0$ for each i. Then the universal covering space of X is $\mathbf{H}^n$, and the decomposition $T_X = \oplus L_i$ lifts to the canonical decomposition $T_{\mathbf{H}^n} = (T_{\mathbf{H}})^{\oplus n}$.*

Proof. Lemma 3.1 gives $c_1(L_i)^2 = 0$ for each i, hence $c_1(X)^2 - 2c_2(X) = 0$. Then Corollary 9.7 of [S] shows that the universal covering space of X is $\mathbf{H}^n$. The assertion about the compatibility of decompositions is not explicitly stated in *loc. cit.*, but follows from the proof; or we can apply Theorem A. □

4. The surface case

Theorem C. *Let X be a compact complex surface. The tangent bundle of X splits as a direct sum of two line bundles if and only if one of the following occurs:*
(a) *The universal covering space of X is a product $U \times V$ of two (simply-connected) Riemann surfaces and the group $\pi_1(X)$ acts diagonally on $U \times V$; in that case the given splitting of T_X lifts to the direct sum decomposition $T_{U \times V} = T_U \oplus T_V$.*
(b) *X is a Hopf surface, with universal covering space $\mathbf{C}^2 - \{0\}$. Its fundamental group is isomorphic to $\mathbf{Z} \oplus \mathbf{Z}/m\mathbf{Z}$, for some integer $m \geq 1$; it is generated by diagonal automorphisms $(x, y) \mapsto (\alpha x, \beta y)$ with $|\alpha| \leq |\beta| < 1$, and $(x, y) \mapsto (\lambda x, \mu y)$ where λ and μ are primitive m-th roots of 1.*

As a corollary, for Kähler surfaces we see that any direct sum decomposition of the tangent bundle gives rise to a splitting of the universal covering, as announced in the introduction.

(4.1). Before starting the proof we will need a few preliminaries. From now on we denote by X a compact complex surface; we assume given a direct sum decomposition

$\Omega_X^1 \cong L \oplus M$. By Lemma 3.1 (or by [BB]) the Chern class $c_1(L) \in H^1(X, \Omega_X^1)$ belongs to the subspace $H^1(X, L)$, and similarly for M. As a consequence, we get:

(4.2). We have $L^2 = M^2 = 0$, and therefore $c_1^2(X) = 2L.M = 2c_2(X)$.

The following consequence is less obvious.

Proposition 4.3. *Let C be a smooth rational curve in X. Then $C^2 \geq 0$.*

Proof. Put $C^2 = -d$ and assume $d > 0$. Since $H^1(C, \mathcal{O}_C(d+2)) = 0$, the exact sequence
$$0 \to \mathcal{O}_C(d) \to \Omega_{X|C}^1 \to \Omega_C^1 \to 0$$
splits, providing an isomorphism $\Omega_{X|C}^1 \cong \mathcal{O}_C(d) \oplus \mathcal{O}_C(-2)$. Thus one of the line bundles L or M, say L, satisfies $L_{|C} \cong \mathcal{O}_C(d)$. Consider the commutative diagram

$$\begin{array}{ccc} H^1(X, L) & \longrightarrow & H^1(X, \Omega_X^1) \\ \downarrow & & \downarrow \\ H^1(C, L_{|C}) & \longrightarrow & H^1(C, \Omega_C^1) \ ; \end{array}$$

since $d > 0$ we have $H^1(C, L_{|C}) = 0$; thus $c_1(L)$ goes to 0 in $H^1(C, \Omega_C^1)$, which means $d = 0$, a contradiction. □

(4.4). We shall come across situations where the vector bundle $\Omega_X^1 = L \oplus M$ appears as an extension
$$0 \to P \to \Omega_X^1 \xrightarrow{p} Q \to 0$$
of two line bundles P and Q. In that case,
– either the restriction of p to one of the direct summands of Ω_X^1, say M, is surjective; then the exact sequence splits, Q is isomorphic to M and P to L;
– or the restriction of p to both L and M is not surjective; then there exists effective (non-zero) divisors A and B, whose supports do not intersect, such that $L \cong Q(-A)$, $M \cong Q(-B)$ and $P \cong Q(-A-B)$; the exact sequence does *not* split.
In particular, if $\mathrm{Hom}(P, Q) = 0$, the exact sequence splits.

(4.5). Finally we will need some classical facts about connections (see [E]). Let $p : M \to B$ be a smooth holomorphic map between complex manifolds, whose fibres are isomorphic to a fixed variety F. A *connection* on p is a splitting of the exact sequence
$$0 \to p^*\Omega_B^1 \to \Omega_M^1 \to \Omega_{M/B}^1 \to 0,$$

that is a sub-bundle $L \subset \Omega^1_M$ mapping isomorphically onto $\Omega^1_{M/B}$; the connection is flat (or integrable) if $dL \subset L \wedge \Omega^1_M$ (this is automatic if B is a curve). In that case the group $\pi_1(B)$ acts on F by complex automorphisms, and M is the fibre bundle on B with fibre F associated to the universal covering $\tilde{B} \to B$, that is the quotient of $\tilde{B} \times F$ by the group $\pi_1(B)$ acting diagonally; the splitting $\Omega^1_M = p^*\Omega^1_B \oplus L$ pulls back to the decomposition $\Omega^1_{\tilde{B} \times F} = \Omega^1_{\tilde{B}} \oplus \Omega^1_F$.

5. Proof of Theorem C

(5.1) Kodaira dimension 2. If $\kappa(X) = 2$, the canonical bundle K_X is ample by Proposition 4.3. The Aubin–Calabi–Yau theorem implies that X admits a Kähler–Einstein metric; we can therefore apply Theorem A.

(5.2) Kodaira dimension 1. If $\kappa(X) = 1$, X admits an elliptic fibration $p : X \to B$. By 4.2 we have $c_2(X) = 0$; this implies that the only singular fibres of p are multiples of smooth elliptic curves (see [B1], VI.4 and VI.5). For $b \in B$, we write $p^*[b] = m_b F_b$, where F_b is a smooth elliptic curve; we have $m_b \geq 1$ and $m_b = 1$ except for finitely many points. Put $\Delta = \sum_b (m_b - 1) F_b$. We have an exact sequence

$$0 \to p^*\Omega^1_B(\Delta) \longrightarrow \Omega^1_X \longrightarrow \omega_{X/B} \to 0 \tag{5.3}$$

where $\omega_{X/B}$ is the relative dualizing line bundle. Since $\chi(\mathcal{O}_X) = 0$ by Riemann–Roch, we deduce from [BPV], V.12.2 and III.18.2, that $\omega_{X/B}$ is a torsion line bundle. Since $K_X = p^*\Omega^1_B(\Delta) \otimes \omega_{X/B}$, the hypothesis $\kappa(X) = 1$ implies $\mathrm{Hom}(p^*\Omega^1_B(\Delta), \omega_{X/B}) = 0$, hence the exact sequence (5.3) splits by 1.1: one of the direct summands of Ω^1_X, say M, maps surjectively onto $\omega_{X/B}$.

Let $\rho : \tilde{B} \to B$ be the orbifold universal covering of $(B, (m_b))$: this is a ramified Galois covering, with $\tilde{B}$ simply-connected, such that the stabilizer of a point $\tilde{b} \in \tilde{B}$ is a cyclic group of order $m_{\rho(\tilde{b})}$ (see for instance [KO], Lemma 6.1; note that because of the hypothesis $\kappa(X) = 1$ and the formula for K_X, there are at least 3 multiple fibers if B is of genus 0). Let $\tilde{X}$ be the normalization of $X \times_B \tilde{B}$. We have a commutative diagram

$$\begin{array}{ccc} \tilde{X} & \xrightarrow{\pi} & X \\ \tilde{p} \downarrow & & \downarrow p \\ \tilde{B} & \xrightarrow{\rho} & B \end{array}$$

where $\tilde{p}$ is smooth and π is étale ([B1], VI.7). The exact sequence

$$0 \to \tilde{p}^*\Omega^1_{\tilde{B}} \to \Omega^1_{\tilde{X}} \to \Omega^1_{\tilde{X}/\tilde{B}} \to 0$$

coincides with the pull back under π of the exact sequence (5.3); therefore p admits an integrable connection, given by the subbundle π^*M of $\Omega^1_{\tilde{X}}$. The result follows from 4.5 and 1.2.

(5.4) Kodaira dimension 0. Assume $\kappa(X) = 0$. By 4.2 and the classification of surfaces, X is either a complex torus, a bielliptic surface, or a Kodaira surface. Complex tori and bielliptic surfaces fall into case (a) of the theorem (a bielliptic surface is the quotient of a product $E \times F$ of elliptic curves by a finite abelian group acting diagonally).

A primary Kodaira surface has trivial canonical bundle and admits a smooth elliptic fibration $p : X \to B$. Thus the exact sequence (5.3) realizes Ω^1_X as an extension of $\mathcal{O}_X$ by $\mathcal{O}_X$. Since $h^{1,0}(X) = 1$, this extension is non-trivial, and it follows from 4.4 that Ω^1_X does not split.

A secondary Kodaira surface admits a primary Kodaira surface as a finite étale cover, hence its tangent bundle cannot split either.

(5.5) Ruled surfaces. We consider the case when X is algebraic and $\kappa(X) = -\infty$. By 4.2 and 4.3, X is a geometrically ruled surface, that is a projective bundle $p : X \to B$ over a curve. We again consider the exact sequence

$$0 \to p^*\Omega^1_B \to \Omega^1_X \to \Omega^1_{X/B} \to 0;$$

since $\Omega^1_{X/B}$ has negative degree on the fibres, we have $\mathrm{Hom}(p^*\Omega^1_B, \Omega^1_{X/B}) = 0$, hence by 4.4 the above exact sequence splits: one of the direct summands of Ω^1_X defines an integrable connection for p. The result follows then from 4.5.

(5.6) Inoue surfaces. We now assume that X is not algebraic and $\kappa(X) = -\infty$, so that X is what is usually called a surface of type VII_0. These surfaces have $b_1 = h^{0,1} = 1$ and therefore $c_1^2 + c_2 = 12\chi(\mathcal{O}_X) = 0$; in our case this gives $c_2 = 0$ in view of 4.2, and finally $b_2 = 0$. Moreover we have $H^0(X, \Omega^1_X \otimes L^{-1}) \neq 0$. The surfaces with these properties have been completely classified by Inoue [I]: they are either Hopf surfaces, or belong to three classes of surfaces constructed by Inoue (*loc. cit.*).

We first consider the Inoue surfaces. The surfaces S_M of the first class are quotients of $\mathbf{H} \times \mathbf{C}$ by a group acting diagonally, hence they fall into case (a) of the theorem.

The surfaces $S^{(+)}_{N,p,q,r;t}$ of the second class are quotients of $\mathbf{H} \times \mathbf{C}$ by a group which does *not* act diagonally. This action leaves invariant the vector field $\partial/\partial z$ on $\mathbf{C}$, which therefore descends to a non-vanishing vector field v on X. This gives rise to an exact sequence

$$0 \to K_X \xrightarrow{i(v)} \Omega^1_X \xrightarrow{i(v)} \mathcal{O}_X \to 0,$$

which does not split since $h^{1,0}(X) = 0$. We have $H^0(X, K_X^{-1}) = 0$, for instance because X contains no curves; we infer from 4.4 that Ω_X^1 does not split.

The surfaces $S_{N,p,q,r}^{(-)}$ of the third class are quotients of certain surfaces of the second class by a fixed point free involution; therefore their tangent bundle does not split either.

(5.7) Primary Hopf surfaces. It remains to consider the class of Hopf surfaces, which are by definition the surfaces of class VII_0 whose universal covering space is $\mathbf{W} := \mathbf{C}^2 - \{0\}$. We consider first the *primary* Hopf surfaces, which are quotients of $\mathbf{W}$ by the infinite cyclic group generated by an automorphism T of $\mathbf{W}$. According to [Ko], §10, there are two cases to consider:

a) $T(x, y) = (\alpha x, \beta y)$ for some complex numbers α, β with $0 < |\alpha| \le |\beta| < 1$;

b) $T(x, y) = (\alpha^m x + \lambda y^m, \alpha y)$ for some positive integer m and non-zero complex numbers α, λ with $|\alpha| < 1$.

As in 2.1, we denote by L_θ, for $\theta \in \mathbf{C}$, the flat line bundle associated to the character of $\pi_1(X)$ mapping T to θ. In case a) we find $\Omega_X^1 = L_\alpha^{-1} \oplus L_\beta^{-1}$, so the tangent bundle splits.

Let us consider case b). The form dy on $\mathbf{W}$ satisfies $T^* dy = \alpha \, dy$, hence descends to a form $\overline{dy}$ in $H^0(X, \Omega_X^1 \otimes L_\alpha)$; similarly the function y descends to a non-zero section of L_α. We have an exact sequence

$$0 \to L_\alpha^{-1} \xrightarrow{\overline{dy}} \Omega_X^1 \longrightarrow L_\alpha^{-m} \to 0.$$

Since L_α has a nonzero section, the space $\mathrm{Hom}(L_\alpha^{-1}, L_\alpha^{-m})$ is zero for $m > 1$. Hence if Ω_X^1 splits, we deduce from 4.4 that the exact sequence splits. This means that there exists a form $\overline{\omega} \in H^0(X, \Omega_X^1 \otimes L_\alpha^m)$ such that $\overline{\omega} \wedge \overline{dy} \ne 0$. Then $\overline{\omega} \wedge \overline{dy}$ is a generator of the trivial line bundle $K_X \otimes L_\alpha^{m+1}$, hence pulls back to $c \, dx \wedge dy$ on $\mathbf{W}$, for some constant $c \ne 0$. Therefore the pull back ω of $\overline{\omega}$ to $\mathbf{W}$ is of the form $c \, dx + f(x, y) dy$ for some holomorphic function f on $\mathbf{C}^2$. The flat line bundle L_α^m carries a flat holomorphic connection ∇; the 2-form $\nabla \overline{\omega}$, which is a global section of $K_X \otimes L_\alpha^m \cong L_\alpha^{-1}$, is zero. This implies $d\omega = 0$, so the function $f(x, y)$ is independent of x; let us write it $f(y)$. Now the condition $T^* \omega = \alpha^m \omega$ reads $\alpha f(\alpha y) + c\lambda m y^{m-1} = \alpha^m f(y)$. Differentiating m times we find $f^{(m)} = 0$, then differentiating $m-1$ times leads to a contradiction. □

(5.8) Secondary Hopf surfaces. A secondary Hopf surface X is the quotient of $\mathbf{W}$ by a group Γ acting freely, containing a central, finite index subgroup generated by an automorphism T of the above type. We assume that Ω_X^1 splits. The primary Hopf surface $Y = \mathbf{W}/T^{\mathbf{Z}}$ is a finite étale cover of X, so Ω_Y^1 also splits; it follows from (5.7) that T is of type a), and that Γ does not contain any transformation of type b). According to [Ka], §3, this implies that after an appropriate change of coordinates, the group Γ acts *linearly* on $\mathbf{C}^2$.

We claim that Γ is contained in a maximal torus of $\mathbf{GL}(2,\mathbf{C})$. This is clear if $\alpha \neq \beta$, because T is central in Γ. If $\alpha = \beta$, the direct sum decomposition of Ω^1_X pulls back to a decomposition $\Omega^1_Y = L_\alpha^{-1} \oplus L_\alpha^{-1}$ (5.7), which for an appropriate choice of coordinates comes from the decomposition $\Omega^1_\mathbf{W} = \mathcal{O}_\mathbf{W} dx \oplus \mathcal{O}_\mathbf{W} dy$. Since Γ must preserve this decomposition, it is contained in the diagonal torus.

Thus we may identify Γ with a subgroup of $(\mathbf{C}^*)^2$; since it acts freely on $\mathbf{W}$, the first projection $\Gamma \to \mathbf{C}^*$ is injective. Therefore the torsion subgroup of Γ is cyclic, and we are in case (b) of the theorem. $\square$

References

[B1] Beauville, A., Surfaces algébriques complexes. Astérisque 54 (1978).

[B2] Beauville, A., Variétés kählériennes dont la première classe de Chern est nulle. J. Differential Geom. 18 (1983), 755–782.

[BB] Baum, P., Bott, R., On the zeros of meromorphic vector-fields. Essays on Topology and Related Topics, 29–47. Springer-Verlag, New York 1970.

[BPV] Barth, W., Peters, C., Van de Ven, A., Compact complex surfaces. Ergeb. Math. Grenzgeb., Springer-Verlag, 1984.

[D] Druel, S., Variétés algébriques dont le fibré tangent est totalement décomposé. Preprint math.AG/9901138, J. Reine Angew. Math., to appear.

[E] Ehresmann, C., Les connexions infinitésimales dans un espace fibré différentiable. Colloque de topologie, Bruxelles (1950), 29–55. G. Thone, Liège 1951.

[I] Inoue, M., On surfaces of class VII_0. Invent. Math. 24 (1974), 269–310.

[Ka] Kato, M., Topology of Hopf surfaces. J. Math. Soc. Japan 27 (1975), 222–238.

[K] Kobayashi, S., First Chern class and holomorphic tensor fields. Nagoya Math. J. 77 (1980), 5–11.

[KO] Kobayashi, S., Ochiai, T., Holomorphic structures modeled after hyperquadrics. Tôhoku Math. J. 34 (1982), 587–629.

[Ko] Kodaira, K., On the structure of compact complex analytic surfaces II. Amer. J. Math. 88 (1966), 682–721.

[S] Simpson, C., Constructing variations of Hodge structure using Yang–Mills theory and applications to uniformization. J. Amer. Math. Soc. 1 (1988), 867–918.

[Y] Yau, S.-T., A splitting theorem and an algebraic geometric characterization of locally hermitian symmetric spaces. Comm. Anal. Geom. 1 (1993), 473–486.

Projections from subvarieties

*Mauro C. Beltrametti, Alan Howard,
Michael Schneider and Andrew J. Sommese*

Contents

1 Introduction   71

2 Background material   73

3 Lower and upper bounds for $h^0(tL)$ 77

4 Some general structure results for projections   82

5 Examples   91

6 The divisorial case   93

7 The linear case   97

8 The linear case in codimension 1   101

1. Introduction

Let $X \subset \mathbb{P}^N$ be an n-dimensional connected projective submanifold of projective space. Let $p : \mathbb{P}^N \to \mathbb{P}^{N-q-1}$ denote the projection from a linear $\mathbb{P}^q \subset \mathbb{P}^N$. Assuming that $X \not\subset \mathbb{P}^q$ we have the induced rational mapping $\psi := p_X : X \to \mathbb{P}^{N-q-1}$. This article started as an attempt to understand the structure of this mapping when ψ has a lower dimensional image. In this case of necessity we have $Y := X \cap \mathbb{P}^q$ is nonempty.

The special case when Y is a point is very classical: X is a linear subspace of $\mathbb{P}^N$. The case when $q = 1$ and $Y = \mathbb{P}^q = \mathbb{P}^1$ was settled for surfaces by the fourth author [18] and by Ilic [12] in general. Beyond this even the special case when $q \geq 2$ and $Y = \mathbb{P}^q$ is open.

We have found it convenient to study a closely related question, which includes many special cases including the case when the center of the projection $\mathbb{P}^q$ is contained in X.

Problem. Let Y be a proper connected k-dimensional projective submanifold of an n-dimensional projective manifold X. Assume that $k > 0$. Let L be a very ample line bundle on X such that $L \otimes \mathcal{J}_Y$ is spanned by global sections, where $\mathcal{J}_Y$ denotes the ideal sheaf of Y in X. Describe the structure of (X, Y, L) under the additional assumption that the image of X under the mapping ψ associated to $|L \otimes \mathcal{J}_Y|$ is lower dimensional.

Let us describe our progress on this problem.

In §3 we study upper and lower bounds for the dimensions of the spaces of sections of powers tL of a very ample line bundle L on a projective manifold X. The need for such bounds arises naturally when we consider line bundles which are multiples of a very ample line bundle. One general result, Proposition (3.8), gives an upper bound for an integer t_0 such that for $t \geq t_0$, $h^0(tL \otimes \mathcal{J}_Y) > 0$.

In §4 we prove a number of general results. For example, Theorem (4.6) shows that $\dim \psi(X) \geq n - k - 1$ with equality only if Y is a complete intersection in X. In particular Corollary (4.7) shows that if Y is a linear $\mathbb{P}^k$, then $\dim \psi(X) \geq n - k - 1$ with equality only if $(X, L) \cong (\mathbb{P}^n, \mathcal{O}_{\mathbb{P}^n}(1))$. Proposition (4.9) further shows that if $\text{rankPic}(X) = 1$ and $\dim \psi(X) \geq n - k$ then $\dim \psi(X) \geq n - k + \frac{k}{n-k} - 1$. Theorem (4.10) shows that if Y is a $\mathbb{P}^k$ (or more generally a projective manifold whose algebraic cohomology is the same as $\mathbb{P}^k$ up to dimension $2(n - k)$), then if $\dim \psi(X) \geq n - k$, it follows that $\dim \psi(X) \geq k$. In particular except for known examples, we have for a wide range of Y including $\mathbb{P}^k$, that $\dim \psi(X) \geq \frac{\dim X}{2}$.

In §5 we give a number of examples showing that the dimensions allowed by the above results do occur. Of particular interest is Example (5.2). This example consists for each positive integer n of an infinite sequence of projective n-folds in $\mathbb{P}^{2n-1}$ which contain a linear $\mathbb{P}^{n-1}$. All degrees of X that are allowed by theory occur.

In §6 we specialize to the case when Y is a divisor. We study bundles of the form $tL - Y$ where t is near $\delta := \deg Y$. One result, Theorem (6.4), implies that if $\delta > 1$ then $|\delta L - Y|$ gives a birational map, which is in fact an isomorphism if $2n \geq \dim \Gamma(L) + 1$.

In §7 we restrict to the case when Y is a linear $\mathbb{P}^k$ and show, among other things, that $\dim \psi(X) \geq n - k$ except when X is a hypersurface in $\mathbb{P}^{n+1}$. In §8 we restrict further to the special case when Y is a linear $\mathbb{P}^{n-1}$. In this case ψ is a morphism. Remmert-Stein factorize $\psi = s \circ \phi$ with $\phi : X \to Z$ a morphism with connected fibers onto a normal projective variety Z, and with s a finite morphism. We know that except for known examples, if $\dim \psi(X) < \dim X$ then $\dim \psi(X) = n - 1$. We show that Z is very well behaved (Cohen–Macaulay, $\mathbb{Q}$-factorial, $\text{Pic}(Z) \cong \mathbb{Z}$). Moreover we examine the possible degrees of s and use adjunction theory to classify the possible (X, L) for extreme values of this degree.

We would like to thank Frank-Olaf Schreyer for his very helpful explanation of how Castelnuovo theory gives lower bounds for the dimensions of spaces of sections of powers of very ample line bundles.

The research in this article was carried out in Bayreuth, the University of Notre Dame, and two sessions of the RiP program at Oberwolfach. All the authors are indebted to the Volkswagen Stiftung, whose generosity allowed us to work together in such an ideal setting. The fourth author thanks the Alexander von Humboldt Stiftung for their generous support.

The final stages of this article were developed during a three-week stay at Oberwolfach in the summer of 1997. Within a few weeks after we separated, Michael Schneider died in a climbing accident. The three remaining authors dedicate this work to his memory. He was our friend and colleague, a person of vibrant energy, keen intelligence, and generous spirit. We feel a deep sense of loss, but are grateful to have had our lives and work enriched by his presence.

2. Background material

We work over the complex numbers $\mathbb{C}$. Through the paper we deal with projective varieties V. We denote by $\mathcal{O}_V$ the structure sheaf of V and by K_V the canonical bundle, for V smooth. For any coherent sheaf $\mathcal{F}$ on V, $h^i(\mathcal{F})$ denotes the complex dimension of $H^i(V, \mathcal{F})$.

Let $\mathcal{L}$ be a line bundle on V. The line bundle $\mathcal{L}$ is said to be *numerically effective* (*nef*, for short) if $\mathcal{L} \cdot C \geq 0$ for all effective curves C on V. $\mathcal{L}$ is said to be *big* if $\kappa(\mathcal{L}) = \dim V$, where $\kappa(\mathcal{L})$ denotes the Kodaira dimension of $\mathcal{L}$. If $\mathcal{L}$ is nef then this is equivalent to $c_1(\mathcal{L})^n > 0$, where $c_1(\mathcal{L})$ is the first Chern class of $\mathcal{L}$ and $n = \dim V$.

2.1. Notation. The notation used in this paper is standard from algebraic geometry. In particular, $\approx$ denotes linear equivalence of line bundles. For a line bundle $\mathcal{L}$ on a compact complex space V, $\chi(\mathcal{L}) := \sum_i (-1)^i h^i(\mathcal{L})$ denotes the Euler characteristic, and $|\mathcal{L}|$ denotes the complete linear system associated with a line bundle. We say that $\mathcal{L}$ is spanned if it is spanned at all points of V by $\Gamma(\mathcal{L})$.

For a compact connected projective manifold V, $h^{2j}(V, \mathbb{Q})_{\text{alg}}$ denotes the dimension of the vector subspace $H^{2j}(V, \mathbb{Q})_{\text{alg}}$ of $H^{2j}(V, \mathbb{Q})$ dual under Kronecker duality to the vector subspace of $H_{2j}(V, \mathbb{Q})$ spanned by the j-dimensional algebraic subvarieties of V.

We denote the ideal sheaf of an irreducible subvariety A of a variety V by $\mathcal{J}_{A/V}$ (or simply $\mathcal{J}_A$ when no confusion can result). For smooth A contained in the smooth locus of V, $N_{A/V}$ denotes the normal bundle of A in V.

Line bundles and divisors are used with little (or no) distinction. Hence we shall freely switch between the multiplicative and the additive notation.

2.2. Conductor formula. Let V be a connected projective manifold of dimension n. Let L be a very ample line bundle on V of degree $d := L^n$ with $|L|$ embedding V into $\mathbb{P}^N$. Then the classical conductor formula for the canonical bundle states that

$$\Delta \in |(d - n - 2)L - K_V|,$$

where Δ is the double point divisor of a projection of V from $\mathbb{P}^N$ to $\mathbb{P}^{n+1}$ (in the degenerate cases when $N = n$ or $n + 1$, Δ is taken to be the empty divisor). In particular the line bundle $(d - n - 2)L - K_V$ is spanned since given any point of V a generic projection can be chosen with the point not in the double point divisor of the projection (see [21], [22, p. 71], and [14]).

The following standard lemma is basic (see also [2, (3.1.8)]).

Lemma 2.3. *Let V be an irreducible normal projective variety with $\mathrm{Pic}(V) \cong \mathbb{Z}$. Let $g : V \to Z$ be a surjective morphism of V to a projective variety Z. Either g is a finite morphism or $g(V)$ is a point. The same conclusion holds for $V \cong \mathbb{P}^n$ and any holomorphic map to a compact complex space Z.*

Proof. Assume that g is not finite and doesn't map V to a point. If Z is projective, then the pullback of an ample line bundle cannot be ample and thus we see that $\mathrm{Pic}(V) \not\cong \mathbb{Z}$. Thus we can assume that $V \cong \mathbb{P}^n$ and Z is not necessarily projective.

Note that $\dim g(\mathbb{P}^n) = n$. If not let F denote a general fiber. Since it is smooth it would have trivial normal bundle. This contradicts the ampleness of the tangent bundle of $\mathbb{P}^n$.

Let F denote a positive dimensional fiber. There is a complex neighborhood U of F which maps generically one-to-one to a Stein space. Since $\mathbb{P}^n$ is homogeneous we have that the translates of F fill out an open set. Since the map g_U must map these positive dimensional subspaces to points we have the contradiction that $\dim g(U) = \dim g(\mathbb{P}^n) < n$. □

We also need the following general fact.

Lemma 2.4. *Let $f : X \to Y$ be a surjective proper map between normal varieties. Assume that X is Cohen–Macaulay and all fibers of f are equal dimensional. Then Y is Cohen–Macaulay.*

Proof. Note that a Cartier divisor on a Cohen–Macaulay variety is Cohen–Macaulay and that if we slice with $\dim X - \dim Y$ sufficiently ample divisors, then the restriction of the map to the slice is finite by a well known theorem of Hironaka [11, (2.1)]. Since a general hyperplane section of a normal variety is normal by Seidenberg's theorem, we can assume without loss of generality that f is finite. Let $n := \dim X = \dim Y$. By using [9, III, (7.6)] we are reduced to showing that for any locally free coherent sheaf $\mathcal{E}$ on Y, we have

$$h^i(\mathcal{E}(-q)) = 0 \text{ for } i < n \text{ and } q \gg 0, \tag{1}$$

where $\mathcal{F}(t)$ for a coherent sheaf $\mathcal{F}$ means $\mathcal{F} \otimes H^{\otimes t}$ for a fixed ample line bundle H on Y.

Since f is finite the pullback of an ample line bundle is ample. Given a coherent sheaf $\mathcal{G}$ on X, $\mathcal{G}(t)$ means $\mathcal{G}$ twisted by the t-th power of the pullback of H. Hence $(f^*\mathcal{E})(t) = f^*(\mathcal{E}(t))$.

Now since X is Cohen–Macaulay we have

$$h^i((f^*\mathcal{E})(-q)) = 0 \text{ for } i < n \text{ and } q \gg 0.$$

By the Leray spectral sequence, the projection formula and vanishing of higher direct images we obtain

$$h^i((f^*\mathcal{E})(-q)) = h^i(\mathcal{E}(-q) \otimes f_*\mathcal{O}_X). \tag{2}$$

Since both X and Y are normal we can use the trace mapping from $f_*\mathcal{O}_X \to \mathcal{O}_Y$ to see that the exact sequence $0 \to \mathcal{O}_Y \to f_*\mathcal{O}_X \to \mathcal{M} \to 0$ splits, where $\mathcal{M}$ denotes the quotient bundle. Thus $f_*\mathcal{O}_X \cong \mathcal{O}_Y \oplus \mathcal{M}$. Thus by combining (1) and (2) we get

$$0 = h^i((f^*\mathcal{E})(-q)) = h^i(\mathcal{E}(-q)) + h^i(\mathcal{E}(-q) \otimes \mathcal{M}),$$

for $i < n$ and $q \gg 0$. Then (1) follows and we are done. $\square$

If X is smooth and f is finite we can say more.

Lemma 2.5. *Let $f : X \to Y$ be a finite surjective map between projective varieties, where X is smooth and Y is normal. Then Y is Cohen–Macaulay and $(\deg f)$-factorial. Moreover, if $-K_X$ is nef and big, the induced map of $\mathrm{Pic}(Y) \to \mathrm{Pic}(X)$ is injective.*

Proof. The fact that Y is Cohen–Macaulay was proved in the previous lemma. To see that it is $(\deg f)$-factorial, let D be a Weil divisor on Y. Since X is smooth, f^*D is a Cartier divisor. We construct a Cartier divisor, $\mathrm{Norm}(f^*D)$, on Y as follows: in a small neighborhood U of any smooth point y in Y over which f is unramified, we define a rational function by multiplying the functions defining f^*D on the connected components of $f^{-1}(U)$; and we construct the divisor determined locally by this construction, first over all smooth points of Y over which f is unramified, and then (since Y is normal) to all of Y by Riemann extension.

From the way $\mathrm{Norm}(f^*D)$ was constructed it is obvious that $\mathrm{Norm}(f^*D) = (\deg f)D$. This shows that Y is $(\deg f)$-factorial. In addition, the same construction shows that if D is a Cartier divisor on Y for which f^*D is trivial, then $(\deg f)D$ is trivial. In particular, the kernel of the induced map of $\mathrm{Pic}(Y) \to \mathrm{Pic}(X)$ consists entirely of torsion elements.

Now suppose that $-K_X$ is nef and big. Then $h^i(\mathcal{O}_X) = 0$ for $i > 0$. Using the direct sum decomposition $f_*\mathcal{O}_X \cong \mathcal{O}_Y \oplus \mathcal{M}$ from the previous lemma together with the Leray spectral sequence applied to the finite map f, we see that $h^i(\mathcal{O}_Y) = 0$ for $i > 0$. Therefore, $\mathrm{Pic}(Y) \cong H^2(Y, \mathbb{Z})$, and it follows that $\mathrm{Pic}(Y) \to \mathrm{Pic}(X)$ is injective unless there is torsion in $H^2(Y, \mathbb{Z})$. We will show this can not occur.

By the universal coefficient theorem, torsion in $H^2(Y, \mathbb{Z})$ is equivalent to torsion in $H_1(Y, \mathbb{Z})$, which in turn implies the existence of a finite unbranched covering $Y' \to Y$. Lifting this to X gives the commutative diagram

$$\begin{array}{ccc} X' & \to & X \\ \downarrow & & \downarrow f \\ Y' & \to & Y \end{array}$$

where Y' is connected, the vertical arrows are branched coverings, and the horizontal arrows are unbranched coverings. Let m be the common sheet number of both of the latter. It is easy to see that X' consists of a finite number of disjoint connected components, each mapping isomorphically onto X; for $h^i(\mathcal{O}_{X'}) = 0$ (because $-K_{X'}$ is big and nef), and $\chi(\mathcal{O}_{X'}) = m\chi(\mathcal{O}_X) = m$, where m is the sheet number. If A is any connected component of X', we thus get a finite surjective map $X \cong A \to Y'$. Arguing as before, we see that $h^i(\mathcal{O}_{Y'}) = 0$ for $i > 0$, so that $\chi(\mathcal{O}_{Y'}) = 1$. On the other hand, we have $\chi(\mathcal{O}_{Y'}) = m\chi(\mathcal{O}_Y) = m$. □

The following general lemma is well known and follows from the results in the introduction of [16] (see also [2, (6.6.1)]).

Lemma 2.6. *Let L be a very ample line bundle on an irreducible projective variety, X. Let $Y \subset X$ be an irreducible subvariety of degree δ relative to L, i.e., $\delta = L^{\dim Y} \cdot Y$. If either Y is smooth or $Y \subset \mathrm{reg}(X)$ and $\mathrm{cod}_X Y = 1$ then $\mathcal{J}_Y(\delta)$ is spanned by global sections, where $\mathcal{J}_Y(\delta)$ denotes the ideal sheaf $\mathcal{J}_Y$ of Y in X tensored with δL.*

The following result we need is a "folklore" result, for which we don't know references.

Proposition 2.7. *Assume that Hartshorne's conjecture [8], that any connected non-degenerate n-dimensional smooth submanifold $X \subset \mathbb{P}^m$ is a complete intersection if $n > \frac{2}{3}m$, is true. Then each vector bundle $\mathcal{E}$ on $\mathbb{P}^m$ of rank $r < \frac{m}{3}$ splits into a direct sum of line bundles.*

Proof. We use induction over r. If $r = 1$ the assertion is true. So, let us assume the assertion true for $r - 1$.

Since the assertion is independent of twisting, we may assume that $\mathcal{E}$ is generated by global sections. Take a general section $s \in H^0(\mathcal{E})$ and let $X := V(s)$, the zero locus of s. Then X is smooth and $\mathrm{cod}_{\mathbb{P}^m} X = r$. The assumption $r < \frac{m}{3}$ is equivalent to $\dim X = m - r > \frac{2}{3}m$ and therefore X is a complete intersection in $\mathbb{P}^m$ by Hartshorne's

conjecture. Thus
$$\mathcal{E}_X \cong N_{X/\mathbb{P}^m} \cong \oplus_{i=1}^r \mathcal{O}_X(a_i),$$
where $\mathcal{E}_X$ denotes the restriction of $\mathcal{E}$ to X. Since $N_{X/\mathbb{P}^m}$ is ample, the a_i's are positive integers and we may assume $a_1 \geq a_2 \geq \cdots \geq a_r > 0$.

Claim. $\mathcal{E}(-a_1)$ has a section without zeros.

Assuming the Claim true, we get an exact sequence (given by that section)
$$0 \to \mathcal{O}_{\mathbb{P}^m} \to \mathcal{E}(-a_1) \to \mathcal{F} \to 0,$$
where the quotient $\mathcal{F}$ is a rank $r - 1$ vector bundle. Then by induction $\mathcal{F}$ splits. Therefore $\mathcal{E}(-a_1)$, and hence $\mathcal{E}$, splits into a direct sum of line bundles.

Thus it remains to show the Claim. Note that
$$\mathcal{E}_X(-a_1) \cong \mathcal{O}_X \oplus \mathcal{O}_X(a_2 - a_1) \oplus \cdots \oplus \mathcal{O}_X(a_r - a_1),$$
where $a_i - a_1 \leq 0$ for $2 \leq i \leq r$. Let $\sigma \in H^0(\mathcal{O}_X)$ be a section of $\mathcal{E}_X(-a_1)$ with no zeros. The obstruction to extending σ to a formal neighborhood $\widehat{X}$ of X belongs to $H^1(X, S^t(N^*) \otimes \mathcal{E}_X(-a_1))$, where $N := N_{X/\mathbb{P}^m}$. Since $S^t(N^*) \otimes \mathcal{E}_X(-a_1)$ is a direct sum of negative line bundles, we have $H^1(S^t(N^*) \otimes \mathcal{E}_X(-a_1)) = 0, t \geq 1$, by Kodaira vanishing. Thus we conclude that there exists a section $\widehat{\sigma} \in H^0(\widehat{X}, \mathcal{E}_{\widehat{X}}(-a_1))$ whose restriction to X coincides with σ. As soon as $\dim X \geq 2$ (which is the case since $m \geq 3$), it is a fairly standard fact, by using results of Barth [1, Proposition 4] and Griffiths [5, Theorems I, III, p. 378, 379] (see also [10, p. 226, 227]), that $\widehat{\sigma}$ extends to a section $\tau \in H^0(\mathcal{E}(-a_1))$. Then the restriction τ_X has no zeros on X. We want to show that τ has no zeros on $\mathbb{P}^m$. Let $Y := V(\tau)$ be the zero locus of τ. If $Y \neq \emptyset$, then $\dim Y \geq m - \text{rank}\mathcal{E} = m - r$. Since $r < \frac{m}{3}$, we have that $\dim(X \cap Y) \geq \dim X + \dim Y - m \geq m - 2r > 0$. Therefore $X \cap Y \neq \emptyset$ in $\mathbb{P}^m$. This contradicts the fact that the restriction τ_X has no zeros on X. □

3. Lower and upper bounds for $h^0(tL)$

We first state some general lower and upper bound formulas for the number of sections of multiples of a given line bundle L.

Lemma 3.1. *Let L be a big and spanned line bundle on an irreducible n-dimensional projective variety X. Then for $t \geq 0$ with $d := \deg_L(X) = L^n$ we have*
$$h^0(tL) \leq \binom{t+n-1}{n-1} \frac{dt+n}{n} = \frac{td+n}{t+n}\binom{t+n}{n}.$$

Proof. If X is a curve then clearly the result is true, i.e., $h^0(tL) \leq \binom{t+1-1}{1-1}$. $\frac{dt+1}{1} = dt + 1$ with equality only if $X \cong \mathbb{P}^1$. Now in general let $A \in |L|$.

Then by using the exact sequence $0 \to (s-1)L \to sL \to sL_A \to 0$ for $1 \le s \le t$ we see that $h^0(tL) \le \sum_{j=0}^{t} h^0(jL_A)$. Thus by induction we have $h^0(tL) \le \sum_{j=0}^{t} \binom{j+n-2}{n-2} \frac{dj+n-1}{n-1} = \binom{t+n-1}{n-1} \frac{dt+n}{n}$. □

Now assume that L is very ample. Then we also have the following lower bound

$$h^0(tL) \ge \binom{t+n+1}{n+1} \quad \text{for } t < d := L^n. \tag{3}$$

To see this note that we can assume $h^0(L) \ge n+2$ since otherwise $(X, L) \cong (\mathbb{P}^n, \mathcal{O}_{\mathbb{P}^n}(1))$ and the assertion is clearly true. Then X is embedded by $|L|$ in $\mathbb{P}^{n+r}$ with $r > 0$, so that we can project X generically one-to-one into $\mathbb{P}^{n+1}$. Now, for any positive integer t,

$$h^0(tL) = h^0(\mathcal{O}_{\mathbb{P}^{n+r}}(t)_X) \ge h^0(\mathcal{O}_{\mathbb{P}^{n+1}}(t)_{X'}),$$

where X' is the image of X in $\mathbb{P}^{n+1}$. But if $t < d := \deg_L(X)$ then $h^0(\mathcal{O}_{\mathbb{P}^{n+1}}(t)_{X'}) \ge h^0(\mathcal{O}_{\mathbb{P}^{n+1}}(t))$ since the kernel of the restriction map has dimension $h^0(\mathcal{O}_{\mathbb{P}^{n+1}}(t-d)) = 0$. Thus we get $h^0(tL) \ge h^0(\mathcal{O}_{\mathbb{P}^{n+1}}(t))$, which is the bound as in (3).

Following Harris' presentation [7] of Castelnuovo theory we can significantly improve the above lower bound. Let us fix some notation. Let X_{n-i} be the $(n-i)$-dimensional subvariety of X obtained as transversal intersection of X with a general $\mathbb{P}^{n+r-i}$, $0 \le i \le n$, and in particular $X_n = X$. Let, for $0 \le i \le n$,

$$h_{X_{n-i}}(t) := \dim(\text{Im}(H^0(\mathbb{P}^{n+r-i}, \mathcal{O}_{\mathbb{P}^{n+r-i}}(t)) \to H^0(X_{n-i}, \mathcal{O}_{X_{n-i}}(tL)))),$$

$h^0(t) := h_{X_0}(t)$. By [7, Lemma (3.1)] one has, for a given integer $t \ge 0$,

$$h_X(t) \ge h_{X_{n-1}}(t) + h_X(t-1).$$

Iterating on t we get $h_{X_{n-i}}(j) \ge \sum_{k=0}^{j} h_{X_{n-i-1}}(k)$ for $0 \le i \le n$. Iterating on n, we get

$$h^0(tL) \ge h_X(t) \ge \sum_{k_{n-1}=0}^{t} \cdots \sum_{k_1=0}^{k_2} \sum_{k=0}^{k_1} h_{X_0}(k). \tag{4}$$

Castelnuovo theory (see [7, p. 94]) gives

$$h_{X_0}(k) \ge h(k) := \min\{d, kr+1\}. \tag{5}$$

(Note that the formula in [7, p. 94] is for a curve in $\mathbb{P}^r$, whereas we are considering a curve in $\mathbb{P}^{r+1}$.) Let $c := \left[\frac{d-1}{r}\right]$, the integral part of $\frac{d-1}{r}$, and let $R := d - 1 - cr$. Then the graph of the function $h(k)$ looks like

Projections from subvarieties 79

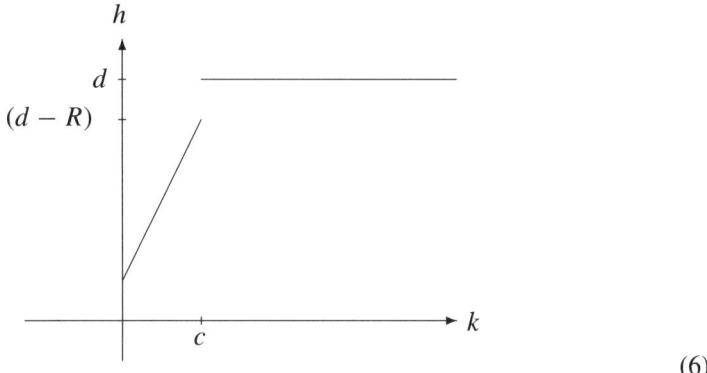

(6)

where the oblique line is the graph of the equation $h = kr + 1$.

Lemma 3.2. *Let X be an n-dimensional irreducible nondegenerate subvariety of $\mathbb{P}^{n+r}$. Let $L := \mathcal{O}_{\mathbb{P}^{n+r}}(1)_X$, and let $d = L^n$ be the degree of X in $\mathbb{P}^{n+r}$. Let $c := \left\lceil \frac{d-1}{r} \right\rceil$. Let R be the remainder defined as $R := d - 1 - cr$. Then for any integer $t \geq 0$ we have the lower bound*

$$h^0(tL) \geq r\binom{n+t}{n+1} + \binom{n+t}{n} - r\binom{n+t-c-1}{n+1} + (R-r)\binom{n+t-c-1}{n}$$
$$= \frac{tr+n+1}{t+n+1}\binom{t+n+1}{n+1} - r\binom{t+n-c}{n+1} + R\binom{t+n-c-1}{n}. \quad (7)$$

Proof. Referring to the inequality in formula (5), we get for any positive integer a

$$\sum_{k=0}^{a} h_{X_0}(k) \geq \sum_{k=0}^{a} h(k) = \sum_{k=0}^{a} kr + 1 - \sum_{k=c+1}^{a} (kr + 1 - d)$$

(see diagram (6)). Iterating the summation as in formula (4) gives

$$h^0(tL) \geq \sum_{k_{n-1}=0}^{t} \cdots \sum_{k_1=0}^{k_2} \sum_{k=0}^{k_1} (kr+1) - \sum_{k_{n-1}=c+1}^{t} \cdots \sum_{k_1=c+1}^{k_2} \sum_{k=c+1}^{k_1} (kr+1-d)$$

$$= \sum_{k_{n-1}=0}^{t} \cdots \sum_{k_1=0}^{k_2} \sum_{k=0}^{k_1} (kr+1) - \sum_{j_{n-1}=0}^{t-c-1} \cdots \sum_{j_1=0}^{j_2} \sum_{j=0}^{j_1} ((j+c+1)r+1-d)$$

$$= r \sum_{k_{n-1}=0}^{t} \cdots \sum_{k_1=0}^{k_2} \sum_{k=0}^{k_1} k + \sum_{k_{n-1}=0}^{t} \cdots \sum_{k_1=0}^{k_2} \sum_{k=0}^{k_1} 1$$

$$- r \sum_{j_{n-1}=0}^{t-c-1} \cdots \sum_{j_1=0}^{j_2} \sum_{j=0}^{j_1} j - \sum_{j_{n-1}=0}^{t-c-1} \cdots \sum_{j_1=0}^{j_2} \sum_{j=0}^{j_1} ((c+1)r+1-d)$$

By repeatedly using the combinatorial identity $\sum_{i=0}^{b} \binom{i+m}{q} = \binom{b+m+1}{q+1} - \binom{m}{q+1}$ for any positive integers b, m, and q, with the usual convention that $\binom{u}{v} = 0$ whenever $v > u$, we get

$$h^0(tL) \geq r\binom{n+t}{n+1} + \binom{n+t}{n} - r\binom{n+t-c-1}{n+1}$$
$$+ (d - 1 - r(c+1))\binom{n+t-c-1}{n}$$
$$= r\binom{n+t}{n+1} + \binom{n+t}{n} - r\binom{n+t-c-1}{n+1} + (R-r)\binom{n+t-c-1}{n}$$
$$= \frac{tr+n+1}{t+n+1}\binom{t+n+1}{n+1} - r\binom{t+n-c}{n+1} + R\binom{t+n-c-1}{n}.$$

□

Remark 3.3. It is easy to see that the right-hand side of this last inequality is minimized when $R = 0$ and $c = 1$, and therefore the bound in Lemma (3.2) yields the simpler form

$$h^0(tL) \geq \frac{tr+n+1}{t+n+1}\binom{t+n+1}{n+1} - r\binom{t+n-1}{n+1}. \tag{8}$$

Note also that if X has nonnegative Kodaira dimension, then $d \geq rn + 2$ (see e.g., [2, (8.1.3)]). Thus $\frac{d-1}{r} > n$, so that $c = \left\lceil \frac{d-1}{r} \right\rceil \geq n$. Therefore in this case we can use the bound in (3.2), with $R = 0$, $c = n$, in the form

$$h^0(tL) \geq \frac{tr+n+1}{t+n+1}\binom{t+n+1}{n+1} - r\binom{t}{n+1}.$$

We have the following general fact.

Proposition 3.4. *Let X be a nondegenerate irreducible n-dimensional subvariety of $\mathbb{P}^{n+r}$. Let $L := \mathcal{O}_{\mathbb{P}^{n+r}}(1)_X$. Let Y be a k-dimensional irreducible subvariety of X of degree $\delta := L^k \cdot Y$. Assume that $k > 0$. Let $t > 0$ be an integer such that*

$$\frac{tr+n+1}{t+n+1}\binom{t+n+1}{n+1} - r\binom{t+n-1}{n+1} > \frac{\delta t + k}{t+k}\binom{t+k}{k}.$$

Then $tL \otimes \mathcal{J}_Y$ has a section not identically zero on X.

Proof. Let L_Y be the restriction of L to Y. By Lemma (3.1) applied to L_Y we have

$$h^0(tL_Y) \leq \frac{\delta t + k}{t+k}\binom{t+k}{k}. \tag{9}$$

Suppose that $h^0(tL \otimes \mathcal{J}_Y) = 0$. Then $h^0(tL) \leq h^0(tL_Y)$. Thus by combining the inequalities (9) and (8) we get

$$\frac{\delta t + k}{t + k}\binom{t+k}{k} \geq \frac{tr + n + 1}{t + n + 1}\binom{t+n+1}{n+1} - r\binom{t+n-1}{n+1},$$

contrary to the inequality assumed in the proposition. $\square$

Let us now make explicit the bound in (3.4) in the case when Y is a divisor on X.

Proposition 3.5. *Let X be a nondegenerate irreducible n-dimensional subvariety of $\mathbb{P}^{n+r}$. Let $L := \mathcal{O}_{\mathbb{P}^{n+r}}(1)_X$. Let D be an irreducible divisor of degree $\delta := L^{n-1} \cdot D > 1$. Thus, for $t \geq 1$, the inequality $t > \frac{n}{r+1}(\delta - 1) - n + 1$ implies that $tL - D$ has a section not identically zero on X.*

Proof. The inequality in (3.4) becomes, in case $k = n - 1$,

$$\frac{tr + n + 1}{t + n + 1}\binom{t+n+1}{n+1} - r\binom{t+n-1}{n+1} > \frac{\delta t + n - 1}{t + n - 1}\binom{t+n-1}{n-1}. \quad (10)$$

Now by a simple calculation (10) gives $-t - rt - 2n + \delta n + 1 - rn + r < 0$, or, solving in t and simplifying, $t > \frac{n}{r+1}(\delta - 1) - n + 1$. $\square$

The following example shows that Proposition (3.5) is sharp.

Example 3.6. Let $X := \mathbb{P}^2 \times \mathbb{P}^2$, with $L := \mathcal{O}(1,1)$, and choose $D \in |\mathcal{O}(2,0)|$. Then $tL - D \approx \mathcal{O}(t-2, t)$ has a non-trivial section if and only if $t \geq 2$. Embed X in $\mathbb{P}^8$ by the Segre mapping. Then $n = r = 4$, and an easy calculation gives $\delta := L^3 \cdot D = 6$.

In this case we see that the hypothesis of (3.5) is satisfied if $t = 2$, and $2L - D$ has a non-zero section; whereas the hypothesis fails if $t = 1$ and $L - D$ has no non-trivial sections. Thus the inequality in (3.5) cannot be weakened.

Remark 3.7. We follow the notation and assumptions of (3.4). In general, it is hard to make the bound in (3.4) explicit in t. If Y has codimension two in X, then, after simplification, the condition for $h^0\left((\delta - 1)L \otimes \mathcal{J}_Y\right)$ to be positive becomes

$$-rn^2 - 3n^2 + \delta n^2 - 2rn\delta + 9n - 4\delta n + 5rn - 6 - 6r + 5\delta + 5r\delta - r\delta^2 - \delta^2 > 0.$$

Let $X \subset \mathbb{P}^N$ be a nondegenerate smooth connected n-fold. Let $\deg(X) = d$ and denote by L the restriction of $\mathcal{O}_{\mathbb{P}^N}(1)$ to X. From Lemma (2.6) we know that $\mathcal{O}_{\mathbb{P}^N}(d) \otimes \mathcal{J}_X$ is spanned by global sections.

Problem. What can we say about the smallest integer $t > 0$ such that $h^0(\mathcal{O}_{\mathbb{P}^N}(t) \otimes \mathcal{J}_X) > 0$?

We define the *lower degree* in $\mathbb{P}^N$, δ_N, of a subvariety $X \subset \mathbb{P}^N$ to be the smallest positive integer t such that $h^0(\mathcal{O}_{\mathbb{P}^N}(t) \otimes \mathcal{J}_X) > 0$. One consequence of the above results is that under modest conditions there must be some form of much lower degree than d vanishing on X.

Proposition 3.8. Let $X \subset \mathbb{P}^N$ be a nondegenerate smooth connected n-fold of degree d. Let $L = \mathcal{O}_{\mathbb{P}^N}(1)_X$ and let δ_N be the degree of the lowest-degree homogeneous form vanishing on X. Then δ_N satisfies the inequality

$$\frac{1}{\delta_N - 1}\left(\frac{(\delta_N + N - 1)\cdots(\delta_N + n - 1)}{N\cdots(n+1)} - n\right) \leq d.$$

Proof. If t is a positive integer for which $h^0(\mathcal{O}_{\mathbb{P}^N}(t) \otimes \mathcal{J}_X) = 0$ then

$$\binom{N+t}{N} = h^0(\mathcal{O}_{\mathbb{P}^N}(t)) \leq h^0(X, tL).$$

By applying Lemma (3.1) we get $\binom{N+t}{N} \leq \frac{td+n}{t+n}\binom{t+n}{n}$. An easy calculation shows that this inequality is equivalent to $\frac{1}{t}\left(\frac{(t+N)\cdots(t+n)}{N\cdots(n+1)} - n\right) \leq d$. From the definition of δ_N it follows that $h^0(\mathcal{O}_{\mathbb{P}^N}(\delta_N - 1) \otimes \mathcal{J}_X) = 0$, and substituting $t = \delta_N - 1$ in the last inequality completes the proof. $\square$

For surfaces here is the explicit bound.

Corollary 3.9. Let X be a nondegenerate smooth connected surface of degree d in $\mathbb{P}^N$. Assume $N \geq 5$. Let δ_N be the degree of the lowest-degree homogeneous form vanishing on X. Then $\delta_N^3 + 11\delta_N^2 + 46\delta_N + 96 \leq 60d$.

Proof. We apply the bound in (3.8) with $n = 2$. Since $N \geq 5$, we get

$$\frac{1}{\delta_N - 1}\left(\frac{(\delta_N + 4)\cdots(\delta_N + 1)}{60} - 2\right) \leq d.$$

After simplifying this becomes $(\delta_N+4)(\delta_N+3)(\delta_N+2)(\delta_N+1) - 120 \leq 60d(\delta_N-1)$. Since $\delta_N \geq 2$ we can divide both sides by $\delta_N - 1$ to obtain the desired result. $\square$

For example, if X is of degree 21 then $\delta_N \leq 6$ and hence there is a form of the sixth degree vanishing on X. Or again, if X is of degree 10,000 there is a form of degree 80 vanishing on X.

Some other special cases are as follows. For threefolds in $\mathbb{P}^N$ with $N \geq 5$ the corresponding bound as in Corollary (3.9) is $\delta_N^2 + 10\delta_N + 36 \leq 20d$. For threefolds in $\mathbb{P}^N$ with $N \geq 6$ the bound becomes $\delta_N^3 + 15\delta_N^2 + 86\delta_N + 240 \leq 120d$.

4. Some general structure results for projections

In this section we discuss some general properties of projections from a k-dimensional subvariety Y of a given polarized variety X. We always assume that $k > 0$. In §6 and

§7 we will present some more refined results in the cases when Y is either a divisor or a linear $\mathbb{P}^k$.

4.1. General set-up of morphisms. Let L be a very ample line bundle on X, a smooth connected variety of dimension $n \geq 2$. Let Y be a k-dimensional connected submanifold of X. *We always assume that $k > 0$.* We will denote by $\mathcal{J}_Y$ the ideal sheaf of Y in X.

Let $\sigma : \overline{X} \to X$ be the blowing up of X along Y and set $E = \sigma^{-1}(Y)$. Let $\psi : \overline{X} \to \psi(\overline{X})$ be the surjective rational map given by $|\sigma^* L - E|$. We refer to the mapping ψ as the *projection from Y associated to L.* If $L \otimes \mathcal{J}_Y$ is spanned by its global sections, then $\sigma^* L - E$ is spanned on $\overline{X}$ and ψ is a morphism. We have the Remmert-Stein factorization $\psi = s \circ \phi$ of $\psi : \overline{X} \to \psi(\overline{X})$, where $\phi : \overline{X} \to Z$ is a morphism with connected fibers onto a normal variety Z and $s : Z \to \psi(\overline{X})$ is a finite morphism. We will refer to $\phi : \overline{X} \to Z$ as *the morphism associated to $L \otimes \mathcal{J}_Y$.* Note there is an ample and spanned line bundle $\mathcal{H}$ on Z such that $\sigma^* L - E \approx \phi^* \mathcal{H}$. We have the following commutative diagram

$$\begin{array}{ccc} E & \hookrightarrow & \overline{X} \\ \downarrow & \downarrow \sigma \searrow^{\phi} & \\ Y & \hookrightarrow X \xrightarrow{\varphi} & Z \end{array}$$

where φ is the connected part of the rational mapping associated to $|L \otimes \mathcal{J}_Y|$.

We need the following technical lemmas.

Lemma 4.2. *Let Y be a connected k-dimensional submanifold of X, a connected projective manifold of dimension $n \geq 2$. Assume that $k > 0$. Let $N := N_{Y/X}$ be the normal bundle of Y in X. Let L be a line bundle on X. Assume that $L \otimes \mathcal{J}_Y$ is spanned by a vector subspace, V, of $\Gamma(L \otimes \mathcal{J}_Y)$ and $c_{n-k}(N^*(L)) = 0$. Then a general element $D \in |V|$ is smooth.*

Proof. A general $D \in |V|$ is smooth on $X \setminus Y$ by Bertini's theorem.

A given $D \in |L \otimes \mathcal{J}_Y|$ is smooth at a point $y \in Y$ if the differential in local coordinates of the defining equation of D is not zero at $y \in Y$. From the exact sequence

$$0 \to \mathcal{J}_Y^2 \otimes L \to \mathcal{J}_Y \otimes L \xrightarrow{\partial} N^*(L) \to 0$$

we see that D is smooth on Y if the image $\partial(s)$ of s defining D in $N^*(L)$ is nowhere zero. Since $c_{n-k}(N^*(L)) = 0$, a general $s \in V$ goes to a nowhere vanishing section $\partial(s)$ in $N^*(L)$. □

Lemma 4.3. *Let Y be a connected k-dimensional submanifold of X, a connected projective manifold of dimension $n \geq 2$. Assume that $k > 0$. Let L be a very ample* line bundle on X. Assume that $L \otimes \mathcal{J}_Y$ is spanned by global sections and let D be a smooth element of $|L \otimes \mathcal{J}_Y|$. Let $\sigma : \overline{X} \to X$ be the blowing up of X along Y and let $\overline{D}$ be the proper transform of D under σ. Let $\phi : \overline{X} \to Z$ be the morphism associated to $L \otimes \mathcal{J}_Y$ as in (4.1). Let $\phi_{\overline{D}}$ be the restriction of ϕ to $\overline{D}$. Then $\phi_{\overline{D}}$ has lower dimensional image if and only if ϕ has lower dimensional image.

Proof. We follow the notation from (4.1). Assume that ϕ has lower dimensional image. We have $\overline{D} \in |\sigma^*L - E|$, $E = \sigma^{-1}(Y)$. Note that $\overline{D}$ is the pullback of some divisor $\mathcal{D} \subset Z$. From $\dim\phi(\overline{D}) < \dim\phi(\overline{X}) < \dim X$ we get $\dim\phi(\overline{D}) < \dim \overline{D}$.

For the converse, note that by definition of $\overline{D}$ one has $\dim\phi(\overline{D}) = \dim\phi(\overline{X}) - 1$. Thus the assumption $\dim\phi(\overline{D}) < \dim \overline{D} = n - 1$ gives the result. □

Let us note some further general properties of the morphism ϕ. The notation is as in (4.1).

1. (Divisorial case) If Y is a divisor, then $\overline{X} \cong X$.

2. (Linear case) Assume that $(Y, L_Y) \cong (\mathbb{P}^k, \mathcal{O}_{\mathbb{P}^k}(1))$ and that $\Gamma(L)$ embeds X in $\mathbb{P}^{n+r}$. Since Y is a linear space it follows that $L \otimes \mathcal{J}_Y$ is spanned by global sections. The mapping given by $\Gamma(L \otimes \mathcal{J}_Y)$ coincides off of Y with the restriction to X of the projection of $\mathbb{P}^{n+r}$ to $\mathbb{P}^{n+r-k-1}$ from Y.

3. (Smooth case) If Y is smooth then $\delta L \otimes \mathcal{J}_Y$, $\delta = L^k \cdot Y$, is spanned by global sections by Lemma (2.6).

We have the following crude structure theorem in the case when the projection has lower dimensional image.

Theorem 4.4. *Let Y be a connected k-dimensional submanifold of X, a connected projective manifold of dimension $n \geq 2$.*

Assume that $k > 0$. Let L be a very ample line bundle on X. Assume that $L \otimes \mathcal{J}_Y$ is spanned by its global sections. Let $\sigma : \overline{X} \to X$ be the blowing up of X along Y. Let $\phi : \overline{X} \to Z$ be the morphism associated to $L \otimes \mathcal{J}_Y$ as in (4.1). Let $E := \sigma^{-1}(Y)$ be the exceptional divisor. Assume $n > \dim Z$. Then we have:

1. $\phi(E) = Z$;

2. Z is uniruled if $\mathrm{cod}_X Y > 1$;

3. Z is unirational if Y is unirational;

4. the restriction, E_F, of E to any fiber F of ϕ is an ample divisor on F.

Proof. To show 1), assume by contradiction that the restriction, $\phi_E : E \to Z$, of ϕ to E is not surjective. Take a point $x \in Z \setminus \phi(E)$ and let $F_x = \phi^{-1}(x)$ be the fiber on x. Then the restriction $(\sigma^*L - E)_{F_x}$ is trivial. But $E_{F_x} \cong \mathcal{O}_x$ since $x \notin \phi(E)$, so that $(\sigma^*L - E)_{F_x} \cong (\sigma^*L)_{F_x} \cong L_{\sigma(F_x)}$ is ample, where the last isomorphism follows from the fact that F_x goes isomorphically to X under σ, since $F_x \cap E = \emptyset$. Thus $(\sigma^*L - E)_{F_x}$ is both ample and trivial, an absurdity that contradicts the assumption $n > \dim Z$.

To show 2), note that since $\mathrm{cod}_X Y > 1$, σ is not an isomorphism and the exceptional divisor is uniruled. This means that there exists an $(n - 2)$-dimensional variety V and a rational map $V \times \mathbb{P}^1 \to E$ which is dominant. Since $\phi(E) = Z$ by 1), we get a dominant map $V \times \mathbb{P}^1 \to Z$, i.e., Z is uniruled.

To show 3), recall that E is birational to $Y \times \mathbb{P}^{n-k-1}$. Since Y is unirational we have a dominant rational map $\mathbb{P}^k \to Y$. Therefore, combining with the surjective map $\phi_E : E \to Z$, we get a dominant rational map $\mathbb{P}^k \times \mathbb{P}^{n-k-1} \to Z$. This implies that Z is unirational.

To show 4), take a fiber F of $\phi : \overline{X} \to Z$. If the restriction σ_F of the blowing up map is finite-to-one then $\sigma_F^* L$ is ample and the assertion is clear. It is easy to see that σ_F is finite. If not, then it follows that there is a positive dimensional fiber, f, of $\sigma_F : F \to \sigma(F)$. This implies that f is contained in a fiber of $E \to Y$. But $\sigma^*L - E$ is ample on fibers of $E \to Y$. On the other hand, since $f \subset F$, the line bundle $\sigma^*L - E$ is trivial on f. □

We need the following lemma.

Lemma 4.5. *Let Y be a connected k-dimensional submanifold of X, a connected projective manifold of dimension $n \geq 2$. Assume that $k > 0$. Let L be a very ample line bundle on X. Let $\mathcal{J}_Y$ be the ideal sheaf of Y in X and let $N := N_{Y/X}$ be the normal bundle of Y in X. Then $L \otimes \mathcal{J}_Y$ is spanned and $N^*(L)$ is trivial if and only if Y is the complete intersection of $n - k$ divisors $D_1, \ldots, D_{n-k} \in |L|$.*

Proof. The "if" part is straightforward. As to the converse, consider the exact sequence

$$0 \to L \otimes \mathcal{J}_Y^2 \to L \otimes \mathcal{J}_Y \xrightarrow{\partial} N^*(L) \to 0.$$

Set $w := n - k$. Since $N^*(L) = \oplus^w \mathcal{O}_Y$ and $L \otimes \mathcal{J}_Y$ is spanned we can find sections $s_1, \ldots, s_w \in \Gamma(L \otimes \mathcal{J}_Y)$ defining w divisors $D_1, \ldots, D_w \in |L \otimes \mathcal{J}_Y|$ on X containing Y.

For each $i = 1, \ldots, w$, D_i is smooth on $X \setminus Y$ by Bertini's theorem.

For each $i = 1, \ldots, w$, D_i is smooth at a point $y \in Y$ if the differential in local coordinates of the defining equation of D_i is not zero at y. From the exact sequence above we see that D_i is smooth on Y if the image $\partial(s_i)$ of s_i defining D_i in $N^*(L)$ is nowhere zero, $1 \leq i \leq w$. Since $N^*(L)$ is trivial we can find the sections $s_1, \ldots, s_w \in \Gamma(L \otimes \mathcal{J}_Y)$ such that $\partial(s_1), \ldots, \partial(s_w)$ are independent in $N^*(L)$. It follows that $D_1, \ldots, D_w$ are smooth as well as the intersection $D_1 \cap \cdots \cap D_w$ is smooth.

Since $\dim X \geq 2$, from the exact sequence $0 \to -L \to \mathcal{O}_X \to \mathcal{O}_{D_i} \to 0$, we see that $h^0(\mathcal{O}_{D_i}) = 1$, so the D_i's are connected, $1 \leq i \leq w$.

Since $D_1 \cap \cdots \cap D_w$ is at least one-dimensional, we know by the Lefschetz hyperplane section theorem that $D_1 \cap \cdots \cap D_w$ is connected. Since it is also smooth of dimension $\dim Y$ and contains Y we conclude that Y is the complete intersection of $D_1, \ldots, D_w$. □

We can prove now the following more refined structure result, which gives a general lower bound for the dimension of the image of the projection.

Theorem 4.6. *Let Y be a connected k-dimensional submanifold of X, a connected projective manifold of dimension $n \geq 2$. Assume that $k > 0$. Let L be a very ample line bundle on X. Assume that $L \otimes \mathcal{J}_Y$ is spanned by its global sections. Let $\sigma : \overline{X} \to X$ be the blowing up of X along Y. Let $\phi : \overline{X} \to Z$ be the morphism associated to $L \otimes \mathcal{J}_Y$ as in (4.1). Then $\dim Z \geq n - k - 1$, with equality if and only if $Z \cong \mathbb{P}^{n-k-1}$ and Y is the complete intersection of $n - k$ divisors $D_1, \ldots, D_{n-k} \in |L|$.*

Proof. Set $w := n - k$. Let $E := \sigma^{-1}(Y)$ be the exceptional divisor. Set $\overline{L} := \sigma^* L$ and $N := N_{Y/X}$. Recall that $E \cong \mathbb{P}(N^*(L))$. Thus $\pi := \sigma_E : E \to Y$ is a $\mathbb{P}^{w-1}$-bundle. Let ξ be the tautological line bundle of $\mathbb{P}(N^*(L))$. Notice that $\xi \cong (\overline{L} - E)_E$. Since $(\overline{L} - E)_{\mathbb{P}^{w-1}} \cong \mathcal{O}_{\mathbb{P}^{w-1}}(1)$ it follows that each fiber $\mathbb{P}^{w-1}$ of $\pi : E \to Y$ maps isomorphically under the map $\psi : \overline{X} \to \psi(\overline{X})$ given by $|\overline{L} - E|$, and hence maps isomorphically into Z under the morphism ϕ associated to $L \otimes \mathcal{J}_Y$. This shows that $\dim Z \geq w - 1$. If $\dim Z = w - 1$, it follows that $Z \cong \mathbb{P}^{w-1}$.

It also follows that Y is a complete intersection. For we have a surjective map of locally free sheaves $\oplus^{\dim Z+1} \mathcal{O}_{\overline{X}} \to \overline{L} - E \to 0$, and, restricting to E, we have a surjection $\oplus^{\dim Z+1} \mathcal{O}_E \to (\overline{L} - E)_E \to 0$. Consider the $\mathbb{P}^{w-1}$-bundle $\pi : E \to Y$. Notice that $(\overline{L} - E)_E \cong \xi$, the tautological line bundle of $E \cong \mathbb{P}(N^*(L))$. By pushing forward under π, we get a surjection

$$\beta : \oplus^{\dim Z+1} \mathcal{O}_Y \to N^*(L) = \pi_* \xi \to 0.$$

By comparing the ranks, since $N^*(L)$ has rank $\mathrm{cod}_X Y = w = \dim Z + 1$, we conclude that β is an isomorphism, i.e., $N^*(L)$ is the trivial bundle. Thus, since $L \otimes \mathcal{J}_Y$ is spanned by global sections, Lemma (4.5) applies to give the result.

Next, we show that if $Y := D_1 \cap \cdots \cap D_w$ is the complete intersection of w divisors $D_1, \ldots, D_w \in |L|$, then $\dim Z = w - 1$. We first observe that $N \cong \oplus^w L_Y$, so that $N^*(L) \cong \oplus^w \mathcal{O}_Y$ is trivial. Consider the exact sequence

$$0 \to L \otimes \mathcal{J}_Y^2 \to L \otimes \mathcal{J}_Y \to N^*(L) \to 0.$$

Since the morphism $L \otimes \mathcal{J}_Y \to N^*(L)$ is surjective at the sheaf level, $L \otimes \mathcal{J}_Y$ is spanned by global sections and $N^*(L)$ is trivial, it follows that the induced map $\alpha : \Gamma(L \otimes \mathcal{J}_Y) \to \Gamma(N^*(L)) \to 0$ is surjective.

For any integer $m \geq 2$, consider the exact sequence

$$0 \to L \otimes \mathcal{J}_Y^{m+1} \to L \otimes \mathcal{J}_Y^m \to S^m(N^*) \otimes L \to 0.$$

Since $S^m(N^*) \otimes L \cong \oplus^w L_Y^{-(m-1)}$ we have $h^0(S^m(N^*) \otimes L) = 0$, for $m \geq 2$, and therefore we get

$$\Gamma(L \otimes \mathcal{J}_Y^2) \cong \cdots \cong \Gamma(L \otimes \mathcal{J}_Y^m), \quad m \geq 2.$$

If $h^0(L \otimes \mathcal{J}_Y^2) \neq 0$ we thus find a section of L vanishing on Y of any given order $m \geq 2$, which is absurd. Therefore we conclude that $h^0(L \otimes \mathcal{J}_Y^2) = 0$ and hence $\Gamma(L \otimes \mathcal{J}_Y)$ injects in $\Gamma(N^*(L))$, i.e., the map α is an isomorphism. Thus $h^0(L \otimes \mathcal{J}_Y) = h^0(\overline{L} - E) = w$. Since $\overline{L} - E$ is spanned and gives the projection $\psi : \overline{X} \to \psi(\overline{X})$, we thus conclude that $\dim Z = w - 1$. □

There are many results from adjunction theory [2] describing all varieties with a given hyperplane section. Combining these results with Theorem (4.6) gives many consequences. By way of illustration we give two useful corollaries.

Corollary 4.7. *Let Y be a connected k-dimensional submanifold of X, a connected projective manifold of dimension $n \geq 2$. Let L be a very ample line bundle on X. Assume that $k > 0$ and Y is a linear $\mathbb{P}^k$ with respect to L, i.e., $L^k \cdot Y = 1$. Let $\sigma : \overline{X} \to X$ be the blowing up of X along Y. Let $\phi : \overline{X} \to Z$ be the morphism associated to $L \otimes \mathcal{J}_Y$ as in (4.1). Then $\dim Z = n - k - 1$ if and only if $(X, L) \cong (\mathbb{P}^n, \mathcal{O}_{\mathbb{P}^n}(1))$.*

Proof. Assume $\dim Z = n - k - 1$. Then, by (4.6), 2), Y is the complete intersection of $n - k$ divisors $D_1, \ldots, D_{n-k} \in |L|$. Since Y is a linear $\mathbb{P}^k$ it thus follows that $(X, L) \cong (\mathbb{P}^n, \mathcal{O}_{\mathbb{P}^n}(1))$.

If $(X, L) \cong (\mathbb{P}^n, \mathcal{O}_{\mathbb{P}^n}(1))$, the projection from $Y = \mathbb{P}^k$ has an $(n - k - 1)$-dimensional image. □

Corollary 4.8. *Let Y be a connected k-dimensional submanifold of X, a connected projective manifold of dimension $n \geq 3$. Assume that $k > 0$. Let L be a very ample line bundle on X. Assume that $L \otimes \mathcal{J}_Y$ is spanned by its global sections. Let $\sigma : \overline{X} \to X$ be the blowing up of X along Y. Let $\phi : \overline{X} \to Z$ be the morphism associated to $L \otimes \mathcal{J}_Y$ as in (4.1). Assume that Y is a $K(\pi, 1)$ and $k \geq 2$. Then $\dim Z \geq n - k$.*

Proof. By (4.6) either we are done or Y is a complete intersection of $n - k$ divisors $D_1, \ldots, D_{n-k} \in |L|$. Since Y is a $K(\pi, 1)$ with $\dim Y \geq 2$, this is not possible by a result of the fourth author [19]. □

Under special conditions on the cohomology of Y, we get stronger lower bounds for the image dimension of the projection. We restrict our attention to the case in which $\dim Z \geq n - k$, since the case $\dim Z = n - k - 1$ was covered in (4.6).

Proposition 4.9. *Let Y be a connected k-dimensional submanifold of X, a connected projective manifold of dimension $n \geq 2$. Assume that $k > 0$. Let L be a very ample line bundle on X. Assume that $L \otimes \mathcal{J}_Y$ is spanned by its global sections. Let $\sigma : \overline{X} \to X$ be the blowing up of X along Y. Let $\phi : \overline{X} \to Z$ be the morphism* associated to $L \otimes \mathcal{J}_Y$ as in (4.1). Assume that $h^2(Y, \mathbb{Q})_{\text{alg}} = 1$ (or equivalently $\text{Pic}(Y) \otimes \mathbb{Q} \cong \mathbb{Q}$). If $\dim Z \geq n - k$, then $\dim Z \geq n - k + \frac{k}{n-k} - 1$. In particular, if $\dim Z = \text{cod}_X Y = n - k$, then $n \geq 2k$.

Proof. Let $w := n - k$ and $N := N_{Y/X}$. As in the proof of (4.6), we have a surjective vector bundle map $\oplus^{\dim Z+1} \mathcal{O}_Y \to N^*(L) \to 0$. This gives a natural map $\rho : Y \to \text{Grass}(w, \dim Z + 1)$ of Y in the Grassmannian of the w-dimensional quotients of $\mathbb{C}^{\dim Z+1}$. We claim that the map ρ is finite. Indeed, to see this, notice that $\det(N^*(L)) = \rho^* \mathcal{P}$, where $\mathcal{P}$ is an ample line bundle, the Plücker bundle, on $\text{Grass}(w, \dim Z + 1)$. Since ρ is a not trivial map, $\det(N^*(L))$ is spanned and not trivial. Since $\text{Pic}(Y) \otimes \mathbb{Q} \cong \mathbb{Q}$, we thus conclude that $\rho^* \mathcal{P}$ is ample. Let F be a connected component of a positive dimensional fiber of ρ. Then $(\rho^* \mathcal{P})_F \cong \mathcal{O}_F$. This contradicts the ampleness of $\rho^* \mathcal{P}$. Thus

$$k = \dim Y \leq \dim \text{Grass}(w, \dim Z + 1) = w(\dim Z + 1 - w)$$

gives the desired inequality.

If $\dim Z = n - k$, we have $k \leq n - k = \text{cod}_X Y$, which is the same as $2k \leq n$. □

Theorem 4.10. *Let Y be a connected k-dimensional proper submanifold of X, a connected projective manifold of dimension $n \geq 2$. Assume that $k > 0$. Let L be a very ample line bundle on X. Assume that $L \otimes \mathcal{J}_Y$ is spanned by its global sections. Let $\sigma : \overline{X} \to X$ be the blowing up of X along Y. Let $\phi : \overline{X} \to Z$ be the morphism associated to $L \otimes \mathcal{J}_Y$ as in (4.1). If $h^{2j}(Y, \mathbb{Q})_{\text{alg}} = 1$ for $j \leq w := n - k$, then either $Z \cong \mathbb{P}^{n-k-1}$ with Y the complete intersection of $n - k$ divisors in $|L|$ or $\dim Z \geq k$.*

Proof. By Theorem (4.6) we can assume that if the theorem is false then

$$k - 1 \geq \dim Z \geq w. \tag{11}$$

Let $E = \sigma^{-1}(Y)$ be the exceptional divisor of $\sigma : \overline{X} \to X$ and let $\phi_E : E \to Z$ be the restriction to E of the morphism $\phi : \overline{X} \to Z$. From (4.4), 1), we know that ϕ_E is surjective. Let $N := N_{Y/X}$ be the normal bundle of Y in X. Set $\overline{L} = \sigma^* L$. Let ξ be the tautological line bundle of $\mathbb{P}(N^*(L)) \cong E$. Notice that $\xi \cong (\overline{L} - E)_E$. Therefore for each fiber $\mathbb{P}^{w-1}$ of the $\mathbb{P}^{w-1}$-bundle $\pi : E \to Y$ we have $(\overline{L} - E)_{\mathbb{P}^{w-1}} \cong \mathcal{O}_{\mathbb{P}^{w-1}}(1)$. This implies that each fiber F of ϕ_E meets $\mathbb{P}^{w-1}$ in at most one point. It thus follows that F goes isomorphically to $\pi(F)$ under π. Since $\xi_F \cong \mathcal{O}_F$, we get a surjective map $(\pi^* N^*(L))_F \to \xi_F \cong \mathcal{O}_F \to 0$. Letting $F' := \pi(F)$, we have by the above $F \cong F'$ and therefore pushing forward under π we get a surjective map

$$N^*(L)_{F'} \to \mathcal{O}_{F'} \to 0. \tag{12}$$

We claim that

$$c_w(N^*(L)) = 0. \tag{13}$$

To see this, let F be a general fiber of ϕ_E and $F' = \pi(F)$. Note that $\dim F' = \dim E - \dim Z \geq n - 1 - (k-1) = w$. In view of this and the assumption that $h^{2j}(Y, \mathbb{Q})_{\text{alg}} = 1, j \leq w$, we see that it is enough to note that from (12) it immediately follows that $c_w(N^*(L)_{F'}) = 0$.

Thus Lemma (4.2) implies that there exists a smooth divisor $D \in |L \otimes \mathcal{J}_Y|$. Since $\dim X \geq 2$, from the exact sequence $0 \to -L \to \mathcal{O}_X \to \mathcal{O}_D \to 0$, we see that $h^0(\mathcal{O}_D) = 1$, so D is connected. Let $H \in |\overline{L} - E|$ be the divisor corresponding to D. I.e., $H = \phi^* Z_1$, where $Z_1 \in |\mathcal{H}|$ and $\overline{L} - E \approx \phi^* \mathcal{H}$ for some ample and spanned line bundle $\mathcal{H}$ on Z. Note that by the generalized Seidenberg theorem (see e.g., [2, (1.7.1)]), Z_1 is irreducible and normal since Z is irreducible and normal. Notice also that $\sigma^* D \approx H + E$.

Let $X_1 := D$. By construction, $Y \subset X_1$. Furthermore the blowing up $\sigma : \overline{X} \to X$ induces a blowing up map $\sigma_1 : \overline{X}_1 \to X_1$ of X_1 along Y. We can also consider the morphism, $\phi_1 : \overline{X}_1 \to Z_1$, associated to $L_{X_1} \otimes \mathcal{J}_Y$, where L_{X_1} is the restriction of L to X_1. Note that ϕ_1 is onto, $\dim X_1 = n - 1$, $\dim Z_1 = \dim Z - 1$. Hence in particular $\dim Z_1 < k$, i.e., (11) is preserved passing from Z to Z_1.

Thus, starting from $X_1 = D, Z_1, \phi_1 : \overline{X}_1 \to Z_1, Y \subset X_1$, we proceed in such a way that from the initial data

$$(k, \dim Z, w)$$

we reach, after $w - 1$ steps, the data (recall that we are working under the initial assumption that $\dim Z \geq w$)

$$(k, \dim Z - w + 1, 1).$$

I.e., Y is a divisor in X_{w-1} with $X_{w-1} \cong \overline{X}_{w-1}$, and the morphism $\phi_{w-1} : \overline{X}_{w-1} \to Z_{w-1}$ has image of dimension $\dim Z - w + 1$. In particular, since $X_{w-1} \cong \overline{X}_{w-1}$, we can restrict ϕ_{w-1} to Y, so that we get a surjective map from Y to Z_{w-1} (see (4.4), 1)). By assumption we have that $h^2(Y, \mathbb{Q})_{\text{alg}} = 1$, and therefore that $\dim Y = \dim Z_{w-1}$. Thus using (11) we have

$$k = \dim Z_{w-1} = \dim Z - w + 1 = \dim Z - n + k + 1$$

which gives that $n = \dim Z + 1$. Combined with (11) we have $n \leq k$, which contradicts the hypothesis that Y is a proper submanifold of X. □

Corollary 4.11. *Let Y be a connected k-dimensional submanifold of X, a connected projective manifold of dimension $n \geq 2$. Assume that $k > 0$. Assume that L is a very ample line bundle on X such that $(Y, L_Y) \cong (\mathbb{P}^k, \mathcal{O}_{\mathbb{P}^k}(1))$. Let $\sigma : \overline{X} \to X$ be the blowing up of X along Y. Let $\phi : \overline{X} \to Z$ be the morphism associated to $L \otimes \mathcal{J}_Y$ as in (4.1). Then either $Z \cong \mathbb{P}^{n-k-1}$ with $(X, L) \cong (\mathbb{P}^n, \mathcal{O}_{\mathbb{P}^n}(1))$ or $\dim Z \geq k$.*

Proof. It immediately follows by combining (4.7) and (4.10). □

Corollary 4.12. *Let Y be a connected k-dimensional submanifold of X, a connected projective manifold of dimension $n \geq 2$. Assume that $k > 0$. Let L be a very ample line bundle on X. Assume that $L \otimes \mathcal{J}_Y$ is spanned by its global sections. Let $\sigma : \overline{X} \to X$ be the blowing up of X along Y. Let $\phi : \overline{X} \to Z$ be the morphism associated to $L \otimes \mathcal{J}_Y$ as in (4.1). Assume that Y is not a complete intersection. Further assume that $h^{2j}(Y, \mathbb{Q})_{\mathrm{alg}} = 1$ for $j \leq n - k$. Then $\dim Z \geq n/2$.*

Proof. From (4.6) and (4.10) it follows that $\dim Z \geq n - k$ and $\dim Z \geq k$. Thus $\dim Z \geq n/2$. □

Let us point out some relations between the results above and Castelnuovo–Mumford regularity theory and the Castelnuovo bound conjecture (see [6]).

Remark 4.13. Let X be an n-dimensional smooth variety in $\mathbb{P}^{n+r}$ and let Y be a k-dimensional subvariety of X of degree $\delta := L^k \cdot Y$ and $L := \mathcal{O}_{\mathbb{P}^{n+r}}(1)_X$. Assume that $k > 0$. Let q be the codimension of Y in the smallest linear subspace $\mathbb{P}^{k+q} \subset \mathbb{P}^{n+r}$ containing Y. The Castelnuovo bound conjecture says that $(\delta - q + 1)L \otimes \mathcal{J}_Y$ is spanned by global sections. The conjecture is related to the question of Castelnuovo–Mumford regularity, and it is known to hold when Y has dimension 1 (see [6]) or 2 (see [15]) and when Y has dimension 3 and $\left\lceil \frac{\delta-1}{q} \right\rceil \geq 6$ (see [17]). Assuming the conjecture true and Y a divisor, we will show that the projection from Y associated to L is birational except in certain specific cases (see §6).

Remark 4.14. Let X be an n-dimensional smooth variety in $\mathbb{P}^{n+r}$ of degree d. Let L be a very ample line bundle on X. From the Castelnuovo–Mumford regularity theory developed in [7] it follows that in case $n = 1$, for $t > \left\lceil \frac{d-1}{r} \right\rceil$, one has $h^1(tL) = 0$ and thus that $h^0(tL) = \chi(\mathcal{O}_X(tL))$. One might hope that this extends in higher dimensions also. Unfortunately this is not true in dimension $n \geq 2$, as the following example shows.

Let C_1 be a smooth plane curve of degree d_1 with L_1 the restriction of the hyperplane section bundle of $\mathbb{P}^2$ to C_1. Let L_2 be a very ample line bundle of degree $d_2 := d' + 2g - 2$, with $d' > 0$, on a smooth curve C_2 of genus $g := g(C_2)$. Let $X := C_1 \times C_2 \subset \mathbb{P}^N$ and let $L := p_1^* L_1 \otimes p_2^* L_2$, where $p_i : X \to C_i$, $i = 1, 2$, are the projections on the two factors. Note that if $g(C_i) \geq 2$ for $i = 1, 2$, then X is a surface of general type. Since $d_2 > 2g - 2$, we have $h^1(L_2) = 0$ and hence

$$h^0(L_2) = d_2 - g + 1 = d' + g - 1.$$

Thus by the Künneth formula we have $h^0(L) = 3(d'+g-1)$, i.e., $N = 3(d'+g-1)-1$. Therefore $r = N - 2 = 3(d' + g - 2)$. Moreover $d = 2d_1 d_2 = 2d_1(d' + 2g - 2)$. Thus the critical value, c, is $c = \left\lceil \frac{2d_1(d'+2g-2)-1}{3(d'+g-2)} \right\rceil$. For a fixed g and taking $d' \gg 0$ and $d_1 \geq 10$ we have $c \sim \frac{2}{3} d_1 < d_1 - 3$. On the other hand, by using again Künneth

formulas we get, for $t = d_1 - 3$,
$$h^1(tL) \geq h^1(\mathcal{O}_{C_1}(t)) = h^1(K_{C_1}) = 1. \tag{14}$$

To show that the equality $h^0(tL) = \chi(\mathcal{O}_X(tL))$ for $t > c = \left\lceil \frac{d-1}{r} \right\rceil$ is not true in general, consider the smooth irreducible curve and the set Γ of d distinct points obtained as transversal intersection of X with a general $\mathbb{P}^{r+1}$ and a general hyperplane $\mathbb{P}^r$ of the $\mathbb{P}^{r+1}$. Look at the exact sequence $0 \to (t-1)L_C \to tL_C \to tL_\Gamma \to 0$, From [7, Theorem (3.7)] we know that $H^0(tL_C) \to H^0(tL_\Gamma)$ is surjective for $t > c$ and therefore $H^1((t-1)L_C)$ injects in $H^1(tL_C)$, for $t > c$. Since $h^1(tL_C) = 0$ for $t \gg 0$, we can conclude that $h^1((t-1)L_C) = 0$ for $t > c$. Thus from the cohomology sequence associated to the exact sequence
$$0 \to (t-2)L \to (t-1)L \to (t-1)L_C \to 0$$
we infer that $h^2((t-2)L) = 0$ for $t > c$. From this we thus conclude that, for $t > c - 2$, the equality $h^0(tL) = \chi(\mathcal{O}_X(tL))$ is equivalent to $h^1(tL) = 0$. We have just shown (see (14)) that this is not the case.

5. Examples

In this section we give some examples to illustrate the results obtained in §4. We use the same notation as in (4.1). The following example shows that the dimension of the image of the projection in Theorem (4.6) can actually reach all possible values.

Example 5.1. Let M be a smooth connected projective variety of dimension $n - s$, $s \geq 0$. Let $X := M \times \mathbb{P}^s$. Let L be a very ample line bundle on X. Let $p : X \to \mathbb{P}^s$ be the product projection and set $H := p^*\mathcal{O}_{\mathbb{P}^s}(1)$. Let Y be the k-dimensional subvariety of X obtained as transversal intersection of $n - k - 1$ general members of $|L|$ and a general $D_{n-k} \in |L - H|$. Assume that $k \geq s$ and $k > 0$. Let $\phi : \overline{X} \to Z$ be the morphism associated to $L \otimes \mathcal{J}_Y$ as in the usual set up (4.1).

Let $N := N_{Y/X}$ be the normal bundle of Y in X. Note that $N \cong (\oplus^{n-k-1} L_Y) \oplus (L - H)_Y$. We let $V := (\oplus^{n-k-1}\mathcal{O}_X) \oplus H$ and $\mathcal{F} := (\oplus^{n-k-1}\mathcal{O}_{\mathbb{P}^s}) \oplus \mathcal{O}_{\mathbb{P}^s}(1)$. Thus $N^*(L) \cong V \cong p^*\mathcal{F}$ and $E \cong \mathbb{P}(p^*\mathcal{F})$, where E is the exceptional divisor of the blowing up, $\sigma : \overline{X} \to X$, of X along Y. Let α, β be the morphisms associated to $|\xi_{p^*\mathcal{F}}|$ and $|\xi_\mathcal{F}|$ respectively, where $\xi_{p^*\mathcal{F}}$ and $\xi_\mathcal{F}$ are the tautological line bundles of $\mathbb{P}(p^*\mathcal{F})$ and $\mathbb{P}(\mathcal{F})$. Consider the projection $p : X \to \mathbb{P}^s$. Since $\mathcal{F}$ is a spanned vector bundle on $\mathbb{P}^s$, it is a general fact that $\alpha : \mathbb{P}(p^*\mathcal{F}) \to \mathbb{P}^{N'}$ factors through $\beta : \mathbb{P}(\mathcal{F}) \to \mathbb{P}^{N'}$. Since $\mathcal{F}$ is the direct sum of a trivial bundle and a very ample line bundle, $\mathcal{O}_{\mathbb{P}^s}(1)$, $\xi_\mathcal{F}$ is big. This implies that $\dim(\mathrm{Im}\beta) = \dim(\mathbb{P}(\mathcal{F})) = n - k + s - 1$. Since $\dim\phi(\overline{X}) \geq \dim\phi(E) = \dim Z$, it follows that $\dim\phi(\overline{X}) \geq n - k + s - 1$.

Consider the Koszul complex
$$0 \to \wedge^{n-k} V \otimes (-(n-k-1)L) \to \cdots \to \wedge^2 V \otimes (-L) \to V \to \mathcal{J}_Y(L) \to 0.$$

Set $T := \oplus^{n-k-1}\mathcal{O}_X$ and note that $\wedge^m(T \oplus H) = \wedge^m T \oplus (\wedge^{m-1}T \otimes H)$ for each $m \geq 1$. Note also that $h^0(H) = s + 1$ and hence $h^0(V) = n - k + s$. From the hypercohomology sequence associated to the Koszul complex above we see that $h^0(\mathcal{J}_Y(L)) = h^0(V) = n - k + s$. This is immediate if $L - H$ is assumed ample, but otherwise requires checking a few cases. Thus we conclude that the image of the morphism, $\phi : \overline{X} \to Z$, associated to $L \otimes \mathcal{J}_Y$ has dimension

$$\dim\phi(\overline{X}) = \dim Z \leq n - k + s - 1 \leq \mathrm{cod}_X Y + s - 1.$$

Thus we conclude that $\dim\phi(\overline{X}) = n - k + s - 1$.

Note that the complete intersection situation corresponds, in our present notation, to the case $s = 0$ with p the constant map.

We have the following three infinite sequences of examples (for one more class of examples see (8.3) in §8).

Example 5.2. (projection from a linear divisor) Let X be an n-dimensional projective submanifold of $\mathbb{P}^{2n-1}$. Assume that there is a linear $\mathbb{P}^{n-1}$, $D \subset X$. Let L denote the restriction of $\mathcal{O}_{\mathbb{P}^{2n-1}}(1)$ to X. Since the morphism, $\psi : X \to \psi(X)$ associated to $|L - D|$ agrees with the restriction of the projection of $\mathbb{P}^{2n-1}$ from D away from D, we see that $\dim\psi(X) \leq n - 1$. From this we conclude that $(L - D)^n = 0$. A calculation given in Proposition (8.1) shows that $d := L^n = \frac{(s+1)^n - 1}{s}$ for $s \geq 1$ and n for $s = 0$, where the normal bundle of D in X is $\mathcal{O}_{\mathbb{P}^{n-1}}(-s)$. Since we have that $(L - D)_D \cong \mathcal{O}_{\mathbb{P}^{n-1}}(s + 1)$ is ample for $s \geq 0$, we conclude that if $s \geq 0$, then the morphism associated to $|L - D|$ has at least an $(n - 1)$-dimensional image.

We now show that such examples occur for all integers $n > 0$ and $s \geq 0$. Fix integers $s \geq 0$ and $n > 0$. Let $\mathcal{P} := \mathbb{P}(\mathcal{O}_{\mathbb{P}^{2n-1}}(1) \oplus \mathcal{O}_{\mathbb{P}^{2n-1}}(s + 1))$ and let $p : \mathcal{P} \to \mathbb{P}^{2n-1}$ denote the bundle projection. Let ξ denote the tautological line bundle on $\mathcal{P}$ such that $p_*\xi \cong \mathcal{O}_{\mathbb{P}^{2n-1}}(1) \oplus \mathcal{O}_{\mathbb{P}^{2n-1}}(s + 1)$. Note that by counting constants we see that the transversal intersection of n general elements of $|\xi|$ is a smooth n-fold X' which maps isomorphically under p to its image X in $\mathbb{P}^{2n-1}$. Let $L := \mathcal{O}_{\mathbb{P}^{2n-1}}(1)_X$. Let $\mathcal{E} := \oplus^n \xi$. From the Koszul complex resolution of the ideal sheaf of X' we get the exact sequence

$$0 \to \det \mathcal{E}^* \to \wedge^{n-1}\mathcal{E}^* \to \cdots \to \wedge^2 \mathcal{E}^* \to \mathcal{E}^* \to \mathcal{O}_\mathcal{P} \to \mathcal{O}_{X'} \to 0.$$

By tensoring the sequence with $p^*\mathcal{O}_{\mathbb{P}^{2n-1}}(1)$ we see that the restriction map gives an isomorphism $H^0(\mathbb{P}^{2n-1}, \mathcal{O}_{\mathbb{P}^{2n-1}}(1)) \cong H^0(X, L)$. Moreover the intersection of X' with the section Σ corresponding to the quotient $\mathcal{O}_{\mathbb{P}^{2n-1}}(1) \oplus \mathcal{O}_{\mathbb{P}^{2n-1}}(s + 1) \to \mathcal{O}_{\mathbb{P}^{2n-1}}(1)$ is a linear $\mathbb{P}^{n-1}$ with respect to $\mathcal{O}_{\mathbb{P}^{2n-1}}(1)$. Thus X' contains a linear $\mathbb{P}^{n-1}$. Denote this by D. Since $N_{\Sigma/\mathcal{P}} \cong \mathcal{O}_{\mathbb{P}^{2n-1}}(-s - 1) \otimes \xi_\Sigma$ and $\xi_\Sigma \cong \mathcal{O}_{\mathbb{P}^{2n-1}}(1)$, and since the normal bundle $N_{D/X}$ of D in X is isomorphic to the restriction of the normal bundle of Σ, we see that $N_{D/X} \cong \mathcal{O}_{\mathbb{P}^{n-1}}(-s)$. As noted above the morphism, $\phi := p_X : X \to \mathbb{P}^{n-1}$, associated to $L \otimes \mathcal{J}_D$ has an $(n - 1)$-dimensional image.

Recall that $L - D \approx \phi^*\mathcal{H}$ for some ample and spanned line bundle $\mathcal{H}$ on $\mathbb{P}^{n-1}$. Then in the example above one has $\mathcal{H}^{n-1} = 1$. Indeed, let $\mathcal{H} = \mathcal{O}_{\mathbb{P}^{n-1}}(h)$. Since

$L - D \approx \phi^* \mathcal{H}$, we see that $h^0(L - D) = \binom{h+n-1}{n-1}$. From the exact sequence $0 \to L - D \to L \to L_D \cong \mathcal{O}_{\mathbb{P}^{n-1}}(1) \to 0$ we infer that $h^0(L) \geq h^0(L - D) + n$. Since $h^0(L) \leq 2n$ we conclude that $h^0(L - D) \leq n$. Thus, since $n \geq 2$, $\binom{h+n-1}{n-1} \leq n$ implies $h = 1$.

The following example is related to Theorem (7.1) in §7.

Example 5.3. We construct here a smooth hypersurface of degree d in $\mathbb{P}^{2k+1}$ containing a linear $\mathbb{P}^k$, such that the projection from the $\mathbb{P}^k$ associated to $L := \mathcal{O}_X(1)$ has a k-dimensional image.

Consider in $\mathbb{P}^{2k+1}$ the degree d hypersurface defined by the equation

$$\sum_{j=0}^{2k+1} x_j x_{2k+1-j}^{d-1} = 0.$$

Then X is smooth and contains the linear $\mathbb{P}^k$ defined by the equations $x_{2k+1} = \cdots = x_{k+1} = 0$. The projection from this $\mathbb{P}^k$ has image $\mathbb{P}^k$.

Example 5.4. Let $X := \mathbb{P}(\mathcal{E} \oplus \mathcal{O}_{\mathbb{P}^k}(1))$, where $\mathcal{E}$ is a rank r vector bundle on $\mathbb{P}^k$ of the form $\mathcal{E} = \oplus_{i=1}^r \mathcal{O}_{\mathbb{P}^k}(a_i)$, $a_i \geq 1$. Then X is of dimension $n = k + r$. Take as $\mathbb{P}^k$ the section of the $\mathbb{P}^r$-bundle $p : X \to \mathbb{P}^k$ corresponding to the quotient

$$\mathcal{E} \oplus \mathcal{O}_{\mathbb{P}^k}(1) \to \mathcal{O}_{\mathbb{P}^k}(1) \to 0.$$

This guarantees that $\xi_{\mathbb{P}^k} \approx \mathcal{O}_{\mathbb{P}^k}(1)$, where $\xi_{\mathbb{P}^k}$ is the restriction to $\mathbb{P}^k$ of the tautological bundle $L := \xi$ of X. Hence in particular $\delta := L^k \cdot \mathbb{P}^k = 1$, i.e., $\mathbb{P}^k$ is linear.

Let $\sigma : \overline{X} \to X$ be the blowing up of X along the $\mathbb{P}^k$. Note that σ induces the blowing up, $\pi : \overline{F} \to \mathbb{P}^r$, at one point, x, of each fiber $F = \mathbb{P}^r$ of p. Consider the morphism $\phi : \overline{X} \to Z$ associated to $L \otimes \mathcal{I}_{\mathbb{P}^k}$. Note that the restriction $\phi_{\overline{F}}$, for each fiber $F = \mathbb{P}^r$, is the morphism given by the line bundle $|\pi^* \mathcal{O}_{\mathbb{P}^r}(1) - \pi^{-1}(x)|$. Therefore $\phi_{\overline{F}}$, being the projection of $\mathbb{P}^r$ from the point x, has lower dimensional image. Since the fibers $F = \mathbb{P}^r$ cover X we thus conclude that ϕ has lower dimensional image.

6. The divisorial case

In this section L always denotes a very ample line bundle on an n-dimensional projective manifold X, such that its global sections, $\Gamma(L)$, embed X in a projective space $\mathbb{P}^{n+r}$. Let $Y = D$ be a smooth connected divisor on X of degree $\delta = L^{n-1} \cdot D$. We assume $n \geq 2$ since the case $n = 1$ is trivial.

Recall that $\delta L - D$ is spanned (see Lemma (2.6)). In the present case we can say considerably more. Let us first show the following fact.

Lemma 6.1. *Let L be a very ample line bundle on a connected projective manifold X of dimension n. Let D be a smooth divisor of degree $\delta = L^{n-1} \cdot D$. Then either $(X, L, \mathcal{O}_X(D)) \cong (\mathbb{P}^n, \mathcal{O}_{\mathbb{P}^n}(1), \mathcal{O}_{\mathbb{P}^n}(\delta))$, or the restriction $(\delta L - D)_D$ is an ample line bundle on D.*

Proof. By the conductor formula (2.2) and the adjunction formula we have that

$$(\delta - n - 1)L_D - K_D \approx (\delta - n - 1)L_D - (K_X + D)_D \qquad (15)$$

is nef. By general adjunction theoretic results (see e.g., [2, (7.2.1)]) we know that $K_X + (n+1)L$ is either ample or $(X, L) \cong (\mathbb{P}^n, \mathcal{O}_{\mathbb{P}^n}(1))$. Therefore we see from (15) that if $(\delta L - D)_D$ is not ample then $K_{X|D} + (n+1)L_D$ is not ample and hence $(X, L) \cong (\mathbb{P}^n, \mathcal{O}_{\mathbb{P}^n}(1))$. In this case $\mathcal{O}_X(D) \cong \mathcal{O}_{\mathbb{P}^n}(\delta)$. □

Next, we recall the following definition.

Definition 6.2. A line bundle, L, on a projective variety, X, is *k-ample* for an integer $k \geq 0$, if mL is spanned for some $m > 0$, and the morphism $X \to \mathbb{P}_{\mathbb{C}}$ defined by $\Gamma(mL)$ for such an m has all fibers of dimension $\leq k$.

Theorem 6.3. *Let L be a very ample line bundle on a connected projective manifold X of dimension $n \geq 2$, such that $\Gamma(L)$ embeds X in $\mathbb{P}^{n+r}$. Let D be a smooth divisor on X of degree $\delta = L^{n-1} \cdot D$. Then $\delta L - D$ is 1-ample except in the case when $(X, L, \mathcal{O}_X(D)) \cong (\mathbb{P}^n, \mathcal{O}_{\mathbb{P}^n}(1), \mathcal{O}_{\mathbb{P}^n}(\delta))$.*

Proof. Let F be a fiber of the morphism associated to $|\delta L - D|$ and assume $\dim F \geq 2$. Then $(\delta L - D)_F \approx \mathcal{O}_F$, so that $D_F \approx \delta L_F$ is ample. This implies that $D \cap F$ contains an effective curve, C, and $D \cdot C > 0$. But $(\delta L - D) \cdot C = 0$ since $\delta L - D$ is trivial on F. If $(X, L, \mathcal{O}_X(D)) \not\cong (\mathbb{P}^n, \mathcal{O}_{\mathbb{P}^n}(1), \mathcal{O}_{\mathbb{P}^n}(\delta))$ this contradicts the ampleness of $(\delta L - D)_D$ (see (6.1)). □

If $\delta > 1$ we can say more.

Theorem 6.4. *Let L be a very ample line bundle on a connected projective manifold X of dimension $n \geq 2$, such that $\Gamma(L)$ embeds X in $\mathbb{P}^{n+r}$. Let D be a smooth divisor on X of degree $\delta = L^{n-1} \cdot D > 1$. Assume that $(X, L, \mathcal{O}_X(D)) \not\cong (\mathbb{P}^n, \mathcal{O}_{\mathbb{P}^n}(1), \mathcal{O}_{\mathbb{P}^n}(\delta))$. Then the morphism associated to $|\delta L - D|$ is birational; moreover, $\delta L - D$ is very ample if $n \geq r + 2$.*

Proof. First assume $n \geq r + 2$, or, equivalently, $2\dim X - (n + r) \geq 2$. Then by the Barth-Lefschetz theorem (see e.g., [2, (2.3.11)]) we conclude that $\text{Pic}(X) \cong \mathbb{Z}$ with generator the restriction of the hyperplane section bundle on projective space. Since $\delta L - D$ is spanned and not trivial unless $(X, L, \mathcal{O}_X(D)) \cong (\mathbb{P}^n, \mathcal{O}_{\mathbb{P}^n}(1), \mathcal{O}_{\mathbb{P}^n}(\delta))$ (see (6.1)), we conclude that $\delta L - D$ is a multiple of the restriction of the hyperplane section bundle on projective space. Thus $\delta L - D$ is very ample.

We next assume that $n \leq r+1$. Then

$$\delta - 1 > \frac{n}{r+1}(\delta - 1) - n + 1,$$

and Proposition (3.5) applies to say that $h^0((\delta - 1)L - D) > 0$, from which it easily follows that the morphism associated to $|\delta L - D|$ is birational. □

Look at the embedding $X \subset \mathbb{P}^{n+r}$ and let q be the codimension of D in the smallest linear subspace $\mathbb{P}^{n-1+q} \subset \mathbb{P}^{n+r}$ containing it. Let us assume that the Castelnuovo bound conjecture holds true, i.e., $(\delta - q + 1)L - D$ is spanned by its global sections (compare with (4.13)). Clearly we have

$$r \geq q - 1. \tag{16}$$

Recall also the usual relations

$$d \geq r+1 \text{ and } \delta \geq q+1. \tag{17}$$

Proposition 6.5. *Let L be a very ample line bundle on a connected projective manifold X of dimension $n \geq 2$, such that $\Gamma(L)$ embeds X in $\mathbb{P}^{n+r}$. Let D be a smooth divisor on X of degree $\delta = L^{n-1} \cdot D > 1$. Let q be the codimension of D in the smallest linear subspace $\mathbb{P}^{n-1+q} \subset \mathbb{P}^{n+r}$ containing it. Assume that $(\delta - q + 1)L - D$ is spanned by its global sections. Then*

1. *If $n \geq r+2$, $(\delta - q + 1)L - D$ is very ample unless $X \cong \mathbb{P}^n$ and $\delta L \approx D$;*

2. *If $n \leq r+1$, then the morphism associated to $|(\delta - q + 1)L - D|$ is birational unless $q = r+1$ and either $n = r+1$ or $n < r+1$ and $\delta = r+2$.*

Proof. Assume $n \geq r+2$. Let $d := L^n$. We have the following fact.
Claim. $(\delta - q + 1)L - D$ is not trivial unless $X \cong \mathbb{P}^n$, $\delta L \approx D$.
Proof of Claim. Assume $D \approx (\delta - q + 1)L$. Dotting with L^{n-1} gives $(\delta - q + 1)d = L^{n-1} \cdot D = \delta$, or $(d-1)\delta = d(q-1)$. Using (17) this gives $(d-1)(q+1) \leq d(q-1)$, or

$$2d \leq q + 1. \tag{18}$$

Since by (17) and (16), $d \geq r+1 \geq q$, we find $q \leq 1$. Thus (18) yields $d = q = 1$ and hence $r = 0$ by (17). Therefore $X \cong \mathbb{P}^n$, $D \approx \delta L$. □
Since $n \geq r+2$ is equivalent to $2\dim X - (n+r) \geq 2$, by the Barth-Lefschetz theorem (see e.g., [2, (2.3.11)]) we have $\text{Pic}(X) \cong \mathbb{Z}$. Since $(\delta - q + 1)L - D$ is spanned and by the Claim we can assume it is not trivial, we conclude that $(\delta - q + 1)L - D$ is very ample. This shows 1).
As for 2), assume $n \leq r+1$. If the morphism associated to $|(\delta - q + 1)L - D|$ is not birational, then $h^0((\delta - q)L - D) = 0$. Thus, by Proposition (3.5), $\delta - q \leq$

$\frac{n}{r+1}(\delta - 1) - n + 1$, or

$$(\delta - 1)\left(1 - \frac{n}{r+1}\right) \leq q - n, \tag{19}$$

or, by using $r \geq q - 1$ from (16),

$$(\delta - 1)\left(\frac{r+1-n}{r+1}\right) \leq r + 1 - n. \tag{20}$$

If $r + 1 = n$, then equality holds in (20) and hence in particular $r = q - 1$, i.e., D spans $\mathbb{P}^{n+r}$. If $r + 1 > n$, inequality (20) yields $\delta - 1 \leq r + 1$, or $\delta \leq r + 2$. Since $q \leq \delta - 1$ by (17), inequality (19) gives

$$(\delta - 1)\left(1 - \frac{n}{r+1}\right) \leq \delta - 1 - n \quad \text{or} \quad n \leq (\delta - 1)\frac{n}{r+1}.$$

This implies $r + 2 \leq \delta$. Thus $\delta = r + 2$. Also, at each step, equalities hold true. Therefore $q = \delta - 1 = r + 1$. □

Example 6.6. Notation as in (6.5). We give here an example in the range $n = r + 1$ where $|(\delta - q + 1)L - D|$ is spanned but the morphism associated to it is not birational, D spans $\mathbb{P}^{n+r}$ and the projection from D has an $(n - 1)$-dimensional image.

Consider the Segre embedding $X = \mathbb{P}^1 \times \mathbb{P}^{n-1} \hookrightarrow \mathbb{P}^{n+r} = \mathbb{P}^{2n-1}$, $r = n - 1$, and let $p_1 : X \to \mathbb{P}^1$, $p_2 : X \to \mathbb{P}^{n-1}$ be the projections on the two factors. Denote $\mathcal{O}(a, b) := p_1^*\mathcal{O}_{\mathbb{P}^1}(a) \otimes p_2^*\mathcal{O}_{\mathbb{P}^{n-1}}(b)$, for given integers a, b. Let $L := \mathcal{O}(1, 1)$, so that $h^0(L) = 2n$. Take a smooth divisor D in the linear system $|\mathcal{O}(2, 1)|$. We have $d := L^n = n$ and $\delta := L^{n-1} \cdot D = n + 1$. Consider the exact sequence

$$0 \to L - D \to L \to L_D \to 0.$$

Note that $L - D = \mathcal{O}(-1, 0)$, so $h^0(L - D) = 0$ and, by using Künneth's formulas, $h^1(L - D) = 0$. Therefore $h^0(L) = h^0(L_D)$. This means that D spans $\mathbb{P}^{n+r} = \mathbb{P}^{2n-1}$, or $q = r + 1 = n$. Then $(\delta - q + 1)L - D = 2L - D = \mathcal{O}(0, 1)$. Thus $(\delta - q + 1)L - D$ is not big, so that the projection from D associated to it is not birational, and has an $(n - 1)$-dimensional image.

Example 6.7. Notation as in (6.5). We give here an example in the range $r = n$, where $(\delta - q + 1)L - D$ is spanned but not ample, in fact is 1-ample, and the morphism associated to it is birational.

Let $X := \mathbb{P}(\oplus^{n-1}\mathcal{O}_{\mathbb{P}^1} \oplus \mathcal{O}_{\mathbb{P}^1}(1))$. Let ξ be the tautological bundle of X and let F be a fiber of the bundle projection $X \to \mathbb{P}^1$. Let $L := \xi + F$ and take a smooth divisor $D \in |\xi + 2F|$. Note that both $\xi + F$ and $\xi + 2F$ are very ample (see e.g., [2, (3.2.4)]).

A standard check shows that $d = L^n = n + 1$, $\delta = L^{n-1} \cdot D = n + 2$ and $h^0(L) = 2n + 1$, $h^0(L - D) = h^0(-F) = 0$, $h^1(L - D) = 0$. Thus $X \subset \mathbb{P}^{2n}$, i.e.,

$q = r + 1 = n + 1$. Then

$$(\delta - q + 1)L - D = 2L - D = \xi.$$

The line bundle ξ is spanned but not ample (see e.g., [2, (3.2.4)]) and the morphism associated to $|\xi|$ is the blowing up $X \to \mathbb{P}^n$ of $\mathbb{P}^n$ along $\mathbb{P}^{n-2}$. Hence in particular ξ is 1-ample.

7. The linear case

Let X be a smooth connected projective variety of dimension n, polarized by a very ample line bundle L. In this section we discuss some further results about the structure of projection maps from a k-dimensional subvariety Y of X, in the case when Y is a linear $\mathbb{P}^k$ with respect to L.

In (7.1) we show that if the morphism, ϕ, associated to $L \otimes \mathcal{J}_Y$ as in (4.1) has image dimension $n - k$, then ϕ has $\mathbb{P}^{n-k}$ as image and X is a hypersurface in $\mathbb{P}^{n+1}$. Next we show in (7.2) that assuming "Hartshorne's conjecture" we have a stronger lower bound for the dimension of the image of ϕ. Finally we prove in (7.4) a spannedness result for the adjoint bundle (see also (8.6) for more adjunction theoretic structure type results in the case when Y is a codimension 1 linear $\mathbb{P}^{n-1}$).

Let us explicitly point out the following fact: if Y is a smooth k-dimensional subvariety of (X, L) of degree $\delta = L^k \cdot Y$, then, since $\delta L \otimes \mathcal{J}_Y$ is spanned by global sections by Lemma (2.6), the morphism associated to $|tL \otimes \mathcal{J}_Y|$ is birational for $t \geq \delta + 1$. In particular, if Y is a linear $\mathbb{P}^k$ with respect to L and the projection from Y associated to tL has lower dimensional image, then necessarily $t = 1$.

In the case when Y is a linear $\mathbb{P}^k$ and the projection has image dimension one bigger than the lowest possible value we have the following result. We recall Theorem (4.6) for a general lower bound for the image dimension of ϕ and we refer back to (5.3) which gives in fact an example of the situation discussed below.

Theorem 7.1. *Let L be a very ample line bundle on X, a connected projective manifold of dimension $n \geq 2$. Let Y be a subvariety of X with $(Y, L_Y) \cong (\mathbb{P}^k, \mathcal{O}_{\mathbb{P}^k}(1))$. Let $\sigma : \overline{X} \to X$ be the blowing up of X along Y. Let $\phi : \overline{X} \to Z$ be the morphism associated to $L \otimes \mathcal{J}_Y$ as in (4.1). Assume that $\dim Z = \text{cod}_X Y = n - k$ and $k \geq 2$. Then X is a hypersurface in $\mathbb{P}^{n+1}$.*

Proof. Set $w := n - k$. Since $\overline{L} - E$ is spanned and gives the projection $\psi : \overline{X} \to \psi(\overline{X})$ and since $\dim Z = w$, we have a surjection of locally free sheaves $\oplus^{w+1} \mathcal{O}_{\overline{X}} \to \overline{L} - E \to 0$. Hence, restricting to E, we have an exact sequence

$$0 \to \mathcal{K} \to \oplus^{w+1} \mathcal{O}_E \to (\overline{L} - E)_E \to 0.$$

Consider the $\mathbb{P}^{w-1}$-bundle map $\pi : E \to Y$. Let $N := N_{X/Y}$ be the normal bundle of Y in X. Notice that $(\overline{L} - E)_E \cong \xi$, the tautological line bundle of $E \cong \mathbb{P}(N^*(L))$.

By pushing forward under π, we get an exact sequence on Y

$$0 \to K \to \oplus^{w+1}\mathcal{O}_Y \to N^*(L) \cong \pi_*\xi \to 0. \tag{21}$$

By comparing the ranks, since $N^*(L)$ has rank $\mathrm{cod}_X Y = w$, we conclude that K is a line bundle.

Since $Y \cong \mathbb{P}^k$, $k \geq 2$, the first cohomology of a line bundle is zero, i.e., $h^1(Y, K) = 0$. This means that the sections of $\oplus^{w+1}\mathcal{O}_Y$ surject onto the sections of $N^*(L)$, so $h^0(N^*(L)) \leq w + 1$.

Notice that $\overline{L} - E \approx \phi^*(\mathcal{H})$ for some ample line bundle $\mathcal{H}$ on Z. Since the restriction $\phi_E : E \to Z$ is onto by (4.4), we have $h^0(N^*(L)) = h^0((\overline{L} - E)_E) \geq h^0(\mathcal{H}) = h^0(\overline{L} - E)$. Thus

$$h^0(L \otimes \mathcal{J}_Y) = h^0(\overline{L} - E) \leq w + 1. \tag{22}$$

Now look at the exact sequence

$$0 \to L \otimes \mathcal{J}_Y \to L \to L_Y \to 0.$$

Recall that $L_Y \cong \mathcal{O}_{\mathbb{P}^k}(1)$ since Y is a linear $\mathbb{P}^k$. Therefore, by (22), $h^0(L) \leq h^0(L \otimes \mathcal{J}_Y) + h^0(\mathcal{O}_{\mathbb{P}^k}(1)) \leq w + k + 2 = n + 2$. Thus, either $\Gamma(L)$ embeds X as hypersurface in $\mathbb{P}^{n+1}$, or else $h^0(L) = n + 1$ and $X \cong \mathbb{P}^n$. However, the latter is ruled out by the assumption $\dim Z \geq n - k$. $\square$

A minor modification of the proof of the theorem above gives us the following result, which states that assuming "Hartshorne's conjecture" (see [8]) the image dimension of ϕ has a stronger lower bound unless X is a complete intersection.

Proposition 7.2. *Let X be a smooth connected projective variety of dimension $n \geq 2$. Let L be a very ample line bundle on X. Let Y be a linear $\mathbb{P}^k$ with respect to the embedding given by $\Gamma(L)$. Let $\sigma : \overline{X} \to X$ be the blowing up of X along Y. Let $\phi : \overline{X} \to Z$ be the morphism associated to $L \otimes \mathcal{J}_Y$ as in (4.1). Assume that Hartshorne's conjecture is true and that X is not a complete intersection. Then $\dim \phi(\overline{X}) \geq \mathrm{cod}_X Y + \frac{k}{3} - 1$.*

Proof. First note that for $k \leq 2$ the bound on $\dim \phi(\overline{X})$ follows from Theorem (4.6) and Corollary (4.7), so we can assume $k \geq 3$.

Set $w := n - k = \mathrm{cod}_X Y$ and $z := \dim \phi(\overline{X})$. Exactly the same argument as in the proof of Theorem (7.1) gives us an exact sequence

$$0 \to K \to \oplus^{z+1}\mathcal{O}_Y \to N^*(tL) \to 0$$

on Y, where K is a vector bundle of rank $z + 1 - w$ and N is the normal bundle of Y in X.

Assume, by contradiction, that $z < w + \frac{k}{3} - 1$, and therefore $\mathrm{rank}(K) = z + 1 - w < \frac{k}{3}$. Thus from (2.7) we know that K splits as a direct sum of line bundles on $\mathbb{P}^k$ (here we are using our present assumption that $k \geq 3$). Then the first cohomology of K is zero. This means that the sections of $\oplus^{z+1}\mathcal{O}_Y$ surject onto the sections of $N^*(L)$, so $h^0(N^*(L)) \leq z + 1$. Again, as in the proof of (7.1), we thus conclude that

$$h^0(L \otimes \mathcal{J}_Y) \leq z + 1. \tag{23}$$

Now look at the exact sequence $0 \to L \otimes \mathcal{J}_Y \to L \to L_Y \to 0$. Recall that $L_Y \cong \mathcal{O}_{\mathbb{P}^k}(1)$ since Y is a linear $\mathbb{P}^k$. Therefore, by (23), $h^0(L) \leq h^0(L \otimes \mathcal{J}_Y) + h^0(\mathcal{O}_{\mathbb{P}^k}(1)) \leq z + k + 2$. Thus $\Gamma(L)$ embeds X in $\mathbb{P}^{z+k+1}$. A direct numerical check shows that the inequality $z < w + \frac{k}{3} - 1$ implies $n > \frac{2}{3}(z + k + 1)$. Since we are assuming that Hartshorne's conjecture is true, we thus conclude that X is a complete intersection. $\square$

We need the following result. The case when $k = 1$ also follows immediately from a result of Ilic [12].

Theorem 7.3. *Let X be a connected projective manifold of dimension $n \geq 2$. Assume that X is a $\mathbb{P}^{n-1}$-bundle $\pi : X \to C$ over a smooth curve C with fibers linear with respect to L, a very ample line bundle on X. Let $Y \subset X$ be a linear $\mathbb{P}^k$ with respect to the embedding given by $\Gamma(L)$. Let $\sigma : \overline{X} \to X$ be the blowing up of X along Y. Let $\phi : \overline{X} \to Z$ be the morphism associated to $L \otimes \mathcal{J}_Y$ as in (4.1). Then $\dim Z < n$ if and only if either*

1. *$\dim \pi(Y) = 1$, $\dim Y = 1$ and Y is a section of π corresponding to a surjection from the vector bundle $\pi_* L$ onto a direct summand $\mathcal{O}_{\mathbb{P}^1}(1)$; or*

2. *$\dim \pi(Y) = 0$, $k = n - 1$, and $(X, L) \cong (\mathbb{P}^{n-1} \times \mathbb{P}^1, \mathcal{O}_{\mathbb{P}^{n-1} \times \mathbb{P}^1}(1, 1))$.*

Proof. We leave the reader to check the straightforward assertion that $\dim Z < n$ in cases 1) and 2). Assume now that $\dim Z < n$.

If $\dim \pi(Y) = 1$, then since $\mathbb{P}^k$ cannot map onto a curve if $k \geq 2$, we conclude that $k = 1$ and $C \cong \mathbb{P}^1$. Since Y and fibers of π are linear, we conclude that Y meets any given fiber transversely in exactly one point. Thus Y corresponds to a surjection $\pi_* L \to \mathcal{O}_{\mathbb{P}^1}(L \cdot Y) \cong \mathcal{O}_{\mathbb{P}^1}(1)$. Using the fact that $\pi_* L$ is very ample and a direct sum of line bundles, it is a simple check that $\pi_* L \to \mathcal{O}_{\mathbb{P}^1}(L \cdot Y) \cong \mathcal{O}_{\mathbb{P}^1}(1)$ splits.

Assume now that $\dim \pi(Y) = 0$. If the codimension of Y is one, then we have $0 = (L - Y)^n = L^n - nL^{n-1} \cdot Y = L^n - n$. From this we see that $\pi_* L$ is a very ample rank n vector bundle of degree n. This immediately implies that $(X, L) \cong (\mathbb{P}^{n-1} \times \mathbb{P}^1, \mathcal{O}_{\mathbb{P}^{n-1} \times \mathbb{P}^1}(1, 1))$.

Now we consider the case when the codimension of Y is greater than one. Since $N_{Y/X} \cong \mathcal{O}_{\mathbb{P}^k} \oplus \left(\oplus^{n-1-k} \mathcal{O}_{\mathbb{P}^k}(1) \right)$, it is a straightforward consequence of Lemma (4.2) and the fact that $N^*(L)$ is spanned, that we can choose $n - k - 1$ smooth divisors $D_1, \ldots, D_{n-k-1}$ in $|L \otimes \mathcal{J}_Y|$ all meeting transversely in a smooth $(k+1)$-dimensional subvariety $X_{k+1} := D_1 \cap \cdots \cap D_{n-k-1}$ containing Y as a divisor. But since it follows from the last paragraph that $(X_{k+1}, L_{X_{k+1}}) \cong (\mathbb{P}^k \times \mathbb{P}^1, \mathcal{O}_{\mathbb{P}^k \times \mathbb{P}^1}(1, 1))$ we infer that

$\pi_{X_{k+1}*}L_{X_{k+1}} \cong \oplus^{k+1}\mathcal{O}_{\mathbb{P}^1}(1)$. Thus we conclude that $\pi_*L \cong (\oplus^{n-k-1}\mathcal{O}_{\mathbb{P}^1}) \oplus (\oplus^{k+1}\mathcal{O}_{\mathbb{P}^1}(1))$. Since π_*L is very ample, we conclude that $n = k+1$. □

The case $k = 1$ of the following spannedness result for the adjoint bundle follows from [12].

Theorem 7.4. *Let X be a smooth connected projective variety of dimension $n \geq 2$. Let L be a very ample line bundle on X. Let $Y \subset X$ be a linear $\mathbb{P}^k$ with respect to the embedding given by $\Gamma(L)$. Let $\sigma : \overline{X} \to X$ be the blowing up of X along Y. Let $\phi : \overline{X} \to Z$ be the morphism associated to $L \otimes \mathcal{J}_Y$ as in (4.1). Assume that $\dim Z < n$. Then $K_X + (n-1)L$ is spanned by global sections unless either*

1. $(X, L) \cong (\mathbb{P}^n, \mathcal{O}_{\mathbb{P}^n}(1))$, $1 \leq k \leq n-1$, with $\dim Z = n-k-1$; or

2. $(X, L) \cong (\mathcal{Q}, \mathcal{O}_{\mathcal{Q}}(1))$, $\mathcal{Q}$ a quadric in $\mathbb{P}^{n+1}$, $1 \leq k \leq [\frac{n}{2}]$, with $\dim Z = n-k$; or

3. (X, L) is a scroll, $\pi : X \to C$, over a smooth curve C, i.e., $K_X + nL \approx \pi^*H$ for some ample line bundle H on C, with either

   (a) $\dim \pi(Y) = 1$, $\dim Y = 1$ and Y is a section of π corresponding to a surjection from the vector bundle π_*L onto a direct summand $\mathcal{O}_{\mathbb{P}^1}(1)$; or

   (b) $\dim \pi(Y) = 0$, $k = n-1$, and $(X, L) \cong (\mathbb{P}^{n-1} \times \mathbb{P}^1, \mathcal{O}_{\mathbb{P}^{n-1} \times \mathbb{P}^1}(1, 1))$.

Proof. From general adjunction theory results we know that $K_X + (n-1)L$ is spanned unless either

(i) $(X, L) \cong (\mathbb{P}^n, \mathcal{O}_{\mathbb{P}^n}(1))$; or

(ii) $(X, L) \cong (\mathcal{Q}, \mathcal{O}_{\mathcal{Q}}(1))$, $\mathcal{Q}$ a quadric in $\mathbb{P}^{n+1}$; or

(iii) (X, L) is a scroll, $\pi : X \to C$, over a smooth curve C, i.e., $K_X + nL \approx \pi^*H$ for some ample line bundle H on C.

In case (i), by looking at the projection of $\mathbb{P}^n$ from $\mathbb{P}^k$, we see that $1 \leq k \leq n-1$ with $\dim Z = n-k-1$.

In case (ii) we see that $1 \leq k \leq [\frac{n}{2}]$ with $\dim Z = n-k$, by looking at the projection of $\mathbb{P}^{n+1}$ from $\mathbb{P}^k$.

In case (iii), use Theorem (7.3). □

Corollary 7.5. *Let X be a smooth connected projective variety of dimension $n \geq 2$. Let L be a very ample line bundle on X. Let $Y \subset X$ be a linear $\mathbb{P}^k$ with respect to the embedding given by $\Gamma(L)$. Let $\phi : \overline{X} \to Z$ be the morphism associated to $L \otimes \mathcal{J}_Y$ as in (4.1). Assume that $\dim Z < n$. Let $N := N_{Y/X}$ be the normal bundle of Y in X. If (X, L) is not as in one of cases 1), 2), 3) of (7.4), one has $c_1(N) \leq n-2-k$.*

Proof. By the assumption, $K_X + (n-1)L$ is spanned. On the other hand,

$$(K_X + (n-1)L)_Y \approx K_Y - \det N + (n-1)L_Y \cong \mathcal{O}_{\mathbb{P}^k}(n-2-k) - \det N.$$

Since $(K_X + (n-1)L)_Y \cong \mathcal{O}_{\mathbb{P}^k}(b)$ for some nonnegative integer b, we thus conclude that $\det N \cong \mathcal{O}_{\mathbb{P}^k}(a)$ for some integer $a \leq n-2-k$. □

8. The linear case in codimension 1

Let X be a smooth connected projective variety of dimension $n \geq 2$. Let L be a very ample line bundle on X. Let P be a linear $\mathbb{P}^{n-1} \subset X$ with respect to L, i.e., $\delta = L^{n-1} \cdot P = 1$. Recall that in this case the line bundle $L - P$ is spanned (see the discussion after Lemma (4.3)). We follow the notation of (4.1), with the exception of denoting Y by P to emphasize its special nature. Thus we let $\psi : X \to \psi(X)$ be the morphism associated to $|L - P|$ and $\psi = \mathfrak{s} \circ \phi$ the Remmert-Stein factorization of ψ with $\phi : X \to Z$ having connected fibers and $\mathfrak{s} : Z \to \psi(X)$ finite.

In this section we study the projection from P, a linear $\mathbb{P}^{n-1}$, under the assumption that $n > \dim \phi(X)$. For shortness, it is convenient to refer to the situation above simply saying that (X, L, P) is a $\mathbb{P}^{n-1}$-*degenerate triple*.

First, let us state the following preliminary facts.

Proposition 8.1. *Let X be a connected n-dimensional manifold and let L be very ample line bundle on X. Assume that (X, L, P) is a $\mathbb{P}^{n-1}$-degenerate triple. Let $N := N_{\mathbb{P}^{n-1}/X} \cong \mathcal{O}_{\mathbb{P}^{n-1}}(-s)$ be the normal bundle of $P := \mathbb{P}^{n-1}$ in X. Then we have:*

1. *$s \geq -1$, with equality only if $(X, L) \simeq (\mathbb{P}^n, \mathcal{O}_{\mathbb{P}^n}(1))$;*

2. *if $s \geq 0$, the morphism $\psi : X \to \psi(X)$ associated to $|L - P|$ has an $(n-1)$-dimensional image with all fibers having dimension one; and ψ_P is finite; and*

3. *the degree of (X, L) is given by $d := L^n = \frac{(s+1)^n - 1}{s}$ for $s \geq 1$ and by n for $s = 0$.*

Proof. Items 1) and 2) follow immediately from Lemma (6.1) and Theorem (6.3). As for 3), note that since $L - P$ is not big we have $(L - P)^n = 0$. Then

$$d = \sum_{j=1}^{n} (-1)^{j-1} \binom{n}{j} L^{n-j} \cdot P^j.$$

By noting that $L^{n-j} \cdot P^j = \mathcal{O}_P(1)^{n-j} \cdot \mathcal{O}_P(-s)^{j-1} = (-1)^{j-1} s^{j-1}$, we find $d = \sum_{j=1}^{n} \binom{n}{j} s^{j-1}$. This gives the result. □

In light of the above results we will assume that $(X, L) \not\cong (\mathbb{P}^n, \mathcal{O}_{\mathbb{P}^n}(1))$, i.e., $N_{P/X} \cong \mathcal{O}_{\mathbb{P}^{n-1}}(-s)$ with $s \geq 0$.

Let $\mathcal{H}$ be the ample line bundle on Z such that $L - P \approx \phi^*(\mathcal{H})$. Set $\mathfrak{h} = \mathcal{H}^{n-1}$ and $t = L \cdot f$ for a general fiber f of ϕ. We have $L_P - P_P \approx \mathcal{O}_{\mathbb{P}^{n-1}}(s+1) \approx \phi_P^*(\mathcal{H})$. Since $t = \deg \phi_P$, we conclude that

$$t\mathfrak{h} = (s+1)^{n-1}. \tag{24}$$

Note that the restriction $\phi_P : \mathbb{P}^{n-1} \to Z$ is a t-to-one finite morphism.

Remark 8.2. Note that by (2.4) and (2.5) applied to the finite map ϕ_P we conclude that Z is Cohen–Macaulay, has t-factorial singularities, and $\operatorname{Pic}(Z) \cong \mathbb{Z}$.

Let us give one more class of examples.

With the notation as above, assume that (X, L, P) is a $\mathbb{P}^{n-1}$-degenerate triple with $s \geq 0$ and $t = 1$. Since the restriction ϕ_P is an isomorphism under this assumption we see that, by using also relation (24), $(Z, \mathcal{H}) \cong (\mathbb{P}^{n-1}, \mathcal{O}_{\mathbb{P}^{n-1}}(s+1))$, and that ϕ is a $\mathbb{P}^1$-bundle (see also [2, (3.2.1)]). We let $V := \phi_* \mathcal{O}_X(P)$ and

$$\mathcal{E} := \phi_* L \cong \phi_*(\mathcal{O}_X(P) \otimes \phi^* \mathcal{H}) \cong V \otimes \mathcal{O}_{\mathbb{P}^{n-1}}(s+1). \tag{25}$$

Then $X \cong \mathbb{P}(\mathcal{E}) \cong \mathbb{P}(V)$.

Proposition 8.3. *If $s \geq 0$ and $t = 1$ then $(X, L) \cong (\mathbb{P}(\mathcal{O}_{\mathbb{P}^{n-1}}(s+1) \oplus \mathcal{O}_{\mathbb{P}^{n-1}}(1))), \xi)$, where ξ denotes the tautological line bundle on $\mathbb{P}(\mathcal{O}_{\mathbb{P}^{n-1}}(s+1) \oplus \mathcal{O}_{\mathbb{P}^{n-1}}(1))$.*

Proof. From the exact sequence $0 \to \mathcal{O}_X \to \mathcal{O}_X(P) \to P_P \cong \mathcal{O}_{\mathbb{P}^{n-1}}(-s) \to 0$, by taking the direct image and since the higher direct image functor $R^i \phi_* \mathcal{O}_X$ is zero for $i > 0$, we get the exact sequence $0 \to \mathcal{O}_{\mathbb{P}^{n-1}} \to V \to \mathcal{O}_{\mathbb{P}^{n-1}}(-s) \to 0$. Since $h^1(\mathcal{O}_{\mathbb{P}^{n-1}}(s)) = 0$ we see that this sequence splits. Thus $\mathcal{E} = \phi_* L \cong V \otimes \mathcal{O}_{\mathbb{P}^{n-1}}(s+1) \cong \mathcal{O}_{\mathbb{P}^{n-1}}(s+1) \oplus \mathcal{O}_{\mathbb{P}^{n-1}}(1)$. From this the result is clear. □

From relation (24) we see that $s = 0$ implies $t = 1$. This gives the following consequence.

Corollary 8.4. *If $s = 0$ then $(X, L) \cong (\mathbb{P}^{n-1} \times \mathbb{P}^1, \mathcal{O}_{\mathbb{P}^{n-1} \times \mathbb{P}^1}(1, 1))$.*

Remark 8.5. Note that the example of a $\mathbb{P}^{n-1}$-degenerate triple given by $\mathbb{P}^n$ blown up at one point z, $p : X \to \mathbb{P}^n$, with $L = p^* \mathcal{O}_{\mathbb{P}^n}(2) - P$, $P = p^{-1}(z)$, fits in Proposition (8.3) with $s = 1$.

By the above, we can work from now on under the extra assumptions that $s \geq 1$ and $t \geq 2$, where $N_{\mathbb{P}^{n-1}/X} \cong \mathcal{O}_{\mathbb{P}^{n-1}}(-s)$ and $t = \deg \phi_{\mathbb{P}^{n-1}}$.

We can now carry out some more adjunction theoretic analysis, improving, in the case of a codimension 1 linear projective space, the results proved in (7.4). We will also assume $n \geq 3$, since the problem is completely solved when $n = 2$ (see [18], [2, §8.4]). For the structure of the first reduction map occurring in the theorem below we refer to [2, Chap. 7].

Theorem 8.6. *Let X be a smooth connected n-dimensional variety, $n \geq 3$, and let L be a very ample line bundle on X. Assume that (X, L, P) is a $\mathbb{P}^{n-1}$-degenerate triple. Let $N := N_{\mathbb{P}^{n-1}/X} \cong \mathcal{O}_{\mathbb{P}^{n-1}}(-s)$. Assume that $s \geq 1$ and $t := \deg \phi_P \geq 2$. Then the first reduction exists, i.e., there exists a map $\pi : X \to X'$ expressing X as the blowup of a projective manifold X' at a finite set B with $K_X + (n-1)L \approx \pi^* H$ for a very ample line bundle H on X'. Moreover it follows that π is an isomorphism unless B is a single point, $s = 1$, and $P := \mathbb{P}^{n-1} = \pi^{-1}(B)$.*

Proof. Set $P := \mathbb{P}^{n-1}$. If $K_X + (n-1)L$ is not spanned, then (X, L) is as in one of cases 1), 3) of (7.4) (notice that case 2) of (7.4) is excluded because we have $\dim P > \lceil \frac{n}{2} \rceil$). In case 1) we have that $K_X + (n+1)L$ is trivial, which implies that $\mathcal{O}_P \approx (K_X + (n+1)L)_P \approx \mathcal{O}_{\mathbb{P}^{n-1}}(s+1)$. Thus $s = -1$. In case 3), we have $t = 1$. Therefore both cases 1), 3) of (7.4) are excluded in view of our present assumptions that $s \geq 1$ and $t \geq 2$.

Therefore we can assume that $K_X + (n-1)L$ is spanned. It follows [2, §7.3] that either $K_X + (n-1)L$ is nef and big or:

1. $K_X \cong -(n-1)L$; or

2. (X, L) is a quadric fibration, $\pi : X \to C$, over a smooth curve C, i.e., $K_X + (n-1)L \cong \pi^* H$ for some ample line bundle H on C; or

3. (X, L) is a scroll, $\pi : X \to S$, over a smooth surface S, i.e., $K_X + (n-1)L \cong \pi^* H$ for some ample line bundle H on S.

In the first case we have that $\mathcal{O}_f \cong (K_X + (n-1)L)_f$ for a general fiber f of ϕ. Since $(1-n)L \cdot f = (1-n)t = K_X \cdot f = \deg(K_f)$ we conclude that $n = 3$ and $t = L \cdot f = 1$, contradicting our present assumption $t \geq 2$.

Since $P = \mathbb{P}^{n-1}$ can't map to a curve by Lemma (2.3), we conclude in the second case that P is a component of a fiber of π. But since $n \geq 3$ fibers are either irreducible quadrics, or two $\mathbb{P}^{n-1}$'s meeting in a $\mathbb{P}^{n-2}$. Indeed multiple fibers don't happen, since otherwise we could slice down to a surface and have $\mathbb{P}^1$ as a multiple fiber, which is a classical standard impossibility. If we are in the case of two $\mathbb{P}^{n-1}$'s meeting in a $\mathbb{P}^{n-2}$, then we have negative normal bundle for each $\mathbb{P}^{n-1}$ and we can contract one $\mathbb{P}^{n-1}$ to get a map of the other $\mathbb{P}^{n-1}$ to a $(n-1)$-dimensional image but with the intersection $\mathbb{P}^{n-2}$ going to a point, which is not possible again by Lemma (2.3).

In the third case we know from a result of the fourth author [20, Theorem (3.3)] that π is a $\mathbb{P}^{n-2}$-bundle. Thus we conclude that P is a section with $n = 3$. Indeed since fibers of π are one dimensional we conclude that P meets a general fiber f of π in a finite nonempty set. Since $L - P$ is nef and $L \cdot f = 1$ we conclude that $P \cdot f = 1$. Since $(L - P) \cdot f = 0$ it is clear that π is the same as ϕ and $t = 1$.

Thus we see that $K_X + (n-1)L$ is big and the first reduction $\pi : X \to X'$ exists. Assume that π is not an isomorphism. Let F be a positive dimensional fiber of π. We know that F is a linear $\mathbb{P}^{n-1}$ with respect to L and $N_{F/X} \cong \mathcal{O}_{\mathbb{P}^{n-1}}(-1)$. If we show that $F = P$ then we see that $s = 1$ and the theorem will be proved. Thus assume that F is not P. Then we see that $F \cap P$ is empty or we would have the absurdity that π maps the positive dimensional subset $F \cap P$ of P to the point $\pi(F)$ without mapping P to the same point. Thus we have $L_F \cong \mathcal{O}_{\mathbb{P}^{n-1}}(1)$. Therefore we see that F is a section of $\phi : X \to Z$. Thus we conclude that ϕ is a $\mathbb{P}^1$-bundle over $\mathbb{P}^{n-1}$. Restricting the bundle to a bundle $\phi_S : S \to R$ on a smooth curve R on Z, we find a $\mathbb{P}^1$-bundle S over R with two disjoint curves, $P \cap S$ and $F \cap S$, each with negative self intersection since both the normal bundles $N_{P/X}$, $N_{F/X}$ are negative. This is absurd. □

We conclude this section by considering the special case of a threefold X.

8.7. The three dimensional case. We use the same notation and assumptions as above. In particular in view of the results above we make the blanket assumption that $s \geq 1$ and $t \geq 2$.

Theorem 8.8. *Let X be a smooth threefold and L a very ample line bundle on X. Assume that (X, L, P) is a $\mathbb{P}^2$-degenerate triple. If $s = 1$ and $t := \deg \phi_P \geq 2$, then $t = 4$. In this case X is the blowing up at one point of the complete intersection of three quadrics in $\mathbb{P}^6$.*

Proof. If $s = 1$ then by Proposition (8.1), 3) we see that $L^3 = 7$. Note that we use the classification of degree 7 manifolds given in [13]. By Theorem (8.6) we can assume that $K_X + 2L$ is nef and big. Thus quadric fibrations over curves and scrolls over curves and surfaces are ruled out. By using the degree 7 classification, two possibilities remain.

1. X is the blowing up at one point, $\pi : X \to X'$, of the complete intersection X' of three quadrics in $\mathbb{P}^6$, with π the first reduction map; or

2. there exists a morphism $\rho : X \to C$ of X to a curve C given by the complete linear system $|m(K_X + L)|$ for $m \gg 0$.

In the first case we know from [13] that L embeds X into $\mathbb{P}^5$. This X contains the positive dimensional fiber of π and thus since projection from this linear $\mathbb{P}^2$ must map to $\mathbb{P}^2$ we conclude that this is an example with $s = 1$. Let $f \cong \mathbb{P}^1$ be a fiber of $\phi : X \to Z$. To see what t is, note that $K_X + L$ being nef yields $t = L \cdot f \geq -K_X \cdot f = 2$. By Theorem (8.6) we know that P coincides with the exceptional divisor of π. Moreover, $-K_{X'} \cong \mathcal{O}_{X'}(1) = L'$, the polarization of the first reduction X', which satisfies the condition $L \cong \pi^*L' - P$. Then

$$K_X \cong \pi^*K_{X'} + 2P \cong -L - P + 2P = -L.$$

Hence we have $K_X \cdot f = \deg(K_f) = 0$. Thus we cannot have $t = 2$ since this would imply f was rational. Since we are assuming $t \geq 2$ we conclude by relation (24) that t must equal 4.

In the second case $\rho(P)$ must be a point by Lemma (2.3) and therefore $(K_X+L)_P \cong \mathcal{O}_P$. Since $(K_X + L)_P \cong \mathcal{O}_P(s - 2)$ we get the contradiction $s = 2$. □

Combining Theorem (8.6) and Theorem (8.8) we have the following result.

Corollary 8.9. *Let X be a smooth threefold and L a very ample line bundle on X. Assume that (X, L, P) is a $\mathbb{P}^2$-degenerate triple. If $s \geq 1$ and $t := \deg \phi_P \geq 2$, then either X is the blowing up at one point of the complete intersection of three quadrics in $\mathbb{P}^6$, or $K_X + 2L$ is very ample.*

Proof. By (8.6) and (8.8) we know that either $s = 1$ and X is the blowing up at one point of the complete intersection of three quadrics in $\mathbb{P}^6$ or X is isomorphic to its own first reduction. □

Theorem 8.10. *Let X be a smooth threefold and L a very ample line bundle on X. Assume that (X, L, P) is a $\mathbb{P}^2$-degenerate triple. Further assume $s \geq 2$. Then the case $t = 2$ does not occur.*

Proof. By Corollary (8.9) we can assume that (X, L) is its own first reduction. A simple check of the list of pairs with $K_X + L$ not nef (see [2, §7.3]) shows that they cannot occur if $s \geq 2$. Thus we can assume that $K_X + L$ is nef. We know that there is a morphism with connected fibers $\rho : X \to W$ of X onto a normal variety W, given by $|m(K_X + L)|$ for $m \gg 0$, with $K_X + L \cong \rho^* H$ for some ample line bundle H on W. Note that if $t = 2$ then the general fiber of $\phi : X \to Z$ is a conic. Thus K_X+L must be trivial on the general fiber of ϕ. Then there exists a surjective morphism $q : Z \to W$ such that $q \circ \phi = \rho$, whence $\dim W \leq 2$. Note also that $\dim W > 0$. Indeed otherwise $K_X + L$ would be trivial and therefore, since $(K_X + L)_P \cong \mathcal{O}_P(s - 2)$, we would have $s = 2$. But $t - s = 2$ contradicts relation (24).

The divisor P can not be in a fiber of ρ. If it was we would have $(K_X+L)_P \cong \mathcal{O}_P$. This would imply $s = 2$. Then again $t = s = 2$ contradicts relation (24). By using Lemma (2.3) we conclude that $\dim W = 2$ and, since P must map onto W, that all fibers of ρ are one dimensional. By the above, (X, L) is a quadric fibration over the surface W. Then by Besana's results [3] we know that W is smooth and thus by Lazarsfeld's theorem (see e.g., [2, (3.1.7)]) we know that W is $\mathbb{P}^2$. We also see that the maps ρ and ϕ are the same.

Note that by pulling back to P we have

$$m(K_X + L)_P \cong \mathcal{O}_P(m(s - 2)) \cong (L - P)_P \cong \phi_P^* \mathcal{H} \cong \mathcal{O}_{\mathbb{P}^2}(s + 1).$$

This gives $s + 1 = m(s - 2)$ and hence either $s = 5, m = 2, L - P \cong 2(K_X + L)$, or $s = 3, m = 4, L - P \cong 4(K_X + L)$. Assume $s = 5$. Then, since $t = 2$, relation (24) gives $\mathfrak{h} = \mathcal{H}^2 = 18$. But since $L - P \cong 2(K_X + L)$ we have the absurdity that $18 = \mathcal{H}^2 = 4H^2$. Assume $s = 3$. Then $\mathfrak{h} = \mathcal{H}^2 = 8$ from relation (24) and $L - P \cong 4(K_X + L)$ gives the absurdity $8 = \mathcal{H}^2 = 16H^2$. □

References

[1] W. Barth, Transplanting cohomology classes in complex projective space, Amer. J. Math. 92 (1970), 951–967.

[2] M. C. Beltrametti and A. J. Sommese, The Adjunction Theory of Complex Projective Varieties, de Gruyter Exp. Math. 16, Walter de Gruyter, Berlin–New York 1995.

[3] G. M. Besana, On the geometry of conic bundles arising in adjunction theory, Math. Nachr. 160 (1993), 223–251.

[4] T. Fujita, Classification Theory of Polarized Varieties, London Math. Soc. Lecture Notes Ser. 155, Cambridge University Press, 1990.

[5] P. Griffiths, The extension problem in complex analysis - II, (Embeddings with positive normal bundle), Amer. J. Math. 88 (1966), 366–446.

[6] L. Gruson, R. Lazarsfeld, and C. Peskine, On a theorem of Castelnuovo and the equation defining space curves, Invent. Math. 72 (1983), 491–506.

[7] J. Harris, Curves in projective space, with the collaboration of D. Eisenbud, Université de Montreal, Montreal (Québec), Canada, 1982.

[8] R. Hartshorne, Varieties of small codimension in projective space, Bull. Amer. Math. Soc. 80 (1974), 1017–1032.

[9] R. Hartshorne, Algebraic Geometry, Grad. Texts in Math. 52, Springer-Verlag, 1977.

[10] R. Hartshorne, Ample Subvarieties of Algebraic Varieties, Lecture Notes in Math. 156, Springer-Verlag, New York 1970.

[11] H. Hironaka, Smoothing algebraic cycles of small dimensions, Amer. J. Math. 90 (1968), 1–54.

[12] B. Ilic, Geometric properties of the double point divisor, Trans. Amer. Math. Soc. 350 (1998), 1643–1661.

[13] P. Ionescu, Embedded projective varieties of small invariants, in: Proceedings of the 1982 Week of Algebraic Geometry, Bucharest, ed. by L. Bădescu and D. Popescu, Lecture Notes in Math. 1056, 142–187, Springer-Verlag, New York 1984.

[14] S. L. Kleiman, The enumerative theory of singularities, in: Real and Complex Singularities, Oslo 1976, ed. by P. Holme, 297–396, Alphen aan den Rijn, Sijthoff and Noordhoff, Rockville, Maryland, 1977.

[15] R. Lazarsfeld, A sharp Castelnuovo bound for smooth surfaces, Duke Math. J. 55 (1987), 423–429.

[16] D. Mumford, Varieties defined by quadratic equations, in: Questions on algebraic varieties, CIME course 1969, Rome (1970), 30–100.

[17] Z. Ran, Local differential geometry and generic projections of threefolds, J. Differential Geom. 32 (1990), 131–137.

[18] A. J. Sommese, Hyperplane sections of projective surfaces, I: The adjunction mapping, Duke Math. J. 46 (1979), 377–401.

[19]   A. J. Sommese, Hyperplane sections, in: Algebraic Geometry, Chicago, 1981, ed. by A. Libgober and P. Wagreich, Lecture Notes in Math. 862, 232–271, Springer-Verlag, New York 1981.

[20]   A. J. Sommese, On the adjunction theoretic structure of projective varieties, in: Proceedings of the Complex Analysis and Algebraic Geometry Conference, Göttingen, 1985, ed. by H. Grauert, Lecture Notes in Math. 1194, 175–213, Springer-Verlag, New York 1986.

[21]   O. Zariski, An Introduction to the Theory of Algebraic Surfaces, Lecture Notes in Math. 83, Springer-Verlag, Berlin 1969.

[22]   O. Zariski, Algebraic Surfaces, with appendices by S.S. Abhyankar, J. Lipman, and D. Mumford, Ergeb. Math. Grenzgeb. (2) 61, Springer-Verlag, Berlin 1971.

Generalized Petersson–Weil metric on the Douady space of embedded manifolds

Indranil Biswas and Georg Schumacher

0. Introduction

An adapted generalized Petersson–Weil metric is the tool of choice to investigate the Kähler geometry of various moduli spaces. Like in the classical case of moduli of compact (or punctured) Riemann surfaces, a natural L^2-inner product of distinguished representatives of Kodaira–Spencer classes defines a hermitian structure on the moduli space and eventually yields a Kähler structure. In various higher dimensional settings a generalized Petersson–Weil form has been constructed. Surprisingly, a generalized Petersson–Weil metric on the Douady space of submanifolds $X \hookrightarrow Z$ of a fixed Kähler manifold (Z, ω_Z) had not yet been constructed. In this note, we give a definition, which also applies to the relative Douady space in the sense of Pourcin [P]. The underlying hermitian inner product in the absolute case is an L^2-product of holomorphic sections of the normal bundle $\mathcal{N}_{X/Z}$. These are represented by differentiable vector fields orthogonal to the reference manifold X with respect to ω_Z. The generalized Petersson–Weil form satisfies a fiber-integral formula, which relates our construction to the results of Varouchas [V], and a construction for embedded Riemann surfaces relying on classical Teichmüller theory was undertaken recently by Maarouf [M]. The moduli space of divisors (divided by the automorphism group of the ambient space) occurred within the theory of moduli of framed manifolds [SCH]. As an application of the fiber integral formula for projective varieties the generalized Petersson–Weil form is understood as the curvature form of a determinant line bundle equipped with a Quillen metric.

1. The relative Douady space

Let C be a fixed complex space, and $An_{/C}$ the category of complex spaces over C whose objects are holomorphic maps $S \to C$ of complex spaces. We consider a fixed complex space $\mathcal{Z} \to C$, and denote the fibers by $(\mathcal{Z}_c, \mathcal{O}_{\mathcal{Z}_c})$, where $c \in C$. The existence of a relative Douady space of all compact complex subspaces follows from a theorem of G. Pourcin [P], which solves a global moduli problem. In the sequel we will compute its tangent space in an intrinsic way as space of infinitesimal deformations. We state the universal property.

For any object S in $An_{/C}$ we set $\mathcal{Z}_S := \mathcal{Z} \times_\mathcal{C} S$ and denote by pr_j, $j = 1, 2$, the canonical projections.

Theorem 1.1 (Pourcin). *Let $\mathcal{Z}$ be a complex space over $\mathcal{C}$. Then there exists a complex space $H \to \mathcal{C}$ together with a closed subspace $\mathcal{X} \subset \mathcal{Z}_H$ satisfying the following properties:*

(i) *the map* $\mathrm{pr}_2 : \mathcal{Z}_H \to H$ *restricted to $\mathcal{X}$ is proper and flat;*

(ii) *for any space $S \to \mathcal{C}$ and any subspace $\mathcal{Y} \subset \mathcal{Z}_S$ such that the canonical map $\mathcal{Y} \to S$ is proper and flat, there exists a unique holomorphic map $S \to H$ over $\mathcal{C}$ such that $\mathcal{X} \times_H S$ equals $\mathcal{Y}$ as a subspace of $\mathcal{Z}_S$.*

Let $p \in H$ be a point with image 0 in $\mathcal{C}$. It corresponds to a compact subspace $X \subset Z$ of the fiber $Z := \mathcal{Z}_0$. We compute $T_p H$ (equipped with a natural map $T_p H \to T_0 \mathcal{C}$) and the Kodaira–Spencer map: The elements of $T_p H$ correspond uniquely to holomorphic maps of the double point D to H (over $\mathcal{C}$) carrying the underlying reduced point to p. General deformations are objects over punctured complex spaces (S, s_0) over $(\mathcal{C}, 0)$ or germs of complex spaces. (Isomorphism classes need not be considered, since all objects come with an embedding onto a fixed space.) Hence, because of the universal property of H, the elements of $T_p H$ are exactly infinitesimal relative deformations of $X \hookrightarrow Z$ over $\mathcal{C}$, which are determined by the following data:

(a) a morphism $g : (D, 0) \to (\mathcal{C}, 0)$

(b) a subspace $\mathcal{Y} \hookrightarrow g^*\mathcal{Z} = \mathcal{Z}_D$.

and a diagram

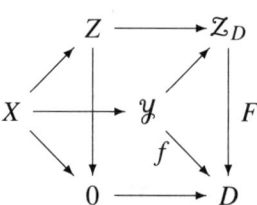

consisting of two embeddings and two Cartesian squares over $0 \hookrightarrow S$. The map F equals $\mathrm{pr}_2 : Z \times_\mathcal{C} D \to D$. In the above situation, we consider the analytic extension of complex spaces together with the canonical epimorphisms:

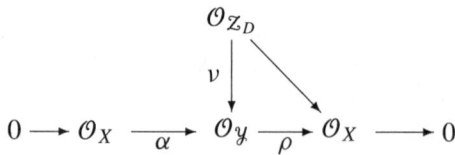

Denote by $\mathcal{J} \subset \mathcal{O}_{Z_D}$ the ideal of X. Then $\nu(\mathcal{J}^2) = 0$, giving rise to a diagram

$$\begin{array}{ccccccccc}
0 & \longrightarrow & \mathcal{J}/\mathcal{J}^2 & \longrightarrow & \mathcal{O}_{Z_D}/\mathcal{J}^2 & \longrightarrow & \mathcal{O}_X & \longrightarrow & 0 \\
& & \varphi \downarrow & & \bar{\nu} \downarrow & & \| & & \\
0 & \longrightarrow & \mathcal{O}_X & \longrightarrow & \mathcal{O}_Y & \longrightarrow & \mathcal{O}_X & \longrightarrow & 0
\end{array}$$

where also the first row is an analytic extension of $\mathcal{O}_X$ and $\varphi \in \operatorname{Hom}_{\mathcal{O}_X}(\mathcal{J}/\mathcal{J}^2, \mathcal{O}_X)$. The map ν as well as $\bar{\nu}$ are $F^{-1}(\mathcal{O}_D)$-linear.

Conversely: Let $(D, 0) \to (C, 0)$ be a double point in our category of relative (punctured) complex spaces, and $\mathcal{J} \subset \mathcal{O}_{Z_D}$ the ideal of X. For any $\mathcal{O}_X$-linear map $\varphi : \mathcal{J}/\mathcal{J}^2 \to \mathcal{O}_X$ we define

$$\mathcal{O}_Y := \left(\mathcal{O}_{Z_D}/\mathcal{J}^2\right)\big/\operatorname{Kernel}(\varphi),$$

which is a fibred sum of a sheaf of $\mathbb{C}$-algebras and an $\mathcal{O}_X$-module.

Proposition 1.2. *The tangent space $T_p H$ is identified with the space of global homomorphisms $\operatorname{Hom}_{\mathcal{O}_X}(\mathcal{J}/\mathcal{J}^2, \mathcal{O}_X)$. Denoting the projection of Z onto C by q, the obvious restriction of global homomorphisms, namely*

$$\operatorname{Hom}_{\mathcal{O}_X}(\mathcal{J}/\mathcal{J}^2, \mathcal{O}_X) \longrightarrow \operatorname{Hom}_{q^{-1}(\mathcal{O}_C)}(q^{-1}(\mathcal{I}_{q(p)}/\mathcal{I}_{q(p)}^2), q^{-1}(\mathcal{O}_C)) = T_{q(p)}C,$$

where $\mathcal{I}_{q(p)} \subset \mathcal{O}_C$ is the ideal sheaf for $q(p)$, coincides with the natural projection $T_p H \longrightarrow T_{q(p)}C$.

From now on we will consider the Douady space of embedded compact complex manifolds, where the family $Z \to C$ is by assumption proper and smooth.

Proposition 1.3. *Let $t \in T_0 C$. Then the fiber of t in $T_p H$ consists of all holomorphic sections of the normal bundle of X in Z which project down to t under $Z \to C$.*

2. The generalized Petersson–Weil metric

We fix an embedding of compact manifolds $i_0 : X \hookrightarrow Z$, and consider a deformation over a one dimensional smooth base $0 \in S \subset \mathbb{C}$. In other words, we have a commutative diagram

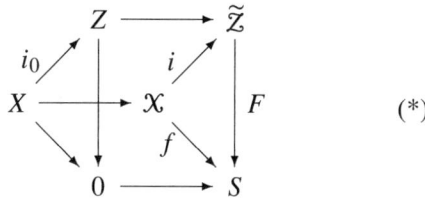

(*)

We need to describe the infinitesimal deformation in adapted local coordinates, namely $\widetilde{\mathcal{Z}} = \{(z, w, s)\}$ where the components of z are z^α; $\alpha = 1, \ldots, n$, and the components of w are w^i; $i = 1, \ldots, m$. These can be chosen such that $\mathcal{X} = \{(z, 0, s)\}$ and $F(z, w, s) = s$. The tangent vector $v = \frac{\partial}{\partial s}|_{s=0} \in T_0 S$ gives rise to a holomorphic section $\rho = \rho(v)$ of the normal bundle $\mathcal{N}_{\mathcal{X}|\widetilde{\mathcal{Z}}}$, which is mapped to v under F_*. In the above local coordinates (z, w, s) it is again given by $\frac{\partial}{\partial s}$. (The holomorphic coefficients of $\frac{\partial}{\partial w^i}$ vanish identically.) Next a real, locally $\partial\bar\partial$-exact $(1, 1)$-form $\tilde\omega$ on $\widetilde{\mathcal{Z}}$ is fixed, which is positive semi-definite and is positive definite in the fiber direction of F. The generalized Petersson–Weil form will be constructed with respect to this form.

We assign to $\rho(\frac{\partial}{\partial s}|_{s=0})$ a differentiable complex vector field $\eta = \eta(\frac{\partial}{\partial s}|_{s=0})$ of type $(1, 0)$ on X with values in $\mathcal{T}_{\widetilde{\mathcal{X}}}$ as follows: For $x \in X$

(i) $\eta(x)$ projects to $\rho(x)$

(ii) $F_*\eta(x) = \frac{\partial}{\partial s}|_{s=0}$

(iii) $\eta(x)$ is perpendicular to X with respect to $\tilde\omega$.

The vector field η is uniquely determined. The form $\tilde\omega$ induces a natural sesquilinear form on all tangent spaces $T_x\widetilde{\mathcal{Z}}$, and by integration over X a semi-norm $\|\eta\|_{PW}$. We denote the coefficients of $\tilde\omega$ by $g_{\alpha\bar\beta}$; $g_{\alpha\bar{j}}$, $g_{\alpha\bar{s}}$ etc. following the above convention on indices. Special attention is paid to $\tilde\omega|X = \sqrt{-1}g_{\alpha\bar\beta}dz^\alpha \wedge dz^{\bar\beta}$. The inverse of $g_{\alpha\bar\beta}$ is denoted by $g^{\bar\beta\alpha}$ and the corresponding volume form on X will be called $g \cdot dv$, where $g = \det(g_{\alpha\bar\beta})$ and dv is the Euclidean volume element. A short calculation gives:

Lemma 2.1. *With respect to local coordinates:*

(i) $\eta = \frac{\partial}{\partial s} - g_{s\bar\beta}g^{\bar\beta\alpha}\frac{\partial}{\partial z^\alpha}$ on X

(ii) $\|\eta(z)\|^2_{PW} = g_{s\bar{s}} - g_{s\bar\beta}g_{\alpha\bar{s}}g^{\bar\beta\alpha} \geq 0$

(iii) $\|\frac{\partial}{\partial s}|_0\|^2_{PW} = \|\eta\|^2_{PW} = \int_X (g_{s\bar{s}} - g_{s\bar\beta}g_{\alpha\bar{s}}g^{\bar\beta\alpha})g\,dv \geq 0$.

The above considerations can be carried out at any point $s \in S$, using for each base point local coordinates that are adapted to the situation. Thus, a positive semi-definite $(1, 1)$-form ω_S^{PW} on S is constructed:

$$\omega_S^{PW} := \sqrt{-1}\|\eta(s)\|^2_{PW}\,ds \wedge \overline{ds},$$

where $\eta(s) = \eta(\frac{\partial}{\partial s}|_s)$ for $s \in S$.

Lemma 2.2. *The Petersson–Weil form on S equals the following fiber integral*

$$\omega_S^{PW} = \frac{1}{(n+1)!} \int_{\mathcal{X}/S} \omega_{\widetilde{\mathcal{Z}}}^{n+1}|\mathcal{X},$$

where $n = \dim X$.

Returning to the general deformations of Section 2 we see that the above construction is functorial and yields a semi-positive form ω^{PW} on the moduli space.

Let $\mathcal{Z} \to \mathcal{C}$ be a holomorphic family of compact complex manifolds over a complex space C, which may also be non-reduced. The total space $\mathcal{Z}$ is equipped with a fixed positive definite real $(1,1)$-form $\omega_{\mathcal{Z}}$, which possesses locally a $\partial\bar{\partial}$-potential of class C^∞, a notion which also makes sense in the non-reduced case.

Let now $S \to C$ denote an arbitrary complex space over C equipped with an embedded family of complex spaces, i.e. we are given a diagram of type $(*)$ with $\tilde{\mathcal{Z}} = \mathcal{Z} \times_C S$. The above construction is compatible with base change and also valid for a one-dimensional base space replaced by a double point D.

Let

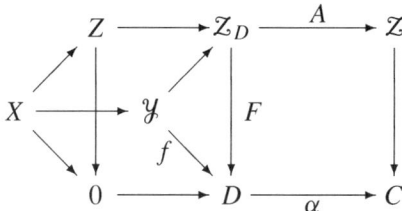

be a deformation over the double point D, represented by η according to Lemma 2.1.

Lemma 2.3. *The Petersson–Weil form is positive definite on the moduli space; i.e.* $\|\eta\|^{PW} = 0$ *holds, if and only if the above infinitesimal deformation is trivial.*

Proof. Let $\|\eta\|^{PW} = 0$. We set $\tilde{\mathcal{Z}} = \mathcal{Z}_D$ and observe that a fiber integral

$$\int_{\mathcal{X}/D} (\omega_{\tilde{\mathcal{Z}}}^{n+1} | x)$$

yields a well defined form on D, which vanishes by assumption. Since $\omega_{\mathcal{Z}}^{(n+1)}$ is a positive $(n+1, n+1)$-form, this means that for all $z \in X$ the normal vector $\eta(z)$ of X in $\mathcal{Z}_D$ at Z is mapped to zero under A_*. According to Section 2 the deformation must be trivial. □

The above results yield the theorem stated below. Let $\mathcal{Z} \to C$ be a family of compact complex manifolds and $\omega_{\mathcal{Z}}$ a Kähler form on $\mathcal{Z}$ (with a local $\partial\bar{\partial}$-potential of class C^∞). Let $H \to C$ be the relative Douady space of embedded submanifolds of dimension n.

Theorem 2.4. *There exists a naturally defined generalized Petersson–Weil form ω^{PW} on H. Let*

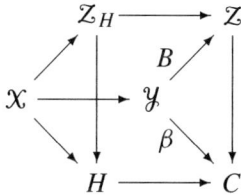

denote the universal family. Then

$$\omega^{PW} = \frac{1}{(n+1)!} \int_{\mathcal{X}/H} (B^*\omega_Z^{n+1}|\mathcal{X})$$

holds. In particular, ω^{PW} is a Kähler form (with a local $\partial\bar{\partial}$-potential).

In the sequel, we restrict ourselves to the absolute Douady space, where $C = 0$. Furthermore we assume that Z is a projective variety equipped with a positive hermitian line bundle $(\mathcal{L}, h)$ and $\omega_Z = c_1(\mathcal{L}, h)$. Let $(\mathcal{L}_\mathcal{X}, h_\mathcal{X})$ be the pull-back to $\mathcal{X}$, and denote by $((\mathcal{L}_\mathcal{X} - \mathcal{L}_\mathcal{X}^{-1})^{n+1}, \tilde{h})$ the induced virtual vector bundle of rank zero. Then

$$\omega^{PW} = \frac{1}{2^{n+1}(n+1)!} \int_{\mathcal{X}/S} \mathrm{ch}((\mathcal{L}_\mathcal{X} - \mathcal{L}_\mathcal{X}^{-1})^{n+1}, \tilde{h})$$

(c.f. also [F-S, B-S]).

The right-hand side integral is invariant under the multiplication with the relative Todd form $\mathrm{td}(\mathcal{X}/S)$. The resp. Kähler metric on $\mathcal{X}$ is immaterial, since only the term in degree zero matters. Invoking a theorem of Bismut, Gillet and Soulé [B-G-S] we get

Theorem 2.5. *The generalized Petersson–Weil form is the curvature form of the determinant bundle* $\lambda = \det((\mathcal{L}_\mathcal{X} - \mathcal{L}_\mathcal{X}^{-1})^{n+1})$ *equipped with the Quillen metric* h_Q.

Acknowledgments. One of us (I.B.) acknowledges the generous hospitality of the Philipps-Universität, Marburg, where the work was carried out. The second named author would like to thank the Tata Institute of Fundamental Research, Bombay, where our collaboration was initiated, for its generous hospitality.

References

[B-G-S] Bismut, J. M., Gillet, H., Soulé, C., Analytic torsion and holomorphic determinant bundles I, II, III. Commun. Math. Phys. 115 (1988), 49–78; 115 (1988), 79–126; 115 (1988), 301–351.

[B-S] Biswas, I., Schumacher, G., Determinant bundle, Quillen metric, and Petersson–Weil form on moduli spaces. Geom. Funct. Anal. 9 (1999), 226–256.

[F-S] Fujiki, A., Schumacher, G., The moduli space of compact extremal Kähler manifolds and generalized Petersson–Weil metrics. Publ. Res. Inst. Math. Sci. 26 (1990), 101–183.

[M] Maarouf, M., Problémes d'hyperbolicité sur l'espace de Douady et ses variantes. Ph.D Thesis, Institut Fourier (1996).

[P] Pourcin, G., Théorème de Douady au-dessus de S. Ann. Scuola Norm. Sup. Pisa, Cl. Sci. Fis. Mat. III. Ser. 23 (1969), 451–459.

[SCH] Schumacher, G., Asymptotics of Kähler-Einstein Metrics on Quasi-Projective Manifolds and an Extension Theorem on Holomorphic Maps. Math. Ann. 311 (1988), 631–645.

[V] Varouchas, J., Stabilité de la classe des variétés Kaehleriennes par certains morphismes propres. Invent. Math. 77 (1984), 117–127.

On simply connected Godeaux surfaces

*Fabrizio Catanese** and Roberto Pignatelli*

Abstract. In this paper we provide a first step towards the classification of the numerical Godeaux surfaces in the still unknown open cases where $\text{Tors}(S) = 0$, or $\text{Tors}(S) = \mathbb{Z}/2\mathbb{Z}$.

Our method works in both cases, but in this paper, after some results which we establish in a greater generality, we mostly restrict ourselves to the case where the Torsion group is zero.

The bicanonical system yields, on a suitable blow up $\tilde{S}$ of the minimal model S, a fibration $f : \tilde{S} \to \mathbb{P}^1$ in curves of genus $2 \leq g \leq 4$, and the invariants of this fibration determine the equations of the image of S under a map which, in the general case, is the product of the tricanonical and of the bicanonical map.

This allows us to subdivide our surfaces into four classes, according to the behaviour of the bicanonical system. For each of these classes we have a complete description but the existence questions are not yet solved.

1991 Mathematics Subject Classification: 14J25, 14J10, 14J29, 14D99, 14B12, 14Q10.

Introduction

Algebraic surfaces with $p_g(S) = q(S) = 0$ were interesting ever since in the theory of algebraic surfaces, when it was first asked whether a surface with such invariants would be rational.

The first counterexample was given by Enriques, who constructed what are by now called the Enriques surfaces (cf. [E1], [E2], and [BPV] for further details and references); these have fundamental group $\mathbb{Z}/2\mathbb{Z}$, and are not of general type.

In the 30's examples of surfaces of general type with $p_g(S) = q(S) = 0$ were constructed by Campedelli ([Cam], these have $K^2 = 2$), and by Godeaux ([G1], [G2], these have $K^2 = 1$).

Later on, Severi asked whether a simply connected surface with $p_g(S) = 0$ would be rational, and again this question had a negative answer by [Dol2], who constructed some simply connected elliptic surfaces, which are nowadays called Dolgachev surfaces.

---

*The research of the author was performed in the realm of the Project AGE HCM, Contract ERBCHRXT 940557.

The interest for surfaces of general type with $p_g(S) = q(S) = 0$, which are considered in Chapter VIII of Enriques' book [E2], with special regard to the question of the good properties of pluricanonical systems, was revived by Bombieri's article [Bo] which left some open problems about their pluricanonical systems.

After that, surfaces of general type with $p_g(S) = q(S) = 0$ and $K^2 = 1$ (for the minimal model) were called numerical Godeaux surfaces, while those with $p_g(S) = q(S) = 0$, $K^2 = 2$ were called numerical Campedelli surfaces.

In the 70's there were several papers devoted to these two classes of surfaces, especially to their tri-and quadri-canonical maps. Reid ([R2]) completely described the geometry of the numerical Godeaux surfaces with Torsion group of order ≥ 3, inverting the method of Godeaux who was constructing these surfaces as quotients by the free action of a cyclic group.

Finally, in the early 80's Barlow constructed ([R3], [Ba2]) a simply connected numerical Godeaux surface. The method here was a clever variant of Godeaux's method, in that the author used a non free action of a non cyclic group.

The Barlow surface was an interesting object both for applications to the differential topology of 4-manifolds ([Kot1], [OVdV]), and recently for problems on Einstein metrics ([CL]).

In fact, a classification of simply connected numerical Godeaux surfaces could produce new simply connected differentiable 4-manifolds with $b^+ = 1$. In this respect, we should point out that another beautiful example of such a surface was produced by Craighero and Gattazzo ([CG]), somehow in the line of thought introduced by Campedelli, i.e. as the minimal resolution of a normal singular surface (that the resulting surface is indeed simply connected, and has K_S ample was recently proved in [DW]). It is yet unclear whether the Barlow surface and the Craighero Gattazzo surface are diffeomorphic.

To conclude our historical motivation, we should point out that another source of interest for the numerical Godeaux surfaces stems from the conjecture of Bloch (cf. [Mu3], [BKL], [Blo]) that for a surface with $q = p_g = 0$ the Chow group of degree zero 0-cycles is trivial (this has been settled only in few very special cases, cf. [IM], [Ba3], [V]).

After all this, the reader might ask why the numerical Godeaux surfaces have not yet been classified. One reason is that the easy lines of the surfaces geography are the lines $K^2 = 2p_g - 4 + m$ with $m = 0, 1$, while here $m = 5$.

In fact, using unramified coverings which make m smaller, Reid was able to show that the numerical Godeaux surfaces with $|\operatorname{Tors}(S)| \geq 3$ form three irreducible families, with fundamental group $\mathbb{Z}/n\mathbb{Z}$, $n = 3, 4, 5$.

The main purpose of the present paper is therefore to attack the classification of the numerical Godeaux surfaces with $\operatorname{Tors}(S) = 0$, or $\mathbb{Z}/2\mathbb{Z}$ (also in the latter case special examples have been constructed, cf. e.g. [Ba1], [Wer1], [Wer2]).

Our method works in both cases, but in this paper, after some results which we establish in a greater generality, we restrict ourselves to the case where the Torsion group (i.e., the Abelianization of the fundamental group) of our surface is zero. We shall consider more amply also the latter case in a sequel to this article.

We would like now to explain what is the new method we are employing in order to classify the numerical Godeaux surfaces.

The first crucial property is that the bicanonical system yields, on a suitable blow up $\tilde{S}$ of the minimal model S, a fibration $f : \tilde{S} \to \mathbb{P}^1$ whose fibres are curves of genus g, where g can only be $= 2, 3, 4$.

To explain further our strategy, we have to subdivide our surfaces into several classes, according to the behaviour of the bicanonical system (cf. 1.1, 1.2).

Writing the bicanonical pencil $|2K_S|$ as $F + |M|$, where F is the fixed part, we see that $KF = 0$, $KM = 2$.

We have then four possibilities:

ia) $M^2 = 4$ $F = 0$ $\quad$ M has genus 4 $\quad$ $|M|$ has 4 base points

ib) $M^2 = 4$ $F = 0$ $\quad$ M has genus 3 $\quad$ $|M|$ has 1 double base point

ii) $M^2 = 2$ $MF = 2$ $F^2 = -2$ $\quad$ M has genus 3

iii) $M^2 = 0$ $MF = 4$ $F^2 = -4$ $\quad$ M has genus 2.

The second ingredient is to consider a product rational mapping $\varphi = \varphi_1 \times \varphi_2 : S \to \mathbb{P}^r \times \mathbb{P}^1$, where φ_2 is the bicanonical map φ_{2K} and φ_1 is

- Case ia): r=3 and φ_1 is the tricanonical morphism φ_{3K}
- Case ib): r =2 and φ_1 is given by the system $|3K_S - P|$, P being the (double) base point of $|2K|$
- Case ii): r=2 and φ_1 is given by the system $|3K_S - F|$
- Case iii): r=1 and φ_1 is given by the system $|3K_S - F|$.

The key idea is simple: namely, that φ_1 induces the complete canonical system on the fibres M of φ_2, and we use the fact that, in all the cases except the last, the general curve M is not hyperelliptic.

Therefore, in these cases φ yields a birational map, and indeed, on a blow up $\tilde{S}$ of S, we get a product morphism $\psi_1 \times f$, which fails to be an embedding exactly when we have a hyperelliptic fibre M of f.

In case iii), which will be treated in the sequel to this paper, we have that the blow up $\tilde{S}$ of S in 5 points is a double covering of $\mathbb{P}^1 \times \mathbb{P}^1$ branched on a curve Δ of bidegree (6,12). The fibration f given by the second projection has precisely 5 non 2-connected fibres and, of these, α are originated by a point of type (3,3) on Δ with horizontal tangency, whereas $5 - \alpha$ are originated by a horizontal fibre in $\mathbb{P}^1 \times \mathbb{P}^1$ contained in Δ, such that Δ has two ordinary quadruple points on it.

We have mentioned this last case because the case where $\alpha = 0$ was already treated in [R4], where Reid showed the existence of such a curve giving rise to a surface with $\text{Tors}(S) = \mathbb{Z}/5\mathbb{Z}$.

In the remaining cases, denoting by Y the image surface, our analysis shows that the blown up canonical model $\tilde{X}$ is precisely the normalization of Y, and the only singular curves of Y correspond to the hyperelliptic fibres of f.

The important conclusion is that the conductor ideal pulls back on $\tilde{S}$ to a divisor which is a sum of the hyperelliptic fibres, counted with a certan multiplicity.

In the cases ib) and ii) Y is a hypersurface, and, after we establish its divisor class, the main question about the existence and geometry of these classes in the moduli space is related to the problem of finding hypersurfaces with certain singularities yielding the right (sub)adjunction conditions.

Case 1a) is more interesting (in this case Y is a subvariety of codimension 2) because then the key idea comes into play.

The bicanonical curves in case 1a) (the most general case) are canonical curves of genus 4, whence the non hyperelliptic ones are proven to be complete intersection curves of type (2,3).

The situation globalizes, in the sense that Y is proven to be a divisor on a hypersurface $\mathcal{Q}$ of bidegree $(2, 7 - 2h)$ in $\mathbb{P}^3 \times \mathbb{P}^1$ (where $0 \leq h \leq 3$ is the number of hyperelliptic fibres in the bicanonical pencil, counted with multiplicity). Clearly, $\mathcal{Q}$ is swept out fibrewise by the quadrics containing the canonical curves, and the cubics also patch together fibrewise to yield a divisor Y.

To calculate the class of this divisor, we introduce a new technique, which is valid more generally for any genus 4 fibration whose fibres have ample canonical system and are 2-connected. The technique consists in defining a notion of multiplicity for the hyperelliptic fibres, and then in showing how this multiplicity determines the divisor classes of the relative quadric and of the relative cubic.

We can partly summarize here our main results as follows:

Theorem 0.1. *Assume that S is a numerical Godeaux surface with torsion $\{0\}$ and of type ia), i.e., such that the bicanonical pencil yields a genus 4 fibration f.*

Let $h = \sum_{C \ hyperelliptic} \mathrm{mult}(C)$: then a priori $0 \leq h \leq 3$.

Moreover $\exists \mathcal{Q} \in |\mathcal{O}_{\mathbb{P}^3 \times \mathbb{P}^1}(2, 7 - 2h)|$, such that $Y := \varphi(S)$ is a divisor in $|\mathcal{O}_\mathcal{Q}(3, 3h - 6)|$ whose singular curves are exactly the twisted cubic curves image of the (honestly) hyperelliptic bicanonical divisors. Moreover, if $\mathcal{C}$ is the conductor ideal, $h^0(\mathcal{CO}_Y(2, h - 3)) > 0$.

Viceversa, assume that $0 \leq h \leq 3$ and that $\mathcal{Q} \in |\mathcal{O}_{\mathbb{P}^3 \times \mathbb{P}^1}(2, 7 - 2h)|$ is an irreducible divisor, and that in turn $Y \in |\mathcal{O}_\mathcal{Q}(3, 3h - 6)|$ is an irreducible divisor whose normalization is a surface $\tilde{X}$ with rational double points as the only singularities. Suppose moreover that the conductor ideal $\mathcal{C}$ defines a divisor on $\tilde{X}$ equal to h fibres (counted with multiplicity). Assume moreover that the singular curves of Y are (irreducible) twisted cubics, and that $h^0(\mathcal{CO}_Y(2, h-3)) > 0$. Then Y is the tri-bicanonical model of a numerical Godeaux surface with torsion $\{0\}$ and of type ia).

A posteriori, the case with three distinct hyperelliptic fibres does not occur. Whereas, for the Barlow surface $h = 2$, while the Craighero Gattazzo surface has exactly two hyperelliptic fibres occurring with multiplicity 1.

The local moduli space of the Barlow and of the Craighero Gattazzo surface is smooth of the expected dimension = 8.

Acknowledgements. The first author acknowledges the hospitality of the MSRI Berkeley in the spring of 1996, where the first ideas of this approach started to develop.

A first description of the proposed approach was then exposed in seminar talks in Warwick and Oberwolfach in the summer of 1996.

Later on, the optimistic hope that there would exist such surfaces as complete intersections in $\mathbb{P}^3 \times \mathbb{P}^1$ having 3 hyperelliptic fibres turned out not to hold.

We are very grateful to F.-O. Schreyer who helped us with the Computer Algebra program Macaulay, and indeed his script allows us to show that the Barlow surface has 2 hyperelliptic fibres.

Added in proof: We were recently able to show that case iii) and case ia) with 3 hyperelliptic fibres cannot occur.

Notation

For the reader's convenience, we enclose here a list of the notation more often used, and of our abbreviations.

In this paper we denote by S the minimal model of a numerical Godeaux surface and (except in section 6) by X its canonical model.

We let moreover $\text{Tors}(S)$ be the torsion subgroup of the first homology group $H_1(S, \mathbb{Z})$ (equivalently, of $H^2(S, \mathbb{Z})$).

For a Gorenstein algebraic variety Z (e.g. S, X) we denote by K_Z a Cartier divisor associated to its dualizing sheaf ω_Z. The rational map associated to a divisor D is denoted by φ_D or $\varphi_{|D|}$; similarly the rational map associated to a line bundle $\mathcal{L}$ is denoted by $\varphi_{\mathcal{L}}$.

$\tilde{S}$ (resp. $\tilde{X}$) is a blow up of S (resp. X): except in case iii) (which is hardly treated here) it is the blow up in the base points of the movable part $|M|$ of $|2K_S|$ (resp. in the smooth base points of $|2K_X|$, we shall prove in Lemma 1.1 that $|2K_X|$ has no fixed part). The induced morphism is denoted by β (resp. $\hat{\beta}$).

We have already defined $\varphi : S \dashrightarrow \mathbb{P}^r \times \mathbb{P}^1$ in the introduction; let us denote by $\hat{\varphi} : X \dashrightarrow \mathbb{P}^r \times \mathbb{P}^1$ the induced map on the canonical model, and set $Y := \varphi(S)$, $\Sigma := \varphi_{3K_S}(S)$.

Moreover, we denote by $\pi_1 : \mathbb{P}^r \times \mathbb{P}^1 \to \mathbb{P}^r$, $\pi_2 : \mathbb{P}^r \times \mathbb{P}^1 \to \mathbb{P}^1$ the natural projections.

This allows us to define the morphisms $g := \varphi \circ \beta : \tilde{S} \to Y$, $\hat{g} := \hat{\varphi} \circ \hat{\beta} : \tilde{X} \to Y$.

Last, we denote by $f := \pi_2 \circ \varphi \circ \beta : \tilde{S} \to \mathbb{P}^1$ the fibration associated to the bicanonical pencil, and by $\hat{f} := \pi_2 \circ \hat{\varphi} \circ \hat{\beta} : \tilde{X} \to \mathbb{P}^1$ the analogous fibration on the canonical model X.

Quite often, given a Cartier divisor D on a scheme Z, by slight abuse of notation we denote also by D the associated invertible sheaf $\mathcal{O}_Z(D)$; and we often write, as shorthand notation, $H^0(D)$ instead of $H^0(\mathcal{O}_Z(D))$.

Moreover, $\mathcal{O}_{\mathbb{P}^r \times \mathbb{P}^1}(a, b)$ is also a quite understandable notation for $\pi_1^* \mathcal{O}_{\mathbb{P}^r}(a) \otimes \pi_2^* \mathcal{O}_{\mathbb{P}^1}(b)$.

1. The canonical ring

Let S be a numerical Godeaux surface, i.e. a minimal surface of general type with $K_S^2 = 1$, $p_g(S) = q(S) = 0$.

Recall that the canonical ring of S (cf. [Mu1]) is defined as the graded ring

$$R(S) := \oplus_{n \in \mathbb{N}} H^0(nK_S).$$

In our case of numerical Godeaux surfaces we have

$$h^0(K_S) = 0; \; \forall n \geq 2 \; h^0(nK_S) = \binom{n}{2} + 1.$$

Let us look for a minimal system of generators of this ring (as a $\mathbb{C}$-algebra).

As usual we denote by 1 the identity of $R(S)$, given by the constant function equal to 1, moreover we fix a basis $\{x_0, x_1\}$ of $H^0(2K_S)$, and a basis $\{y_0, y_1, y_2, y_3\}$ of $H^0(3K_S)$.

We remark that $x_0^2, x_0 x_1, x_1^2$ are independent in $H^0(4K_S)$, since $R(S)$ is an integral domain; whence, we can complete these elements to a basis $\{x_0^2, x_0 x_1, x_1^2, v_0, v_1, v_2, v_3\}$ of $H^0(4K_S)$.

Let $X := \text{Proj}(R(S))$ be the canonical model of S, and let $\pi : S \to X$ be the natural map; X has an invertible dualizing sheaf and, as customary, we denote by K_X an associated Cartier divisor. Since $\pi^*(K_X) = K_S$, one has a natural isomorphism between the canonical rings $R(S)$ and $R(X)$.

Lemma 1.1. *The fixed part F of the bicanonical pencil $|2K_S|$ is supported on the fundamental cycles of S (normal crossing configurations of smooth rational curves with self-intersection -2).*

In particular $|2K_X|$ has no fixed part.

Proof. We can write $|2K_S| = |M| + F$ where M is a linear pencil without fixed components; since K_S is nef and the only curves with $K_S C = 0$ are the finitely many smooth rational (-2) curves, building the so called fundamental cycles (cf. [Bo], [BPV]), we know that $K_S M > 0$, $K_S F \geq 0$.

Since $K_S M + K_S F = 2K_S^2 = 2$ we get $0 < K_S M \leq 2$, and clearly our purpose is to show that $K_S F = 0$, equivalently $K_S M = 2$.

Assume by contradiction that $K_S M = 1$. M being a pencil without fixed part, we have $M^2 \geq 0$, but $M^2 + K_S M = 0 \pmod 2$. It follows then that $M^2 = 1$, whence equality holds in the inequality given by algebraic index theorem.

Our conclusion is thus that M is numerically equivalent to K_S, and since $h^1(\mathcal{O}_S) = 0$ but $h^0(K_S) = 0 \neq h^0(M) = 2$, $M - K_S = \mu$ yields a non zero torsion element μ in Pic S.

An easy calculation $\chi(M) = \chi(K_S) = 1$, $h^0(M) = 2 \Rightarrow 1 \leq h^1(M) = h^1(K_S + \mu) = h^1(-\mu)$ shows that the covering of S induced by μ, yields an irregular covering of S. This is a contradiction, because the equality $K_S^2 = \chi(S)$ holds for S, hence for all its unramified coverings, whereas for minimal irregular surfaces Y we have the inequality $K_Y^2 \geq 2\chi(Y)$ (cf. [Bo]). $\square$

Remark 1.2. We have seen that $KF = 0$ $KM = 2$. So, we have three possibilities for F and M, namely
  i) $M^2 = 4$ $F = 0$
  ii) $M^2 = 2$ $MF = 2$ $F^2 = -2$
  iii) $M^2 = 0$ $MF = 4$ $F^2 = -4$.

In the second case F is precisely a fundamental cycle, i.e., on the canonical model X, we get in the base point scheme a reduced singular point.

Lemma 1.3. $H^0(2K) \otimes H^0(3K) \to H^0(5K)$ *is injective.*

Proof. Otherwise we would have a relation $x_0 y = x_1 y'$ for suitable elements y, y' in $H^0(3K)$. By Lemma 1.1, on X $\text{Min}(\text{div}(x_0), \text{div}(x_1)) = 0$; whence, $\text{div}(x_0) < \text{div}(y')$ and therefore the rational section y'/x_0 of $3K_X - 2K_X = K_X$ is a regular section, contradicting $p_g(X) = 0$. $\square$

Corollary 1.4. *We can fix a basis of $H^0(5K)$ of the form*

$$\{x_i y_j, w_1, w_2, w_3\}.$$

Let us now consider the polynomial ring $A := \mathbb{C}[y_0, y_1, y_2, y_3]$, and let us look for a set of generators of $R(S)$ as A-module, ($R(S)$ is an A-algebra via the natural homomorphism $A \to R(S)$).

Define

$$R^{(0)} = \oplus_{n \geq 0} H^0(3n K_S),$$

$$R^{(1)} = \oplus_{n \geq 0} H^0((3n + 1) K_S),$$

$$R^{(2)} = \oplus_{n \geq 0} H^0((3n + 2) K_S),$$

Of course, there is a splitting (as A-modules) $R(S) = R^{(0)} + R^{(1)} + R^{(2)}$.

Theorem 1.5. *There are three resolutions*

$$0 \to A(-3)^7 \xrightarrow{\alpha} A \oplus A(-2)^6 \to R^{(0)} \to 0$$

$$0 \to A(-4) \oplus A(-2)^6 \xrightarrow{\alpha^t} A(-1)^7 \to R^{(1)} \to 0$$

$$0 \to A(-3)^2 \oplus A(-2)^3 \xrightarrow{\beta} A^2 \oplus A(-1)^3 \to R^{(2)} \to 0$$

where $\beta = \beta^t$.

Proof. It is an easy exercise following the same argument of [Cat3]. $\square$

Corollary 1.6. $R(S) = R(X)$ *is generated in degree ≤ 6 as an A-module.*

Remark 1.7. In [Ci] is shown the weaker result that $R(S) = R(X)$ is generated in degree ≤ 6 as a ring.

From now on, let us assume $\mathrm{Tors}(S) = 0$ or $\mathbb{Z}/2\mathbb{Z}$. In this last case, denote by μ the nonzero torsion element in $\mathrm{Pic}(S)$.

Under this assumption, we can prove

Proposition 1.8. $R(S) = R(X)$ *is generated as a ring in degree ≤ 5.*

Proof. By Corollary 1.6, we must only prove that every section of $H^0(6K_X)$ can be written as sum of products of sections of degree ≤ 5.

Take an effective divisor $C \in |2K_X|$, and let $C = \mathrm{div}(c)$; then

$$H^0(6K_X) \supset W_C := cH^0(4K_X) + S^2(H^0(3K_X)).$$

We will prove that there exists some C such that this inclusion is an equality. Since $H^1(4K) = 0$ we get the following exact sequence

$$0 \to H^0(\mathcal{O}_X(4K)) \xrightarrow{c \cdot} H^0(\mathcal{O}_X(6K)) \xrightarrow{\pi} H^0(\omega_C^2) \to 0.$$

By definition W_C contains $\mathrm{Ker}\,\pi$; whence, it suffices to show that there exists C such that $\pi : S^2(H^0(3K_X)) \to H^0(\omega_C^2)$ is surjective; this is equivalent to the surjectivity of $\pi|_C : S^2(H^0(\omega_C)) \to H^0(\omega_C^2)$, that is verified for every C irreducible and non hyperelliptic by Noether's theorem.

It is clear that the general C is irreducible (since we have no fixed part, and $|M|$ is a linear pencil with $h^0(\mathcal{O}_S(M)) = 2$). That the general C is non-hyperelliptic follows from the forthcoming Lemma 1.10. $\square$

Lemma 1.9. *Let C be a 3-connected genus 4 Gorenstein curve, let ω be the dualizing sheaf of C. Assume that φ_ω embeds C. Then $\varphi_\omega(C)$ is a complete intersection of type $(2, 3)$.*

Proof. C has genus 4, so $h^0(C, \omega) = 4$, $h^0(C, \omega^2) = 9$, $h^0(C, \omega^3) = 15$. So the natural map $S^2(H^0(C, \omega)) \to H^0(C, \omega^2)$ has a non-trivial kernel.

Assume, by contradiction, that this kernel has dimension greater than 1.

Thus, there exist two distinct quadrics Q_1, Q_2 containing the degree 6 curve $\varphi_\omega(C)$. If Q_1, Q_2 have no common components, their intersection is a curve of degree 4, a contradiction.

Therefore there do exist linear forms L_0, L_1, L_2 such that $Q_1 = L_0 L_1$, $Q_2 = L_0 L_2$, and $\varphi_\omega(C) \subset L_0 L_1 \cap L_0 L_2 = L_0 \cup (L_1 \cap L_2)$.

But $\varphi_\omega(C)$ is non degenerate, so we can write $C = C_1 + C_2$, with $\varphi_\omega(C_1) \subset L_0$ of degree 5, $\varphi_\omega(C_2) = L_1 \cap L_2$ a line.

Now, recalling that C is assumed to be 3-connected, we can compute (note that $C_1 C_2$ is defined as: $\deg_{C_1}(\omega_C) - \deg_{C_1}(\omega_{C_1})$, cf. [CFHR])

$$4 = g(C) = g(C_1) + g(C_2) - 1 + C_1 C_2 \geq 6 + 1 - 1 + 3 = 9,$$

hence we derive a contradiction.

Therefore there is only one quadric containing $\varphi_\omega(C)$, let us denote it by Q.

Now, by a dimension count, the map $S^3(H^0(C, \omega)) \to H^0(C, \omega^3)$ has a kernel of dimension at least 5. In particular we get at least one cubic surface G containing $\varphi_\omega(C)$ and not having Q as a component.

If G and Q have no common components, their intersection is a degree 6 curve containing $\varphi_\omega(C)$, so $\varphi_\omega(C) = Q \cap G$ and we are done.

Otherwise, there must exist linear forms L_0, L, and a quadratic form Q', such that $Q = L_0 L$, $G = L_0 Q'$, and $\varphi_\omega(C) \subset L_0 \cup (L \cap Q')$. Again, we can decompose C as $C_1 + C_2$, with $\varphi_\omega(C_1) \subset L_0$, $\varphi_\omega(C_2) \subset L \cap Q'$. If $\varphi_\omega(C_2) \neq L \cap Q'$, we have decomposed $\varphi_\omega(C)$ as the union of a plane quintic and of a line, and we have already excluded this case. Else, $\varphi_\omega(C_1)$ is a plane quartic, $\varphi_\omega(C_2)$ a plane conic, and again we get

$$4 = g(C) = g(C_1) + g(C_2) - 1 + C_1 C_2 \geq 3 + 1 - 1 + 3 = 6,$$

a contradiction. $\square$

Let us recall (cf. [Cat2]) that a honestly hyperelliptic curve is a finite covering of degree 2 of $\mathbb{P}^1$.

Lemma 1.10. *Let $C \in |2K_X|$; one of the following holds:*

a) C is embedded by ω_C, $\varphi_{|3K|}(C) = \varphi_{\omega_C}(C)$ is the complete intersection of a quadric and a cubic; moreover, if $\varphi_{\omega_C}(C)$ is reducible, it decomposes as the union of two plane cubics intersecting (with multiplicity) in three points;

b) C is honestly hyperelliptic, $\varphi_{|3K|}(C) = \varphi_{\omega_C}(C)$ is a double twisted cubic curve;

c) $C = 2D$; in this case $\mathrm{Tors}(S) = \mathbb{Z}/2\mathbb{Z}$, $D \in |K + \mu|$, and $\varphi_{|3K|}(C) = \varphi_{\omega_C}(C)$ is a sextuple line.

Case a) is the general one.

Proof. Let us consider first the case where C is not 3-connected. Then, by [CFHR], Lemma 4.2 and its proof, $\pi^* C$ is not 3-connected, and we have a decomposition $\pi^*(C) = D_1 + D_2$ with $D_1 D_2 \leq 2$ and with $K_S D_i = 1$.

We get then $D_1^2 + D_2^2 = (2K_S)^2 - 2D_1D_2 \geq 0$, so we can assume D_1^2 non negative, whence positive because it must be odd; by the algebraic index theorem $D_1^2 = 1$, and $D_1 = K_S + \varepsilon$, $D_2 = K_S - \varepsilon$, $\varepsilon \in \text{Tors}(S)$.

Since $H^0(K_S) = 0$, by our hypothesis on the torsion group follows that $\text{Tors}(S) = \mathbb{Z}/2\mathbb{Z}$, and that $\varepsilon = \mu = -\mu$.

Remark that $h^0(K + \mu) = 1$, whence $D_1 = D_2 \in |K_S + \mu|$.

Since $h^0(3K - (K + \mu)) = h^0(2K - \mu) = h^0(2K + \mu) = 2$, then $\varphi_{|3K|}(D_1)$ is a curve of degree $(K_S + \mu)3K_S = 3$ contained in a line, thus it is a triple line. This gives case c).

If C is 3-connected, by [CFHR], Theorem 3.6, either $\omega_C = 3K_{|C}$ is very ample or C is honestly hyperelliptic. Note that if C is honestly hyperelliptic and reducible, then C consists of two smooth rational curves intersecting (with multiplicity) in 5 points. In this case $\varphi_{\omega_C}(C)$ is an irreducible non degenerate curve of degree 3 in $\mathbb{P}^3$, so its schematic image is a double structure on a twisted cubic curve.

Assume now that C is canonically embedded: by Lemma 1.9 $\varphi_{\omega_C}(C)$ is a complete intersection of type $(2, 3)$.

Finally, if C is reducible, $C = C_1 + C_2$, where C_1, C_2 are irreducible and $C_1 \neq C_2$ by the hypothesis of 3-connectedness (else, $C_1 C_2 = 1$). Since $K_S \pi^*(C_i) = 1$, $\varphi_{\omega_{\pi^*(C)}}(C_i)$ is a plane curve of degree $3K_S \pi^*(C_i) = 3$, and we get thus two distinct irreducible plane cubics intersecting in three points.

Remark that case a) is the general one because (under our assumption about the torsion of S), $\varphi_{|3K_S|}$ is a birational morphism, as proved in [Cat1]. $\square$

2. The tri-bicanonical map

As we recalled in the proof of Lemma 1.10, by [Cat1], for a numerical Godeaux surface with torsion 0 or $\mathbb{Z}/2\mathbb{Z}$ the tricanonical system defines a birational morphism onto a surface $\Sigma \subset \mathbb{P}^3$ of degree 9.

We consider the rational map $\Phi : S \to \mathbb{P}^3 \times \mathbb{P}^1$ whose components are the tricanonical and the bicanonical maps. This is not a morphism in the base points of the movable part of the bicanonical system; its image Y is a birational model of our S that dominates Σ.

We know that the general bicanonical curve is a complete intersection of a quadric and a cubic; then, as mentioned in the introduction, our aim would be to construct two hypersurfaces in $\mathbb{P}^3 \times \mathbb{P}^1$ of bidegree respectively $(2, m)$ and $(3, m')$ such that their complete intersection is Y.

Let $\beta : \tilde{S} \to S$ be a minimal sequence of ordinary blow ups such that $\varphi \circ \beta$ is a morphism. Denote by π_1, π_2 the two respective projections of Y on $\mathbb{P}^3$ and $\mathbb{P}^1$, set $f := \pi_2 \circ \varphi \circ \beta : \tilde{S} \to \mathbb{P}^1$.

Lemma 2.1. *Let B be a smooth curve and $f : \tilde{S} \to B$ be a genus 4 fibration whose generic fibre is non hyperelliptic. Let F be a fibre of f, set $\omega = F + K_{\tilde{S}}$.*

Suppose moreover that the nonhyperelliptic fibres are 3-connected (equivalently, such that their canonical image is a complete intersection of type (2, 3)).

Consider the homomorphisms of sheaves

$$S^n(f_*(\omega)) \xrightarrow{\sigma_n} f_*(\omega^n),$$

and set $\mathcal{L}_n = \ker \sigma_n$ and $\mathcal{T}_n = \operatorname{coker} \sigma_n$. Then

i) $\mathcal{T}_n$ is a torsion sheaf supported on the image of the hyperelliptic fibres.

ii) Let $p \in \mathbb{P}^1$ be the image of a honestly hyperelliptic fibre (whose canonical image is a twisted cubic); then there is a positive integer s such that

$$\forall k \geq 2 \quad \operatorname{length}(\mathcal{T}_k, p) = s(3k - 4).$$

Proof. Let $\mathcal{M}_p$ be the maximal ideal sheaf of the point p in $\mathcal{O}_B$.

By Grauert's base change theorem (cf. e.g. [BPV]) $\forall p \in B$, $\frac{(f_*\omega^k)}{(\mathcal{M}_p f_*\omega^k)} \cong H^0(F_p, \omega^k)$; mod $\mathcal{M}_p$ the morphism σ_k acts on the stalks as $S^k(H^0(F_p, \omega)) \to H^0(F_p, \omega^k)$, whence it is surjective when F_p is non hyperelliptic.

By Nakayama's lemma, if p corresponds to a non hyperelliptic fibre, then $(\sigma_n)_p$ is surjective, and the first part of the teorem is proved.

Let us now assume that F_p is a honestly hyperelliptic fibre, so that F_p is a double cover of $\mathbb{P}^1$ branched, by Hurwitz formula, on a divisor of degree 10. We can embed F_p in $\mathbb{P}(5, 1, 1)$ as the hypersurface defined by the equation $w^2 = P(t_0, t_1)$, where P is the homogeneus polynomial of degree 10 whose divisor is the branch divisor of the canonical map.

Following the same line of [ML] it is easy to prove that the canonical ring of F_p is generated in degree 2; this ring can be described as a subring of the ring $\mathcal{A} = \mathbb{C}[t_0, t_1, w]/\langle w^2 = P(t_0, t_1)\rangle$.

In fact, generators for $H^0(\omega)$ are $y_0 = t_0^3$, $y_1 = t_0^2 t_1$, $y_2 = t_0 t_1^2$, $y_3 = t_1^3$; the kernel of the map $S^2(H^0(F_p, \omega)) \to H^0(F_p, \omega^2)$ has dimension 3 (three independent quadrics through a twisted cubic), so the cokernel has dimension $9 - 10 + 3 = 2$, and we can see that it is generated by $v_0 = t_0 w$, $v_1 = t_1 w$.

It follows that, if we choose 3 degree 4 polynomials $P_{00}(y_i)$, $P_{01}(y_i)$, $P_{11}(y_i)$, such that in the ring $\mathcal{A}$ is $P_{00} = t_0^2 P$, $P_{01} = t_0 t_1 P$, $P_{11} = t_1^2 P$, we get the following 9 relations:

$$r_1 := y_1^2 - y_0 y_2 \qquad r_2 := y_2^2 - y_1 y_3 \qquad r_3 := y_0 y_3 - y_1 y_2$$
$$r_4 := v_0 y_1 - v_1 y_0 \qquad r_5 := v_0 y_2 - v_1 y_1 \qquad r_6 := v_0 y_3 - v_1 y_2$$
$$r_7 := v_0^2 - P_{00} \qquad r_8 := v_0 v_1 - P_{01} \qquad r_9 := v_1^2 - P_{11}.$$

So we can describe the canonical ring R of F_p as a quotient of the graded ring $\mathbb{C}[y_0, y_1, y_2, y_3, v_0, v_1]/\langle r_1, \ldots, r_9\rangle$, where $\deg y_i = 1$, $\deg v_i = 2$. We have $H^0(F_p, \omega) = 4$, and $\forall k \geq 2$ $H^0(F_p, \omega^k) = 6k - 3$ but on the other hand an easy calculation yields that the homogeneus part of degree k of our ring has at most the same dimension. Therefore follows that $R = \mathbb{C}[y_0, y_1, y_2, y_3, v_0, v_1]/\langle r_1, \ldots, r_9\rangle$.

Let us denote by R_k the homogeneous part of R of degree k. Remark that

$$(f_*\omega^k)_p \otimes_{\mathcal{O}_p} \mathbb{C} = R_k,$$

so, by flatness, $\oplus_k(f_*\omega^k)_p = \mathcal{O}_p[y_0, y_1, y_2, y_3, v_0, v_1]/\langle \bar{r}_1, \ldots, \bar{r}_9 \rangle$, where the $\bar{r}_i$'s are lifts to $\mathcal{O}_p$ of the r_i's.

Moreover, every syzygy of R lifts to a syzygy of the $\mathcal{O}_p$-module $\oplus_k(f_*\omega^k)_p$.

Let t be a local parameter for $\mathcal{O}_p$, and write

$$\begin{pmatrix} \bar{r}_1 \\ \bar{r}_2 \\ \bar{r}_3 \end{pmatrix} = \begin{pmatrix} \tilde{q}_1(y_j, t) \\ \tilde{q}_2(y_j, t) \\ \tilde{q}_3(y_j, t) \end{pmatrix} + \begin{pmatrix} \tilde{\alpha}_{11}(t) & \tilde{\alpha}_{12}(t) \\ \tilde{\alpha}_{21}(t) & \tilde{\alpha}_{22}(t) \\ \tilde{\alpha}_{31}(t) & \tilde{\alpha}_{32}(t) \end{pmatrix} \begin{pmatrix} v_0 \\ v_1 \end{pmatrix},$$

where $\tilde{q}_i(y_j, 0) = r_i$, and $\tilde{\alpha}_{i,j}(0) = 0$. Let s be the minimum of the orders of vanishing of the $\tilde{\alpha}_{ij}(t)$'s; we can then find a new basis for the respective vectors spaces generated by v_0 and v_1, and by $\bar{r}_1, \bar{r}_2, \bar{r}_3$, and a new local parameter t, so that we can write our relations in the following simpler form:

$$\begin{pmatrix} \bar{r}_1 \\ \bar{r}_2 \\ \bar{r}_3 \end{pmatrix} = \begin{pmatrix} q_1(y_j, t) \\ q_2(y_j, t) \\ q_3(y_j, t) \end{pmatrix} + \begin{pmatrix} t^s & t^{s+1}\alpha_{12}(t) \\ t^{s+1}\alpha_{21}(t) & t^s \alpha_{22}(t) \\ t^{s+1}\alpha_{31}(t) & t^s \alpha_{32}(t) \end{pmatrix} \begin{pmatrix} v_0 \\ v_1 \end{pmatrix}.$$

Clearly then the linear space of conics generated by the $q_i(y_j, 0)$'s coincides with the space generated by the r_i's ($i = 1, 2, 3$).

Lifting the syzygy $y_3 r_1 + y_1 r_2 + y_2 r_3$, by degree reasons we get a syzygy of the form $L_3(y_j, t)\bar{r}_1 + L_1(y_j, t)\bar{r}_2 + L_2(y_j, t)\bar{r}_3 + f_4(t)\bar{r}_4 + f_5(t)\bar{r}_5 + f_6(t)\bar{r}_6$, where the $L_i(y_j, 0)$'s are three independent linear forms, and $\bar{r}_4, \bar{r}_5, \bar{r}_6$ are lifts of r_4, r_5, r_6.

Working modulo the ideal generated by t^{s+1} and by the monomials of degree 3 in the (y_j)'s we get

$$t^s(L_3(y_j, 0)v_0 + (\alpha_{22}(0)L_1(y_j, 0) + \alpha_{23}(0)L_2(y_j, 0))v_1) \in (\bar{r}_4, \bar{r}_5, \bar{r}_6)$$

But in fact, there are no constant coefficients syzygies among r_4, r_5, r_6, thus we conclude that

$$L_3(y_j, 0)v_0 + (\alpha_{22}(0)L_1(y_j, 0) + \alpha_{23}(0)L_2(y_j, 0))v_1 \in (r_4, r_5, r_6)$$

which excludes the possibility that $\alpha_{22}(0) = \alpha_{23}(0) = 0$.

Therefore, choosing new bases for the respective $\mathcal{O}_p$-modules generated by v_0 and v_1, and by $\bar{r}_1, \bar{r}_2, \bar{r}_3$, we can write our relations in the following even simpler form:

$$\begin{pmatrix} \bar{r}_1 \\ \bar{r}_2 \\ \bar{r}_3 \end{pmatrix} = \begin{pmatrix} q_1(y_j, t) \\ q_2(y_j, t) \\ q_3(y_j, t) \end{pmatrix} + \begin{pmatrix} t^s & 0 \\ 0 & t^s \\ 0 & 0 \end{pmatrix} \begin{pmatrix} v_0 \\ v_1 \end{pmatrix}.$$

This allows us to compute, using the lifts of r_7, r_8, r_9 to eliminate the multiples of $v_0^2, v_0 v_1, v_1^2$, and the lifts of r_4, r_5, r_6 to eliminate the multiples of v_0 as much as possible, that there exists a nonzero linear form $L_0(y_j)$ such that the set $\{t^i v_0 L_0^{k-2}(y_j), t^i v_1 q_l(y_j) \mid i < s\}$ is a basis for $(\mathcal{T}_k)_p$, when $\{q_l\}$ is a basis for the homogeneus degree $k-2$ part of $\mathbb{C}[y_0, y_1, y_2, y_3]/\langle r_1, r_2, r_3\rangle$. But this is the projective coordinate ring of a twisted cubic, whose homogeneus part of degree d has dimension $3d+1$; whence the dimension of $(\mathcal{T}_k)_p$ equals $s(1+3(k-2)+1) = s(3k-4)$. □

The integer s arising in Lemma 2.1 can in fact be defined as follows:

Definition 2.2. Let C be a honestly hyperelliptic curve of genus g, occurring as a fibre of a genus g fibration $f : \tilde{S} \to B$ where $\tilde{S}$ is smooth. Define the *multiplicity* of C (or mult(C)), as the multiplicity of C in the conductor ideal of f (recall that the conductor is a divisorial ideal).

Proposition 2.3. *The integer s associated to a honestly hyperelliptic fibre C as in Lemma 2.1 equals the multiplicity.*

Proof. Let $p \in B$ be a point such that C is the fibre of p and U a sufficiently small affine open neighbourhood of p. Let $Y \subset \mathbb{P}^3 \times U$ be the image of the relative canonical map φ of f.

By abuse of notation let us still denote by $\tilde{S} = f^{-1}(U)$. The sheaf of double points Δ, supported on the image Γ of C is defined via the exact sequence

$$0 \to \mathcal{O}_Y \to \varphi_* \mathcal{O}_{\tilde{S}} \to \Delta \to 0.$$

Twisting the exact sequence by $\mathcal{O}_{\mathbb{P}^3 \times \mathbb{P}^1}(n, 0)$, and observing that $\varphi^* \mathcal{O}_{\mathbb{P}^3 \times \mathbb{P}^1}(1, 0) \cong \omega$, from the definition of $\mathcal{T}_n$ we get that

$$(\mathcal{T}_n)_p \cong H^0(\Gamma, \Delta(n)).$$

From Lemma 2.1 we conclude that the length of Δ at the generic point of Γ equals s. Since (as we shall also see in Lemma 2.6) at the general point of Γ we have a singularity consisting of two smooth branches, we conclude immediately that s equals the multiplicity of C in the conductor divisor. □

The geometric meaning of the definition and proposition above is that s should be interpreted as the intersection multiplicity of the curve B with the hyperelliptic locus inside the moduli space of the curves of genus 4.

Remark 2.4. The fibration f we had already defined is a genus 4 fibration if and only if (see Remark 1.2) $F = 0$ and the map β is a sequence of blow-ups in smooth points, possibly infinitely near, of the generic bicanonical divisor; i.e., if and only if $F = 0$ and the base locus of $|2K_S|$ is not consisting of a single point where every bicanonical divisor has multiplicity 2.

The above condition is of course an open condition; in fact we shall prove later that the Craighero Gattazzo surface enjoys such a property.

Theorem 2.5. *Assume that S is a numerical Godeaux surface with torsion $\{0\}$, such that f is a genus 4 fibration. Let $h = \sum_{C \text{ hyperelliptic}} \text{mult}(C)$.*

Then $\exists Q \in |\mathcal{O}_{\mathbb{P}^3 \times \mathbb{P}^1}(2, 7 - 2h)|$, such that $Y := \varphi(S)$ is a divisor in
$$|\mathcal{O}_Q(3, 3h - 6)|.$$

Proof. By Remark 2.4, the map β we had already defined at the beginning of this section, is a sequence of blow-ups of smooth points of the generic bicanonical divisor.

Let E_i, $i = 1, \ldots, 4$ be the corresponding exceptional divisors of the first kind, and set $E = \sum E_i$. Then $K_{\tilde{S}} = \beta^*(K_S) + E$, and if F is a generic fibre of f (F is the strict transform of a generic bicanonical divisor on S by β), $F = \beta^*(2K_S) - E = 2K_{\tilde{S}} - 3E$ is a genus 4 curve.

The pull-back of the tricanonical system is given by $\omega := \beta^*(3K_S) = 3K_{\tilde{S}} - 3E = K_{\tilde{S}} + F$. In view of Lemma 1.10 the hypotheses of Lemma 2.1 are satisfied.

Consider now the exact sequences

$$0 \to \mathcal{L}_2 \to S^2(f_*\omega) \xrightarrow{\sigma_2} f_*\omega^2 \to \mathcal{T}_2 \to 0; \tag{1}$$

$$0 \to \mathcal{L}_3 \to S^3(f_*\omega) \xrightarrow{\sigma_3} f_*\omega^3 \to \mathcal{T}_3 \to 0. \tag{2}$$

$\forall F$, the map $H^0(\tilde{S}, \omega) \to H^0(F, \omega)$ is an isomorphism, therefore $f_*\omega \cong \mathcal{O}_{\mathbb{P}^1}^4$.

In particular $\mathcal{L}_2$ is a subsheaf of $S^2(f_*\omega) \cong \mathcal{O}^{10}$, while $\mathcal{L}_3$ is a subsheaf of $S^3(f_*\omega) \cong \mathcal{O}^{20}$.

Moreover, since $\mathcal{T}_k$ is a torsion sheaf, the rank of $\mathcal{L}_k$ equals the difference $\dim S^k(H^0(F, \omega)) - h^0(F, \omega^k)$. Therefore rank $\mathcal{L}_2 = 1$, rank $\mathcal{L}_3 = 5$ and we can write $\mathcal{L}_2 \cong \mathcal{O}_{\mathbb{P}^1}(-m)$.

Since $\mathcal{L}_2$ is a subsheaf of $\mathcal{O}^{10}$, $m \geq 0$, and the injection $\mathcal{L}_2 \to \mathcal{O}^{10}$ defines a hypersurface $Q \in |\mathcal{O}_{\mathbb{P}^3 \times \mathbb{P}^1}(2, m)|$ that contains Y (in particular, then, $m \geq 1$).

Let us now compute the Euler characteristic of the exact sequence (1).

We get

$$\chi(\mathcal{L}_2) = 1 - m;$$

$$\chi(S^2(f_*\omega)) = \chi(\mathcal{O}^{10}) = 10;$$

$$R^1 f_*\omega^2 = 0 \Rightarrow \chi(f_*\omega^2) = \chi(\omega^2) = 16;$$

$$\chi(\mathcal{T}_2) = \text{length}(\mathcal{T}_2) = 2h;$$

so $1 - m + 16 = 10 + 2h$, i.e. $m = 7 - 2h$.

The splitting surjective homomorphism $S^2(f_*\omega) \otimes f_*\omega \to S^3(f_*\omega)$, induces a homomorphism $\mathcal{L}_2^4 \to \mathcal{L}_3$; it is easy to see that this is injective and that its cokernel is a subsheaf of the locally free sheaf $(f_*\omega^2)^4$.

So, $\mathcal{L}_3$ being a locally free sheaf of rank $20 - 15 = 5$, we can write the following exact sequence

$$0 \to \mathcal{O}(2h - 7)^4 \to \mathcal{L}_3 \to \mathcal{O}(-m') \to 0. \tag{3}$$

But then

$$\chi(\mathcal{L}_3) = 8h - m' - 23;$$

$$\chi(S^3(f_*\omega)) = \chi(\mathcal{O}^{20}) = 20;$$

$$R^1 f_*\omega^3 = 0 \Rightarrow \chi(f_*\omega^3) = \chi(\omega^3) = 37;$$

$$\chi(\mathcal{J}_3) = \text{length}(\mathcal{J}_3) = 5h;$$

so by the exact sequence (2) we get $8h - m' - 23 + 37 = 20 + 5h$, i.e. $m' = 3h - 6$.

Remark now that the injection $\mathcal{O}_{\mathbb{P}^1}(6 - 3h) = \mathcal{L}_3/\mathcal{L}_2^4 \to S^3(f_*\omega)/\mathcal{L}_2 \cdot H^0(f_*\omega)$ defines a divisor $\mathcal{D} \in |\mathcal{O}_{\mathcal{Q}}(3, 3h-6)|$ containing Y. $\mathcal{D}$ and Y have the same dimension; $\forall H_2 \in |\mathcal{O}_{\mathbb{P}^3 \times \mathbb{P}^1}(0, 1)|$, $H_2 \cap Y$ has degree 6. But if $H_2 \cap \mathcal{D}$ is a curve, it has the same degree and contains $H_2 \cap Y$, so they coincide. So, if $\mathcal{D} \neq Y$, $\exists H_2$ such that $H_2 \cap \mathcal{D}$ is a component of $H_2 \cap \mathcal{Q}$; if $\mathcal{Q} \cap H_2 = \mathcal{D} \cap H_2$ we get some torsion element in coker $\mathcal{L}_3 \to S^3(f_*\omega)$, that is a subsheaf of the locally free sheaf $f_*\omega^3$, a contradiction. Otherwise $\mathcal{Q} \cap H_2$ is the union of two planes, and $\mathcal{D} \cap H_2$ is exactly one of the two planes; but we have already excluded this case in the proof of Lemma 1.9. □

Now, let us understand the local behaviour of Y near the image of every honestly hyperelliptic curve occuring as fibre of $f : \tilde{S} \to \mathbb{P}^1$.

Proposition 2.6. *Let C be a honestly hyperelliptic genus 4 fibre of f, $\Gamma = \varphi(C)$. Assume that the multiplicity of C equals s: then, in the neighbourhood of a general point $p \in \Gamma$ there exist local coordinates (y_1, y_2, y_3, t), such that Y is defined by the equations $y_2 = y_1(y_1 - t^s) = 0$, Γ by $y_1 = y_2 = t = 0$, and the projection π_2 is (still) given by the coordinate t.*

Proof. For the general $p \in \Gamma$ there exists a neighbourhood U of p in $\mathbb{P}^3 \times \mathbb{P}^1$ such that $\varphi^{-1}(U)$ has two smooth connected components, and φ identifies the two smooth holomorphic curves corresponding to C.

So, for a first suitable choice of local coordinates in the source and in the target we can assume that $\Gamma = \{y_1 = y_2 = t = 0\}$, the projection π_2 is given by the coordinate t, and the two branches of Y are parametrized as follows

$(u_1, t_1) \to (0, 0, u_1, t_1)$

$(u_2, t_2) \to (t_2\phi_1(u_2, t_2), t_2\phi_2(u_2, t_2), u_2, t_2).$

So, for a suitable local analytic coordinate change that fixes t, we get the simpler form

$(u_1, t_1) \to (0, 0, u_1, t_1)$

$(u_2, t_2) \to (t_2^a, 0, u_2, t_2).$

And Y is described by the equations $y_2 = y_1(y_1 - t^a) = 0$.

Finally, remarking that the conductor ideal is generated by y_1, t^a, we get $a = s$. □

Corollary 2.7. *Assume that a fibre $F = \{t = 0\}$ appears in the conductor divisor with multiplicity s. Thus, if $Q(y_i, t)$ represents a divisor in $\mathbb{P}^3 \times \mathbb{P}^1$ such that $\varphi^* \operatorname{div}(Q(y_i, t)) \geq 2sF$, then $t^s | Q(y_i, t) \pmod{\mathcal{J}_Y}$.*

Proof. By our assumption $\operatorname{div} Q(y_i, t)$ pulls back to a divisor $\geq 2sF$.

Since we are interested in $Q \bmod \mathcal{J}_Y$, this means, using the local coordinates introduced above, that we can look at Q modulo y_2, and writing $Q'(y_1, y_3, t) = Q(y_1, 0, y_3, t)$ we get as a first condition that

1) $\quad Q' \in (y_1, t^{2s}) \Leftrightarrow Q' = y_1 q' + t^{2s} g$

and it suffices therefore to prove that $t^s | q'$.

The condition imposed by the second branch is that $t^{2s} | t^s q'(t^s, v, t) \Leftrightarrow q' \in (y_1, t^s)$. Thus, $\bmod \mathcal{J}_Y$, $Q' \equiv y_1^2 a + y_1 t^s b + t^{2s} g$.

But $\mathcal{J}_Y \ni y_1(y_1 - t^s)$ and thus $Q' \equiv t^s(y_1(a+b) + t^s g)$. $\square$

Consider now the case where $F = 0$, but f is not a genus 4 fibration. In this case we can consider the blow up $\beta' : \tilde{S}' \to S$ of S in the single base point P of $|2K_S|$. If E is the exceptional divisor of β', the strict transform of the bicanonical system is given by $\beta'^* 2K_S - 2E$.

Let $g' : \tilde{S}' \to \mathbb{P}^2 \times \mathbb{P}^1$ be the morphism obtained from $|\beta'^* 3K_S - E| \times |\beta'^* 2K_S - 2E|$, let $\pi'_2 : \mathbb{P}^2 \times \mathbb{P}^1 \to \mathbb{P}^2$ be the second projection, set $f' = \pi'_2 \circ g'$.

Recall, to understand the statement of the following Lemma, that a curve C is said to be hyperelliptic if the canonical map is not birational.

Lemma 2.8. *Let $f' : \tilde{S} \to \mathbb{P}^1$ be a genus 3 fibration whose fibres are 2-connected and whose generic fibre is non hyperelliptic. Let F be the fibre of f', set $\omega = F + K_{\tilde{S}}$. Consider the homomorphisms of sheaves*

$$S^n(f'_* \omega) \xrightarrow{\sigma_n} f'_*(\omega^n),$$

and denote by $\mathcal{L}_n = \ker \sigma_n$ and $\mathcal{J}_n = \operatorname{coker} \sigma_n$. Then

i) $\mathcal{J}_n$ is a torsion sheaf supported on the image of the hyperelliptic fibres.

ii) Let $p \in \mathbb{P}^1$ be the image of some hyperelliptic fibre; then $\exists s > 0, s \in \mathbb{N}$, such that

$$\forall k \geq 2 \quad \operatorname{length}(\mathcal{J}_k, p) = s(2k-3).$$

Proof. i) This point follows since if C is not hyperelliptic the canonical image of C is a plane quartic (in fact, the hypothesis of 2-connectedness ensures that the canonical system has no base points, see [CFHR], Lemma 3.3.b).

ii) Recall that, by [ML], the canonical ring of a hyperelliptic fibre has the form

$$R = \mathbb{C}[x_1, x_2, x_3, y]/\langle r_1 := Q(x_i), r_2 := y^2 - F(x_i) \rangle,$$

where $\deg x_i = 1$, $\deg y = 2$, $\deg Q = 2$, $\deg F = 4$.

Acting as in Lemma 2.1, if we choose a suitable local parameter t in $\mathcal{O}_p$, we can write a lift of r_1 as

$$\bar{r}_1 = \overline{Q}(x_i, t) + t^s y$$

for some $s > 0$, $\overline{Q}(x_i, 0) = Q(x_i)$.

This allows us to compute, using the lift of r_2 to eliminate the multiples of y^2, that the set $\{t^i q_j y \mid i < s\}$ is a basis for $\mathcal{J}_k$ when the set $\{q_j\}$ is a basis for the homogeneus part of degree $k - 2$ of the quotient ring $\mathbb{C}[x_1, x_2, x_3]/Q$. □

Theorem 2.9. *For a numerical Godeaux surface with torsion $\{0\}$, and of type* ib) *(bicanonical system without fixed part possessing a double base point) f' is a genus 3 fibration, and g' yields fibrewise the canonical map of the fibres. Moreover, f' has exactly 7 hyperelliptic fibres (counted with multiplicity according to 2.8) and the image of g' is a divisor in $|\mathcal{O}_{\mathbb{P}^2 \times \mathbb{P}^1}(4, 8)|$.*

We need the following

Lemma 2.10. *Under the above assumption, all the fibres are 2-connected and the generic fibre F is non hyperelliptic.*

Proof. Let F be a fibre of f'. Since $EF = 2$ and E is irreducible, it follows that if we have a decomposition $F = A + B$, then $AE \geq 0$, and similarly $BE \geq 0$, whence

$$\begin{cases} AE = 0, BE = 2 \text{ or} \\ AE = BE = 1. \end{cases}$$

Assume that $AB = r$: then in the first case we get $A(B+2E) = r$, in the second we get $(A+E)(B+E) = r+1$. In both cases we obtain a decomposition $(A' + B') \in |2K_S|$ where $A'B' \leq (r + 1)$. We conclude that $r \geq 2$ because under our assumptions (see [Bo], Lemma 2, page 181) $A'B' \geq 2$ and is equal to 2 only if, say, $A'K_S = 0$, what excludes $AE = 1$ (since otherwise there would be a fixed part of the bicanonical system).

Assume by contradiction that every F is hyperelliptic. We observe that the first component of g' restricts to every fibre F to the complete canonical system of F. Therefore we obtain that g' is $2 : 1$ so it defines a (birational) involution σ on $\tilde{S}$ that is the hyperelliptic involution on every fibre. Since σ acts biregularly on the minimal model S and clearly fixes P, σ also acts biregularly on $\tilde{S}$ leaving E invariant.

In particular on every F the hyperelliptic involution induces a involution on the corresponding bicanonical divisor in S, so every bicanonical divisor on S is hyperelliptic. This is a contradiction because the rational map induced by $|3K_S|$ is birational. □

Proof of Theorem 2.9. Consider the exact sequences

$$0 \to \mathcal{L}_2 \to S^2(f'_*\omega) \xrightarrow{\sigma_2} f'_*\omega^2 \to \mathcal{T}_2 \to 0; \tag{4}$$

$$0 \to \mathcal{L}_3 \to S^3(f'_*\omega) \xrightarrow{\sigma_3} f'_*\omega^3 \to \mathcal{T}_3 \to 0; \tag{5}$$

$$0 \to \mathcal{L}_4 \to S^4(f'_*\omega) \xrightarrow{\sigma_4} f'_*\omega^4 \to \mathcal{T}_4 \to 0. \tag{6}$$

Argueing as in Theorem 2.5, we get that $\mathcal{L}_2, \mathcal{L}_3, \mathcal{L}_4$ are locally free sheaves of respective ranks 0, 0, 1, and that the $\mathcal{T}_j$ are torsion sheaves supported on the points corresponding to hyperelliptic fibres. By Lemma 2.8 for every such point $p \in \mathbb{P}^1$ there is a multiplicity s_p such that

$$\text{length}(\mathcal{T}_2, p) = s_p,$$
$$\text{length}(\mathcal{T}_3, p) = 3s_p,$$
$$\text{length}(\mathcal{T}_4, p) = 5s_p.$$

Computing the Euler characteristics in the sequences (4), (5) and (6), we get $s = \sum s_p = 7$ and $\mathcal{L}_4 = \mathcal{O}_{\mathbb{P}^1}(-8)$.

So we can conclude that $g(\tilde{S}) \in |\mathcal{O}_{\mathbb{P}^2 \times \mathbb{P}^1}(4, 8)|$. □

Remark that, in the case of torsion $\mathbb{Z}/2\mathbb{Z}$, everything works almost identically, except that we have to calculate the hyperelliptic multiplicity corresponding to the non 1-connected (double) fibre: this will be done in the sequel to the present paper.

3. The genus 4 fibration cannot have three distinct hyperelliptic fibres

The goal of this section is to prove the following

Proposition 3.1. *Let S a numerical Godeaux surface with torsion $\{0\}$ such that $|2K_S|$ has 4 base points possibly infinitely near (equivalently, such that f is a genus 4 fibration). Then the bicanonical pencil cannot contain three distinct honestly hyperelliptic fibres.*

We shall argue by contradiction and assume by Theorem 2.5 that $h = 3$ and that Y is a divisor in $|\mathcal{O}_Q(3, 3)|$, with $Q \in |\mathcal{O}_{\mathbb{P}^3 \times \mathbb{P}^1}(2, 1)|$.

Remark 3.2. *Y is the complete intersection of Q with a hypersurface $\mathcal{G}$, where $\mathcal{G} \in |\mathcal{O}_{\mathbb{P}^3 \times \mathbb{P}^1}(3, 3)|$.*

In fact, exact sequence 3 splits because $\text{Ext}^1(\mathcal{O}(-3), \mathcal{O}(-1)^4) = H^1(\mathcal{O}(2)^4) = 0$; so exact sequence 2 induces a divisor $\mathcal{G}$ in $|\mathcal{O}_{\mathbb{P}^3 \times \mathbb{P}^1}(3,3)|$ that cuts out Y on $\mathcal{Q}$.

Lemma 3.3. *If the torsion group is $\{0\}$ and $h = 3$, then there exists a quadric Q'' in $\mathbb{P}^3$ containing each twisted cubic, image of a honestly hyperelliptic fibre. Moreover, the pull-back of Q'' to $\tilde{S}$ yields an effective divisor which is greater than the adjoint (conductor) divisor.*

Proof. Let X be the canonical model of S, and let $\hat{\beta} : \hat{X} \to X$ be the blow up in the base points of $|2K_X|$ (they are smooth points of X by our hypothesis).

Let ψ be given by the relative canonical map of $\hat{f} : X \to \mathbb{P}^1$.

We have the following diagram

$$\begin{array}{ccccc}
\tilde{X} & & \Sigma & \subset & \mathbb{P}^3 \\
{\scriptstyle \hat{\beta}}\downarrow & {\scriptstyle \hat{g}}\swarrow & \uparrow{\scriptstyle \pi_1} & & \\
X & \overset{\hat{\varphi}}{\dashrightarrow} & Y & \subset & \mathbb{P}^3 \times \mathbb{P}^1 \\
& & \downarrow{\scriptstyle \pi_2} & & \\
& & \mathbb{P}^1 & &
\end{array}$$

and recall that $\hat{g} = \hat{\varphi} \circ \hat{\beta}$ is a birational morphism factoring through a possible contraction $\epsilon : \tilde{X} \to \hat{X}$ of strings of (-2) curves (to rational double point singularities), and a finite birational map $\tilde{g} : \hat{X} \to Y$.

By [H], ex. III.6.10 and III.7.2, $\hat{g}_*(K_{\tilde{X}}) = \mathcal{H}om_{\mathcal{O}_Y}(\hat{g}_*\mathcal{O}_{\tilde{X}}, K_Y)$. Moreover $\omega_Y = \mathcal{O}_Y(2 + 3 - 4, 1 + 3 - 2) = \mathcal{O}_Y(1, 2)$, whence $\hat{g}_*(K_{\tilde{X}}) = \mathcal{C} \otimes \mathcal{O}_Y(1, 2)$, $\mathcal{C}$ being the conductor ideal of $\tilde{g}$.

The pull-back to $\tilde{X}$ of the conductor ideal $\mathcal{C}$ is an invertible sheaf $\mathcal{O}_{\tilde{X}}(-D)$, D is here the adjunction divisor. We have $K_{\tilde{X}} + D = \hat{g}^*(\mathcal{O}_Y(1,2))$, so $D \equiv 3F$ (as we already know). More generally, since $\mathcal{C} = \hat{g}_*\mathcal{O}_{\tilde{X}}(-D)$, the nth adjoint ideal $\hat{g}_*\mathcal{O}_{\tilde{X}}(-nD)$ equals $\mathcal{C}^n$.

Whence
$$h^0(S, nK_S) = h^0(\tilde{X}, nK_{\tilde{X}}) = h^0(Y, \hat{g}_*(nK_{\tilde{X}})) = h^0(Y, \mathcal{C}^n \mathcal{O}_Y(n, 2n)).$$

In particular, a global section of $g_*(K_{\tilde{S}})$ is a global section of $\mathcal{O}_Y(1, 2)$, whose divisor pulls back to an effective divisor containing the honestly hyperelliptic fibres with their multiplicity; in particular its divisor contains the special twisted cubics. Since no plane contains a twisted cubic curve, we recover the basic assumption $h^0(K_S) = 0$.

Moreover, letting E be the sum of the four (-1) divisors of the blow-up, since $|\varphi^*(\mathcal{O}_Y(0,1))| = |2K_S - E| = |2K_{\tilde{S}} - 3E| = |\varphi^*(\mathcal{O}_Y(2,4)) - 2D - 3E|$, there exists $Q' \in |\mathcal{O}_Y(2,3)|$ whose pull-back on $\tilde{S}$ is a divisor consisting of $3E$ plus the sum of the honestly hyperelliptic fibres counted each $2s$ times (s being their respective multiplicity).

By Corollary 2.7 Q' belongs to the sheaf of ideals $(\mathcal{Q}, \mathcal{G}, P)$, where P is a polynomial of degree 3 on $\mathbb{P}^1$ such that its divisor pulls back to the adjunction divisor on $\tilde{S}$.

Since (Q, G, P) form a regular sequence, it follows easily that there exists a quadratic polynomial $Q''(y_i)$ such that $Q' = Q''P$.

Since the pull-back of Q' contains the adjunction divisor D doubly, while P pulls back to D, it follows that the pull-back of div Q'' is at least D. □

Let us devote our analysis to the case of torsion $\{0\}$ and let us write down explicitly the equations of the two divisors whose complete intersection gives our image surface Y.

Let $Q_\lambda = \lambda_0 Q_0 + \lambda_1 Q_1$; $G_\lambda = \lambda_0^3 G_{000} + \lambda_0^2 \lambda_1 G_{100} + \lambda_0 \lambda_1^2 G_{011} + \lambda_1^3 G_{111}$.

We can assume that for $\lambda = (1, 0), (0, 1)$ or $\mu = (\mu_0, \mu_1)$ (fixed), $Q_\lambda \cap G_\lambda$ is a double twisted cubic.

Lemma 3.4. Q_0, Q_1, Q_μ *are quadric cones of rank 3.*

Proof. If one of these quadrics, say Q_0, were smooth then Q_0 would be isomorphic to $(\mathbb{P}^1 \times \mathbb{P}^1)$. Then the cubic G_{000} would cut on Q_0 a divisor in the linear system $(3, 3)$, while we know that this intersection must be twice an irreducible twisted cubic curve (t.c.c. for short), a contradiction (observe that a t.c.c. lies in a linear system of type $(2, 1)$ or $(1, 2)$). Moreover, since the t.c.c. is irreducible and non degenerate, rank $Q_0 =$ rank $Q_1 =$ rank $Q_\mu = 3$. □

By Lemma 3.4, we know that Q_0, Q_1, Q_μ, are quadric cones. Let V_0, V_1, V_μ be their respective vertices, $\Gamma_0, \Gamma_1, \Gamma_\mu$ the corresponding twisted cubic curves. The tricanonical image Σ of S is the hypersurface of $\mathbb{P}^3$ defined by $\Sigma = \{Q_1^3 G_{000} - Q_1^2 Q_0 G_{100} + Q_1 Q_0^2 G_{011} - Q_0^3 G_{111} = 0\}$.

Lemma 3.5. *If* $V_0 = V_1 \Rightarrow Q_0, Q_1, Q''$ *have a common line* L.

Proof. Let us consider the lines l_0, l_1, l_μ residual to the twisted cubics in the respective intersections of the three quadratic cones with the "adjoint" quadric Q''. I.e., we have $Q_0 \cap Q'' = \Gamma_0 \cup l_0$, $Q_1 \cap Q'' = \Gamma_1 \cup l_1$, $Q_\mu \cap Q'' = \Gamma_\mu \cup l_\mu$.

Observe that, since $V_0 = V_1 = V_\mu$, then clearly $V_0 \in l_0 \cap l_1 \cap l_\mu$.

$Q'' \supset \Gamma_0 \Rightarrow$ rank $Q'' \geq 3$.

If Q'' is smooth, then $Q'' \cong \mathbb{P}^1 \times \mathbb{P}^1$ and every line in Q'' is contained in one of the two rulings. So, at least two of the above lines are in the same ruling, and since they intersect, they do coincide.

This line is in the base locus of the pencil Q_λ hence our assertion follows.

If Q'' is a quadric cone, denote by V'' its vertex. Every t.c.c. in a quadric cone passes trough the vertex, so $\forall i \ V'' \in \Gamma_i \subset Q_i$; let $V = V_0 = V_1$, and observe that $V \neq V''$ (else the two quadric cones would intersect in 4 lines), whence the line $\overline{VV''}$ is contained in all these quadrics. □

Lemma 3.6. $V_0 \neq V_1$.

Proof. Observe preliminarily that the previous lemma implies that the three twisted cubics $\Gamma_0, \Gamma_1, \Gamma_\mu$ are distinct (otherwise there would be a twisted cubic Γ contained in each Q_λ: but then $Q_0 = Q_1$, since they have the same vertex V and they are the join of V with Γ).

Observe now that Σ must be singular in our three twisted cubics: in fact Σ is the image of Y under the birational morphism given by the first projection, and Y is singular along the three twisted cubics.

Thus, $Q'' \cap \Sigma \geq 2\Gamma_0 + 2\Gamma_1 + 2\Gamma_\mu$.

Both Q'' and Σ are irreducible, so their intersection must be a curve of degree 18 and equality must hold.

But, by Lemma 3.5, we have a line in $Q'' \cap Q_0 \cap Q_1$, which is a fortiori also in $Q'' \cap \Sigma$, whence a contradiction. $\square$

So, we can assume $V_0 \neq V_1$.

Lemma 3.7. $\forall \lambda \in \mathbb{P}^1$, Q_λ is a quadric cone and the line $\overline{V_0 V_1}$ is contained in Q_λ.

Proof. Recall that $Q_0 \cap G_{000} = 2\Gamma_0$. But Q_0 is singular in V_0, so G_{000} must be smooth in V_0, thus also in a general point of Γ_0.

Observe that $V_0 \in \Gamma_0 \subset \text{Sing }\Sigma$, so, by inspecting the equation of Σ, we infer that $V_0 \in Q_1$. Similarly, $V_1 \in Q_0$, and the line $\overline{V_0 V_1} \subset Q_\lambda$ $\forall \lambda$.

Let us now fix coordinates such that $V_0 = (0, 0, 0, 1)$, $V_1 = (0, 0, 1, 0)$; $\overline{V_0 V_1} = \{x_0 = x_1 = 0\}$.

Thus the matrix of the quadric Q_λ has the following form:

$$Q_\lambda = \begin{pmatrix} * & * & \lambda_0 * & \lambda_1 * \\ * & * & \lambda_0 * & \lambda_1 * \\ \lambda_0 * & \lambda_0 * & 0 & 0 \\ \lambda_1 * & \lambda_1 * & 0 & 0 \end{pmatrix},$$

whence the determinant of the matrix of $|Q_\lambda|$ equals the square p_2^2, of a homogeneous polynomial p_2 of degree 2 in the λ_i's; since we know that it has at least three distinct roots, we conclude that $p_2 = 0$, therefore Q_λ is a pencil of quadric cones. $\square$

So, after a suitable change of coordinates in $\mathbb{P}^1$ and in $\mathbb{P}^3$, we may assume that

$$Q_\lambda = \begin{pmatrix} 0 & 0 & -\frac{\lambda_0}{2} & -\frac{\lambda_1}{2} \\ 0 & \lambda_0 + \lambda_1 & 0 & 0 \\ -\frac{\lambda_0}{2} & 0 & 0 & 0 \\ -\frac{\lambda_1}{2} & 0 & 0 & 0 \end{pmatrix},$$

i.e., $Q_0 = x_1^2 - x_0 x_2$, $Q_1 = x_1^2 - x_0 x_3$.

Remark that this choice imposes that it cannot be $\mu_0 = -\mu_1$, because otherwise we get a t.c.c. Γ_μ contained in a reducible quadric Q_μ.

End of the proof.

The vertices V_0, V_1 and V_μ of the quadric cones Q_0, Q_1 and Q_μ must be respectively contained in the twisted cubics Γ_0, Γ_1 and Γ_μ, therefore also in Q''. However, these three points lie on the same line $\overline{V_0 V_1}$; in particular, we get a line intersecting a quadric in three distinct points. The conclusion is that $\overline{V_0 V_1} \subset Q''$.

Recall that Σ must be singular in our three twisted cubics, and that by inspecting its equation, it follows easily that Σ is triple on the complete intersection of the two quadrics Q_0, Q_1, which contains the line $\overline{V_0 V_1}$.

Let us write the complete intersection $Q_0 \cap Q_1$ as $\overline{V_0 V_1} + T$, where T is thus a 1-cycle of degree 3.

Only two cases can occur:

- Γ_0, Γ_1 and Γ_μ are distinct

- $\Gamma_0 = \Gamma_1 = \Gamma_\mu = T$

In the first case, the schematic intersection $\Sigma \cap Q''$ has degree 18, however it contains Γ_0, Γ_1 and Γ_μ with multiplicity two and $\overline{V_0 V_1}$ with multiplicity three: this is clearly a contradiction, since $18 + 3 = 21 > 18$.

In the second case, the irreducible twisted cubic T would intersect the line $\overline{V_0 V_1}$ in the three distinct points V_0, V_1, V_μ, which is well known not to be possible. □

4. The Barlow surface

Up to now there are only two known explicit constructions of numerical Godeaux surfaces with torsion $\{0\}$ (and indeed simply connected), respectively due to Barlow ([Ba2]), and Craighero and Gattazzo ([CG]): let us consider first Barlow's example.

For the Barlow surface, we can study the bicanonical and tricanonical system according to the manuscript [R3], where Reid describes the canonical ring of the Barlow surface as follows.

Let A the symmetric matrix

$$A = \begin{pmatrix} -2x_4 & x_2 - x_0 - x_4 & x_0 - x_1 - x_4 & x_3 - x_2 - x_4 & x_1 - x_3 - x_4 \\ & -2x_0 & x_3 - x_1 - x_0 & x_1 - x_2 - x_0 & x_4 - x_3 - x_0 \\ & & -2x_1 & x_4 - x_2 - x_1 & x_2 - x_3 - x_1 \\ & & & -2x_2 & x_0 - x_3 - x_2 \\ & & & & -2x_3 \end{pmatrix}.$$

Let A_{ij} the ij-th entry of A, B_{ij} the ij-th entry of the adjoint matrix B of A.

Let us consider the authomorphism β of $\mathbb{C}[x_0, \ldots, x_4, y_0, \ldots, y_4]$ that acts as $\beta(x_i) = x_{i+1}$, $\beta(y_i) = y_{i+1}$, and the automorphism α that acts as $\alpha(x_i) = x_{a(i)}$ ($a = (25)(34)$ in S_5), $\alpha(y_i) = -y_{4-i}$, where all indices are to be taken in $\mathbb{Z}/5\mathbb{Z}$.

They generate a subgroup G of the group of automorphisms of $\mathbb{C}[x_0, \ldots, x_4, y_0, \ldots, y_4]$. One can indeed check that $G \cong D_{10}$.

Let $R = \mathbb{C}[x_0, \ldots, x_4, y_0, \ldots, y_4]/I$, where the ideal I is generated by

$$\sum x_i = 0,$$

$$\forall 1 \leq i \leq 5, \quad \sum_{1}^{5} A_{ij} y_{j-1} = 0,$$

$$\forall 1 \leq i, j \leq 5, \quad y_{i-1} y_{j-1} - B_{ij} = 0.$$

We consider the ring R as a graded ring via the following grading which makes I a homogeneus ideal: deg $x_i = 1$, deg $y_i = 2$.

One can check that the ideal I is G-invariant, whence G acts on R. Since the action acts only with isolated fixed points, it follows (cf. [R3]) that the canonical ring of the Barlow surface can be described as the ring of the invariants of R for the action of G.

In order to simplify the computations, one can choose as generators for G, β and $\alpha' = \beta\alpha$; $\alpha'(x_i) = x_{a'(i)}$, with $a' = (12)(35)$, and $\alpha'(y_i) = -y_{-i}$.

So we can easily compute that there are no nontrivial invariants in R_1, while the subspace of invariants in R_2 is generated by

$$\xi_0 = x_1 x_2 + x_2 x_3 + x_3 x_4 + x_4 x_0 + x_0 x_1,$$

$$\xi_1 = x_1 x_3 + x_2 x_4 + x_3 x_0 + x_4 x_1 + x_0 x_2,$$

$$\xi_2 = x_1^2 + x_2^2 + x_3^2 + x_4^2 + x_0^2.$$

Moreover, the relation $\sum x_i = 0$ induces the relation

$$2\xi_0 + 2\xi_1 + \xi_2 = 0.$$

So we can take ξ_0, ξ_1 as generators of the bicanonical system.

The tricanonical system needs more computations.

We know that R_3 is generated by $x_i x_j x_k$ and $x_i y_j$; the invariants must have the same decomposition.

The subspace of invariants in the span of the monomials $x_i x_j x_k$ is generated by the invariants:

$$\eta_0 = x_1 x_2 x_3 + x_2 x_3 x_4 + x_3 x_4 x_0 + x_4 x_0 x_1 + x_0 x_1 x_2,$$

$$\eta_1 = x_1 x_2 x_4 + x_2 x_3 x_0 + x_3 x_4 x_1 + x_4 x_0 x_2 + x_0 x_1 x_3,$$

$$\eta_2 = x_1^2(x_2 + x_0) + x_2^2(x_3 + x_1) + x_3^2(x_4 + x_2) + x_4^2(x_0 + x_3) + x_0^2(x_1 + x_4),$$

$$\eta_3 = x_1^3 + x_2^3 + x_3^3 + x_4^3 + x_0^3,$$

$$\eta_4 = x_1^2(x_3 + x_4) + x_2^2(x_4 + x_0) + x_3^2(x_0 + x_1) + x_4^2(x_1 + x_2) + x_0^2(x_2 + x_3).$$

The relation $\sum x_i = 0$ induces the three linear relations

$$\begin{cases} 2\eta_0 + \eta_1 + \eta_2 = 0 \\ \eta_0 + 2\eta_1 + \eta_4 = 0 \\ \eta_3 + \eta_2 + \eta_4 = 0. \end{cases}$$

Thus the above subspace is generated by two independent generators, say η_0, η_1.

Now we have to find two more independent generators for the subspace of invariants in the span of the monomials $\{x_i y_j\}$.

Here the β invariants are generated by $\zeta_j = \sum_i x_i y_{i+j}$, where the indices $0 \leq j \leq 4$ are again to be understood as elements of $\mathbb{Z}/5\mathbb{Z}$.

The ζ_j verify the trivial relation $\sum \zeta_j = 0$, and the sum of the five linear relations $\forall 1 \leq i \leq 5 \sum_1^5 A_{ij} y_{j-1} = 0$. An easy calculation shows that this sum yields exactly $(-6)\zeta_1$. Whence, we have only the other relation $\zeta_1 = 0$.

Another easy calculation shows that $\alpha'(\zeta_0) = -\zeta_2, \alpha'(\zeta_1) = -\zeta_1, \alpha'(\zeta_3) = -\zeta_4$, and we can easily conclude that a system of independent generators for the tricanonical system of the Barlow surface is given by $\eta_0, \eta_1, \zeta_0 - \zeta_2, \zeta_3 - \zeta_4$.

In order to understand how many hyperelliptic divisors (with multiplicity) there are in the bicanonical system of the Barlow surface, we have only to check what is the minimal m such that there exists a non trivial element in

$$(S^m(\langle \xi_0, \xi_1 \rangle)) \otimes S^2(\langle \eta_0, \eta_1, \zeta_0 - \zeta_2, \zeta_3 - \zeta_4 \rangle)) \cap I.$$

We are indebted to F.-O. Schreyer who wrote a Macaulay script that verifies that this minimal number m is indeed equal to 3, and that the relation is given by the following polynomial

$$1728\xi_0^3\eta_0^2 + 1872\xi_0^2\xi_1\eta_0^2 - 1296\xi_0\xi_1^2\eta_0^2 - 1584\xi_1^3\eta_0^2 + 5472\xi_0^3\eta_0\eta_1$$
$$+ 5184\xi_0^2\xi_1\eta_0\eta_1 - 5184\xi_0\xi_1^2\eta_0\eta_1 - 5472\xi_1^3\eta_0\eta_1 + 1584\xi_0^3\eta_1^2$$
$$+ 1296\xi_0^2\xi_1\eta_1^2 - 1872\xi_0\xi_1^2\eta_1^2 - 1728\xi_1^3\eta_1^2 - 13\xi_0^3(\zeta_0 - \zeta_2)^2$$
$$- 22\xi_0^2\xi_1(\zeta_0 - \zeta_2)^2 - 10\xi_0\xi_1^2(\zeta_0 - \zeta_2)^2 + \xi_1^3(\zeta_0 - \zeta_2)^2$$
$$+ 14\xi_0^3(\zeta_0 - \zeta_2)(\zeta_3 - \zeta_4) + 24\xi_0^2\xi_1(\zeta_0 - \zeta_2)(\zeta_3 - \zeta_4)$$
$$+ 24\xi_0\xi_1^2(\zeta_0 - \zeta_2)(\zeta_3 - \zeta_4) + 14\xi_1^3(\zeta_0 - \zeta_2)(\zeta_3 - \zeta_4) - \xi_0^3(\zeta_3 - \zeta_4)^2$$
$$+ 10\xi_0^2\xi_1(\zeta_3 - \zeta_4)^2 + 22\xi_0\xi_1^2(\zeta_3 - \zeta_4)^2 + 13\xi_1^3(\zeta_3 - \zeta_4)^2.$$

Afterwards we wrote a Macaulay2 script (available upon request) that obtains the same result in characteristic 0.

We can therefore summarize the main result of the foregoing section in the following

Theorem 4.1. *The bicanonical system of the Barlow surface has exactly 4 distinct base points and contains two hyperelliptic fibres (counted with multiplicity).*

The same result was obtained independently by [Lee].

5. The Craighero Gattazzo surface

Let us now compute what happens for the Craighero Gattazzo surface. As the Barlow surface, this is a numerical Godeaux surface with torsion {0} (and indeed simply connected, as shown in [DW]).

The Craighero Gattazzo surface S is constructed in [CG] as the minimal resolution of the quintic $X \in \mathbb{P}^3$ defined by the equation F_5

$$F_5 = (x + my + az)^2 t^3 + \left[a^2 x^3 + xy(bx + cy) + m^2 y^3 + (ex^2 + fxy + cy^2)z\right.$$
$$\left. + (bx + ey)z^2 + z^3\right]t^2 + \left[2ax^3 y + ex^2 y^2 + 2amxy^3\right.$$
$$+ (2amx^3 + fx^2 y + fxy^2 + 2my^3)z + (cx^2 + fxy + by^2)z^2$$
$$+ 2(mx + ay)z^3\right]t + x^3 y^2 + a^2 x^2 y^3 + xy(2mx^2 + bxy + 2ay^2)z$$
$$+ (m^2 x^3 + cx^2 y + exy^2 + y^3)z^2 + (mx + ay)^2 z^3 = 0.$$

where r is a root of the polynomial $t^3 + t^2 - 1$ and where the various coefficients are defined as follows :

$$a = r^2 \qquad\qquad b = -\tfrac{1}{7}(2r^2 - 13r - 18)$$
$$c = \tfrac{1}{49}(73r^2 + 75r + 92) \qquad e = -\tfrac{1}{7}(r^2 - 24r - 9)$$
$$f = \tfrac{1}{49}(181r^2 + 241r + 163) \qquad m = \tfrac{1}{7}(3r^2 + 5r + 1).$$

In [CG] are given different expressions for the coefficients a, e, b, m, f, c, expressed as rational functions of r; we have computed the equivalent expression as $\mathbb{Q}$-linear combinations of $1, r, r^2$ in order to simplify the calculations (we have done this both by hand and via a calculation using MAPLE).

This quintic surface X is invariant for the $\mathbb{Z}/4\mathbb{Z}$-action on $\mathbb{P}^3$ induced by the cyclical permutation of the coordinates $x \mapsto y \mapsto z \mapsto t$; the singular locus of X is the set of coordinate points $\{(1, 0, 0, 0), (0, 1, 0, 0), (0, 0, 1, 0), (0, 0, 0, 1)\}$.

It is possible to show, as we shall do shortly, that in the neighbourhood of every singular point the singularity can be represented as a double cover of the plane branched on a curve with a singularity of type (3,3) (a triple point that has an infinitely near ordinary triple point). Therefore our singular points are simple elliptic (-1)-singularities (for which the exceptional curve in the minimal resolution is a smooth elliptic curve with self-intersection -1).

It follows that the adjoint divisor on the resolution is precisely the elliptic exceptional curve counted with multiplicity one, whence the bicanonical system of S is cut by the quadrics in $\mathbb{P}^3$ whose pull-back on S yields a divisor containing the exceptional locus twice, and the tricanonical system is cut by the cubics in $\mathbb{P}^3$ whose pull-back on S contains the exceptional locus with multiplicity three.

Craighero and Gattazzo compute explicitly both systems, but we found that their computation is different (and non-equivalent) to ours. It is possible that some misprint occurred, so let us sketch our calculation.

Let us look at the equation of X in a neighbourhood of $(0, 0, 0, 1)$. Setting $w = (x + my + az)$ we can write the Taylor development of the equation F_5 in affine coordinates as follows:

$$w^2 + m^2 y^3 + w f_2(w, y, z) + f_4(w, y, z) + f_5(w, y, z) = 0$$

with f_i homogeneus of degree i.

In local analytic coordinates (u, y, z), where $u = w + 1/2 f_2(w, y, z)$, the equation takes the form

$$u^2 + m^2 y^3 + g_4(u, y, z) + \cdots = 0.$$

Whence $y = 0$ is the equation (in the plane of coordinates (y, z)) of the direction of the tangent cone of the branching locus, and therefore the pull-back on S of the divisor $\operatorname{div}(y)$ is easily shown to contain the exceptional curve E at least twice.

Of course the multiplicity in the exceptional curve of w and z is at least one. But, since w^2 belongs to the cube of the maximal ideal, it follows that $\operatorname{div}(w) \geq 2E$.

Again, writing

$$f_2(w, y, z) = \alpha z^2 + w F_1(w, y, z) + y G_1(w, y, z)$$

$$f_4(w, y, z) = \beta z^4 + w F_3(w, y, z) + y G_3(w, y, z)$$

we are able to rewrite our equation in a slightly different way as follows:

$$w^2 + m^2 y^3 + [\alpha w z^2 + w^2 F_1(w, y, z) + w y G_1(w, y, z)]$$
$$+ [\beta z^4 + w F_3(w, y, z) + y G_3(w, y, z)] + f_5(w, y, z) = 0.$$

From the above remarks follows that the function

$$w^2 + \alpha w z^2 + \beta z^4$$

has a divisor which is greater than $5E$.

But a tedious calculation shows that $w^2 + \alpha w z^2 + \beta z^4 = (w - \frac{1}{7}(6r^2 + 3r - 5)z^2)^2$. Whence the multiplicity of $w - \frac{1}{7}(6r^2 + 3r - 5)z^2$ is at least 3.

It is in fact obvious that $p_g(S) = 0$; moreover it is also clear that $|2K_S|$ contains the divisors corresponding to the quadrics $Q_0 = xz$ and $Q_1 = yt$: on the other hand these two quadrics generate a fixed part free pencil on X, therefore the corresponding pencil in $|2K_S|$ has no rational curve in its fixed part; whence S is minimal. Since $K_S^2 = 1$, it follows that the bigenus $P_2(S) = 2$, hence the bicanonical system is precisely the above pencil.

We can proceed further by using the $\mathbb{Z}/4\mathbb{Z}$-invariance of F_5, since then $|3K_S|$ is generated by the $\mathbb{Z}/4\mathbb{Z}$ orbit of the cubic $C_0 = a(x + my + az)t^2 + txy + a^3yzt + \frac{1}{7}(6r^2 + 3r - 5)xzt$.

If σ is the generator of the $\mathbb{Z}/4\mathbb{Z}$ action such that $\sigma(x) = y$, let us set $C_1 = \sigma(C_0)$, $C_2 = \sigma^2(C_0)$, $C_3 = \sigma^3(C_0)$.

In this way, also for the Craighero Gattazzo surface, we can calculate using the computer algebra program Macaulay2 what is the minimal number m such that the kernel of the map $S^m(H^0(2K)) \otimes S^2(H^0(3K)) \to H^0((2m+6)K)$ is not trivial; and again the answer we get is $m = 3$.

At the moment we cannot yet determine whether there do exist numerical Godeaux surfaces with bigger values of $m = 5$ or 7; we hope to address this question in a sequel to this paper.

The explicit equation of this polynomial is

$$-(3r^2 + 5r + 1)Q_0^3 C_0 C_2 + Q_0^2 Q_1 [-7r(C_0^2 + C_2^2)$$
$$- 14(r+1)(C_0C_1 + C_2C_3) - 7(r+1)(C_1^2 + C_3^2) + (r^2 + 4r - 9)C_0C_2$$
$$- 7(r^2 + r + 1)(C_1C_2 + C_0C_3) - (11r^2 + 16r + 6)C_1C_3]$$
$$+ Q_0 Q_1^2 [7(r+1)(C_0^2 + C_2^2) + 7(r^2 + r + 1)(C_0C_1 + C_2C_3) + 7r(C_1^2 + C_3^2)$$
$$+ (11r^2 + 16r + 6)C_0C_2 + 14(r+1)(C_1C_2 + C_0C_3) - (r^2 + 4r - 9)C_1C_3]$$
$$+ (3r^2 + 5r + 1)Q_1^3 C_1 C_3$$

We can combine the results of our calculations above with the previous results of Craighero and Gattazzo ([CG]) and Dolgachev and Werner ([DW]),

Theorem 5.1. *The Craighero Gattazzo surface is a simply connected numerical Godeaux surface with ample canonical bundle. The bicanonical system has exactly 4 distinct base points and contains exactly two hyperelliptic fibres with multiplicity 1.*

Proof. We need only to verify the last two assertions.

Recall that by [DW], S does not contain (-2)-curves; so, by Lemma 1.1, $2K_S$ has no fixed part.

Restricting to the line $x = y = 0$ the equation F_5, we get the polynomial $az^2t^3 + z^3t^2$. So the smooth point of X of coordinates $(0, 0, -a, 1)$ is a base point of the bicanonical system of $2K_S$; since its orbit by the $\mathbb{Z}/4\mathbb{Z}$ action consists of four distinct points, we have gotten 4 distinct base points. These build up the whole base locus because $(2K_S)^2 = 4$.

We have shown before that the minimal m such that the kernel of the map $S^m(H^0(2K)) \otimes S^2(H^0(3K)) \to H^0((2m + 6)K)$ is not trivial, is 3. This allows us to conclude, by Theorem 2.5, that there are two hyperelliptic bicanonical divisors (counted with multiplicity).

But the $\mathbb{Z}/4\mathbb{Z}$ action on X induces a $\mathbb{Z}/2\mathbb{Z}$ action on the bicanonical system (since the bicanonical sections are invariant by σ^2). So, if there were only one hyperelliptic bicanonical divisor (with multiplicity two), it would be cut by a σ-invariant quadric in the pencil generated by Q_0 and Q_1, i.e. by $Q_0 + Q_1$ or $Q_0 - Q_1$.

But we have written down explicitly the tricanonical system, so we can explicitly write the tricanonical images of these two divisors. We can prove that neither of them is hyperelliptic, because otherwise we would find three quadrics containing the image of one of them, whereas we have checked with the program Macaulay2 that in both cases there is only one such a quadric. □

6. The local moduli space of the Craighero Gattazzo surface

It is known that the local moduli space of the Barlow surface is smooth of dimension 8 (cf. [CL], and also [Lee]). The main scope of this section is to prove that the same holds for the Craighero Gattazzo surface:

Theorem 6.1. *The local moduli space of the Craighero Gattazzo surface is smooth of dimension 8.*

Let X be the quintic constructed by Craighero and Gattazzo, and $\pi : S \to X$ its minimal resolution.

By Kodaira and Spencer's first main result in deformation theory (cf. [KS], also [KM]) our claim will be stablished if we show that $h^1(\Theta_S) = 8$, $h^2(\Theta_S) = 0$.

In fact, $h^0(\Theta_S) = 0$, since S is of general type, and moreover $h^1(\Theta_S) - h^2(\Theta_S) = -\chi(\Theta_S) = 10\chi(\mathcal{O}_S) - 2K_S^2 = 8$. Therefore, it suffices to prove that $h^1(\Theta_S) = 8$.

Applying to the standard exact sequence

$$0 \to \mathcal{O}_X(-5) \to \Omega^1_{\mathbb{P}|X} \to \Omega^1_X \to 0$$

the functor $\mathrm{Hom}_{\mathcal{O}_X}(\cdot, \mathcal{O}_X)$, we get the standard long exact sequence

$$H^0(\Theta_{\mathbb{P}^3|X}) \to H^0(\mathcal{O}_X(5)) \to \mathrm{Ext}^1_{\mathcal{O}_X}(\Omega^1_X, \mathcal{O}_X) \to H^1(\Theta_{\mathbb{P}^3|X}) \to H^1(\mathcal{O}_X(5))$$

$$\to \mathrm{Ext}^2_{\mathcal{O}_X}(\Omega^1_X, \mathcal{O}_X) \to H^2(\Theta_{\mathbb{P}^3|X}).$$

However, taking the restriction to X of the Euler exact sequence

$$0 \to \mathcal{O}_X \to \mathcal{O}_X(1)^4 \to \Theta_{\mathbb{P}^3|X} \to 0$$

we can easily compute that $H^1(\Theta_{\mathbb{P}^3|X}) = H^2(\Theta_{\mathbb{P}^3|X}) = 0$.

Therefore, keeping also in mind that $H^1(\mathcal{O}_X(5)) = 0$, we find that the map $H^0(\mathcal{O}_X(5)) \xrightarrow{f} \mathrm{Ext}^1_{\mathcal{O}_X}(\Omega^1_X, \mathcal{O}_X)$ is surjective and that $\mathrm{Ext}^2_{\mathcal{O}_X}(\Omega^1_X, \mathcal{O}_X) = 0$.

In turn, applying the Ext spectral sequence, we obtain the following exact sequence:

$$0 \to H^1(\Theta_X) \to \mathrm{Ext}^1_{\mathcal{O}_X}(\Omega^1_X, \mathcal{O}_X) \xrightarrow{g} H^0(\mathcal{E}xt^1_{\mathcal{O}_X}(\Omega^1_X, \mathcal{O}_X)) \to H^2(\Theta_X) \to 0.$$

We are now going to show the vanishing of $H^1(\Theta_X)$.

Let us denote by p the natural projection $H^0(\mathcal{O}_{\mathbb{P}^3}(5)) \to H^0(\mathcal{O}_X(5))$, and consider the map $g \circ f \circ p : H^0(\mathcal{O}_{\mathbb{P}^3}(5)) \to H^0(\mathcal{E}xt^1_{\mathcal{O}_X}(\Omega^1_X, \mathcal{O}_X))$.

The $\mathbb{Z}/4\mathbb{Z}$ action on X allows us to choose a basis in $H^0(\mathcal{O}_{\mathbb{P}^3}(5))$, say $v_1, \ldots, v_{56}$, such that, if σ is the generator of the action given in the previous section,

$$\sigma(v_j) = \begin{cases} v_j & \text{if } 1 \leq v_j \leq 14 \\ iv_j & \text{if } 15 \leq v_j \leq 28 \\ -v_j & \text{if } 29 \leq v_j \leq 42 \\ -iv_j & \text{if } 43 \leq v_j \leq 56 \end{cases}$$

(notice in fact that σ acts freely on the set of monomials of degree 5).

We observe that $H^0(\mathcal{E}xt^1_{\mathcal{O}_X}(\Omega^1_X, \mathcal{O}_X))$, as a representation of $\mathbb{Z}/4\mathbb{Z}$, is isomorphic to the direct sum of the quotients of $\mathcal{O}_X$ by the jacobian ideal in the 4 singular points of X, and these addenda are permuted by σ, since the 4 singular points are a orbit for σ.

Thus the map $H^0(\mathcal{O}_{\mathbb{P}^3}(5)) \to H^0(\mathcal{E}xt^1_{\mathcal{O}_X}(\Omega^1_X, \mathcal{O}_X))$ is given via a matrix of the following form:

$$\begin{pmatrix} A & B & C & D \\ A & iB & -C & -iD \\ A & -B & C & -D \\ A & -iB & -C & iD \end{pmatrix}$$

where every block is a matrix of size 10×14. We observe immediately that the above matrix has the same rank of the matrix

$$\begin{pmatrix} A & 0 & 0 & 0 \\ 0 & B & 0 & 0 \\ 0 & 0 & C & 0 \\ 0 & 0 & 0 & D \end{pmatrix}.$$

We have explicitly checked with the program Macaulay2 that the matrices A, B, C, D have maximal rank, so that g is a surjective map; since

$$\dim \mathrm{Ext}^1_{\mathcal{O}_X}(\Omega^1_X, \mathcal{O}_X) = \dim H^0(\mathcal{E}xt^1_{\mathcal{O}_X}(\Omega^1_X, \mathcal{O}_X)) = 40,$$

it follows that g is an isomorphism and therefore $H^1(\Theta_X) = H^2(\Theta_X) = 0$.

By [BW] $\pi_*(\Theta_S) = \Theta_X$. So, by the Leray spectral sequence we get $H^1(\Theta_S) \cong H^0(R^1\pi_*\Theta_S)$, and the last vector space equals, by the theorem on formal functions ([H]) to

$$\varprojlim H^1(\Theta_{S|nD}),$$

where D is the exceptional locus of π.

Since D consists of the sum of the four elliptic curves $D_1, \ldots, D_4$, corresponding to the 4 singular points of X, we can conclude that

$$h^1(\Theta_S) = 4 \dim \varprojlim H^1(\Theta_{S|nC}),$$

where C is a smooth elliptic curve with $C^2 = -1$, $K_S C = 1$.

So we are left with proving the following lemma:

Lemma 6.2. *Let S a smooth surface containing a smooth elliptic curve with normal bundle of degree -1. Then*

$$\dim \varprojlim H^1(\Theta_{|nC}) = 2.$$

Proof. Since a simple elliptic singularity is analytically isomorphic to the blow down of the 0-section in the normal bundle to the exceptional curve (cf. [R1], [Lau]), we can assume, w.l.o.g., that S the total space of a line bundle over C of degree -1, that is, $\mathcal{O}_C(-p)$ for some $p \in C$.

By the exact sequence

$$0 \to \Theta_C \to \Theta_{S|C} \to \mathcal{O}_C(-p) \to 0,$$

where C is a smooth elliptic curve (thus $\Theta_C = \mathcal{O}_C$), we get $h^1(\Theta_{S|C}) = 2$.

Tensoring this exact sequence by $\mathcal{O}_C(mp)$, we obtain, $\forall m > 0$, $h^1(\Theta_{S|C}(mp)) = h^1(\mathcal{O}_C((m-1)p))$, whence we get 0 if $m \geq 2$, 1 for $m = 1$.

Applying this result to the exact sequence

$$0 \to \Theta_{S|C}(-(n-1)C) \to \Theta_{S|nC} \to \Theta_{S|(n-1)C} \to 0$$

we get that for $n \geq 3$, the restriction map $H^1(\Theta_{S|nC}) \to H^1(\Theta_{S|(n-1)C})$ is an isomorphism, therefore

$$\varprojlim H^1(\Theta_{|nC}) \cong H^1(\Theta_{|2C})$$

and $2 \leq h^1(\Theta_{S|2C}) \leq 3$.

Let us now consider the canonical projection $q: S \to C$; for every line bundle L on S, $h^0(R^1 q_* L) = 0$, so $h^1(q_* L) = h^1(L)$. Moreover, $q_*(\mathcal{O}_S) \cong \bigoplus_{n \geq 0} \mathcal{O}_C(np)$.

Consider the exact sequence

$$(\#) \quad 0 \to q^* \mathcal{O}_C(-p) \to \Theta_S \to q^* \Theta_C (\cong \mathcal{O}_S) \to 0.$$

Tensoring this sequence by $\mathcal{O}_S(-2C) \cong q^* \mathcal{O}_C(2p)$, since

$$H^i(q^* \Theta_C \otimes \mathcal{O}_S(-2C)) = H^i(q_* \mathcal{O}_S \otimes \mathcal{O}_C(2p))$$
$$= H^i((\bigoplus_{n \geq 0} \mathcal{O}_C(np)) \otimes \mathcal{O}_C(2p)) = \bigoplus_{n \geq 2} H^i(\mathcal{O}_C(np)),$$

we get $h^1(\Theta_S)(-2C) = h^2(\Theta_S)(-2C) = 0$, so $h^1(\Theta_{S|2C}) = h^1(\Theta_S)$.

Again by (#), since

$$h^1(q^*\Theta_C) = h^1(q_*\mathcal{O}_S) = \sum_{n\geq 0} h^1(\mathcal{O}_C(np)) = 1$$

$$h^1(q^*\mathcal{O}_C(-p)) = \sum_{n\geq -1} h^1(\mathcal{O}_C(np)) = 2$$

remembering that we have shown that $2 \leq h^1(\Theta_{S|2C}) = h^1(\Theta_S) \leq 3$, we see that we have to prove that the projection map

$$H^0(\Theta_S) \to H^0(q^*\Theta_C)$$

is not surjective.

We claim that we can write $S = (\mathbb{C}^* \times \mathbb{C})/\sim$, where $\sim$ is the equivalence relation generated by $(z, w) \sim (\mu^2 z, \mu w z)$. $C \subset S$ is defined by the equation $w = 0$, so that $C \cong \mathbb{C}^*/\langle z \sim \mu^2 z \rangle$.

In fact we can assume the point p to be the origin of the elliptic curve C, and we observe that every elliptic curve occurs as a quotient of $\mathbb{C}^*$ as above. Since the functional equation of the Riemann theta function is then $f(\mu^2 z) = \mu^{-1} z^{-1} f(z)$, we obtain the desired assertion.

We shall prove now that the global holomorphic never vanishing section of $q^*\Theta_C$ defined by $z\frac{\partial}{\partial z}$ is not a projection of a global section of Θ_S.

In fact, a global holomorphic vector field on S can be written as $a(z,w)\frac{\partial}{\partial z} + b(z,w)\frac{\partial}{\partial w}$ with a, b global holomorphic functions on $\mathbb{C}^* \times \mathbb{C}$ satisfying the following functional equations: $\forall z, w \in \mathbb{C}^* \times \mathbb{C}$

$$a(z, w) = \mu^2 a(\frac{z}{\mu^2}, \mu\frac{w}{z})$$

$$b(z, w) = \mu w a(\frac{z}{\mu^2}, \mu\frac{w}{z}) + \mu z b(\frac{z}{\mu^2}, \mu\frac{w}{z}).$$

If there were a global holomorphic vector field on S whose projection on $q^*\Theta_C$ is $z\frac{\partial}{\partial z}$, then there would be a global holomorphic function b in $\mathbb{C}^* \times \mathbb{C}$ such that

$$b(z, w) = \mu^{-1} wz + \mu z b(\frac{z}{\mu^2}, \mu\frac{w}{z}).$$

Let us write b as a power series

$$b(z, w) = \sum_{n \in \mathbb{Z}, i \in \mathbb{N}} b_{ni} z^n w^i.$$

Then our condition can be written as :

$$\mu^{-1} wz = b(z, w) - \mu z b(\frac{z}{\mu^2}, \mu\frac{w}{z}) = \sum_{n,i} b_{ni}(z^n w^i - \mu z (\frac{z}{\mu^2})^n (\mu\frac{w}{z})^i) =$$

$$= \sum_{n,i} b_{ni}(z^n w^i - \mu^{i+1-2n} z^{n+1-i} w^i);$$

looking at the coefficient of wz we get $\mu^{-1} = b_{11}(1 - \mu^0) = 0$, a contradiction. □

7. End of the proof of the main theorem

In this section we summarize some of the previous results, in order to prove Theorem 0.1. The first two assertions are already proven in Theorem 2.5.

That the curves in Y which are images of the hyperelliptic bicanonical divisors are irreducible twisted cubic curves was proved in Lemma 1.10, part b); the nature of the singularity along these curves was explained in Proposition 2.6.

Remark that $\omega_Y = \mathcal{O}_Y(2 + 3 - 4, 7 - 2h + 3h - 6 - 2) = \mathcal{O}_Y(1, h - 1)$. Moreover, recall that $g^*\mathcal{O}_Y(0, 1)$ gives the movable part of the bicanonical system, and that $g_*\omega_{\tilde{S}}^2 = \mathcal{C}^2 \omega_Y^2$. So we have a non trivial section Q' in $H^0(\mathcal{C}^2\mathcal{O}_Y(2, 2h - 3))$, which, by Proposition 2.6, induces a non trivial section Q'' in $H^0(\mathcal{C}\mathcal{O}_Y(2, h - 3))$.

Let us denote by H_1 the class of a divisor in $|\mathcal{O}_{\mathbb{P}^3 \times \mathbb{P}^1}(1, 0)|$ and by H_2 the class of a divisor in $|\mathcal{O}_{\mathbb{P}^3 \times \mathbb{P}^1}(0, 1)|$.

The divisor associated to the non trivial section Q' of $H^0(\mathcal{C}^2\mathcal{O}_Y(2, 2h - 3))$ gives a curve in $\mathbb{P}^3 \times \mathbb{P}^1$ of class $(2, 7 - 2h)(3, 3h - 6)(2, h - 3) = 12H_1^3 + 12hH_1^2H_2$. It must contain doubly the h singular twisted cubic curves, so we can consider the residual curve E'', of class $12H_1^3$.

The bicanonical system of $\tilde{S}$ has $2(K_{\tilde{S}} - \beta^*K_S)$ as fixed part, a divisor which is easily shown to contain (with multiplicity) 12 (-1)-curves that are not contracted by g. So their image in Y is precisely E''.

Viceversa, let $Y \subset \mathbb{P}^3 \times \mathbb{P}^1$ be as described. Let us consider the normalization $\varepsilon : \tilde{X} \to Y$, and a minimal resolution of singularities $\delta : \tilde{S} \to \tilde{X}$; we have $\varepsilon_*\omega_{\tilde{X}} \cong \mathcal{C}\omega_Y$. By our assumptions, $\omega_Y = \mathcal{O}_Y(1, h - 1)$.

First, we claim that $p_g(\tilde{X}) = 0$.

In fact, an easy computation shows that the restriction maps

$$H^0(\mathcal{O}_{\mathbb{P}^3 \times \mathbb{P}^1}(1, h - 1)) \to H^0(\mathcal{O}_Q(1, h - 1)) \to H^0(\mathcal{O}_Y(1, h - 1))$$

are isomorphisms.

So, if $h = 0$, $p_g(\tilde{X}) = h^0(\mathcal{O}_{\mathbb{P}^3 \times \mathbb{P}^1}(1, -1)) = 0$, while, if $h > 0$, a non trivial section of $H^0(\varepsilon_*\omega_{\tilde{X}})$ induces a non trivial section of $H^0(\mathcal{O}_{\mathbb{P}^3 \times \mathbb{P}^1}(1, h-1))$ containing some of the singular twisted cubic curves; since a plane in $\mathbb{P}^3$ cannot contain a twisted cubic curve, we derive a contradiction.

Let us now denote by Q'' a non trivial section in $H^0(\mathcal{C}\mathcal{O}_Y(2, h - 3))$; let moreover F' be a non trivial section (unique up to scalar multiplication) in $H^0(\mathcal{O}_Y(0, h))$ whose pull-back in $\tilde{X}$ gives the conductor divisor. Let us set $Q' = F'Q'' \in H^0(\mathcal{C}^2\mathcal{O}_Y(2, 2h - 3))$, $\overline{Q}' = F'Q' \in H^0(\mathcal{C}^3\mathcal{O}_Y(2, 3h - 3))$.

The sections Q' and $\overline{Q}'$ define two injective homomorphisms of sheaves

$$\mathcal{O}_Y(0,1) \to \mathcal{C}^2 \mathcal{O}_Y(2, 2h-2) \cong \varepsilon_* \omega_{\tilde{X}}^2$$

$$\mathcal{O}_Y(1,0) \to \mathcal{C}^3 \mathcal{O}_Y(3, 3h-3) \cong \varepsilon_* \omega_{\tilde{X}}^3.$$

In particular we can conclude that the morphisms $\pi_2 \circ \varepsilon : \tilde{X} \to \mathbb{P}^1$ and $\pi_1 \circ \varepsilon : \tilde{X} \to \mathbb{P}^3$ are induced by some subsystem of the bicanonical, respectively of the tricanonical system. It follows that $\tilde{X}$ is of general type.

Since $\tilde{X}$ has only R.D.P.'s as singularities, $\tilde{S}$ is a surface of general type with geometric genus $p_g = 0$; in particular $q = 0$ and $\chi = 1$.

Let us denote by S the minimal model of $\tilde{S}$; then $K_S^2 \geq 1$. In order to prove that S is a numerical Godeaux surface, we need only to prove that $K_S^2 = 1$.

Observe that the divisor associated to Q'' gives a curve in $\mathbb{P}^3 \times \mathbb{P}^1$ of class $12 H_1^3 + 6h H_1^2 H_2$.

The assumption $Q'' \in H^0(\mathcal{C}\mathcal{O}_Y(2, h-3))$ ensures that such a divisor contains h fibres; so we can consider the residual curve E'' of class $12 H_1^3$ (thus consisting with multiplicity of exactly 12 fibres of the projection over $\mathbb{P}^3$). Let us denote by E' and by E the respective divisors in $\tilde{X}$ and $\tilde{S}$ given by the difference between the pull-back of $\text{div}(Q'')$ and the h fibres corresponding to the conductor divisor.

We have

$$(2K_{\tilde{S}}) = ((\varepsilon \circ \delta)^* \mathcal{O}_Y(0,1) + E)^2 = 24 + E^2$$

$$(3K_{\tilde{S}}) = ((\varepsilon \circ \delta)^* \mathcal{O}_Y(1,0) + E)^2 = 9 + E^2.$$

In particular $K_{\tilde{S}}^2 = (9 + E^2 - 24 - E^2)/5 = -3$.

The morphism $\beta : \tilde{S} \to S$ is a sequence of n blow ups. Since S is of general type and $K_{\tilde{S}}^2 = -3$, it follows that $n = K_{\tilde{S}}^2 - K_S^2 \geq 4$.

An easy computation shows that, if we denote by $\overline{E}$ the difference $K_{\tilde{S}} - \beta^* K_S$, $\overline{E}$ contains, with multiplicity, at least n (-1)-curves. Remark that the morphism $\tilde{S} \to Y$ is composition of a finite map $(\tilde{X} \to Y)$ and of the minimal resolution of the singularities of $\tilde{X}$. By hypotheses, $\tilde{X}$ has only R.D.P., so the only curves contracted are (-2)-curves, and our (-1)-curves cannot be contracted to Y.

Now we only need to remark that the fixed part of $3K_{\tilde{S}}$ contains $3\overline{E}$, whence at least $3n$ (-1)-curves; and the corresponding divisor maps on Y to E'', which has 12 components.

Since $n \geq 4$, $3\overline{E}$ is exactly the fixed part of $3K_{\tilde{S}}$; in particular $n = 4$, $K_S^2 = 1$ and S is a numerical Godeaux surface.

Thus $3\overline{E}$ is the fixed part of both $2K_{\tilde{S}}$ and $3K_{\tilde{S}}$; the rational map $S \dashrightarrow Y$ is the tri-bicanonical morphism, $3K_S$ has no base points, whence (as shown in [Cat1], [Mi1]) the torsion group of S is either 0 or $\mathbb{Z}/2\mathbb{Z}$. But if the torsion were $\mathbb{Z}/2\mathbb{Z}$, by

Lemma 1.10, part c), in the singular locus we would obtain a fibre consisting of a line with multiplicity 6, a contradiction.

Since the bicanonical system yields a genus 4 fibration, we are in case 1a).

We proved that the case with three distinct hyperelliptic fibres cannot occur in Proposition 3.1; the computations of the number of hyperelliptic fibres for the Barlow and the Craighero Gattazzo surface are given in sections 4 and 5.

Finally, the local moduli space of the Barlow surface is computed in [CL], whereas the assertion concerning the local moduli space of the Craighero Gattazzo surface is the contents of Theorem 6.1.

References

[Ba1]   Barlow, R., Some new surfaces with $p_g = 0$, Duke Math. J. 51:4 (1984), 889–904.

[Ba2]   Barlow, R., A simply connected surface of general type with $p_g = 0$, Invent. Math. 79 (1985), 293–301.

[Ba3]   Barlow, R., Rational equivalence of zero cycles for some more surfaces with $p_g = 0$, Invent. Math., 79 (1985), 303–308.

[BPV]   Barth, W., Peters, C., Van de Ven, A., Compact complex surfaces, Springer-Verlag, 1984.

[Blo]   Bloch, S., Lectures on algebraic cycles, Duke University Mathematics Series IV, Durham 1980.

[BKL]   Bloch, S., Kas, A., Lieberman, D., Zero cycles on surfaces with $p_g = 0$, Compositio Math. 33 (1976), 135–145.

[Bo]   Bombieri, E., Canonical models of surfaces of general type, Inst. Hautes Études Sci. Publ. Math. 42 (1973), 173–219.

[BC]   Bombieri, E., Catanese, F., The tricanonical map of a surface with $K^2 = 2$, $p_g = 0$, in: C. P. Ramanujam – A Tribute, Stud. Math. 8, Tata Inst., Bombay (1978), 279–290.

[Bu]   Burniat, P., Sur les surfaces de genre $P_{12} > 0$, Ann. Mat. Pura e Appl. (4) 71 (1966), 1–24.

[BW]   Burns, D. M., Jr., Wahl, J. M., Local contributions to global deformations of surfaces, Invent. Math. 26 (1974), 67–88.

[Cam]   Campedelli, L., Sopra alcuni piani doppi notevoli con curve di diramazione del decimo ordine, Atti Acc. Naz. Lincei 15 (1932), 536–542.

[Cas]   Castelnuovo, G., Sulle superficie di genere zero, Memorie Soc. Ital. Scienze (3) 10 (1896), 103–123.

[Cat1]   Catanese, F., Pluricanonical mappings of surfaces with $K^2 = 1, 2, q = p_g = 0$, in: C.I.M.E., Algebraic surfaces, Liguori Editore, Napoli 1981, 247–266.

[Cat2]   Catanese, F., Pluricanonical Gorenstein curves, in: Enumerative geometry and classical algebraic geometry, Proc. Nice 1981, Prog. Math. 24, Birkhäuser, 1982, 50–96.

[Cat3] Catanese, F., Commutative algebra methods and equations of regular surfaces, in: Algebraic Geometry Bucharest 1982, Lecture Notes in Math. 1056, Springer-Verlag 1984, 68–111.

[CF] Catanese F., Franciosi, M., Divisors of small genus on algebraic surfaces and projective embeddings, in: Proc. of the 1993 Hirzebruch 65 Conference on Alg. Geom., Israel Math. Conf. Proc. 9, Contemp. Math., Amer. Math. Soc., 1996, 109–140.

[CFHR] Catanese F., Franciosi, M., Hulek, K., Reid, M., Embedding of curves and surfaces, Nagoya Math. J. 154 (1999), 185–220.

[CL] Catanese, F., Le Brun, C., On the scalar curvature of Einstein manifolds., Math. Res. Letters 4 (1997), 843–854.

[Ci] Ciliberto, C., Sul grado dei generatori dell'anello canonico di una superficie di tipo generale, Rend. Sem. Mat. Univ. Polit. Torino 41:3 (1983), 83–111.

[CG] Craighero, P., Gattazzo, R., Quintic surfaces of $\mathbb{P}^3$ having a non singular model with $q = p_g = 0$, $P_2 \neq 0$, Rend. Sem. Mat. Univ. Padova 91 (1994), 187–198.

[Dol1] Dolgachev, I., On rational surfaces with a pencil of elliptic curves, Izv. Akad. Nauk SSSR (Ser. Math.) 30 (1966), 1073–1100 (in Russian).

[Dol2] Dolgachev, I., On Severi's conjecture on simply connected algebraic surfaces, Soviet Math. Dokl. 7 (1966), 1169–1172.

[Dol3] Dolgachev, I., Algebraic surfaces with $p_g = q = 0$, in: C.I.M.E., Algebraic surfaces, Liguori Editore, Napoli 1981, 97–215.

[DW] Dolgachev, I., Werner, C., A simply connected numerical Godeaux surface with ample canonical class, to appear in J. Alg. Geom.

[E1] Enriques, F., Un' osservazione relativa alle superficie di bigenere uno, Rend. Acad. Sci. Bologna, 1908, 40–45.

[E2] Enriques, F., Le superficie algebriche, Zanichelli, Bologna, 1949.

[G1] Godeaux, L., Sur une surface algébriques de genre zero et de bigenre deux, Atti Acad. Naz. Lincei 14 (1931), 479–481.

[G2] Godeaux, L., Les surfaces algébriques non rationnelles de genres arithmétique et géométriques nuls, Actualité Scientifiques et Industrielles 123, Exposés de Géométrie, IV, Hermann, Paris 1934.

[GH] Griffiths, P., Harris, J., Principles of algebraic geometry, John Wiley & Sons, 1978.

[H] Hartshorne, R., Algebraic Geometry, Grad. Texts in Math. 52, Springer-Verlag, 1977.

[KM] Kodaira, K., Morrow, J., Complex Manifolds, New York, Holt, Rinehart and Winston, 1974.

[KS] Kodaira, K., Spencer, D. C., On deformations of complex analytic structures I–II, Ann. Math. 67 (1958), 328–466.

[Kot1] Kotschick, D., On manifolds homeomorphic to $\mathbb{CP}^2 \# 8\overline{\mathbb{CP}}^2$, Invent. Math. 95:3 (1989), 591–600.

[Kot2] Kotschick, D., On the pluricanonical maps of Godeaux and Campedelli surfaces, Int. J. Math., 5:1 (1994), 53–60.

[IM]　　　Inose, H., Mizukami, M., Rational equivalence of 0-cycles on some surfaces of general type with $p_g = 0$, Math. Ann. 244 (1979), 205–217.

[Ino]　　Inoue, M., Some new surfaces of general type, Tokyo J. Math. 17:2 (1994), 295–319.

[Lau]　　Laufer, Henry B., On minimally elliptic singularities, Amer. J. Math. 99 (1977), 1257–1295.

[Lee]　　Lee, Y., Bicanonical pencil of a determinantal Barlow surface, to appear in Trans. Amer. Math. Soc.

[L]　　　Lipman, J., Dualizing sheaves, differentials and residues on algebraic varieties, Astérisque 117, 1984.

[ML]　　Mendes Lopes, M. The relative canonical algebra for genus three fibrations, Thesis, University of Warwick, 1989.

[Mi1]　　Miyaoka, Y., Tricanonical maps of numerical Godeaux surfaces, Invent. Math. 34 (1976), 99–111.

[Mi2]　　Miyaoka, Y., On numerical Campedelli surfaces, in; Complex analysis and algebraic geometry, Cambridge Univ. Press, 1977, 113–118.

[Mu1]　　Mumford, D., The canonical ring of an algebraic surface, Ann. Math II 76 (1962), 612–615.

[Mu2]　　Mumford, D. Lectures on curves on an algebraic surface, Ann. of Math. Stud. 59, Princeton University Press, 1966.

[Mu3]　　Mumford, D., Rational equivalence of O-cycles on surfaces, J. Math. Kyoto Univ. 9 (1969), 195–204.

[N]　　　Naie, D., Surfaces d'Enriques et une construction de surfaces de type général avec $p_g = 0$, Math. Z. 215 (1994), 269–280.

[OVdV]　Okonek, C., Van de Ven, A., Γ-type-invariants associated to PU(2)-bundles and the differentiable structure of Barlow's surface, Invent. Math. 95:3 (1989), 601–614.

[OP]　　Oort, F., Peters, C., A Campedelli surface with torsiongroup $\mathbb{Z}/2\mathbb{Z}$, Indag. Math. 43 (1981), 399–407.

[Pet]　　Peters, C., On two types of surfaces of general type with vanishing geometric genus, Invent. Math. 32 (1976), 33–47.

[R1]　　Reid, M., Elliptic Gorenstein singularities of surfaces, Manuscript ca. 1975.

[R2]　　Reid, M., Surfaces with $p_g = 0$, $K^2 = 1$, Jour. Fac. Sc. Univ. Tokyo, IA 25:1 (1978), 75–92.

[R3]　　Reid, M., A simply connected surface of general type with $p_g = 0$, $K^2 = 1$ due to Rebecca Barlow, Manuscript ca. 1981.

[R4]　　Reid, M., Campedelli versus Godeaux, in: Problems in the theory of surfaces and their classification, Cortona 1988, ed. by Catanese, F. et al., Symp. Math. 32, Academic Press., London 1991, 309–365.

[Se]　　Severi, F. Colloque de géométrie algébrique Liège 1949, George Thome, Liège, Paris 1950, p. 9.

[St]	Stagnaro, E., On Campedelli Branch Loci, Ann. Univ. Ferrara sez. VII 43 (1997), 1–26.
[V]	Voisin, C. , Sur les zero-cycles de certaines hypersurfaces munies d'un automorphisme, Ann. Sc. Norm. Super. Pisa, Cl. Sci. IV. Ser. 19:4 (1992), 473–492.
[Wer1]	Werner, C., A surface of general type with $p_g = q = 0$, $K^2 = 1$, Manuscripta Math. 84:3-4 (1994), 327–341.
[Wer2]	Werner, C., A four-dimensional deformation of a numerical Godeaux surface, Trans. Amer. Math. Soc. 349:4 (1997), 1515–1525.

On the Wahl map of plane nodal curves

Ciro Ciliberto*, Angelo Felice Lopez* and Rick Miranda

1991 Mathematics Subject Classification: Primary 14H10. Secondary 14C20, 14H99.

1. Introduction

Let C be a smooth irreducible curve of genus g and let $\Phi_{\omega_C} : \bigwedge^2 H^0(\omega_C) \to H^0(\omega_C^3)$ be the Wahl map of C. Since Wahl's introduction in 1987, the corank of Φ_{ω_C} has been related to many geometrical properties of the curve, such as the possibility of embedding C on a K3 surface [W1] and on Fano varieties [CLM1], [CLM2], the Clifford index and the existence of linear series on C [BEL], [P], the fact that C has general moduli [CHM], [V], just to mention a few. A recent addition to the above list was made by Wahl in 1995 [W2], where he remarked the importance of the cohomology of the square of the ideal sheaf of a canonical curve and its connection with Green's conjecture. Wahl proved that if $C \subset \mathbb{P}^{g-1}$ is a canonical curve satisfying
(∗) $H^1(\mathcal{I}_C^2(t)) = 0$ for every $t \geq 3$, then C is extendable if and only if Φ_{ω_C} is not surjective. In the same article he conjectured that (∗) holds for large g and for any curve with Clifford index at least 3 and remarked that if Green's conjecture holds for $p = 3$ (that is if Cliff $C > p$ then $C \subset \mathbb{P}^{g-1}$ is projectively normal, its ideal is generated by quadrics and the syzygies are generated by linear polynomials up to the p-th module) then $H^1(\mathcal{I}_C^2(4)) = 0$ for every curve C with Cliff $C > 3$. Inspired by this, in a recent article the first two authors [CL] considered the problem of finding families of curves in $\mathcal{M}_g$ of large dimension, for example $\frac{3}{2}g + 15$, with nonsurjective Wahl map, as their general element would be a counterexample to Wahl's conjecture. As [CL] shows this task appears to be one of not so easy solution, and in the present article we confirm this intuition by studying the behaviour of the Wahl map in a very natural family of curves, i.e. plane curves with nodes.

Let $d \geq 1, 0 \leq \delta \leq \binom{d-1}{2}$ be given integers and let $D \subset \mathbb{P}^2$ be an irreducible curve of degree d with δ nodes $P_1, \ldots, P_\delta \in \mathbb{P}^2$ and no other singularity. Denote by $V_{d,\delta}$ the Severi variety parametrizing such curves. For a real number x let $[x]$ be its integer part. Our result is

---

* Research partially supported by the MURST national project 'Geometria Algebrica';

Theorem 1.1. *Let $D \subset \mathbb{P}^2$ be an irreducible curve of degree $d \geq 15$ with δ nodes and no other singularity, C its normalization. Then the Wahl map of C is surjective if one of the following holds:*

(1.1) *D is a general member of the Severi variety $V_{d,\delta}$ and $10 \leq \delta \leq \frac{1}{2}[\frac{d}{3}]([\frac{d}{3}]+3)-5$*

or

(1.2) *D is any member of the Severi variety $V_{d,\delta}$ such that the nodes $P_1, \ldots, P_\delta$ do not lie on a cubic and $10 \leq \delta \leq [\frac{d}{3}] + 5$.*

The surjectivity of the Wahl map of C is obtained in the following way. Let $\epsilon : X \to \mathbb{P}^2$ be the blow-up of $\mathbb{P}^2$ at the nodes $P_1, \ldots, P_\delta$ with exceptional divisors $E_1, \ldots, E_\delta$. Let l be the divisor of a line in $\mathbb{P}^2$ and $H = \epsilon^* l$. By the standard diagram

(1.1)
$$\begin{array}{ccc} \bigwedge^2 H^0(X, \mathcal{O}_X(K_X + C)) & \xrightarrow{\Phi_{K_X+C}} & H^0(X, \Omega^1_X(2K_X + 2C)) \\ & & \downarrow \phi \\ \downarrow p & & H^0(C, \Omega^1_X(2K_X + 2C)_{|C}) \\ & & \downarrow \psi \\ \bigwedge^2 H^0(C, \omega_C) & \xrightarrow{\Phi_{\omega_C}} & H^0(C, \omega_C^3) \end{array}$$

it follows that Φ_{ω_C} is surjective as soon as the same holds for $\psi \circ \phi$ and Φ_{K_X+C}. In section two we show that the surjectivity of the first map is related to the cohomology of a certain sheaf of differentials on X with logarithmic poles along C and we prove that it does hold in most cases. The Gaussian map Φ_{K_X+C} is studied in section three, employing a technique introduced in [CLM3], mainly an application of the Kawamata–Viehweg vanishing theorem. We show that this method is successful if the line bundle $[\frac{d}{3}]H - \sum_{j=1}^\delta E_j$ is very ample on X, hence giving the bound on δ in Theorem (1.1).

It should be remarked that this bound on the number of nodes forces the image in $\mathcal{M}_g$ of the Severi variety to have in fact not so large dimension (of the order of g). When the number of nodes is higher our technique does not apply and, as far as we know, the problem of computing the corank of the Wahl map for such curves is still open.

2. The role of differentials with logarithmic poles

The aim of this section is to study the composition map $\psi \circ \phi$ in diagram (1.1). We start by recording a general result.

Lemma 2.1. *Let S be a smooth irreducible surface, $C \subset S$ a smooth irreducible curve and L a line bundle on S. Consider the maps $\phi : H^0(S, \Omega_S^1 \otimes L^2) \to H^0(C, \Omega_S^1 \otimes L_{|C}^2)$ and $\psi : H^0(C, \Omega_S^1 \otimes L_{|C}^2) \to H^0(C, \omega_C \otimes L_{|C}^2)$.*
Then $\operatorname{cork} \psi \circ \phi \leq \operatorname{cork} \psi + h^1(S, \Omega_S^1(\log C) \otimes L^2(-C))$.

Proof. The diagram

$$\begin{array}{ccccccc}
& & 0 & & 0 & & 0 \\
& & \downarrow & & \downarrow & & \downarrow \\
0 \to & \Omega_S^1 \otimes L^2(-C) & \to & \Omega_S^1(\log C) \otimes L^2(-C) & \to & \mathcal{O}_C \otimes L^2(-C) & \to 0 \\
& \downarrow \text{id} & & \downarrow & & \downarrow & \\
0 \to & \Omega_S^1 \otimes L^2(-C) & \to & \Omega_S^1 \otimes L^2 & \to & \Omega_S^1 \otimes L_{|C}^2 & \to 0 \\
& \downarrow & & \downarrow & & \downarrow & \\
& 0 & \to & \omega_C \otimes L_{|C}^2 & \xrightarrow{\text{id}} & \omega_C \otimes L_{|C}^2 & \to 0 \\
& & & \downarrow & & \downarrow & \\
& & & 0 & & 0 &
\end{array}$$

induces in cohomology

$$\begin{array}{c}
0 \\ \downarrow \\
\operatorname{Ker} \psi = H^0(\mathcal{O}_C \otimes L^2(-C)) \xrightarrow{\beta} H^1(\Omega_S^1 \otimes L^2(-C)) \to H^1(\Omega_S^1(\log C) \otimes L^2(-C)) \\
\downarrow \quad \searrow \alpha \quad \searrow \text{id} \\
H^0(\Omega_S^1 \otimes L^2) \xrightarrow{\phi} H^0(\Omega_S^1 \otimes L_{|C}^2) \to \operatorname{Coker} \phi \subseteq H^1(\Omega_S^1 \otimes L^2(-C)) \\
\downarrow \psi \\
H^0(\omega_C \otimes L_{|C}^2)
\end{array}$$

Therefore $\operatorname{cork} \psi \circ \phi - \operatorname{cork} \psi = \operatorname{cork} \alpha \leq \operatorname{cork} \beta \leq h^1(S, \Omega_S^1(\log C) \otimes L^2(-C))$. $\square$

We apply the above lemma to our case $L = K_X + C$ where X is the blow-up of $\mathbb{P}^2$ at the nodes of D and $C \in |\mathcal{O}_X(dH - 2\sum_{j=1}^{\delta} E_j)|$ is its normalization.

Proposition 2.1. *Suppose $d \geq 7, \delta \geq 10$ and that the nodes $P_1, \ldots, P_\delta$ do not lie on a plane cubic. Then $h^1(X, \Omega_X^1(\log C) \otimes \omega_X^2(C)) = 0$ and the map $\psi \circ \phi$ in diagram (1.1) is surjective.*

Proof. Let x, y be local coordinates on $\mathbb{P}^2$ so that the local equation of D is $xy = 0$. On X we have local coordinates ξ, η with $x = \xi, y = \eta\xi$ where $x \neq 0$ and ξ', η' with $y = \xi', x = \eta'\xi'$ where $y \neq 0$. The sheaf $\Omega_X^1(\log C)$ is then locally generated by $d\xi, d\eta, \frac{d\eta}{\eta}$ and $d\xi', d\eta', \frac{d\eta'}{\eta'}$ on the two open subsets. Let $\mathcal{F}$ be the subsheaf of $\Omega^1_X(\log C)$ which coincides with $\Omega^1_X(\log C)$ away from the exceptional divisors and in a neighborhood of a point of an exceptional divisor is generated by $d\xi, \xi d\eta$ and $\xi \frac{d\eta}{\eta}$ (and $d\xi', \xi' d\eta', \xi' \frac{d\eta'}{\eta'}$). By restricting to a local equation of C we get an exact sequence

(2.3) $$0 \to \epsilon^* \Omega^1_{\mathbb{P}^2} \to \mathcal{F} \to \mathcal{O}_C \to 0$$

Note that $2K_X + C \sim (d-6)H$ on X, hence $H^1(\mathcal{O}_X(-K_X - C)) = H^1(\mathcal{O}_X((d-6)H))^* = 0$. Tensoring (2.3) by $\mathcal{O}_X((d-6)H)$ we deduce $H^1(\mathcal{F}((d-6)H)) = 0$ since $H^1(\Omega^1_{\mathbb{P}^2}(d-6)) = 0$ and $H^1(\mathcal{O}_C((d-6)H)) = H^0(\mathcal{O}_C(-K_X))^* = H^0(\mathcal{O}_X(-K_X))^* = 0$ by hypothesis. Now restriction to the exceptional divisors gives

$$0 \to \mathcal{F} \to \Omega^1_X(\log C) \to (\mathcal{O}_{\mathbb{P}^1} \oplus \mathcal{O}_{\mathbb{P}^1}(3))^{\oplus \delta} \to 0$$

and therefore $H^1(\Omega^1_X(\log C)((d-6)H)) = 0$. Finally by the definition of ψ we get Coker $\psi \subseteq H^1(\mathcal{O}_C(2K_X + C)) = H^1(\mathcal{O}_C((d-6)H)) = 0$, hence we conclude applying Lemma (2.1). □

Remark 2.1. The hypothesis that the nodes do not lie on a cubic is of course necessary, otherwise the map ψ is not surjective in most cases, hence the same holds for Φ_{ω_C}, by diagram (1.1). This is for example the case of a smooth plane curve [CM] (see also [K] for the case of one or two singular points).

3. An application of Kawamata–Viehweg's vanishing theorem

Let S be a smooth irreducible surface and L a line bundle on it. We start by recalling the techniques of [CLM3] to ensure the surjectivity of the Gaussian map $\Phi_L : \bigwedge^2 H^0(S, L) \to H^0(S, \Omega^1_S \otimes L^2)$. Let Y be the blow-up of $S \times S$ along its diagonal Δ, E the exceptional divisor and for every sheaf $\mathcal{G}$ on S let us denote by $\mathcal{G}_i, i = 1, 2$, its pull-back via the map $Y \to S \times S \xrightarrow{p_i} S$ where p_i is the i-th projection. As is well-known [W3], a sufficient condition for the surjectivity of Φ_L is the vanishing of $H^1(S \times S, p_1^* L \otimes p_2^* L \otimes \mathcal{I}_\Delta^2) \cong H^1(Y, L_1 + L_2 - 2E)$. As $L_1 + L_2 - 2E \sim K_Y + (L - K_S)_1 + (L - K_S)_2 - 3E$ the idea is to apply Kawamata–Viehweg's vanishing theorem to the line bundle $(L - K_S)_1 + (L - K_S)_2 - 3E$, provided this is big and nef. However in our case on the blow-up X of $\mathbb{P}^2$ we have $L = K_X + C$ and it is easily seen that $C_1 + C_2 - 3E$ has a negative intersection with curves contained on the strict transform of $E_i \times E_i$, so Kawamata–Viehweg's theorem does not apply as it is and we will need some extra work to prove our result on the surjectivity of

Φ_{K_X+C}. The proof will be divided in two parts, reducing a cohomological statement to one of a more geometrical nature.

Lemma 3.1. *Let Y be the blow-up of $X \times X$ along its diagonal Δ, E the exceptional divisor and $Y_i \subset Y$ the transform of $E_i \times X$, for $i = 1, \ldots, \delta$. Let $M = dH - 3\sum_{j=1}^{\delta} E_j$ and suppose that*

(3.1) $M_1 + M_2 - 3E$ *is big and nef on Y;*

(3.2) $[(C - \sum_{j=1}^{i} E_j)_1 + (C - \sum_{j=1}^{i} E_j)_2 - 3E]_{|Y_i}$ *is big and nef on Y_i for $i = 1, \ldots, \delta$.*

Then the Gaussian map Φ_{K_X+C} is surjective.

Proof. Consider the divisors $Z_i \subset Y$ transform of $X \times E_i$, $F_i \sim (E_i)_1 + (E_i)_2 \sim Y_i + Z_i$ and $F = \sum_{i=1}^{\delta} F_i$. Set $\mathcal{L} = K_Y + C_1 + C_2 - 3E$, so that $\mathcal{L} - F = K_Y + M_1 + M_2 - 3E$ and $H^1(Y, \mathcal{L} - F) = 0$ by Kawamata–Viehweg's vanishing theorem and (3.1). Set $F_0 = 0$. We claim that

(3.3) $$H^1(Y, \mathcal{L} - \sum_{j=0}^{i-1} F_j) = 0 \text{ for } i = 1, \ldots, \delta + 1.$$

Of course (3.3) implies the lemma since the case $i = 1$ gives the required surjectivity of Φ_{K_X+C} (by [W3]). To see (3.3) we proceed by induction on $\delta + 1 - i \geq 0$, the first case having already being done. Suppose then $\delta - i \geq 0$ and consider the exact sequence

$$0 \to \mathcal{O}_Y(\mathcal{L} - \sum_{j=0}^{i} F_j) \to \mathcal{O}_Y(\mathcal{L} - \sum_{j=0}^{i-1} F_j) \to \mathcal{O}_{F_i}(\mathcal{L} - \sum_{j=0}^{i-1} F_j) \to 0$$

By induction (3.3) follows if we show

(3.4) $$H^1(F_i, \mathcal{O}_{F_i}(\mathcal{L} - \sum_{j=0}^{i-1} F_j)) = 0 \text{ for } i = 1, \ldots, \delta.$$

Let us denote by U_{ij} the strict transform on Y of $E_i \times E_j$. Notice that $U_{ij} \cong E_i \times E_j$. By the notation introduced above, we have $F_i = Y_i \cup Z_i$ and if we set $W_i = Y_i \cap Z_i \sim U_{ii} + E_{|Y_i}$ we have

(3.5) $$0 \to \mathcal{O}_{Y_i}(\mathcal{L} - W_i - \sum_{j=0}^{i-1} F_j) \to \mathcal{O}_{F_i}(\mathcal{L} - \sum_{j=0}^{i-1} F_j) \to \mathcal{O}_{Z_i}(\mathcal{L} - \sum_{j=0}^{i-1} F_j) \to 0.$$

As $W_i \cap F_j = \emptyset$ for $j \leq i-1$, on Y_i we also have the exact sequence

(3.6) $\quad 0 \to \mathcal{O}_{Y_i}(\mathcal{L} - W_i - \sum_{j=0}^{i-1} F_j) \to \mathcal{O}_{Y_i}(\mathcal{L} - \sum_{j=0}^{i-1} F_j) \to \mathcal{O}_{W_i}(\mathcal{L}) \to 0$

therefore by (3.6), (3.5), the definition of Z_i and symmetry, we will be done if we prove

(3.7) $\quad H^1(\mathcal{O}_{W_i}(\mathcal{L})) = 0$ for $i = 1, \ldots, \delta$

and

(3.8) $\quad H^1(\mathcal{O}_{Y_i}(\mathcal{L} - W_i - \sum_{j=0}^{i-1} F_j)) = 0$ for $i = 1, \ldots, \delta$.

Set $E'_i = E_{|Y_i}$, so that $E'_i \cong \mathbb{P}\mathcal{E}$ with $\mathcal{E} \cong N^*_{\Delta_{E_i}/E_i \times X}$. From the normal bundle sequence

$$0 \to N_{\Delta_{E_i}/E_i \times E_i} \to N_{\Delta_{E_i}/E_i \times X} \to N_{E_i \times E_i/E_i \times X | \Delta_{E_i}} \to 0$$

and the isomorphisms $N_{\Delta_{E_i}/E_i \times E_i} \cong \mathcal{O}_{\mathbb{P}^1}(2)$, $N_{E_i \times E_i/E_i \times X | \Delta_{E_i}} \cong \mathcal{O}_{\mathbb{P}^1}(-1)$ we easily see that $\mathcal{E} \cong \mathcal{O}_{\mathbb{P}^1}(1) \oplus \mathcal{O}_{\mathbb{P}^1}(-2)$ and $\mathcal{L}_{|E'_i} \cong \mathcal{O}_{\mathbb{P}\mathcal{E}}(2C_0 + 2f)$ where C_0 is a divisor in $|\mathcal{O}_{\mathbb{P}\mathcal{E}}(1)|$ and f is a fiber of $\mathbb{P}\mathcal{E} \to \mathbb{P}^1$. Also the intersection B_i between U_{ii} and E'_i is isomorphic to Δ_{E_i}, whence it is a divisor of type $C_0 + bf$ on $\mathbb{P}\mathcal{E}$, for some b. Therefore $h^1(\mathcal{O}_{E'_i}(\mathcal{L})) = 1$, $h^2(\mathcal{O}_{E'_i}(\mathcal{L} - B_i)) = 0$. From the exact sequence

$$0 \to \mathcal{O}_{E'_i}(\mathcal{L} - B_i) \to \mathcal{O}_{W_i}(\mathcal{L}) \to \mathcal{O}_{U_{ii}}(\mathcal{L}) \to 0$$

we deduce $H^2(\mathcal{O}_{W_i}(\mathcal{L})) = 0$ since $\mathcal{O}_{U_{ii}}(\mathcal{L}) \cong \mathcal{O}_{\mathbb{P}^1 \times \mathbb{P}^1}(-1, -1)$. Now applying this to

$$0 \to \mathcal{O}_{U_{ii}}(\mathcal{L} - B_i) \to \mathcal{O}_{W_i}(\mathcal{L}) \to \mathcal{O}_{E'_i}(\mathcal{L}) \to 0$$

we deduce (3.7) because $\mathcal{O}_{U_{ii}}(\mathcal{L} - B_i) \cong \mathcal{O}_{\mathbb{P}^1 \times \mathbb{P}^1}(-2, -2)$ hence $h^1(\mathcal{O}_{U_{ii}}(\mathcal{L} - B_i)) = 0$, $h^2(\mathcal{O}_{U_{ii}}(\mathcal{L} - B_i)) = 1$.

To prove (3.8) observe that, as a divisor on Y_i, we have $W_i \sim [(E_i)_2]_{|Y_i}$, hence $\mathcal{O}_{Y_i}(\mathcal{L} - W_i - \sum_{j=0}^{i-1} F_j) \cong K_{Y_i} + [(C - \sum_{j=0}^{i} E_j)_1 + (C - \sum_{j=0}^{i} E_j)_2 - 3E]_{|Y_i}$ and therefore (3.8) and the lemma follow by (3.2) and Kawamata–Viehweg's vanishing theorem. $\square$

To take care of the geometrical statements (3.1) and (3.2) we will use an idea of L. Ein, contained in [CLM3]. We denote by X_i the blow-up of $\mathbb{P}^2$ at $P_1, \ldots, P_i$, for $i = 1, \ldots, \delta$, $X = X_\delta$ and we use the same notation H for the pull-back of a line and E_j for the exceptional divisors on X_i.

Lemma 3.2. *Suppose that $\delta \geq 1$ and that the line bundle $[\frac{d}{3}]H - \sum_{j=1}^{i} E_j$ is very ample on X_i, for $i = 1, \ldots, \delta$. Then (3.1) and (3.2) hold.*

Proof. As $M \sim dH - 3\sum_{j=1}^{\delta} E_j$, by hypothesis there are three very ample line bundles A_1, A_2, A_3 of type $a_k H - \sum_{j=1}^{\delta} E_j$, $a_k \geq [\frac{d}{3}]$ such that $M \sim A_1 + A_2 + A_3$ and $h^0(A_k) \geq 4$ for $k = 1, 2, 3$. We recall from [CLM3] that if A_k is very ample then the linear system $|A_{k1} + A_{k2} - E|$ on Y has a sublinear system defining the morphism $Y \to \mathbb{G}(1, \mathbb{P}H^0(A_k)^*)$ associating to $(x, y) \in Y$ the linear span of $\phi_{A_k}(x)$ and $\phi_{A_k}(y)$ (note that this still makes sense if $(x, y) \in E$ since we can think of (x, y) as a pair with $x \in X$, $y \in \mathbb{P}T_{X|x}$). Therefore $A_{k1} + A_{k2} - E$ is nef and also big since the image of X in $\mathbb{P}H^0(A_k)^*$ is non degenerate. Hence we get (3.1) as $M_1 + M_2 - 3E \sim \sum_{k=1}^{3}(A_{k1} + A_{k2} - E)$. Notice now that we can write $C - \sum_{j=1}^{i} E_j \sim B_1 + B_2 + B_3$ with $B_1 = B_2 = [\frac{d}{3}]H - \sum_{j=1}^{\delta} E_j$, and $B_3 = (d - 2[\frac{d}{3}])H - \sum_{j=1}^{i} E_j$. By hypothesis B_1 is very ample on X hence the restriction of $|B_{11} + B_{12} - E|$ on Y_i is certainly nef. But it is also big since B_1 embeds X so that E_i is a line, hence if $|B_{11} + B_{12} - E|$ were not big on Y_i, then the restriction to Y_i of the map $Y \to \mathbb{G}(1, \mathbb{P}H^0(B_1)^*)$ would have no finite fiber, hence every chord joining a point of the line $\phi_{B_1}(E_i)$ and a point of the image $\phi_{B_1}(X)$ would be contained in $\phi_{B_1}(X)$, that is $\phi_{B_1}(X)$ is a plane, a contradiction. Finally we are left to see that $B_{31} + B_{32} - E$ is nef on Y_i. Of course it can fail to be nef only on a curve contained in the indeterminacy locus of the map $Y_i \to \mathbb{G}(1, \mathbb{P}H^0(B_3)^*)$, that is a curve $Z \subset \{(x, y) \in Y_i : \phi_{B_3}(x) = \phi_{B_3}(y)\}$. By hypothesis B_3, as a divisor on X, defines an isomorphism off E_j, $j > i$, hence certainly $i < \delta$ and $Z \subset \bigcup_{j>i} U_{jj}$, but this is a contradiction since $\bigcup_{j>i} U_{jj} \cap Y_i = \emptyset$. □

Remark 3.1. It would be nice to have a more geometrical interpretation of the two lemmas just proved, in terms of the ideal of the nodes. By pushing down to X it can be seen that the vanishing of $H^1(Y, L_1 + L_2 - 2E)$ that we used to get the surjectivity of Φ_{K_X+C}, is in fact related to the cohomology of the normal bundle of the image of X via the linear system $K_X + C$. Therefore a good knowledge of the resolution of the ideal of this image would probably give a better result.

Proof of Theorem 1.1. If we assume (1.2) the map $\psi \circ \phi$ in diagram (1.1) is surjective by Proposition 2.1. The same is true under hypothesis (1.1) as in that case the nodes are generic [AC]. Whence the surjectivity of the Wahl map of C follows by diagram (1.1) (since p is surjective) as soon as we prove that the Gaussian map Φ_{K_X+C} is surjective. By Lemmas (3.1) and (3.2) we just need the very ampleness of $[\frac{d}{3}]H - \sum_{j=1}^{i} E_j$ on X_i. The latter follow by [AH] in case (1.1) and by [DG] in case (1.2). □

Remark 3.2. It is not difficult to see that, by Theorem (1.1) one gets an explicit example of a smooth irreducible curve with surjective Wahl map for all integers g such that $g \geq 149$ or $g \in [76, 81] \cup [90, 95] \cup [105, 110] \cup [114, 126] \cup [131, 143]$, thus reproving the main result of [CHM] for these genera.

References

[AC] Arbarello, E., Cornalba, M., Footnotes to a paper of Beniamino Segre. The number of g_d^1's on a general d-gonal curve, and the unirationality of the Hurwitz spaces of 4-gonal and 5-gonal curves, Math. Ann. 256 (1981), 341–362.

[AH] d'Almeida, J., Hirschowitz, A., Quelques plongements non-speciaux de surfaces rationelles, Math. Z. 211 (1992), 479–483.

[BEL] Bertram, A., Ein, L., Lazarsfeld, R., Surjectivity of Gaussian maps for line bundles of large degree on curves, in: Algebraic Geometry, Proceedings Chicago 1989, Lecture Notes in Math. 1479, Springer-Verlag, Berlin–New York 1991, 15–25.

[CHM] Ciliberto, C., Harris, J., Miranda, R., On the surjectivity of the Wahl map, Duke Math. J. 57 (1988), 829–858.

[CL] Ciliberto, C., Lopez, A. F., On the number of moduli of extendable canonical curves, preprint 1998.

[CLM1] Ciliberto, C., Lopez, A. F., Miranda, R., Projective degenerations of K3 surfaces, Gaussian maps and Fano threefolds, Invent. Math. 114 (1993), 641–667.

[CLM2] Ciliberto, C., Lopez, A. F., Miranda, R., Classification of varieties with canonical curve section via Gaussian maps on canonical curves, Amer. J. Math. 120 (1998), 1–21.

[CLM3] Ciliberto, C., Lopez, A. F., Miranda, R., On the corank of Gaussian maps for general embedded K3 surfaces, Israel Mathematical Conference Proceedings, Papers in honor of Hirzebruch's 65th birthday, Amer. Math. Soc. Publ. 9, 1996, 141–157.

[CM] Ciliberto, C., Miranda, R., Gaussian maps for certain families of canonical curves, in: Complex Projective Geometry, Proc. Trieste–Bergen 1989, London Math. Soc. Lecture Notes Ser. 179, Cambridge University Press, 1992, 106–127.

[DG] Davis, E. D., Geramita, A. V., Bese's very ampleness theorem and punctured complete intersections, The curves seminar at Queen's, Vol. IV, Queen's Papers in Pure and Appl. Math. 76, Exp. No. G. Queen's Univ., Kingston, ON, 1986.

[K] Kang, E., On the corank of a Wahl map on smooth curves with a plane model, Kyushu J. Math. 50 (1996), 471–492.

[P] Pareschi, G., Gaussian maps and multiplication maps on certain projective varieties, Compositio Math. 98 (1995), 219–268.

[V] Voisin, C., Sur l'application de Wahl des courbes satisfaisant la condition de Brill–Noether–Petri, Acta Math. 168 (1992), 249–272.

[W1] Wahl, J., The Jacobian algebra of a graded Gorenstein singularity, Duke Math. J. 55 (1987), 843–871.

[W2] Wahl, J., On cohomology of the square of an ideal sheaf, J. Algebraic Geom. 6, (1997), 481–511.

[W3] Wahl, J., Introduction to Gaussian maps on an algebraic curve, in: Complex Projective Geometry, Trieste–Bergen 1989, London Math. Soc. Lecture Notes Ser. 179, Cambridge University Press, 1992, 304–323.

A remark on projective embeddings of varieties with non-negative cotangent bundles

Lawrence Ein*, Bo Ilic** and Robert Lazarsfeld***

Introduction

The purpose of this note is to establish an elementary but somewhat unexpected lower bound on the degrees of projective embeddings of varieties with numerically effective cotangent bundles, including in particular quotients of bounded domains.

In recent years, there has been interest in understanding the geometry of complex projective varieties whose tangent or cotangent bundles satisfy various positivity properties. In this note, we shall be concerned with smooth complex projective varieties X satisfying the following non-negativity property:

(NCB). *The cotangent bundle Ω_X^1 of X is numerically effective (nef).*

By definition, the condition means that the Serre line bundle $\mathcal{O}_{\mathbf{P}(\Omega_X^1)}(1)$ on the projectivization $\mathbf{P}(\Omega_X^1)$ is numerically effective, or equivalently that for any non-constant map $\nu : C \longrightarrow X$ from a smooth curve C to X, any quotient bundle of $\nu^*\Omega_X^1$ has non-negative degree. Property (NCB) is satisfied, for example, by smooth subvarieties of abelian varieties, by varieties uniformized by the ball or other irreducible Hermitian symmetric spaces (cf. [Mok], §1), and by products and submanifolds thereof. More generally, a very nice theorem of Kratz ([Kratz], Theorem 2) states that any projective manifold whose universal covering is a bounded domain in $\mathbf{C}^n$ satisfies (NCB).

Our result is that if X satisfies (NCB), then the degree of X in any projective embedding must grow essentially exponentially in the dimension of X. Specifically, given a positive integer n, define

$$\delta(n) = 2^{[\sqrt{n}]},$$

where as usual $[x]$ denotes the integer part of x.

---

* Partially supported by NSF Grant DMS 96-22540
** Partially supported by an NSERC postdoctoral fellowship
*** Partially supported by NSF Grant DMS 97-13149

Theorem. *Let X be a smooth projective variety of dimension n which satisfies Property (NCB), and let*
$$f : X \longrightarrow \mathbf{P}^n$$
be any finite surjective mapping. Then $\deg(f) \geq \delta(n)$. *In particular, the degree of X in any projective embedding* $X \subset \mathbf{P}^r$ *must be at least* $\delta(n)$.

We suspect that these statements are not optimal, and that there should be genuinely exponential, or even factorial, bounds on the degree. It would be interesting to prove results along these lines. More philosophically, these results suggest that the complexity of the projective geometry associated to varieties satisfying (NCB) grows exponentially with their dimension. It would be interesting to know if one could make this viewpoint precise, and whether it has any other manifestations.

The proof of the theorem requires only a few lines, and in fact the two ingredients that enter into the argument are at least implicitly quite well known. One simply notes that the hypothesis (NCB) forces the presence of points where the derivative of f drops rank substantially, and that this in turn leads to a lower bound on $\deg(f)$. Nonetheless, the conclusion came as something of a surprise to us: while linear bounds on the degree are very familiar (e.g. [GL], Theorem 2), the existence of essentially exponential statements seems to have been overlooked.

The third author had the opportunity to discuss some of these matters with Michael Schneider about a year before his death, and as always Michael was enthusiastic and encouraging. We hope therefore that the present note might not be out of place in this volume dedicated to his memory. Schneider contributed a lot to algebraic geometry on both a personal and a professional level, and he will be greatly missed.

The proof of the main result occupies in §1. Some applications and variants appear in §2. We are grateful to D. Burns and N. Mok for some valuable discussions. We particularly wish to thank Mohan Ramachandran for informing us of Kratz's theorem.

1. Proof of the theorem

We start with a lemma on degrees and singularities of branched coverings. It was suggested by some examples of Flenner and Ran alluded to in [Ran 2].

Lemma 1.1. *Let* $f : X \longrightarrow Y$ *be a finite surjective map of smooth complex varieties of dimension n. Fix a point* $x \in X$, *let* $y = f(x) \in Y$, *and denote by* $e_f(x)$ *the local degree of f at x, i.e. the multiplicity of x in its fibre* $f^{-1}f(x)$. *Suppose that derivative* $df_x : T_x X \longrightarrow T_y Y$ *of f at x has rank* $n - k$. *Then* $e_f(x) \geq 2^k$, *and consequently* $\deg(f) \geq 2^k$.

Proof. By hypothesis, the co-derivative $df_x^* : T_y^* Y \longrightarrow T_x^* X$ has a k-dimensional kernel. Denoting by $m_x \subset \mathcal{O}_x X$ and $m_y \subset \mathcal{O}_y Y$ the maximal ideals of x and y

respectively, we can therefore choose a system of parameters $u_1, \ldots, u_n \in m_y$ in such a way that $f^*u_1, \ldots, f^*u_k \in m_x^2$. Now

$$e_f(x) = \dim_{\mathbf{C}} \mathcal{O}_x X / f^* m_y,$$

i.e. $e_f(x)$ is alternatively the intersection multiplicity at x of the (germs of) divisors defined by the f^*u_i. On the other hand, it is well known (cf [Fult, 12.4]) that this intersection multiplicity is at least the product of the multiplicities $\mathrm{ord}_x f^*u_i$ of the individual divisors. Since by construction $\mathrm{ord}_x f^*u_i \geq 2$ for $1 \leq i \leq k$, the stated lower bound on $e_f(x)$ follows. The inequality on $\deg(f)$ is then a consequence the fact that for fixed $y \in Y$,

$$\sum_{f(x)=y} e_f(x) = \deg(f).$$

$\square$

The plan is to apply the lemma to branched coverings of projective space. The following well-known fact, which we include for the convenience of the reader, will let us apply theorems on degeneracy loci to guarantee the existence of singularities.

Lemma 1.2. *Let X be a projective variety, and let E and F be vector bundles on X. If E is nef and F is ample, then $E \otimes F$ is ample.*

Sketch of Proof. The statement is a consequence of Kleiman's criterion (cf. [Hart]) that the nef cone is the closure of the ample cone, and the argument is most easily stated using the language of vector bundles twisted by **Q**-divisors, as in [Myka]. First, one verifies the statement when F is a line bundle, or more generally an ample Q-divisor: we leave this to the reader. Next, fix an ample line bundle H on X. Since E is nef, it follows that $E(\frac{1}{N}H)$ is ample for any $N > 0$, and since F is ample, $F(-\frac{1}{N}H)$ is ample for $N \gg 0$. Therefore $E \otimes F = E(\frac{1}{N}H) \otimes F(-\frac{1}{N}H)$ is ample. $\square$

Now we turn to the

Proof of the theorem. Assume that X is smooth projective variety of dimension n whose cotangent bundle Ω_X^1 is nef, and suppose given a branched covering $f : X \longrightarrow \mathbf{P}^n$. Let

$$S_i(f) = \{x \in X \mid \mathrm{rank}\, df_x \leq n - i\}.$$

This is an algebraic subset of X whose expected dimension is $n - i^2$ (cf. [Fult], Chapter 14). In particular, setting $k = [\sqrt{n}]$, $S_k(f)$ has non-negative postulated dimension. The asserted bound on $\deg(f)$ will follow from Lemma 1.1 as soon as we show that $S_k(f) \neq \emptyset$. But this is a consequence of [FL1] or [L, §2] or [FL2]. In fact, since the tangent bundle $T\mathbf{P}^n$ (and hence also $f^*T\mathbf{P}^n$) is ample, the hypothesis (NCB) implies by Lemma 1.2 that $\Omega_X^1 \otimes f^*T\mathbf{P}^n$ is an ample vector bundle on X. The cited results then guarantee that the vector bundle map $df : TX \longrightarrow f^*T\mathbf{P}^n$ must actually drop rank whenever it is dimensionally predicted to do so. Finally, given an embedding

$X \subset \mathbf{P}$ of X into some projective space, we get by projection a branched covering $f : X \longrightarrow \mathbf{P}^n$ whose degree is the degree of X in $\mathbf{P}$, and so $\deg(X) \geq \delta(n)$. □

Remark. Given a smooth variety X with nef cotangent bundle, and an ample line bundle L on X which is generated by its global sections, the theorem is equivalent to the assertion that $\int c_1(L)^n \geq \delta(n)$. It is perhaps worth noting that this bound can fail if L is not globally generated. For example, fixing n, let C be a smooth curve of genus $g \gg n$ which carries no g_n^1, and let $X = \operatorname{Sym}^n(C)$ be the nth symmetric product of C. The Abel–Jacobi map $X \longrightarrow \operatorname{Jac}^n(C)$ is an embedding, so X satisfies (NCB). On the other hand, upon choosing a base-point $P \in C$, $\operatorname{Sym}^{n-1}(C)$ embeds as a divisor in X (via $D \mapsto D + P$), and the corresponding line bundle $L = \mathcal{O}_X(\operatorname{Sym}^{n-1}(C))$ is ample (cf. [FL1, §2]). But $\int_X c_1(L)^n = 1$, as one sees from the fact that there is a unique effective divisor of degree n containing n given points of C.

2. Applications and variants

We begin with a simple application of the theorem:

Corollary 2.1. *Let A be an abelian variety of dimension m, and let $X \subset A$ be a smooth subvariety of dimension n. Assume that X is of general type. Then the top self-intersection of the canonical bundle of X satisfies the inequality:*

$$\int c_1(\mathcal{O}_X(K_X))^n \geq \delta(n).$$

Proof. The embedding $X \subset A$ gives rise to a Gauss mapping $\gamma : X \longrightarrow \mathbf{G}$ of X into the Grassmannian $\mathbf{G} = \mathbf{G}(n, m)$ of n-dimensional subspaces of $T_0 A$, which is generically finite since X is of general type (cf. [Mori, §3]). A theorem of Ran [Ran1] implies that then γ is actually finite. On the other hand, the Plücker line bundle $\mathcal{O}_\mathbf{G}(1)$ on $\mathbf{G}$ pulls back to the canonical bundle on X. Therefore the canonical bundle $\mathcal{O}_X(K_X)$ is ample and globally generated. But X – like any submanifold of A – satisfies Property (NCB), and the desired inequality then follows from the theorem. □

We next prove a variant of the theorem for certain smooth subvarieties of projective space:

Proposition 2.2. *Let $X \subset \mathbf{P}^{n+e} = \mathbf{P}$ be a smooth subvariety of projective space having dimension n and codimension e, and denote by $N = N_{X/\mathbf{P}}$ the normal bundle to X in $\mathbf{P}$. If $N(-1)$ is ample, then*

$$\deg(X) \geq \min\left\{2^e, \delta(n)\right\}.$$

Recall that the hypothesis on $N(-1)$ is equivalent to requiring that every hyperplane tangent to X be tangent at only finitely many points. Note that we do not assume here that X satisfies (NCB). Observe also that if $e^2 \leq n$, then the stated bound $\deg(X) \geq 2^e$ is best possible for a complete intersection of quadrics.

Proof of Proposition 2.2. Fix a linear space L^{e-1} disjoint from X, and project from L to get a finite mapping $f : X \longrightarrow \mathbf{P}^n$. Setting $k = \min\{e, [\sqrt{n}]\}$, we will show that the singularity locus $S_k(f)$ appearing in the proof of the theorem is non-empty, and then the result will follow as above from Lemma 1.1. To this end, recalling that $\mathbf{P}^{n+e} - L$ is the total space of $\mathcal{O}_{\mathbf{P}^n}(1)^{\oplus e}$, one finds the exact sequence of bundles on X:

$$0 \longrightarrow \mathcal{O}_X(1)^{\oplus e} \longrightarrow T\mathbf{P}^{n+e}|X \longrightarrow f^*T\mathbf{P}^n \longrightarrow 0.$$

Combining this with the sequence

$$0 \longrightarrow TX \longrightarrow T\mathbf{P}^{n+e}|X \longrightarrow N_{X/\mathbf{P}} \longrightarrow 0,$$

we arrive at a mapping of vector bundles

$$u : \mathcal{O}_X(1)^{\oplus e} \longrightarrow N_{X/\mathbf{P}}$$

on X whose degeneracy loci are the same as the degeneracy loci of the derivative $df : TX \longrightarrow f^*T\mathbf{P}^n$. Since the bundle $N_{X/\mathbf{P}}(-1)$ is ample, the results cited in the proof of the theorem imply that $S_k(u) \neq \emptyset$, as desired. □

Exercise 2.3. Suppose that $X \subset \mathbf{P}^{n+e}$ is a smooth subvariety having the property that for some $x \in X$ the embedded tangent space $T_x X \subset \mathbf{P}^{n+e}$ meets X at only finitely many points (so that in particular $e \geq n$). Then $\deg(X) \geq 2^n$.

An argument similar to the one proving Proposition 2.2 also leads to the following generalization of the Main Theorem:

Proposition 2.4. *Let X be a smooth variety of dimension n satisfying (NCB), and let E be an ample vector bundle of rank e on X which is generated by its global sections. Then*

$$\int_X s_n(E) \geq \min\{2^n, \delta(n+e-1)\},$$

where $s_n(E)$ denotes the nth Segre class of E.

Outline of proof. In brief, consider the projective bundle $\pi : \mathbf{P}(E) \longrightarrow X$, and fix a general subspace $V \subset H^0(X, E)$ of dimension $n + e$ generating E. This gives rise to a finite mapping $f : \mathbf{P}(E) \longrightarrow \mathbf{P}(V) = \mathbf{P}^{n+e-1}$ whose degree is equal to $\int s_n(E)$. Setting $k = \min\{n, [\sqrt{n+e-1}]\}$, it is enough as above to show that the singularity locus $S_k(f)$ is non-empty. To this end, let M be the vector bundle of rank n on X

defined by the exact sequence

(*) $$0 \longrightarrow M \longrightarrow V \otimes_{\mathbf{C}} \mathcal{O}_X \longrightarrow E \longrightarrow 0,$$

the homomorphism on the right being the canonical evaluation map. Now f factors through the embedding $\mathbf{P}(E) \subset \mathbf{P}(V \otimes_{\mathbf{C}} \mathcal{O}_X) = X \times \mathbf{P}(V)$ determined by (*), and as in the proof of the proposition, the degeneracy loci of df coincide with those of the resulting vector bundle map

$$u : \pi^* TX \longrightarrow N_{\mathbf{P}(E)/X \times \mathbf{P}(V)}.$$

But $\mathbf{P}(E)$ is cut out in $X \times \mathbf{P}(V)$ by a section of $pr_1^* M^* \otimes pr_2^* \mathcal{O}_{\mathbf{P}(V)}(1)$, and consequently $N_{\mathbf{P}(E)/X \times \mathbf{P}(V)} = \pi^* M^* \otimes \mathcal{O}_{\mathbf{P}(E)}(1)$, which by Lemma 1.2 is ample thanks to the amplitude of E and the fact that M^* is globally generated. As $\pi^* \Omega_X^1$ is nef by assumption, it follows that $\pi^* \Omega_X^1 \otimes (\pi^* M^* \otimes \mathcal{O}_{\mathbf{P}(E)}(1))$ is ample. But then [FL1] or the other references cited above guarantee that u must actually drop rank whenever it is dimensionally predicted to do so. □

Remark 2.5. The inequalities established in this note all spring via Lemma 1.1 from producing singularities of a branched covering of projective space. It would be interesting to know whether one can recover or improve these statements by applying positivity theorems to some well-chosen Chern class calculations. It is natural to wonder in particular whether the inequalities of [BSS] might not be relevant here.

References

[BSS] M. Beltrametti, M. Schneider, A. Sommese, Chern inequalities and spannedness of adjoint bundles. In: Proceedings of the Hirzebruch 65 Conference on Algebraic Geometry (Ramat Gan, 1993), Israel Math. Conf. Proc. 9, 1996, 165–198.

[Fult] W. Fulton, Intersection Theory, Ergeb. Math. Grenzgeb. (3) 2, Springer-Verlag, New York–Berlin–Heidelberg 1984.

[FL1] W. Fulton, R. Lazarsfeld, On the connectedness of degeneracy loci and special divisors. Acta Math. 146 (1981), 271–283.

[FL2] W. Fulton, R. Lazarsfeld, Positive polynomials for ample vector bundles. Ann. of Math. 118 (1983), 35–60.

[GL] T. Gaffney, R. Lazarsfeld, On the ramification of branched coverings of $\mathbf{P}^n$. Invent. Math. (1980), 53–38.

[Hart] R. Hartshorne, Ample subvarieties of algebraic varieties. Lecture Notes in Math. 156, Spinger-Verlag, 1970.

[Kratz] H. Kratz, Compact complex manifolds with numerically effective cotangent bundles. Documenta Math. 2 (1997), 183–193.

[L] R. Lazarsfeld, Some applications of the theory of positive vector bundles. In: Complete Intersections (Acriele 1983). Lecture Notes in Math. 1092 (1984), 29–61.

[Myka] Y. Miyaoka, The Chern classes and Kodaira dimension of a minimal variety. In: Algebraic geometry, Sendai 1985. Adv. Stud. Pure Math. 10 (1987), 449–476.

[Mok] N. Mok, Uniqueness theorems of Hermitian metrics of seminegative curvature on quotients of bounded symmetric domains. Ann. of Math. 125 (1987), 105–152.

[Mori] S. Mori, Classification of higher dimensional varieties. In: Algebraic Geometry, Bowdoin 1985. Proc. Sympos. Pure Math. 46 (1987), 269–331.

[Ran1] Z. Ran, The structure of Gauss-like maps. Compositio Math. 52 (1984), 171–177.

[Ran2] Z. Ran, The (dimension +2)-secant lemma. Invent. Math. 106 (1991), 65–71.

The fundamental group's structure of the complement of some configurations of real line arrangements

David Garber and Mina Teicher

Abstract. In this paper, we give a fully detailed exposition of computing fundamental groups of complements of line arrangements, using the Moishezon–Teicher technique for computing the braid monodromy of a curve and the Van Kampen theorem which induces a presentation of the fundamental group of the complement from the braid monodromy of the curve. For example, we treated the cases where the arrangement has t multiple intersection points and the rest are simple intersection points. In this case, the fundamental group of the complement is a direct sum of infinite cyclic groups and t free groups. Hence, the fundamental groups in these cases are "big". These calculations will be useful in computing the fundamental group of Hirzebruch covering surfaces.

Contents

1 **Introduction**   174

2 **Preliminaries**   176
   2.1 Some background 176
   2.2 Definition of g-base 177
   2.3 Braid group and braid monodromy 178
   2.4 The braid monodromy of a real line arrangement 181
   2.5 The algorithm of Moishezon–Teicher 183
   2.6 The Van Kampen theorem 184
   2.7 An application of the Van Kampen theorem 187
   2.8 Outline of the computation of the fundamental group of the complement of line arrangements 188

3 **Arrangements with t non-collinear multiple points**   189
   3.1 The affine case 189
   3.2 The projective case 190

4 Arrangements with t collinear multiple points   193
   4.1 The affine case 194
   4.2 Proof of Lemma 4.2 194
      4.2.1 First case – with the restriction. 194
      4.2.2 Second case – without the restriction 200
   4.3 Proof of Lemma 4.3 209
   4.4 Proof of Theorem 4.1 214
   4.5 The projective case 217

5 Arrangements with more than one equivalence class   218
   5.1 The definition of the equivalence relation 218
   5.2 The affine case 218
   5.3 The projective case 220

6 Results concerning the bigness of the fundamental group   221

1. Introduction

In this paper, we give a fully detailed exposition of calculations of fundamental groups of the complements of certain configurations of real line arrangements, using the Moishezon–Teicher algorithm (which calculates the braid monodromy of curves), the Van Kampen theorem (which induces a finite presentation, in terms of generators and relations, of the fundamental group of curves' complements, from its braid monodromy), and some group computations.

In particular, we got:

1. Let $\mathcal{L}$ be a real line arrangement which is a union of t subsets of lines each of which consists of $k_i + 1$ lines meeting in a single point, and any two lines belonging to different subsets meet in a simple point. Then:

$$\pi_1(\mathbb{C}^2 - \mathcal{L}, u_0) \cong \left(\bigoplus_{i=1}^{t} \mathbb{F}^{k_i}\right) \oplus \mathbb{Z}^t$$

and

$$\pi_1(\mathbb{CP}^2 - \mathcal{L}, u_0) \cong \left(\bigoplus_{i=1}^{t} \mathbb{F}^{k_i}\right) \oplus \mathbb{Z}^{t-1}.$$

2. Let $\mathcal{L}$ be a real line arrangement which consists of t subsets of lines each of which consists of $k_i + 1$ lines meeting in a single point and all the t multiple

points lie on the same line $L \in \mathcal{L}$. Then:

$$\pi_1(\mathbb{C}^2 - \mathcal{L}, u_0) \cong \left(\bigoplus_{i=1}^{t} \mathbb{F}^{k_i}\right) \oplus \mathbb{Z}$$

and

$$\pi_1(\mathbb{CP}^2 - \mathcal{L}, u_0) \cong \bigoplus_{i=1}^{t} \mathbb{F}^{k_i}.$$

3. Generalizations: Let $\mathcal{L}$ be a real line arrangement in $\mathbb{CP}^2$ consists of n lines. We choose the line at infinity such that all the lines are intersected in $\mathbb{C}^2$. Assume that there are k multiple intersection points $p_1, \ldots, p_k$ with multiplicities $m_1, \ldots, m_k$ respectively. Assume also that all the multiple intersection points in every equivalence class (of multiple points) are collinear, i.e. in every equivalence class (of multiple points) there is a unique line of $\mathcal{L}$ which all the multiple points of that class lie on it. Then:

$$\pi_1(\mathbb{C}^2 - \mathcal{L}, u_0) \cong \bigoplus_{i=1}^{k} \mathbb{F}^{m_i-1} \oplus \mathbb{Z}^{n-(\sum_{i=1}^{k}(m_i-1))}$$

and

$$\pi_1(\mathbb{CP}^2 - \mathcal{L}, u_0) \cong \bigoplus_{i=1}^{k} \mathbb{F}^{m_i-1} \oplus \mathbb{Z}^{n-1-(\sum_{i=1}^{k}(m_i-1))}$$

The number of infinite cyclic groups in the affine case is a sum of two numbers: the number of equivalence classes (see definitions in Section 5) and the number of lines which have only simple intersection points.

4. Therefore, in all the above cases, the fundamental group is "big".

We will organize the paper as follows: in Section 2, we introduce the needed background for the techniques which will be used, and we give a detailed description of the Moishezon–Teicher algorithm for the case of line arrangements and the Van Kampen theorem.

In Section 3, we compute the structure of the fundamental group of the complement of a line arrangement which consists of t subsets of lines and the multiple points are not collinear.

In Section 4, we compute the structure of the fundamental group of the complement of a line arrangement which consists of t subsets of lines and the multiple points are collinear.

In Section 5, we generalize the results of the calculations of Sections 3 and 4.

In Section 6, we discuss the bigness of the groups which have been treated.

2. Preliminaries

2.1. Some background

This topic starts with Zariski, who proved in [Z, p. 317] that:

Proposition 2.1 (Zariski). *The fundamental group of the complement of n lines in general position is abelian.*

Among the modern works on this topic, one can mention [Fa1], [Fa2], [OS], [Sa], [Ra] and more.

Moishezon and Teicher developed an algorithm for computing fundamental groups of complements of branch curves of generic projection of surfaces of general type (see [MoTe1], [MoTe2]). This algorithm can be used also for computing fundamental groups of complement of line arrangements. In this paper we give a detailed exposition of this technique in some configurations of line arrangements.

Simultaneously and independently, by entirely different methods, Fan proved in [Fa1], [Fa2] the following results for the projective case:

Proposition 2.2 (Fan). *Let $\Sigma = \bigcup l_i$ be a line arrangement in $\mathbb{CP}^2$ and assume that there is a line L of Σ such that for any singular point S of Σ with multiplicity ≥ 3, we have $S \in L$. Then: $\pi_1(\mathbb{CP}^2 - \Sigma)$ is isomorphic to a direct product of free groups.*

Proposition 2.3 (Fan). *Let Σ be an arrangement of n lines and $S = \{a_1, \ldots, a_k\}$ be the set of all singularities of Σ with multiplicity ≥ 3. Suppose that $\beta(\Sigma) = 0$, where $\beta(\Sigma)$ is the first Betti number of the subgraph of Σ which contains only the higher singularities (i.e. with multiplicity ≥ 3) and their edges. Then:*

$$\pi_1(\mathbb{CP}^2 - \Sigma) \cong \mathbb{Z}^r \oplus \mathbb{F}^{m(a_1)-1} \oplus \cdots \oplus \mathbb{F}^{m(a_k)-1}$$

where $r = n + k - 1 - m(a_1) - \cdots - m(a_k)$.

It has to be noted that the assumption $\beta(\Sigma) = 0$ is equivalent to the assumption that Σ is a union of trees. The r in the last proposition is actually a sum of two combinatorial ingredients: the number of the trees in Σ minus 1 and the number of lines which are intersected only in simple intersection points.

Oka and Sakamoto proved in [OS] the following theorem, which will be a useful tool in some of our calculations:

Theorem 2.4 (Oka–Sakamoto). *Let C_1 and C_2 be algebraic plane curves in $\mathbb{C}^2$. Assume that the intersection $C_1 \cap C_2$ consists of distinct $d_1 \cdot d_2$ points, where d_i ($i = 1, 2$) are the respective degrees of C_1 and C_2.*
Then:

$$\pi_1(\mathbb{C}^2 - (C_1 \cup C_2)) \cong \pi_1(\mathbb{C}^2 - C_1) \oplus \pi_1(\mathbb{C}^2 - C_2)$$

Our computations on the fundamental groups of complements of line arrangements have applications to the fundamental groups of complements of branch curves, which is an important invariant of surfaces [Te2] (when we degenerate a surface to a union of planes, the branch curve degenerates to a union of lines). Moreover, the methods of this paper are important tools in the computations of the fundamental groups of Hirzebruch covering surfaces.

2.2. Definition of g-base

Here, we will present the required definitions and results for the presentation of the algorithm of Moishezon–Teicher. We follow the presentation of [MoTe1].

In this section, we will define the notion of *g-base (good geometric base)* for $\pi_1(D - K, *)$, where K is a finite set in a disk D. For this definition, we have to define:

Definition 2.1 ($l(\gamma)$). Let D be a disk. Let w_i, $i = 1, \ldots, n$, be small disks in $\text{Int}(D)$ such that:

$$w_i \cap w_j = \emptyset, \quad \forall i \neq j.$$

Let $u \in \partial D$. Let γ be a simple path connecting u with one of the w_i's, say w_{i_0}, which does not meet any other w_j, $j \neq i_0$.

We assign to γ a loop $l(\gamma)$ (actually an element of $\pi_1(D - K, u)$) as follows: let c be a simple loop equal to the (oriented) boundary of a small neighbourhood V of w_{i_0} chosen such that $\gamma' = \gamma - V \cap \gamma$ is a simple path.

Then: $l(\gamma) = \gamma' \cup c \cup (\gamma')^{-1}$ (we will not distinguish between $l(\gamma)$ and its representative in $\pi_1(D - K, u)$).

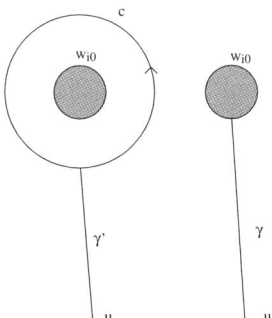

Definition 2.2 (Bush, g-base (good geometric base)). Let D be a disk, $K \subset D, \#K < \infty$. Let $u \in D - K$. A set of simple paths $\{\gamma_i\}$ is a *bush* in (D, K, u), if $\forall i, j$, $\gamma_i \cap \gamma_j =$

u; $\forall i$, $\gamma_i \cap K$ = one point, and γ_i are ordered counterclockwise around u. Let $\Gamma_i = l(\gamma_i) \in \pi_1(D - K, u)$ be a loop around $K \cap \gamma_i$ determined by γ_i. $\{\Gamma_i\}$ is called a *g-base* of $\pi_1(D - K, u)$.

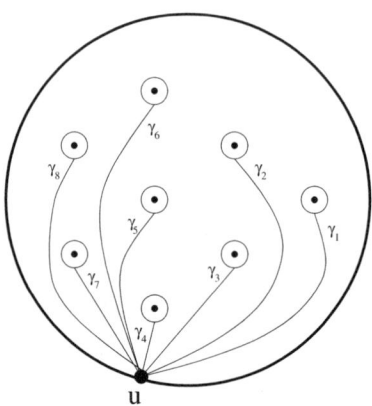

2.3. Braid group and braid monodromy

Let D be a closed disk in $\mathbb{R}^2$, $K \subset D$ a finite set, $u \in \partial D$. In such a case, we can define the *braid group* $B_n[D, K]$ ($n = \#K$):

Definition 2.3 (Braid group $B_n[D, K]$). Let $\mathcal{B}$ be the group of all diffeomorphisms β of D such that $\beta(K) = K$, $\beta|_{\partial D} = \text{Id}|_{\partial D}$. Such diffeomorphism acts naturally on $\pi_1(D - K, u)$. We say that two such diffeomorphisms are equivalent if they define the same automorphism on $\pi_1(D - K, u)$. The quotient of $\mathcal{B}$ by this equivalence relation is called the *braid group* $B_n[D, K]$. An element of $B_n[D, K]$ is called a *braid*. A composition of braids is from *left to right*.

Let us now define the concept of a *half-twist braid*. After fixing an orientation on $\mathbb{R}^2$, we can define a simple path σ such that $[\sigma] \subseteq (D - \partial D - K) \cup \{a, b\}$, σ connects a with b ($a, b \in K$). Choose now a small regular neighbourhood U of σ, and an orientation preserving diffeomorphism $f : \mathbb{R}^2 \to \mathbb{C}$ ($\mathbb{C}$ is taken with the usual "complex" orientation) such that $f(\sigma) = [-1, 1]$, $f(U) = \{z \in \mathbb{C} \mid |z| < 2\}$. Let $\alpha(x)$ be any real smooth monotone function such that

$$\alpha(x) = \begin{cases} 1 & x \in [0, \frac{3}{2}] \\ 0 & x \geq 2 \end{cases}.$$

With this function, we define a diffeomorphism $h : \mathbb{C} \to \mathbb{C}$ as follows: for any $z = re^{i\varphi} \in \mathbb{C}$, we define: $h(z) = re^{i(\varphi + \alpha(r)\pi)}$. It is clear that $\forall z, |z| \leq \frac{3}{2}$, $h(z)$ is a positive rotation on 180° and $h(z) = \text{Id } \forall z, |z| \geq 2$. After these preparations, we can define:

Definition 2.4 ($H(\sigma)$ – (positive) half-twist defined by σ). $H(\sigma)$ is the braid defined by $(f^{-1} \cdot h \cdot f)|_D$.

We have also another way to look at braids – via *motions* of K.

Definition 2.5 (Motion of K' to K). Let K and K' be the two sets $\{a_1, \ldots, a_n\}$ and $\{a'_1, \ldots, a'_n\}$. A motion of K' to K in D is n continuous functions $m_i : [0, 1] \to D$, $i = 1, \ldots, n$, such that:
(a) $\forall i, m_i(0) = a'_i, m_i(1) = a_i$.
(b) $\forall i \neq j, m_i(t) \neq m_j(t) \ \forall t \in [0, 1]$.

According to the following proposition, we can define a family of diffeomorphisms induced from the motion (under the condition that $K = K'$).

Proposition 2.5. *Given a motion $\mathcal{R}$, there exists a continuous family of diffeomorphisms $D_{\mathcal{R},t} : D \to D$, $t \in [0, 1]$, such that:*
(a) $D_{\mathcal{R},t}|_{\partial D} = \text{Id}|_{\partial D}$.
(b) $\forall t, i, D_{\mathcal{R},t}(a'_i) = m_i(t)$.

Definition 2.6 ($b_\mathcal{R}$ (braid induced from a motion $\mathcal{R}$)). When $K = K'$, $b_\mathcal{R}$ is the braid defined by the diffeomorphism $D_{\mathcal{R},1}$.

We define another important notion:

Definition 2.7 (Skeleton in (D, K, K'')). Let $K'' \subset K$, $K'' = \{b_1, \ldots, b_m\}$. A skeleton in (D, K, K'') is represented by a consecutive sequence of simple paths $(p_1, \ldots, p_{m-1})$ in $D - \partial D$ such that each p_i connects b_i to b_{i+1}. We say that two such sequences, say $(p_1, \ldots, p_{m-1}), (\tilde{p}_1, \ldots, \tilde{p}_{m-1})$, represent the same skeleton, if $H(p_i) = H(\tilde{p}_i)$, $i = 1, \ldots, m - 1$.

Before introducing the definition of *braid monodromy*, we have to make some more constructions. From now, we will work in $\mathbb{C}^2$. Let E (resp. D) be a closed disk on x-axis (resp. y-axis), and let C be a part of an algebraic curve in $\mathbb{C}^2$ located in $E \times D$. Let $\pi_1 : E \times D \to E$ and $\pi_2 : E \times D \to D$ be the canonical projections, and let $\pi = \pi_1|_C : C \to E$. Assume π is a proper map, and $\deg \pi = n$. Let $N = \{x \in E \mid \#\pi^{-1}(x) < n\}$, and assume $N \cap \partial E = \emptyset$. Now choose $M \in \partial E$ and let $K = K(M) = \pi^{-1}(M)$. By the assumption that $\deg \pi = n$ ($\Rightarrow \#K = n$), we can write: $K = \{a_1, a_2, \ldots, a_n\}$. Under these constructions, from each loop in $E - N$, we can define a braid in $B_n[M \times D, K]$ in the following way:

(1) Because $\deg \pi = n$, we can lift any loop in $E - N$ with a base point M to a system of n paths in $(E - N) \times D$ which start and finish at $\{a_1, a_2, \ldots, a_n\}$.

(2) Project this system into D (by π_2), to get n paths in D which start and end at the image of K in D (under π_2). These paths actually form a motion.

(3) Induce a braid from this motion, as we did in Definition 2.6.

To conclude, we can match a braid to each loop. Therefore, we get a map $\varphi : \pi_1(E - N, M) \to B_n[M \times D, K]$, which is also a group homomorphism which is called the *braid monodromy of C with respect to $E \times D, \pi_1, M$*.

For the next definitions, let us assume $M_0, M_1 \in E - N$ and $T : [0, 1] \to E - N$ be a path which connects M_0 with M_1. We know that there exists a continuous family of diffeomorphisms $\psi_{(t)} : M_0 \times D \to T(t) \times D$, $\forall t \in [0, 1]$, such that:
(a) $\psi_{(0)} = \text{Id}|_{M_0 \times D}$.
(b) $\forall t \in [0, 1]$, $\psi_{(t)}(\pi_1^{-1}(M_0) \cap C) = \pi_1^{-1}(T(t)) \cap C$.
(c) $\forall y \in \partial D$, $\psi_{(t)}(M_0, y) = (T(t), y)$.

In this situation, we can define the *Lefschetz diffeomorphism induced by T*:

Definition 2.8 (ψ_T, Lefschetz diffeomorphism induced by T).

$$\psi_T = \psi_{(1)} : M_0 \times D \tilde{\to} M_1 \times D.$$

Let $s = (x(s), y(s)) \in C$ be a singular point of π (i.e. $x(s) \in N$). Let $D'(s)$ be such a small disk on y-axis centered at $y(s)$ that $(x(s) \times D'(s)) \cap C = s$, i.e. there are no other branches of C which intersect $D'(s)$. Therefore, for any sufficiently small neighbourhood U of $x(s)$ on the x-axis centered at $x(s)$ such that $\forall x \in U - x(s)$, $\#(x \times \text{Int}(D'(s))) \cap C$ is independent of x (we call this number the *local degree of π at s* and denote it by $\deg_s \pi$). Let $k = \deg_s \pi$ and E' be a small closed disk on the x-axis centered at $x(s)$, such that $\forall x \in E' - x(s)$, $\#(x \times \text{Int}(D'(s))) \cap C = k$. Choose a point $a(s) \in \partial E'$ and let $T : [0, 1] \to \mathbb{C}$ be a path in $E - N - \text{Int}(E')$ connecting $a(s)$ to a point $M' \in E - N$. Let $K_{a,s} = (a(s) \times D'(s)) \cap C$.

Definition 2.9 ($\tilde{\psi}_T$, Lefschetz embedding induced by T). Let ψ_T be the Lefschetz diffeomorphism as defined above. Let T be as above, $a = a(s)$, $D' = D'(s)$. Then:

$$\tilde{\psi}_T = \psi_T|_{a \times D'} : a \times D' \to M' \times D.$$

Remark 2.10. Take k liftings of T to C starting at the different points of $K_{a,s} = (a \times D') \cap C$. These liftings are real curves in $T \times D$. We can think of $\tilde{\psi}_T$ as "pulling" of $a \times D'$ in $T \times D$ along these real curves.

Definition 2.11 ($\mathcal{L}_{T,s}$, Lefschetz injection induced by T). Consider $\tilde{\psi}_T : a \times D' \to M' \times D$, Lefschetz embedding induced by T.
Let $K(M') = (M' \times D) \cap C$. We have

$$\tilde{\psi}_T(K_{a,s}) \subset K(M'), (K(M') - \tilde{\psi}_T(K_{a,s})) \cap \tilde{\psi}_T(\text{Int}(D')) = \emptyset.$$

Therefore, the following canonical injection is well-defined:
$$\mathcal{L}_{T,s} = \psi_T^\vee : B_k[a \times D', K_{a,s}] \hookrightarrow B_n[M' \times D, K].$$

In order to define the *Lefschetz vanishing cycle*, we need the following definition:

Definition 2.12 (Linear frame of a braid group $B_n[D, K]$). Let K be the set $\{a_1, a_2, \ldots, a_n\}$. Let $\{\xi_1, \xi_2, \ldots, \xi_{n-1}\}$ be a system of straight line segments in $D - \partial D$ such that each ξ_i connects a_i with a_{i+1} (and does not intersect any other ξ_j except of end points). Let $H_i = H(\sigma_i)$. The ordered system of positive half-twists $(H_1, H_2, \ldots, H_{n-1})$ is called a *linear frame of* $B_n[D, K]$ *defined by* $\{\xi_1, \xi_2, \ldots, \xi_{n-1}\}$.

Now, we come to one of the most important definitions:

Definition 2.13 ($\mathcal{L}.\text{V.C.}(T, H')$, Lefschetz vanishing cycle induced by T). We call $\mathcal{L}.\text{V.C.}(T, H')$ a skeleton $\langle \xi_1, \ldots, \xi_{k-1} \rangle$ in $(M' \times D, K, \tilde{\psi}_T(K_{a,s}))$ corresponding $\mathcal{L}_{T,s}$ and a linear frame $(H') = (H'_1, \ldots, H'_{k-1})$ of $B_k[a \times D', K_{a,s}]$, that is $\mathcal{L}_{T,s}(H'_i) = H(\xi_i)$, $i = 1, \ldots, k-1$.

Because of the fact that such a linear frame is unique only when all the points of K are on a straight line in $D \subset \mathbb{R}^2$, $\mathcal{L}.\text{V.C.}(T, H')$ will be well-defined if all the points of $K_{a,s}$ are on a straight line in $a \times \mathbb{C}$. If all the points of $K_{a,s}$ are real, we will choose the unique linear frame $(H'_1, \ldots, H'_{k-1})$ determined by an increasing sequence of consecutive real segments on the real axis of $a \times \mathbb{C}$.

2.4. The braid monodromy of a real line arrangement

Definition 2.14 (Line arrangement in $\mathbb{CP}^2$). A *line arrangement in* $\mathbb{CP}^2$ is an algebraic curve in $\mathbb{CP}^2$ which is a union of projective lines.

If the lines are given by the linear forms $l_1, l_2, \ldots, l_k$, the union of the lines is the reducible curve defined by
$$l_1 l_2 \ldots l_k = 0.$$

We say that the arrangement is *real* if each line can be defined by an equation with real coefficients (i.e. each linear form l_i has real coefficients).

Let $\mathbb{C}^2 = \mathbb{CP}^2 -$ (projective line) be an affine part of $\mathbb{CP}^2$. Let E (resp. D) be a closed disk on x-axis (resp. on y-axis) with the center on the real part of x-axis (resp. y-axis). Let $\pi_1 : E \times D \to E$, $\pi_2 : E \times D \to D$ be the canonical projections.

Definition 2.15 (Real line arrangement in a polydisk $E \times D$). We say that C is a *real line arrangement in a polydisk* $E \times D$ (as above), if there exists a real line arrangement $\hat{C}$ in $\mathbb{CP}^2$, such that:
(a) $C = \hat{C} \cap (E \times D)$.
(b) $\forall x \in E$, $\pi_1^{-1}(x) \cap C \subset x \times \text{Int}(D)$.

Let $\pi = \pi_1|_C$, $n = \deg \pi$ (=number of lines in C), $N = \{x \in E \mid \#\pi^{-1}(x) < n\}$, $K_x = \pi^{-1}(x)$. Therefore, for any real $x \notin N$, we have n distinct real points $(x, y_i(x))$, $1 \leq i \leq n$, in K_x. We choose a numeration in $\{y_1(x), \ldots, y_n(x)\}$, such that $y_1(x) < y_2(x) < \cdots < y_n(x)$.

Let $\tilde{D} = \{z \in \mathbb{C} \mid |z - \frac{n+1}{2}| \leq \frac{n+1}{2}\}$, $\tilde{K} = \{1, 2, \ldots, n\} \subset \tilde{D}$ ($\tilde{D}$ is a model which simplifies the treatment with the theoretic calculations of the braid monodromy). Let $\tilde{H} = (\tilde{H}_1, \tilde{H}_2, \ldots, \tilde{H}_{n-1})$ be the linear frame of $B_n[\tilde{D}, \tilde{K}]$ defined by the sequence of real segments $\tilde{\xi} = ([1, 2], [2, 3], \ldots, [n-1, n])$, i.e. $\tilde{H}_j = H([j, j+1])$.

For the set $E'_{\mathbb{R}} = \{x \in E - N \mid x \text{ real}\}$, we can construct a set of diffeomorphisms $\{\beta_x \mid x \times D \tilde{\to} \tilde{D}\}$ with the following properties:

(a) $\beta_x(K_x) = \tilde{K}$.

(b) $\beta_x(x \times \text{real part of } D) = \text{real part of } \tilde{D}$ (order preserved).

(c) $\forall x, x' \in E'_{\mathbb{R}}, y \in \partial D$, $\beta_x(x, y) = \beta_{x'}(x', y)$.

(d) On each connected component $\tilde{\mathcal{L}}$ of $E'_{\mathbb{R}}$, $\{\beta_x \mid x \in \tilde{\mathcal{L}}\}$ is a continuous family of diffeomorphisms.

Let $\xi_x = \{\xi_{x,1}, \xi_{x,2}, \ldots, \xi_{x,n-1}\}$ ($x \in E'_{\mathbb{R}}$) be the sequence of real segments $[y_i(x), y_{i+1}(x)]$, $1 \leq i \leq n-1$, in $x \times D$ and let $H_x = (H_{x,1}, H_{x,2}, \ldots, H_{x,n-1})$ be the linear frame of $B_n[x \times D, K_x]$ defined by ξ_x.

Now, we assume that $\forall x_j \in N$, there is only one singular point of C over x_j.

Let $x_j \in N$. Choose $x'_j = x_j + \epsilon$, $\epsilon > 0$ a very small number. Let A_j be the singularity of C over x_j (i.e. $x(A_j) = x_j$), and let Y_j be the union of irreducible components of C containing A_j. In $\{y_1(x'_j), \ldots, y_n(x'_j)\}$, there is a subsequence with consecutive indices $\{y_{k_j}(x'_j), y_{k_j+1}(x'_j), \ldots, y_{l_j}(x'_j)\}$ which is equal to $K'_{x'_j} = Y_j \cap (x'_j \times D)$.

In this situation, we can define the following notions:

Definition 2.16 (Local Lefschetz vanishing cycle ($\mathcal{L}$.V.C.) of A_j). A skeleton in $(x'_j \times D, K_{x'_j}, K'_{x'_j})$ represented by the sequence of real segments

$$[y_{k_j+r-1}(x'_j), y_{k_j+r}(x'_j)], \quad 1 \leq r \leq l_j - k_j$$

is called a *local $\mathcal{L}$.V.C. of A_j*.

Definition 2.17 ((k_j, l_j), Lefschetz pair of A_j). The smallest and biggest indices k_j, l_j in the sequence considered above form a pair (k_j, l_j), which is called the *Lefschetz pair of A_j*.

Obviously, the local $\mathcal{L}$.V.C. of A_j is uniquely defined by the Lefschetz pair (k_j, l_j).

Definition 2.18 ($\langle k_j, l_j \rangle$, skeleton representing local $\mathcal{L}$.V.C. of A_j). Denote by $\langle k_j, l_j \rangle$ the skeleton in $(\tilde{D}, \tilde{K}, (k_j, k_j+1, \ldots, l_j))$ represented by consecutive real segments connecting points of $(k_j, k_j+1, \ldots, l_j)$.

Lemma 2.6. *Let γ be a simple path in $E - N$ connecting x_j with $M (\in \partial E)$, $[x_j, x'_j] \subset \gamma$. Let γ' be the part of γ from x'_j to M. Let*

$$\varphi : \pi_1(E - N, M) \to B_n[M \times D, K_M]$$

be the braid monodromy of C w.r.t. $E \times D, \pi_1, M$. Let Γ be the element represented by $l(\gamma)$.

Then:

$$\varphi(\Gamma) = \Delta^2 \langle \mathcal{L}.V.C.(\gamma', H(\langle \xi_x \rangle)) \rangle$$

(where, intuitively, $\Delta \langle$skeleton$\rangle$ is a generalized half-twist which is defined according to the skeleton, and $\Delta^2 \langle$skeleton$\rangle$ is applying this half-twist twice).

2.5. The algorithm of Moishezon–Teicher

Following Lemma 2.6, in order to calculate the braid monodromy, we have to find the appropriate Lefschetz vanishing cycles. This is given by the following theorem [MoTe1]:

Theorem 2.7 (Moishezon–Teicher). *Let $N = \{x_1, x_2, \ldots, x_q\}$ with $x_q < x_{q-1} < \cdots < x_1$, $M \in \partial E \cap$ (real axis), with $M > x_1$, and $\epsilon > 0$ a very small number. Let $T_j (1 \leq j \leq q)$ be the path from $x_j - \epsilon$ to $x_j + \epsilon$ along the semicircle below real axis centered at x_j.*

Let γ_j be the path from x_j to M defined by

$$\gamma_j = [x_j, x_{j-1} - \epsilon] \cdot T_{j-1} \cdot [x_{j-1} + \epsilon, x_{j-2} - \epsilon] \cdot T_{j-2} \ldots T_1 \cdot [x_1, M]$$

$$(\gamma_j = [x_j, x_{j-1} - \epsilon] \cdot T_{j-1} \cdot \left(\prod_{r=j-1}^{2} [x_r + \epsilon, x_{r-1} - \epsilon] \cdot T_{r-1} \right) \cdot [x_1, M]).$$

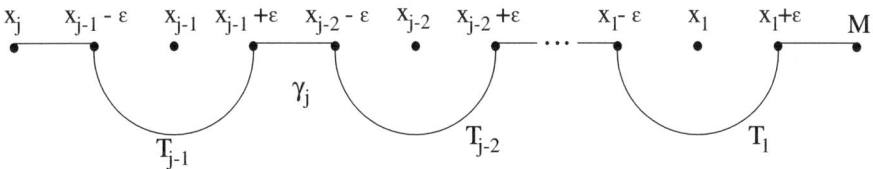

Considering $l(\gamma_j)$'s, we get a g-base $\{\delta_1, \delta_2, \ldots, \delta_q\}$ in $\pi_1(E - N, M)$.

Assume that for all x_j, $1 \leq j \leq q$, there is only one singular point A_j with $x(A_j) = x_j$. Let (k_j, l_j) be the Lefschetz pair of A_j, and $\langle k_j, l_j \rangle$ be the skeleton in $(\tilde{D}, \tilde{K}, (k_j, k_j + 1, \ldots, l_j - 1, l_j))$ representing local $\mathcal{L}$.V.C. of A_j. Let γ'_j be the part of γ_j from $x'_j = x_j + \epsilon$ to M.

Then:

$$\mathcal{L}.\text{V.C.}(\gamma_j') = \beta_M^{-1}(\langle k_j, l_j \rangle \cdot \prod_{m=j-1}^{1} \Delta \langle k_m, l_m \rangle)$$

(where $\prod_{m=j-1}^{1} \Delta \langle k_m, l_m \rangle = \Delta \langle k_{j-1}, l_{j-1} \rangle \cdot \Delta \langle k_{j-2}, l_{j-2} \rangle \ldots \Delta \langle k_1, l_1 \rangle \in B_n[\tilde{D}, \tilde{K}]$) and

$$\mathcal{L}.\text{V.C.}(\gamma_1') = \beta_M^{-1}(\langle k_1, l_1 \rangle).$$

According to this theorem, in order to compute the braid monodromy of a line arrangement, we have to do the following steps:

1. Check that the line arrangement fulfills the assumption that there are no more than one intersection point with the same x-coordinate (so we can apply the theorem).

2. Find the Lefschetz pairs of all the intersection points.

3. Calculate the Lefschetz vanishing cycle of every intersection point according to the last Theorem (2.7).

4. The braid monodromy is the Δ^2 of this $\mathcal{L}$.V.C.

2.6. The Van Kampen theorem

The Van Kampen theorem induces a finite presentation of the fundamental group of complements of curves by meaning of generators and relations. From this finite presentation, we will calculate the structure of the group in our cases (the original theorem is in [VK], other versions can be found at [Mo, pp. 127–130], [MoTe3], [MoTe4, ch. 13], [Te1]. The theorems presented here are from [MoTe3],[MoTe4] and [Te1]).

Let S be an algebraic curve in $\mathbb{C}^2$ ($p = \deg S$). Let $\pi = \pi_1 : \mathbb{C}^2 \to \mathbb{C}$ be the canonical projection on the first coordinate. Let $\mathbb{C}_x = \pi^{-1}(x)$, and now define: $K_x = \mathbb{C}_x \cap S$ (By assumption $\deg S = p$, we know $\#K_x \leq p$).

Let $N = \{x \mid \#K_x < p\}$. Choose now $u \in \mathbb{C}$, u real, such that $x \ll u$, $\forall x \in N$, and define: $B_p = B_p[\mathbb{C}_u, \mathbb{C}_u \cap S]$. Let $\varphi_u : \pi_1(\mathbb{C} - N, u) \to B_p$ be the braid monodromy of S w.r.t π, u. Also choose $u_0 \in \mathbb{C}_u$, $u_0 \notin S$, u_0 below real line far enough such that B_p does not move u_0. It is known that the group $\pi_1(\mathbb{C}_u - S, u_0)$ is free. There exists an epimorphism $\pi_1(\mathbb{C}_u - S, u_0) \to \pi_1(\mathbb{C}^2 - S, u_0)$, so a set of generators for $\pi_1(\mathbb{C}_u - S, u_0)$ determines a set of generators for $\pi_1(\mathbb{C}^2 - S, u_0)$.

In this situation, Van Kampen's theorem says:

Theorem 2.8 (Van Kampen's Theorem – classical version). *Let S be an algebraic curve, u, u_0, φ_u defined as above. Let $\{\delta_i\}$ be a g-base of $\pi_1(\mathbb{C} - N, u)$. Let $\{\Gamma_j \mid 1 \leq j \leq p\}$ ($p = \deg S$) be a g-base for $\pi_1(\mathbb{C}_u - S, u_0)$.*

Then, $\pi_1(\mathbb{C}^2 - S, u_0)$ is generated by the images of Γ_j in $\pi_1(\mathbb{C}^2 - S, u_0)$ and we get a complete set of relations from those induced from

$$(\varphi_u(\delta_i))(\Gamma_j) = \Gamma_j; \quad \forall i \forall j.$$

Here we present also the classical Van Kampen theorem for the projective case. The only difference between the affine case and the projective case is that there is one additional relation in the projective case – the multiplication of all the generators is equal to the identity of the group.

Theorem 2.9 (Classical Van Kampen Theorem for projective case). *Let S be an algebraic curve, u, u_0, φ_u defined as above. Let $\{\delta_i\}$ be a g-base of $\pi_1(\mathbb{C} - N, u)$. Let $\{\Gamma_j \mid 1 \leq j \leq p\}$ ($p = \deg S$) be a g-base for $\pi_1(\mathbb{C}_u - S, u_0)$.*

Then, $\pi_1(\mathbb{CP}^2 - S, u_0)$ is generated by the images of Γ_j in $\pi_1(\mathbb{C}^2 - S, u_0)$ and we get a complete set of relations from those induced from

$$(\varphi_u(\delta_i))(\Gamma_j) = \Gamma_j; \quad \forall i \forall j$$

with one additional relation:

$$\Gamma_p \Gamma_{p-1} \ldots \Gamma_1 = 1.$$

Oka [O] proved the following connection between the fundamental group of the affine case and the fundamental group of the projective case:

Theorem 2.10 (Oka). *Let C be a curve in $\mathbb{CP}^2$ and let L be a general line to C. Then, we have a central extension:*

$$1 \to \mathbb{Z} \to \pi_1(\mathbb{CP}^2 - (C \cup L)) \to \pi_1(\mathbb{CP}^2 - C) \to 1.$$

Due to the fact that L is in a general position to C, we can say:

$$\pi_1(\mathbb{CP}^2 - (C \cup L)) \cong \pi_1((\mathbb{CP}^2 - L) - C) \cong \pi_1(\mathbb{C}^2 - C)$$

(by choosing L as the line at infinity). Therefore, we get the following short exact sequence (see also [OS]):

$$1 \to \mathbb{Z} \to \pi_1(\mathbb{C}^2 - C) \to \pi_1(\mathbb{CP}^2 - C) \to 1.$$

We will show that in the cases which we treat, we get:

$$\pi_1(\mathbb{C}^2 - C) \cong \pi_1(\mathbb{CP}^2 - C) \oplus \mathbb{Z}$$

and therefore, this short exact sequence splits.

Now we return to the affine case. In order to give a more precise version of Van Kampen's theorem for cuspidal curves, i.e. for curves with only nodes and cusps as singularities, we need the following two lemmas.

Lemma 2.11. *Let V be a half-twist in $B_p[D, K]$, $u_0 \notin K$. Then: there exists $A_V, B_V \in \pi_1(D - K, u_0)$, such that:*
(a) *$\{A_V, B_V\}$ can be extended to a g-base of $\pi_1(D - K, u_0)$.*
(b) *$V(A_V) = B_V$.*

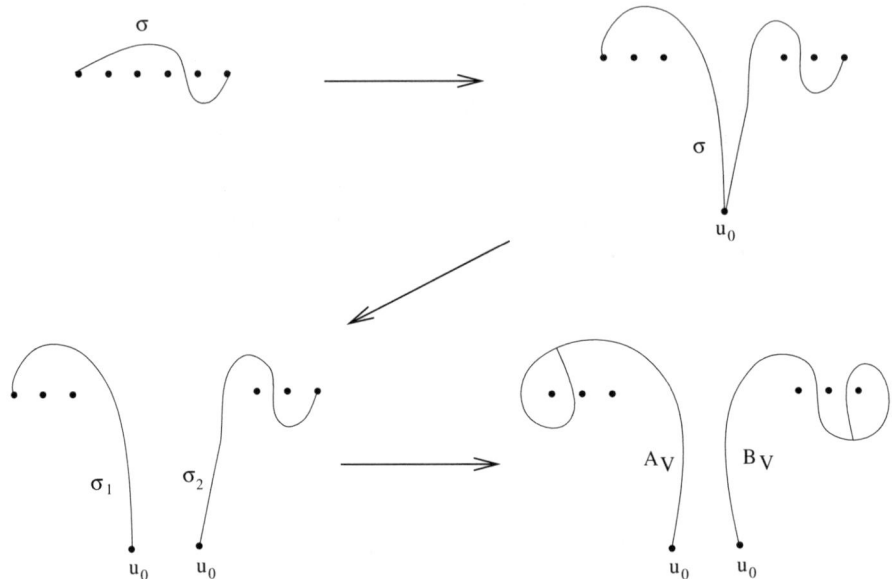

Let S be a cuspidal curve in $\mathbb{C}^2$ ($p = \deg S$). We assume that for every $x \in N$ (N as above), there is only one singular point over it (in $\mathbb{C}^2$). Thus, for every $x \in N$, let x' be the singular point over x. Because S is a cuspidal curve, the point x' is either a branch point, a node or a cusp.

Lemma 2.12. *Let $\{\delta_i\}$ be a g-base for $\pi_1(\mathbb{C} - N, u)$. For every δ_i, there exists V_i and ν_i, where V_i is a half-twist and ν_i is a number such that $\varphi_u(\delta_i) = V_i^{\nu_i}$. Moreover, $\nu_i = 1, 2, 3$ if c'_i (the singular point) = a branch point, a node or a cusp respectively.*

We denote:

$$[A, B] = ABA^{-1}B^{-1},$$

$$\langle A, B \rangle = ABAB^{-1}A^{-1}B^{-1}.$$

Now, we can give the precise version of the Van Kampen theorem for cuspidal curves:

Theorem 2.13 (Van Kampen's theorem for cuspidal curves). *Let S be a cuspidal curve, u, u_0, φ_u, A_{V_i}, B_{V_i} defined as above. Let $\{\delta_i\}$ be a g-base of $\pi_1(\mathbb{C} - N, u)$. Let $\varphi_u(\delta_i) = V_i^{\nu_i}$, V_i is a half-twist, $\nu_i = 1, 2, 3$ (as above).*

Let $\{\Gamma_j \mid 1 \leq j \leq p\}$ ($p = \deg S$) be a g-base for $\pi_1(\mathbb{C}_u - S, u_0)$. Then: $\pi_1(\mathbb{C}^2 - S, u_0)$ is generated by the images of Γ_j in $\pi_1(\mathbb{C}^2 - S, u_0)$ and we get a complete set of relations from those induced from $\varphi_u(\delta_i) = V_i^{\nu_i}$, as follows (when A_{V_i}, B_{V_i} are expressed in terms of $\{\Gamma_j\}$):

(a) $A_{V_i} = B_{V_i}$, when $\nu_i = 1$.

(b) $[A_{V_i}, B_{V_i}] = 1$, when $\nu_i = 2$.

(c) $\langle A_{V_i}, B_{V_i} \rangle = 1$, when $\nu_i = 3$.

What do we get from this theorem? After we calculate the appropriate braid monodromy, we can get a finite presentation of the desired fundamental group.

Note that it is easy to see that the relation, which is induced from the braid monodromy, is uniquely determined by the half-twist V, and is independent of the choice of A_V, B_V.

Now, we will present the version of Van Kampen's theorem for an arrangement with a single multiple point, i.e. an arrangement where all the lines meet in one point (the proof is easy, and can be found, for example, in [Ga, p. 25]):

Lemma 2.14 (Van Kampen's theorem for a single multiple point). *Let $l_1, \ldots, l_k$ be k real lines in $\mathbb{CP}^2$ meeting in a single point p. Let δ be a loop in $\pi_1(E - N, u_0)$ around $x(p)$. Let $\{\Gamma_1, \ldots, \Gamma_k\}$ be a g-base of $\pi_1(\mathbb{C}_{u_0} - \bigcup_{i=1}^k l_i)$.*
Then, the relations which are induced from this intersection point are:

$$\Gamma_k \Gamma_{k-1} \ldots \Gamma_1 = \Gamma_1 \Gamma_k \ldots \Gamma_3 \Gamma_2 = \cdots = \Gamma_{k-1} \Gamma_{k-2} \ldots \Gamma_1 \Gamma_k.$$

2.7. An application of the Van Kampen theorem

Here, we will prove a simple proposition, which will help us in the future. We denote $[x, y] = xyx^{-1}y^{-1}$ for x, y in a group G.

Proposition 2.15. *Let p be an intersection point of k real lines $l_{j_1}, \ldots, l_{j_k}$ in $\mathbb{CP}^2$. Let δ be a loop in $\pi_1(E - N, u_0)$ around $x(p)$. Let $\{\Gamma_{j_1}, \ldots, \Gamma_{j_k}\}$ be a g-base of $\pi_1(\mathbb{C}_{u_0} - \bigcup_{i=1}^k l_{j_i})$.*
Then, the relations which are induced from this intersection point are:

$$[\Gamma_{j_k} \Gamma_{j_{k-1}} \ldots \Gamma_{j_1}, \Gamma_{j_i}] = 1; \quad 1 \leq i \leq k.$$

Proof. By the Van Kampen version for a multiple point (2.14), the following set of relations is induced from the intersection point p:

$$\Gamma_{j_k} \Gamma_{j_{k-1}} \ldots \Gamma_{j_1} = \Gamma_{j_{k-1}} \ldots \Gamma_{j_1} \Gamma_{j_k} = \cdots = \Gamma_{j_1} \Gamma_{j_k} \ldots \Gamma_{j_2}.$$

We will prove now that this set of relations is equivalent to the set of relations in the formulation of the proposition.

($\Rightarrow$) Let $1 \leq i \leq k$. We have to show that
$$\Gamma_{j_k}\Gamma_{j_{k-1}}\cdots\Gamma_{j_1}\Gamma_{j_i} = \Gamma_{j_i}\Gamma_{j_k}\Gamma_{j_{k-1}}\cdots\Gamma_{j_1}.$$
We know (from the first set of relations) that

(*) $\quad \Gamma_{j_k}\Gamma_{j_{k-1}}\cdots\Gamma_{j_1} = \Gamma_{j_i}\Gamma_{j_{i-1}}\cdots\Gamma_{j_1}\Gamma_{j_k}\cdots\Gamma_{j_{i+1}},$

(**) $\quad \Gamma_{j_k}\Gamma_{j_{k-1}}\cdots\Gamma_{j_1} = \Gamma_{j_{i-1}}\Gamma_{j_{i-2}}\cdots\Gamma_{j_1}\Gamma_{j_k}\cdots\Gamma_{j_i}.$

Now:
$$(\Gamma_{j_k}\Gamma_{j_{k-1}}\cdots\Gamma_{j_1})\Gamma_{j_i} \stackrel{(*)}{=} (\Gamma_{j_i}\Gamma_{j_{i-1}}\cdots\Gamma_{j_1}\Gamma_{j_k}\cdots\Gamma_{j_{i+1}})\Gamma_{j_i}$$
$$= \Gamma_{j_i}(\Gamma_{j_{i-1}}\cdots\Gamma_{j_1}\Gamma_{j_k}\cdots\Gamma_{j_{i+1}}\Gamma_{j_i}) \stackrel{(**)}{=} \Gamma_{j_i}(\Gamma_{j_k}\Gamma_{j_{k-1}}\cdots\Gamma_{j_1}).$$

($\Leftarrow$) From the first relation we have:
$$[\Gamma_{j_k}\Gamma_{j_{k-1}}\cdots\Gamma_{j_1},\Gamma_{j_1}] = 1$$
i.e. $\Gamma_{j_k}\Gamma_{j_{k-1}}\cdots\Gamma_{j_1}\Gamma_{j_1} = \Gamma_{j_1}\Gamma_{j_k}\Gamma_{j_{k-1}}\cdots\Gamma_{j_1}$. Now, multiply it by $\Gamma_{j_1}^{-1}$ from the right to get:

(***) $\quad \Gamma_{j_k}\Gamma_{j_{k-1}}\cdots\Gamma_{j_1} = \Gamma_{j_1}\Gamma_{j_k}\cdots\Gamma_{j_2}.$

From the second relation we have: $(\Gamma_{j_k}\Gamma_{j_{k-1}}\cdots\Gamma_{j_1})\Gamma_{j_2} = \Gamma_{j_2}(\Gamma_{j_k}\Gamma_{j_{k-1}}\cdots\Gamma_{j_1})$, but from (***) we get: $(\Gamma_{j_1}\Gamma_{j_k}\cdots\Gamma_{j_2})\Gamma_{j_2} = \Gamma_{j_2}(\Gamma_{j_1}\Gamma_{j_k}\cdots\Gamma_{j_2})$. Now, multiply it by $\Gamma_{j_2}^{-1}$ from the right to get:
$$\Gamma_{j_1}\Gamma_{j_k}\cdots\Gamma_{j_2} = \Gamma_{j_2}\Gamma_{j_1}\Gamma_{j_k}\cdots\Gamma_{j_3}.$$
Applying the same argument together with the rest of the commutative relations give us the requested cyclic relations. $\square$

2.8. Outline of the computation of the fundamental group of the complement of line arrangements

Let us summarize the steps we have to follow in order to compute the fundamental group of the complement of a given real line arrangement $\mathcal{L}$:

(1) Calculation of the braid monodromy of $\mathcal{L}$:
- Check that the line arrangement fulfills the assumption that there are no more than one intersection point with the same x-coordinate (so we can apply the theorem).

- Find the Lefschetz pairs of all the intersection points.
- Calculate the Lefschetz vanishing cycle of every intersection point according to the Moishezon–Teicher theorem.

(2) Calculation of the relations induced on $\pi_1(\mathbb{C}^2 - \mathcal{L})$ from the braid monodromy:
- Choose u as in Section 2.6.
- Choose a g-base for $\pi_1(\mathbb{C}_u - \mathcal{L})$: $\{\Gamma_1, \ldots, \Gamma_n\}$.
- Calculate the A_{V_i}, B_{V_i} from the $\mathcal{L}$.V.C. for every singular point in terms of $\Gamma_i, i = 1, \ldots, n$.
- Find the induced relations according to the Van Kampen theorem.

(3) Computing the structure of $\pi_1(\mathbb{C}^2 - \mathcal{L})$ from the relations in (2). This step contains some group calculations and combinatorics.

3. Arrangements with t non-collinear multiple points

In this section, we are going to calculate the fundamental group of the complement of line arrangements where there is no line on which there are two multiple points. Thus, we can divide the arrangement into t subsets of lines where all the lines in each subset intersect at a single (multiple) point and any two such subsets intersect in simple points only. We define:

Definition 3.1 (Simple point, multiple point, multiplicity of a point). A *simple point* in a line arrangement is a point where two lines meet. A *multiple point* in a line arrangement is a point where more than two lines meet. The *multiplicity of a point* is the number of lines which meet in the point.

Definition 3.2 (An arrangement with t non-collinear multiple points). An *arrangement with t non-collinear multiple points* is an arrangement where there is no line on which there are two multiple points and we can divide it into t subsets of lines where all the lines in each subset intersect in a single multiple point.

We denote by $\mathbb{F}^k$ the free group with k generators.

3.1. The affine case

We calculate the affine case:

Theorem 3.1. *Let $\mathcal{L}$ be a real line arrangement in $\mathbb{CP}^2$ with t non-collinear multiple points. Let $k_i + 1$ be the multiplicity of the multiple point P_i, $1 \leq i \leq t$. Then:*

$$\pi_1(\mathbb{C}^2 - \mathcal{L}) \cong \left(\bigoplus_{i=1}^{t} \mathbb{F}^{k_i}\right) \oplus \mathbb{Z}^t.$$

Proof. Randell [Ra] showed that the fundamental group of the complement of a real line arrangement which consists of n lines meet in a single point is $\mathbb{F}^{n-1} \oplus \mathbb{Z}$.

We can observe $\mathcal{L}$ as a union of t subsets of lines $\mathcal{L}_i$, $1 \leq i \leq t$, where every such subset $\mathcal{L}_i$, $1 \leq i \leq t$, consists of $k_i + 1$ lines which are passing through the multiple point P_i (there is no $l \in \mathcal{L}_i \cap \mathcal{L}_j$, because then l connects P_i and P_j, a contradiction to the assumption). The degree of each $\mathcal{L}_i$ is exactly $k_i + 1$, because there are $k_i + 1$ lines which pass through the point P_i. Moreover, $\mathcal{L}_i \cap \mathcal{L}_j = (k_i + 1)(k_j + 1)$ points, because every line in $\mathcal{L}_i$ meets every line in $\mathcal{L}_j$.

Every $\mathcal{L}_i$, $1 \leq i \leq t$, consists of $k_i + 1$ lines which pass through the multiple point P_i. This is the configuration of Randell. Therefore:

$$\pi_1(\mathbb{C}^2 - \mathcal{L}_i) = \mathbb{F}^{k_i} \oplus \mathbb{Z}$$

Now we can use the Oka–Sakamoto theorem (see Section 2.1), in order to compute the fundamental group of the complement of $\mathcal{L}$:

$$\pi_1(\mathbb{C}^2 - \mathcal{L}) = \pi_1\left(\mathbb{C}^2 - \bigcup_{i=1}^{t}\mathcal{L}_i\right) \stackrel{(O-S)}{\cong} \bigoplus_{i=1}^{t}(\pi_1(\mathbb{C}^2 - \mathcal{L}_i))$$

$$= \bigoplus_{i=1}^{t}(\mathbb{F}^{k_i} \oplus \mathbb{Z}) = \left(\bigoplus_{i=1}^{t}\mathbb{F}^{k_i}\right) \oplus \mathbb{Z}^t. \qquad \square$$

The Oka–Sakamoto theorem gives us a new inductive approach to prove Zariski's proposition:

Proposition 3.2 (Zariski). *The fundamental group of the complement of n lines in general position is abelian.*

Proof. It is known that for a line L:

$$\pi_1(\mathbb{C}^2 - L) \cong \mathbb{Z}$$

Due to the general position of the lines in the arrangement, we can use the Oka–Sakamoto theorem (see Section 2.1) inductively in the following way:

$$\pi_1(\mathbb{C}^2 - \mathcal{L}) = \pi_1\left(\mathbb{C}^2 - \bigcup_{i=1}^{n}l_i\right) \stackrel{(O-S)}{\cong} \bigoplus_{i=1}^{n}(\pi_1(\mathbb{C}^2 - l_i)) \cong \bigoplus_{i=1}^{n}\mathbb{Z} \cong \mathbb{Z}^n$$

And $\mathbb{Z}^n$ is an abelian group (see [O] too). $\qquad \square$

3.2. The projective case

Now, we will investigate the projective case.

Theorem 3.3. *Let $\mathcal{L}$ be a real line arrangement in $\mathbb{CP}^2$ with t non-collinear multiple points. Let $k_i + 1$ be the multiplicity of the multiple point P_i, $1 \leq i \leq t$.*

Then:
$$\pi_1(\mathbb{CP}^2 - \mathcal{L}) \cong \left(\bigoplus_{i=1}^{t} \mathbb{F}^{k_i}\right) \oplus \mathbb{Z}^{t-1}.$$

Proof. First, we will prove this theorem for $t = 1$, i.e. if $\mathcal{L}$ is a real line arrangement in $\mathbb{CP}^2$ which consists of $k+1$ lines meeting in one point P, then $\pi_1(\mathbb{CP}^2 - \mathcal{L}) \cong \mathbb{F}^k$.

Let $\{\Gamma_1, \ldots, \Gamma_{k+1}\}$ be a g-base of $\pi_1(\mathbb{C}_u - \mathcal{L})$ (see Section 2.8). In this line arrangement, we have only one singular point - P. Therefore, according to Lemma 2.14 and Proposition 2.15, this singular point induced the following set of relations:

$$[\Gamma_{k+1}\Gamma_k \ldots \Gamma_1, \Gamma_i] = 1, \quad i = 1, \ldots, k+1.$$

Hence, the fundamental group of its affine complement has the following presentation:

$$\pi_1(\mathbb{C}^2 - \mathcal{L}) = \langle \Gamma_1, \ldots, \Gamma_{k+1} \mid [\Gamma_{k+1}\Gamma_k \ldots \Gamma_1, \Gamma_i] = 1, \ i = 1, \ldots, k+1 \rangle$$

We will compute now another presentation for this group.

Let us modify the set of generators $g = \{\Gamma_1, \ldots, \Gamma_{k+1}\}$ by replacing the generator Γ_1 by the generator

$$\Gamma' = \Gamma_{k+1}\Gamma_k \ldots \Gamma_1.$$

Then, we have to check that after the modifications we get an equivalent set of generators, and we have to calculate the new set of relations.

Claim 3.4. *After replacing Γ_1 by Γ' (which was defined above) in g, we again get a set of generators. We denote this set of generators by $\tilde{g}$.*

Proof. We have to show that $\Gamma_1 \in \langle \tilde{g} \rangle$. But this is obvious, because:

$$\Gamma_1 = \Gamma_2^{-1}\Gamma_3^{-1} \ldots \Gamma_{k+1}^{-1}\Gamma'. \qquad \square$$

The next step is the calculation of the new set of relations for $\tilde{g}$.

Claim 3.5. *The set of relations:*

$$\{[\Gamma', \Gamma] = 1 \mid \forall \Gamma \in \tilde{g}\}$$

is a complete set of relations for $\tilde{g}$.

Proof. We have to show that

$$(*) \qquad \{[\Gamma', \Gamma] = 1 \mid \forall \Gamma \in \tilde{g}\}$$

is an equivalent set of relations to

$$(**) \qquad \{[\Gamma_{k+1}\Gamma_k \ldots \Gamma_1, \Gamma_i] = 1 \mid 1 \leq i \leq k+1\}$$

under the assignment: $\Gamma' = \Gamma_{k+1}\ldots\Gamma_1$.

Let us assume (∗). All the relations are equal except the first one. We have to show that:
$$[\Gamma_{k+1}\ldots\Gamma_1, \Gamma_1] = 1.$$

But:
$$\Gamma'\Gamma_1 = \Gamma'(\Gamma_2^{-1}\ldots\Gamma_{k+1}^{-1}\Gamma') \stackrel{(*)}{=} {}^{+\ ab=ba \Rightarrow ab^{-1}=b^{-1}a} (\Gamma_2^{-1}\ldots\Gamma_{k+1}^{-1}\Gamma')\Gamma' = \Gamma_1\Gamma'.$$

Now, if we assume (∗∗), all the relations in (∗) are equal except of $\Gamma'\Gamma' = \Gamma'\Gamma'$ which is trivial. □

Hence we got the following presentation for the fundamental group of the affine complement of $\mathcal{L}$:
$$\pi_1(\mathbb{C}^2 - \mathcal{L}) = \langle \Gamma', \Gamma_2, \ldots, \Gamma_{k+1} \mid [\Gamma_i, \Gamma'] = 1,\ 2 \leq i \leq k+1 \rangle.$$

Now, when we are going to the projective case, we add one additional relation, according to Theorem 2.9:
$$\Gamma_{k+1}\ldots\Gamma_1 = 1.$$

In terms of the new generator Γ', this relation gets the following form:
$$\Gamma' = 1.$$

Therefore, we can copmute the structure of the fundamental group in the projective case with $t = 1$:
$$\pi_1(\mathbb{CP}^2 - \mathcal{L}) = \langle \Gamma', \Gamma_2, \ldots, \Gamma_{k+1} \mid [\Gamma_i, \Gamma'] = 1,\ 2 \leq i \leq k+1;\ \Gamma' = 1 \rangle$$
$$\cong \langle \Gamma_2, \ldots, \Gamma_{k+1} \rangle \oplus \langle \Gamma' \mid \Gamma' = 1 \rangle \cong \mathbb{F}^k.$$

Now we continue to the general case ($t > 1$). For simplicity of the proof, we will prove it for two multiple points and the proof for t multiple points uses exactly the same arguments.

From the last theorem, we get for a line arrangement $\mathcal{L}$ with two multiple points:
$$\pi_1(\mathbb{C}^2 - \mathcal{L}) \cong \mathbb{F}^{k_1} \oplus \mathbb{F}^{k_2} \oplus \mathbb{Z}^2.$$

Let $l_1, \ldots, l_{k_1+1}$ be $k_1 + 1$ lines which pass through P_1 and let $l_{k_1+2}, \ldots, l_{k_1+k_2+2}$ be $k_2 + 1$ lines which pass through P_2. We choose $\{\Gamma_1, \ldots, \Gamma_{k_1+k_2+2}\}$, a g-base of $\pi_1(\mathbb{C}_u - \mathcal{L})$ (see Section 2.8) where Γ_i corresponds to the line l_i.

Similarly to the first part of the proof, we can write the following presentation for $\pi_1(\mathbb{C}^2 - \mathcal{L})$:

Generators: $g = \{\Gamma_1, \ldots, \Gamma_{k_1}, \Gamma', \Gamma_{k_1+2}, \ldots, \Gamma_{k_1+k_2+1}, \Gamma''\}$.

Relations: $\mathcal{R} = \{\Gamma_i\Gamma_j = \Gamma_j\Gamma_i, 1 \leq i \leq k_1, k_1 + 2 \leq j \leq k_1 + k_2 + 1;\ [\Gamma', \Gamma] = 1, \forall \Gamma \in g;\ [\Gamma'', \Gamma] = 1, \forall \Gamma \in g\}$, where:

$$\Gamma' = \Gamma_{k_1+1}\ldots\Gamma_1;\quad \Gamma'' = \Gamma_{k_1+k_2+2}\ldots\Gamma_{k_1+2}$$

Now, when we are going to the projective case, we add one additional relation, according to Theorem 2.9:

$$\Gamma_{k_1+k_2+2} \ldots \Gamma_1 = 1.$$

In terms of the new generators Γ', Γ'', this relation gets the following form:

$$\Gamma''\Gamma' = 1.$$

Now, we can finish to compute the structure the fundamental group in the projective case:

$$\begin{aligned}
\pi_1(\mathbb{CP}^2 - \mathcal{L}) &= \langle g \mid \mathcal{R}, \Gamma''\Gamma' = 1 \rangle \\
&\cong \langle \Gamma_1, \ldots, \Gamma_{k_1} \rangle \oplus \langle \Gamma_{k_1+2}, \ldots, \Gamma_{k_1+k_2+1} \rangle \oplus \langle \Gamma', \Gamma'' \mid \Gamma''\Gamma' = 1 \rangle \\
&\cong \mathbb{F}^{k_1} \oplus \mathbb{F}^{k_2} \oplus \mathbb{Z}.
\end{aligned}$$

□

As a consequence of the last theorem, we get:

Corollary 3.6.

$$\pi_1(\mathbb{C}^2 - \mathcal{L}) \cong \pi_1(\mathbb{CP}^2 - \mathcal{L}) \oplus \mathbb{Z}.$$

Therefore, the short exact sequence which was proved by Oka (Theorem 2.10):

$$1 \to \mathbb{Z} \to \pi_1(\mathbb{C}^2 - \mathcal{L}) \to \pi_1(\mathbb{CP}^2 - \mathcal{L}) \to 1$$

splits.

4. Arrangements with t collinear multiple points

In this section, we are going to calculate the fundamental group of the complement of line arrangements which consist of t subsets of lines where all the lines in each subset intersect at a single (multiple) point, all the t multiple intersection points lie on a single line which belongs to all the subsets and any two subsets of lines intersect in that line and in simple points out of that line. We define:

Definition 4.1 (An arrangement with t collinear multiple points). An *arrangement with t collinear multiple points* is a line arrangement which contains a line where all the t multiple points lie on it.

4.1. The affine case

Theorem 4.1. *Let $\mathcal{L}$ be a real line arrangement in $\mathbb{CP}^2$ with t collinear multiple points $P_1, \ldots, P_t$ with multiplicities $k_1 + 1, \ldots, k_t + 1$, respectively. Then:*

$$\pi_1(\mathbb{C}^2 - \mathcal{L}) \cong \bigoplus_{i=1}^{t} \mathbb{F}^{k_i} \oplus \mathbb{Z}.$$

It has to be noted that this theorem has a similar result to what we have got in the previous section in the non-collinear case. In both cases, the multiple points induced the free groups. The difference between the cases is that the connected line of the collinear case degenerates all the infinite cyclic groups of the non-collinear case into one infinite cyclic group.

Let L be the line on which all the multiple points lie. We choose $\{\Gamma_1, \ldots, \Gamma_n\}$ ($n = \#\{l \in \mathcal{L}\}$), a g-base of $\pi_1(\mathbb{C}_{u_0} - \mathcal{L})$ (see Section 2.8), where Γ_i corresponds to the line l_i in $\mathcal{L}$. The proof of the theorem is based on the following two lemmas:

Lemma 4.2. *In the situation of the theorem, let $\mathcal{L}_i$ be the subset of lines meet in P_i apart from L. Then: $[\Gamma_i, \Gamma_j] = 1$ where $l_i \in \mathcal{L}_i$, $l_j \in \mathcal{L}_j$ and $1 \le i < j \le t$.*

Lemma 4.3. *Let $\mathcal{L}_i \cup L = \{l_{p_1}, \ldots, l_{p_{k_i+1}}\}$ be the $k_i + 1$ lines that meet in the multiple point P_i. Then, the relations that are induced from this multiple point are:*

$$[\Gamma_{p_{k_i+1}} \ldots \Gamma_{p_1}, \Gamma_{p_j}] = 1, \quad 1 \le j \le k_i + 1$$

The proof of Lemma 4.2 is in Section 4.2. The proof of Lemma 4.3 is in Section 4.3. The proof of the Theorem (4.1) is in Section 4.4.

4.2. Proof of Lemma 4.2

For simplicity, we prove the lemma only for two multiple points, and the proof for t multiple points uses exactly the same arguments.

We will split the proof of this lemma into two cases: with the restriction that all the simple intersection points are to the right of the multiple points, and without this restriction. This restriction simplifies the proof significantly, and help to understand the proof of the general case.

4.2.1. First case – with the restriction. In this case, all the simple points are to the right of the multiple points.

Let $N = \{x \in \mathbb{C} \mid (x, y) \text{ is an intersection point}\}$, and let $u_0 \in \mathbb{R}$ such that $x \ll u_0$ for all $x \in N$. Let $\mathbb{C}_{u_0} = \{(u_0, y) \mid y \in \mathbb{C}\}$. We numerate the lines according to their intersection with $\mathbb{C}_{u_0}$. By a proper choosing of the line in infinity and homotopic movements of the lines, we can assume that the line arrangement has the following property: for $1 \leq i < j \leq k_1$,

$$x(l_i \cap l_t) < x(l_j \cap l_s), \quad k_1 + 1 \leq t, s \leq k_1 + k_2.$$

Therefore, we get the following line arrangement:

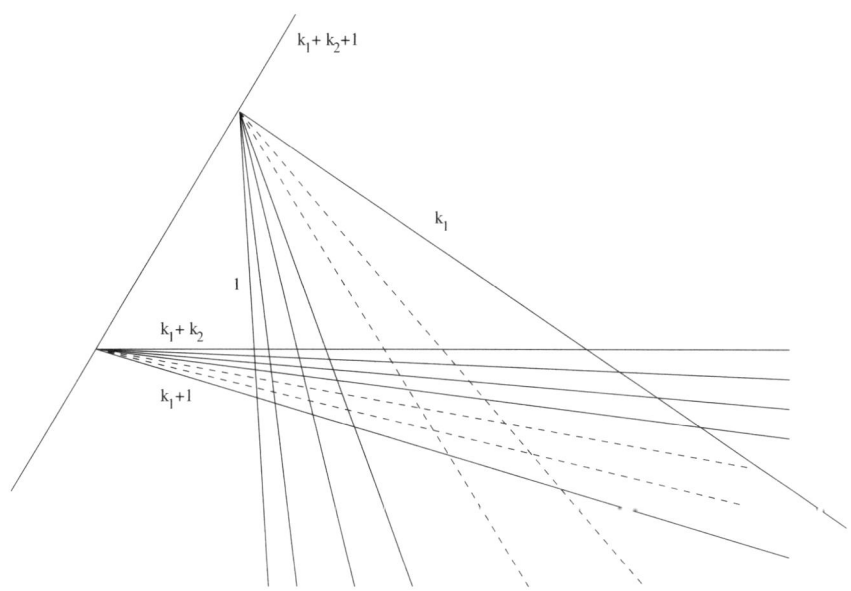

Let $g = \{\Gamma_1, \ldots, \Gamma_{k_1+k_2+1}\}$ be a g-base of $\pi_1(\mathbb{C}_{u_0} - \mathcal{L})$. By abuse of notations, let us denote the images of Γ_i in $\pi_1(\mathbb{C}^2 - \mathcal{L})$ by the same notation.

Now, we prove this lemma using the braid monodromy techniques (2.7) and the Van Kampen theorem (2.13). First, let us calculate the skeletons representing the $\mathcal{L}$.V.C.s of the braid monodromy.

According to this line arrangement, we have the following set of Lefschetz pairs:

j	λ_{x_j}
1	(k_1, k_1+1)
2	(k_1+1, k_1+2)
3	(k_1+2, k_1+3)
$\vdots$	$\vdots$
k_2	(k_1+k_2-1, k_1+k_2)
k_2+1	(k_1-1, k_1)
k_2+2	(k_1, k_1+1)
$\vdots$	$\vdots$
$2k_2$	(k_1+k_2-2, k_1+k_2-1)
$\vdots$	$\vdots$
$(k_1-1)k_2+1$	$(1, 2)$
$(k_1-1)k_2+2$	$(2, 3)$
$\vdots$	$\vdots$
$k_1 k_2$	(k_2, k_2+1)
$k_1 k_2+1$	(k_2+1, k_1+k_2+1)
$k_1 k_2+2$	$(1, k_2+1)$

Let $\{\delta_i \mid 1 \le i \le k_1 k_2 + 2\}$ be a g-base for $\pi_1(\mathbb{C}^X - N, u_0)$ (where $\mathbb{C}^X$ is the x-axis). Let φ be the braid monodromy of $\mathcal{L}$ w.r.t. π_1, u_0.

Now, using the table of Lefschetz pairs, we can calculate the skeletons representing the $\mathcal{L}$.V.C.s for the braids $\varphi(\delta_i)$ (according to Moishezon–Teicher's algorithm (2.7)). Here, we will calculate the $\mathcal{L}$.V.C.s of the two general cases.

Skeleton representing the $\mathcal{L}$.V.C. of $\varphi(\delta_{lk_2+1})$, $0 \le l \le k_1 - 1$: The Lefschetz pair is (k_1-l, k_1-l+1). So the skeleton representing the local $\mathcal{L}$.V.C. is:

●   ●   ●   · · ·   ●   ●━━━●   ●   · · ·   ●   ●
1   2   3 k_1-l-1 k_1-l k_1-l+1 k_1-l+2 k_1+k_2 k_1+k_2+1

According to the algorithm, we have to apply on the skeleton the composition of the following l sequences of braids:

$$\Delta\langle k_1+k_2-l, k_1+k_2-l+1\rangle \Delta\langle k_1+k_2-l-1, k_1+k_2-l\rangle$$
$$\ldots \Delta\langle k_1-l+2, k_1-l+3\rangle \Delta\langle k_1-l+1, k_1-l+2\rangle$$
$$\Delta\langle k_1+k_2-l+1, k_1+k_2-l+2\rangle \Delta\langle k_1+k_2-l, k_1+k_2-l+1\rangle$$
$$\ldots \Delta\langle k_1-l+3, k_1-l+4\rangle \Delta\langle k_1-l+2, k_1-l+3\rangle$$
$$\vdots$$
$$\Delta\langle k_1+k_2-1, k_1+k_2\rangle \Delta\langle k_1+k_2-2, k_1+k_2-1\rangle \ldots \Delta\langle k_1, k_1+1\rangle$$

Fundamental group's structure

In every sequence, only the last braid of the sequence affects the skeleton (because the region of the others has no intersection with the region of the skeleton). Therefore, we get the following skeleton:

[Skeleton diagrams with transformations labeled $\Delta\langle k_1-l+1, k_1-l+2\rangle$, $\Delta\langle k_1-l+2, k_1-l+3\rangle$, $\Delta\langle k_1-2, k_1-1\rangle$, $\Delta\langle k_1-1, k_1\rangle$, $\Delta\langle k_1, k_1+1\rangle$]

Skeleton representing the $\mathcal{L}.\mathcal{V}.\mathcal{C}$. of $\varphi(\delta_{lk_2+i})$, $0 \leq l \leq k_1 - 1$, $2 \leq i \leq k_2$: The Lefschetz pair is $(k_1 - l + i - 1, k_1 - l + i)$. So the skeleton representing local $\mathcal{L}.\mathcal{V}.\mathcal{C}$. is:

[Skeleton diagram with nodes labeled 1, 2, 3, ..., k_1-l+i-2, k_1-l+i-1, k_1-l+i, k_1-l+i+1, ..., k_1+k_2, k_1+k_2+1]

According to the algorithm, we have to apply on the skeleton the composition of the following $l + 1$ sequences of braids:

$$\Delta\langle k_1 - l + i - 2, k_1 - l + i - 1\rangle \Delta\langle k_1 - l + i - 3, k_1 - l + i - 2\rangle$$
$$\ldots \Delta\langle k_1 - l + 1, k_1 - l + 2\rangle \Delta\langle k_1 - l, k_1 - l + 1\rangle$$
$$\Delta\langle k_1 + k_2 - l, k_1 + k_2 - l + 1\rangle \Delta\langle k_1 + k_2 - l - 1, k_1 + k_2 - l\rangle$$
$$\ldots \Delta\langle k_1 - l + 2, k_1 - l + 3\rangle \Delta\langle k_1 - l + 1, k_1 - l + 2\rangle$$
$$\Delta\langle k_1 + k_2 - l + 1, k_1 + k_2 - l + 2\rangle \Delta\langle k_1 + k_2 - l, k_1 + k_2 - l + 1\rangle$$
$$\ldots \Delta\langle k_1 - l + 3, k_1 - l + 4\rangle \Delta\langle k_1 - l + 2, k_1 - l + 3\rangle$$
$$\vdots$$
$$\Delta\langle k_1 + k_2 - 1, k_1 + k_2\rangle \Delta\langle k_1 + k_2 - 2, k_1 + k_2 - 1\rangle \ldots \Delta\langle k_1, k_1 + 1\rangle$$

The first sequence causes the following effect to the skeleton:

198 D. Garber and M. Teicher

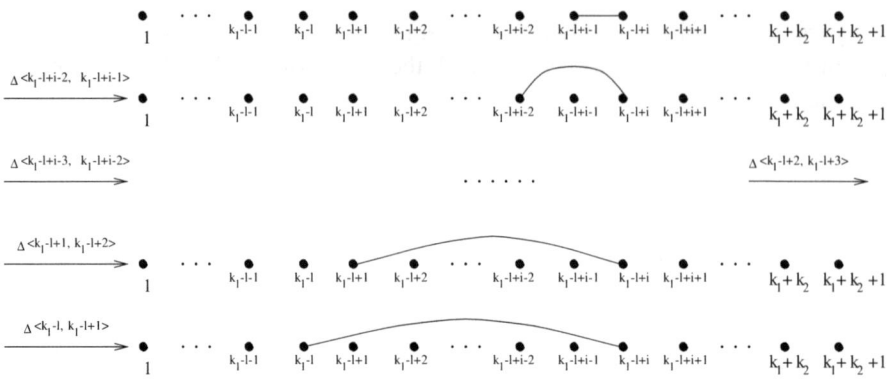

Only the last part of the second sequence affects the skeleton as follows:

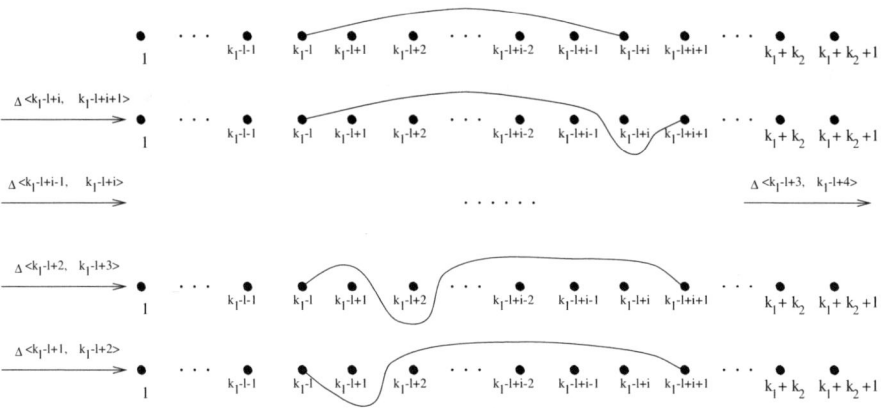

In the other $l-1$ sequences of braids, only the second part of the sequence affects, i.e. only the braids whose region intersects the region of the skeleton. Therefore, we get the following skeleton representing the $\mathcal{L}$.V.C.:

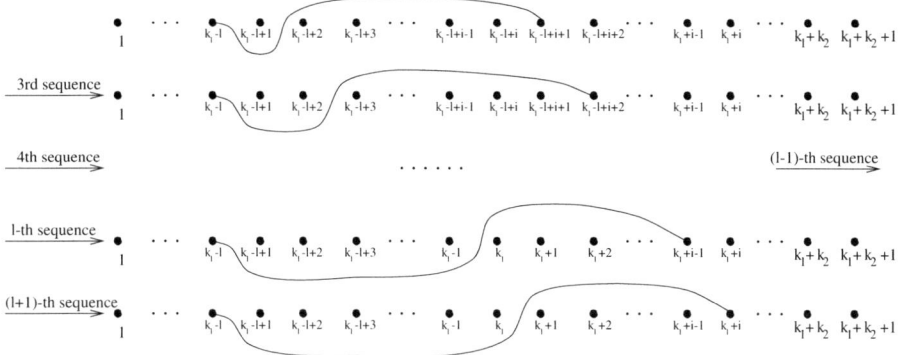

After we have calculated the skeletons representing $\mathcal{L}$.V.C.s for the braid monodromy, we can calculate the relations that they induced. As we have introduced in Section 2, according to Van Kampen theorem (2.13), every $\mathcal{L}$.V.C. induces a relation. Now, we will calculate the general relations which are induced from the general $\mathcal{L}$.V.C.s.

The relation which is induced from $\varphi(\delta_{lk_2+1})$, $0 \leq l \leq k_1 - 1$:

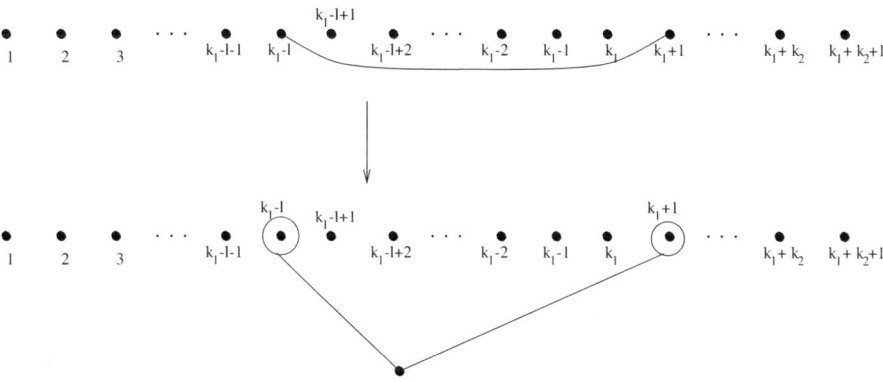

Therefore, the relation is:

$$\Gamma_{k_1-l}\Gamma_{k_1+1} = \Gamma_{k_1+1}\Gamma_{k_1-l}$$

The relation which is induced from $\varphi(\delta_{lk_2+i})$, $0 \leq l \leq k_1 - 1$, $2 \leq i \leq k_2$:

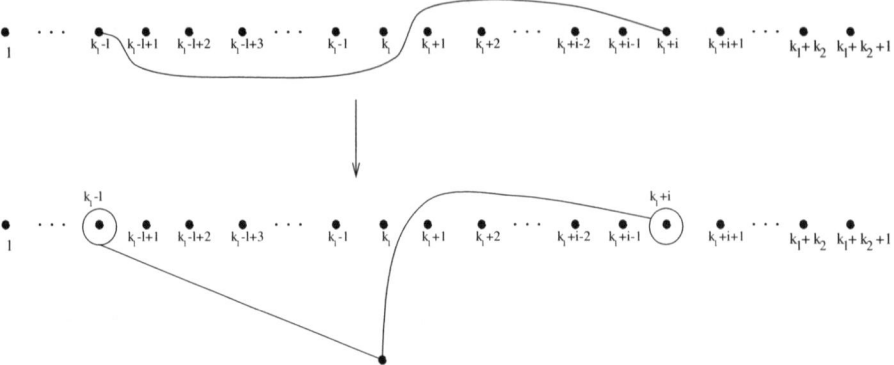

Therefore, the relation is:

$$\Gamma_{k_1-l}\Gamma^{-1}_{k_1+1}\cdots\Gamma^{-1}_{k_1+i-1}\Gamma_{k_1+i}\Gamma_{k_1+i-1}\cdots\Gamma_{k_1+1}$$

$$= \Gamma^{-1}_{k_1+1}\cdots\Gamma^{-1}_{k_1+i-1}\Gamma_{k_1+i}\Gamma_{k_1+i-1}\cdots\Gamma_{k_1+1}\Gamma_{k_1-l}$$

Therefore, we got the following set of relations: for all $0 \le l \le k_1-1$, $1 \le i \le k_2$,

$$\Gamma_{k_1-l}\Gamma^{-1}_{k_1+1}\cdots\Gamma^{-1}_{k_1+i-1}\Gamma_{k_1+i}\Gamma_{k_1+i-1}\cdots\Gamma_{k_1+1}$$

$$= \Gamma^{-1}_{k_1+1}\cdots\Gamma^{-1}_{k_1+i-1}\Gamma_{k_1+i}\Gamma_{k_1+i-1}\cdots\Gamma_{k_1+1}\Gamma_{k_1-l}$$

Now, it is easy to see that this set of relations is equivalent to the following set of relations (see [Ga]):

$$\Gamma_i\Gamma_j = \Gamma_j\Gamma_i; \quad 1 \le i \le k_1, \; k_1+1 \le j \le k_1+k_2$$

and this finishes the proof of the first case of the first Lemma (4.2).

4.2.2. Second case – without the restriction Let $N = \{x \in \mathbb{C} \mid (x, y)$ is an intersection point$\}$, and let $u_0 \in \mathbb{R}$ such that $x \ll u_0$ for all $x \in N$. Let $\mathbb{C}_{u_0} = \{(u_0, y) \mid y \in \mathbb{C}\}$. We numerate the lines according to their intersection with $\mathbb{C}_{u_0}$. We organized this line arrangement in such a way that the following property holds: for $1 \le i < j \le l$ and $k_1+l+1 \le i < j \le k_1+k_2$,

$$x(L_i \cap L_t) < x(L_j \cap L_s), \quad l+1 \le s, t \le k_1+l$$

It is easy to see that this is the general case, i.e. every line arrangement is homotopic to this situation by homotopic rotations and a proper choosing of the line at infinity.

Therefore, we get the following line arrangement:

Fundamental group's structure

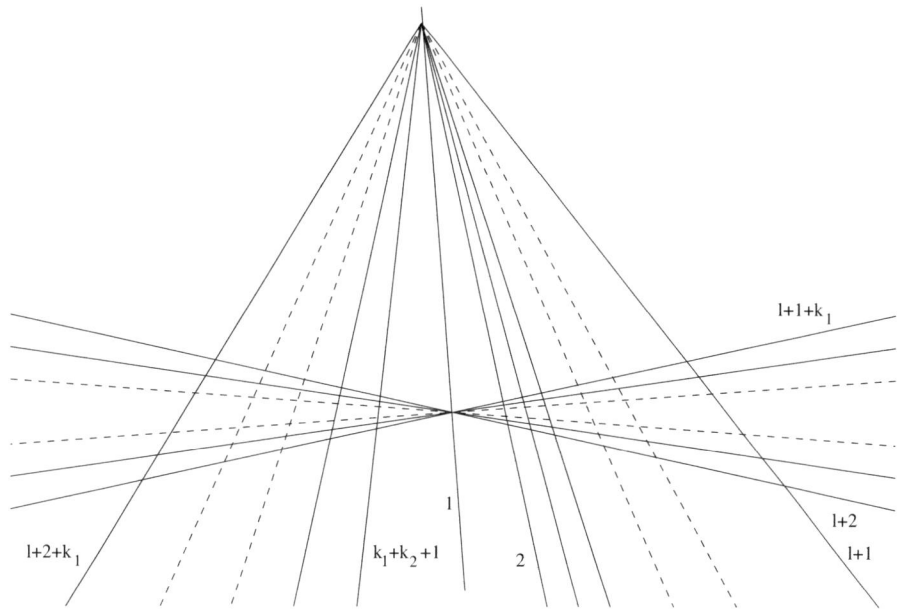

Let $g = \{\Gamma_1, \ldots, \Gamma_{k_1+k_2+1}\}$ be a g-base of $\pi_1(\mathbb{C}_{u_0} - \mathcal{L})$. By abuse of notations, let us denote the images of Γ_i in $\pi_1(\mathbb{C}^2 - \mathcal{L})$ by the same notation.

Now, we prove this lemma using the braid monodromy techniques (2.7) and the Van Kampen theorem (2.13). First, let us calculate the skeletons representing the $\mathcal{L}$.V.C.s of the braid monodromy.

According to this line arrangement, we have the following set of Lefschetz pairs:

j	λ_{x_j}
1	$(l+1, l+2)$
2	$(l+2, l+3)$
$\vdots$	$\vdots$
k_1	(k_1+l, k_1+l+1)
k_1+1	$(l, l+1)$
k_1+2	$(l+1, l+2)$
$\vdots$	$\vdots$
$2k_1$	(k_1+l-1, k_1+l)
$\vdots$	$\vdots$
$(l-1)k_1+1$	$(2, 3)$
j	λ_{x_j}
$(l-1)k_1+2$	$(3, 4)$
$\vdots$	$\vdots$

j	λ_{x_j}
lk_1	(k_1+1, k_1+2)
lk_1+1	$(1, k_1+1)$
lk_1+2	(k_1+1, k_1+k_2+1)
$(lk_1+2)+1$	(k_1, k_1+1)
$(lk_1+2)+2$	(k_1-1, k_1)
$\vdots$	$\vdots$
$(lk_1+2)+k_1$	$(1, 2)$
$(lk_1+2)+k_1+1$	(k_1+1, k_1+2)
$(lk_1+2)+k_1+2$	(k_1, k_1+1)
$\vdots$	$\vdots$
$(lk_1+2)+2k_1$	$(2, 3)$
$\vdots$	$\vdots$
$(lk_1+2)+(k_2-l-1)k_1+1$	(k_1+k_2-l-1, k_1+k_2-l)
$(lk_1+2)+(k_2-l-1)k_1+2$	$(k_1+k_2-l-2, k_1+k_2-l-1)$
$\vdots$	$\vdots$
$(lk_1+2)+(k_2-l)k_1 [= k_1k_2+2]$	(k_2-l, k_2-l+1)

Let $\{\delta_i \mid 1 \leq i \leq k_1k_2+2\}$ be a g-base for $\pi_1(\mathbb{C}^X - N, u_0)$ (where $\mathbb{C}^X$ is the x-axis). Let φ be the braid monodromy of $\mathcal{L}$ w.r.t. π_1, u_0.

Now, using the table of the Lefschetz pairs, we can calculate the skeletons representing $\mathcal{L}$.V.C.s for the braids $\varphi(\delta_i)$ (according to the Moishezon–Teicher algorithm (2.7)).

Until singular point number lk_1 we have almost the same configuration as in the first case of the lemma, hence the general skeleton, which represents the $\mathcal{L}$.V.C., which we have found there is identical (but its center is shifted one point left) to the general skeleton in this case of the lemma until point number lk_1. Therefore:

Skeleton representing the $\mathcal{L}$.V.C. of $\varphi(\delta_{ik_1+1})$, $0 \leq i \leq l-1$:

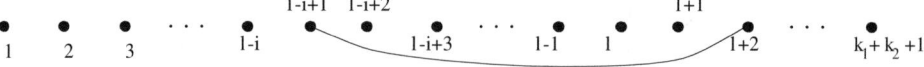

Skeleton representing the $\mathcal{L}$.V.C. of $\varphi(\delta_{ik_1+j})$, $0 \leq i \leq l-1$, $2 \leq j \leq k_1$:

Fundamental group's structure

We skip the calculations of the braid monodromy of the two multiple points (which will be done in the proof of the next Lemma (4.3)), and we continue with the rest of the simple points and we pass directly to the general case:

Skeleton representing the $\mathcal{L}.\mathcal{V}.\mathcal{C}.$ of $\varphi(\delta_{(lk_1+2)+ik_1+1}), 0 \leq i \leq (k_2 - l - 1)$: The Lefschetz pair is

$$(k_1 + i, k_1 + i + 1),$$

therefore the skeleton representing the local $\mathcal{L}.\mathcal{V}.\mathcal{C}.$ is:

●   ●   ●   · · ·   ●   ●———●   ●   · · ·   ●   ●
1   2   3 $k_1{+}i{-}1$ $k_1{+}i$ $k_1{+}i{+}1$ $k_1{+}i{+}2$ $k_1{+}k_2$ $k_1{+}k_2{+}1$

We have to apply on this skeleton the following sequences of braids:

$$\Delta\langle i, i+1 \rangle \Delta\langle i+1, i+2 \rangle \ldots \Delta\langle k_1+i-1, k_1+i \rangle$$

$$\vdots$$

$$\Delta\langle 1, 2 \rangle \Delta\langle 2, 3 \rangle \ldots \Delta\langle k_1, k_1+1 \rangle$$
$$\Delta\langle k_1+1, k_1+k_2+1 \rangle \Delta\langle 1, k_1+1 \rangle$$
$$\Delta\langle k_1+1, k_1+2 \rangle \Delta\langle k_1, k_1+1 \rangle \ldots \Delta\langle 2, 3 \rangle$$

$$\vdots$$

$$\Delta\langle k_1+l, k_1+l+1 \rangle \Delta\langle k_1+l-1, k_1+l \rangle \ldots \Delta\langle l+1, l+2 \rangle$$

In the first $i - 1$ sequences, only the last braid in each sequence affects the skeleton, hence we get:

●   ●   ●   · · ·   ●   ●   ●   · · ·   ●   ●———●   · · ·   ●
1   2   3 k_1 $k_1{+}1$ $k_1{+}2$ $k_1{+}i{-}1$ $k_1{+}i$ $k_1{+}i{+}1$ $k_1{+}k_2{+}1$

$\xrightarrow{\Delta\langle k_1+i-1, k_1+i \rangle}$

●   ●   ●   · · ·   ●   ●   ●   · · ·   ●⌒●   ●   · · ·   ●
1   2   3 k_1 $k_1{+}1$ $k_1{+}2$ $k_1{+}i{-}1$ $k_1{+}i$ $k_1{+}i{+}1$ $k_1{+}k_2{+}1$

$\xrightarrow{\Delta\langle k_1+i-2, k_1+i-1 \rangle}$

· · · · · ·

$\xrightarrow{\Delta\langle k_1+2k_1+3 \rangle}$

$\xrightarrow{\Delta\langle k_1+1, k_1+2 \rangle}$

●   ●   ●   · · ·   ●   ●⌒———⌒●   · · ·   ●   ●   ●   · · ·   ●
1   2   3 k_1 $k_1{+}1$ $k_1{+}2$ $k_1{+}i{-}1$ $k_1{+}i$ $k_1{+}i{+}1$ $k_1{+}k_2{+}1$

$\xrightarrow{\Delta\langle k_1, k_1+1 \rangle}$

●   ●   ●   · · ·   ●⌒————————⌒●   · · ·   ●   ●   ●   · · ·   ●
1   2   3 k_1 $k_1{+}1$ $k_1{+}2$ $k_1{+}i{-}1$ $k_1{+}i$ $k_1{+}i{+}1$ $k_1{+}k_2{+}1$

Next, the action of the braids $\Delta\langle k_1+1, k_1+k_2+1\rangle$ and $\Delta\langle 1, k_1+1\rangle$ is as follows:

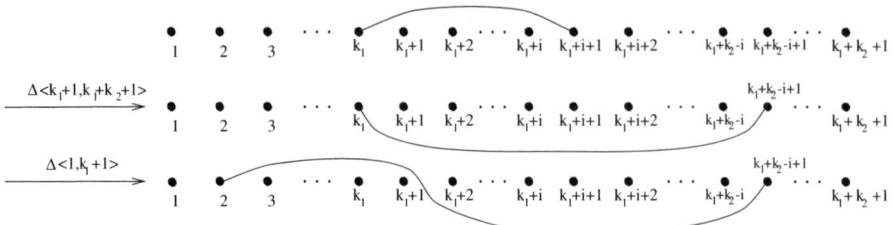

Then, the l sequences of braids move the left side of the skeleton l points right:

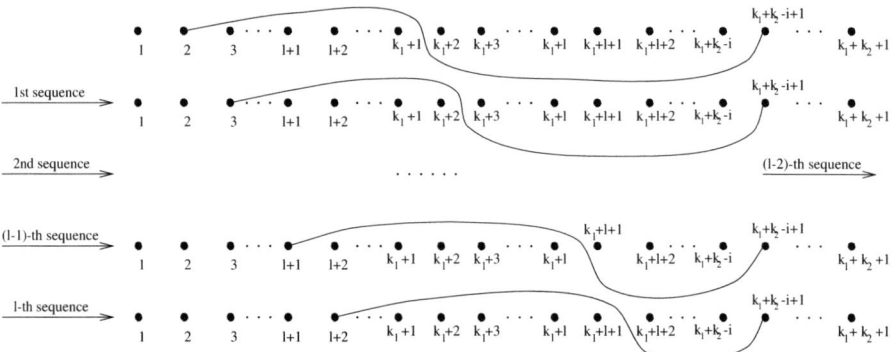

Skeleton representing the $\mathcal{L}$.V.C. of $\varphi(\delta_{(lk_1+2)+ik_1+j})$; $0 \le i \le k_2 - l - 1$, $2 \le j \le k_1$: The Lefschetz pair is

$$(k_1 + i - j + 1, k_1 + i - j + 2)$$

therefore the skeleton representing the local $\mathcal{L}$.V.C. is:

• • • ⋯ •——• • ⋯ • •
1 2 3 k_1+i-j+1 k_1+i-j+2 k_1+i-j+3 k_1+k_2 k_1+k_2+1

(with k_1+i-j labeled above)

We have to apply on this skeleton the following sequences of braids:

$$\Delta\langle k_1+i-j+2, k_1+i-j+3\rangle \Delta\langle k_1+i-j+3, k_1+i-j+4\rangle$$
$$\ldots \Delta\langle k_1+i-1, k_1+i\rangle \Delta\langle k_1+i, k_1+i+1\rangle$$

Fundamental group's structure

$$\Delta\langle i, i+1 \rangle \Delta\langle i+1, i+2 \rangle \ldots \Delta\langle k_1+i-1, k_1+i \rangle$$
$$\vdots$$
$$\Delta\langle 1, 2 \rangle \Delta\langle 2, 3 \rangle \ldots \Delta\langle k_1, k_1+1 \rangle$$
$$\Delta\langle k_1+1, k_1+k_2+1 \rangle \Delta\langle 1, k_1+1 \rangle$$
$$\Delta\langle k_1+1, k_1+2 \rangle \Delta\langle k_1, k_1+1 \rangle \ldots \Delta\langle 2, 3 \rangle$$
$$\vdots$$
$$\Delta\langle k_1+l, k_1+l+1 \rangle \Delta\langle k_1+l-1, k_1+l \rangle \ldots \Delta\langle l+1, l+2 \rangle$$

The first sequence acts as follows:

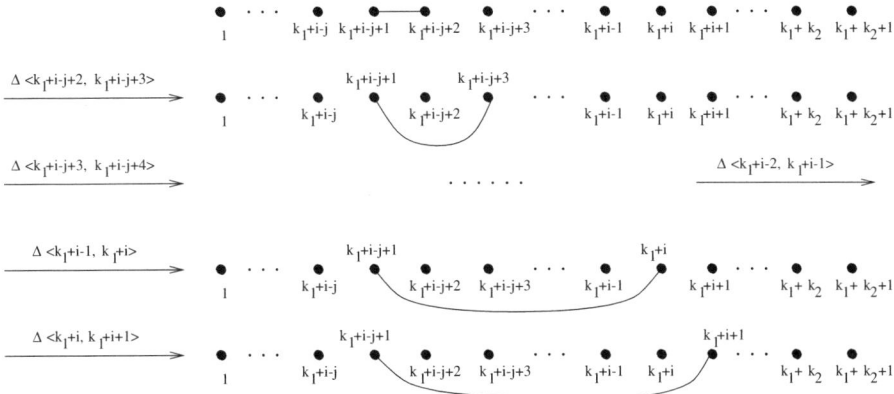

The second sequence moves the left side of the skeleton one point left (the first part of the sequence does not affect the skeleton):

Each of the next $i - 1$ sequences moves the left side of the skeleton another step left, so we get the following:

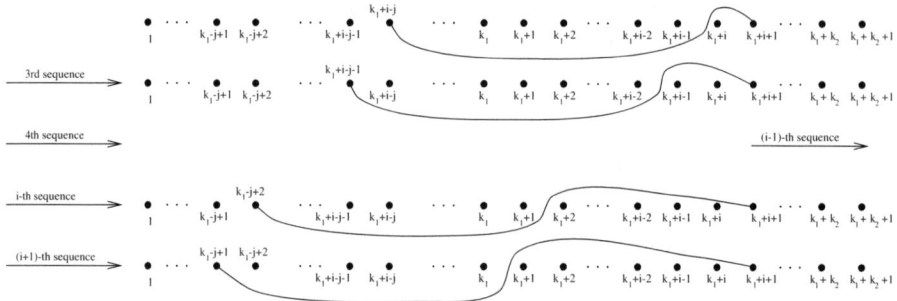

Next, the action of the braids $\Delta \langle k_1 + 1, k_1 + k_2 + 1 \rangle$ and $\Delta \langle 1, k_1 + 1 \rangle$ is as follows:

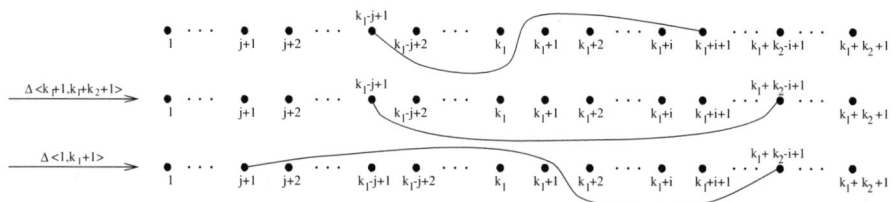

Then, the l sequences of braids move the left side of the skeleton l points right:

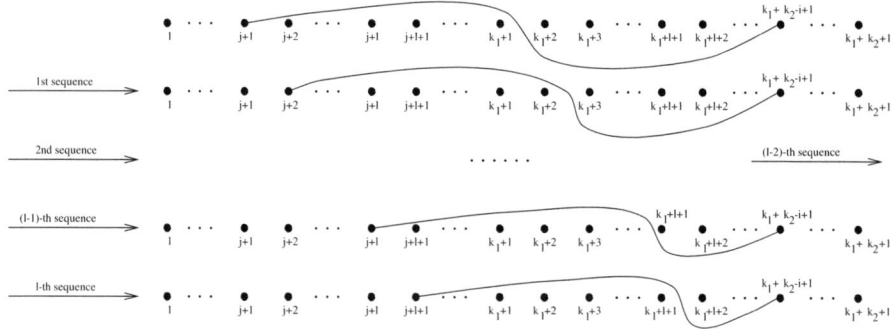

After we have calculated the skeletons representing $\mathcal{L}$.V.C.s for the braid monodromy, we can calculate the relations that they induced. As we have introduced in

Section 2, according to Van Kampen's theorem (2.13), every $\mathcal{L}$.V.C. induces a relation. Now, we will calculate the general relations which are induced from the general $\mathcal{L}$.V.C.s.

The relation which is induced from $\varphi(\delta_{ik_1+1})$, $0 \leq i \leq l-1$:

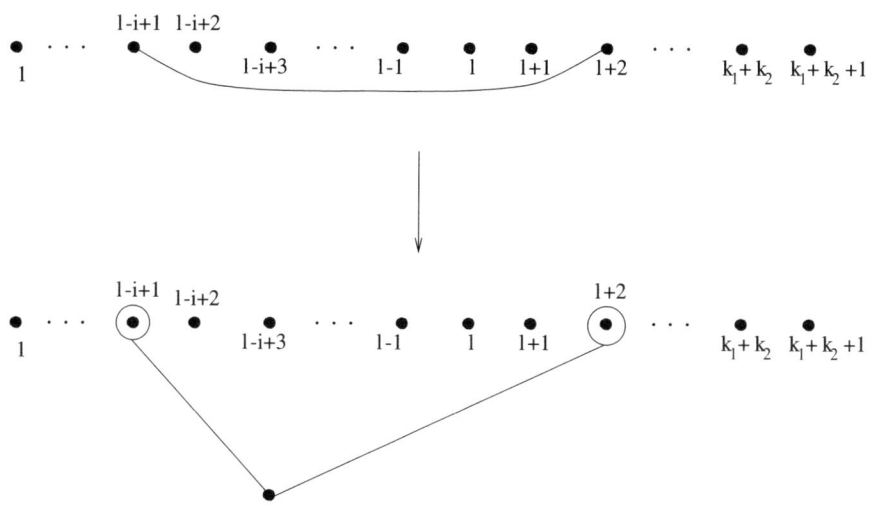

Therefore, the relation is:

$$\Gamma_{l-i+1}\Gamma_{l+2} = \Gamma_{l+2}\Gamma_{l-i+1}$$

The relation which is induced from $\varphi(\delta_{ik_1+j})$, $0 \leq i \leq l-1$, $2 \leq j \leq k_1$:

Therefore, the relation is:

$$\Gamma_{l-i+1}\Gamma_{l+2}^{-1}\dots\Gamma_{l+j}^{-1}\Gamma_{l+j+1}\Gamma_{l+j}\dots\Gamma_{l+2} = \Gamma_{l+2}^{-1}\dots\Gamma_{l+j}^{-1}\Gamma_{l+j+1}\Gamma_{l+j}\dots\Gamma_{l+2}\Gamma_{l-i+1}.$$

The relation which is induced from $\varphi(\delta_{(lk_1+2)+ik_1+1})$, $0 \leq i \leq (k_2-l-1)$:

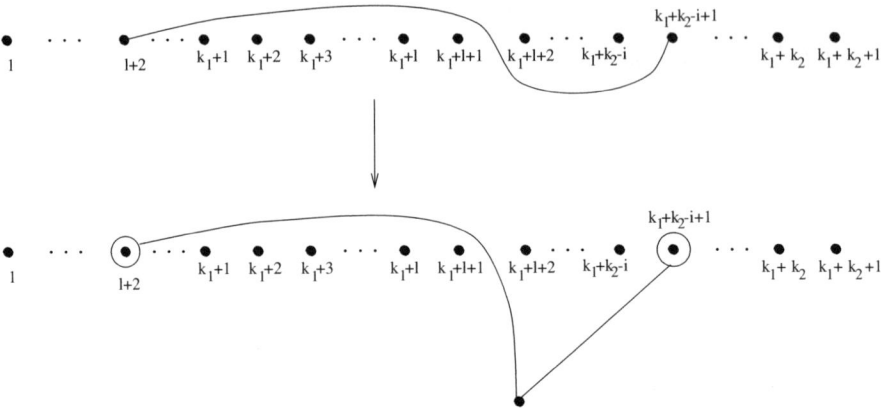

Therefore, the relation is:

$$\Gamma_{k_1+l+1}\ldots\Gamma_{l+3}\Gamma_{l+2}\Gamma_{l+3}^{-1}\ldots\Gamma_{k_1+l+1}^{-1}\Gamma_{k_1+k_2-i+1}$$
$$=\Gamma_{k_1+k_2-i+1}\Gamma_{k_1+l+1}\ldots\Gamma_{l+3}\Gamma_{l+2}\Gamma_{l+3}^{-1}\ldots\Gamma_{k_1+l+1}^{-1}.$$

The relation which is induced from $\varphi(\delta_{(lk_1+2)+ik_1+j})$, $0 \leq i \leq (k_2-l-1)$, $2 \leq j \leq k_1$:

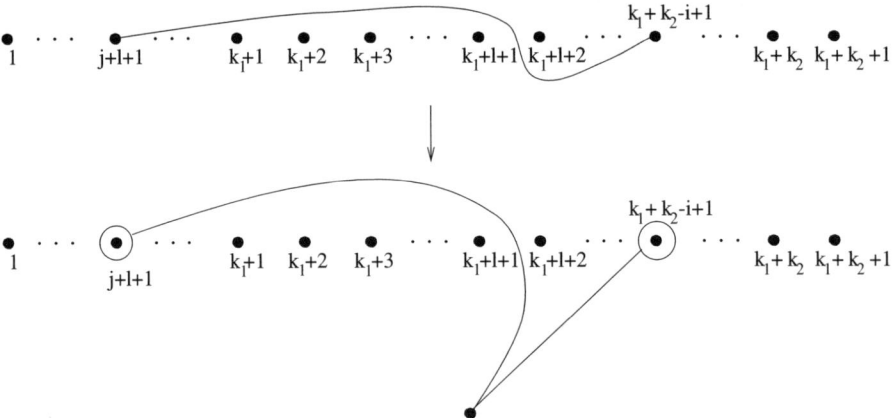

Therefore, the relation is:

$$\Gamma_{k_1+l+1}\ldots\Gamma_{l+j+2}\Gamma_{l+j+1}\Gamma_{l+j+2}^{-1}\ldots\Gamma_{k_1+l+1}^{-1}\Gamma_{k_1+k_2-i+1}$$
$$=\Gamma_{k_1+k_2-i+1}\Gamma_{k_1+l+1}\ldots\Gamma_{l+j+2}\Gamma_{l+j+1}\Gamma_{l+j+2}^{-1}\ldots\Gamma_{k_1+l+1}^{-1}.$$

Therefore, we got the following two sets of relations: for all $0 \leq i \leq l-1$, $1 \leq j \leq k_1$:

$$\Gamma_{l-i+1}\Gamma_{l+2}^{-1}\ldots\Gamma_{l+j}^{-1}\Gamma_{l+j+1}\Gamma_{l+j}\ldots\Gamma_{l+2}$$
$$= \Gamma_{l+2}^{-1}\ldots\Gamma_{l+j}^{-1}\Gamma_{l+j+1}\Gamma_{l+j}\ldots\Gamma_{l+2}\Gamma_{l-i+1}$$

and for all $0 \leq i \leq k_2 - l - 1$, $1 \leq j \leq k_1$:

$$\Gamma_{k_1+l+1}\ldots\Gamma_{l+j+2}\Gamma_{l+j+1}\Gamma_{l+j+2}^{-1}\ldots\Gamma_{k_1+l+1}^{-1}\Gamma_{k_1+k_2-i+1}$$
$$= \Gamma_{k_1+k_2-i+1}\Gamma_{k_1+l+1}\ldots\Gamma_{l+j+2}\Gamma_{l+j+1}\Gamma_{l+j+2}^{-1}\ldots\Gamma_{k_1+l+1}^{-1}.$$

Now, it is easy to see that these two sets of relations are equivalent to the following two sets of relations:

$$\Gamma_i\Gamma_j = \Gamma_j\Gamma_i;\ 2 \leq i \leq l+1,\ l+2 \leq j \leq l+k_1+1$$

and

$$\Gamma_i\Gamma_j = \Gamma_j\Gamma_i;\ l+2 \leq i \leq l+k_1+1,\ l+k_1+2 \leq j \leq k_1+k_2+1$$

and this finishes the proof of the second case of the first Lemma (4.2).

4.3. Proof of Lemma 4.3

As in the first lemma, we prove this lemma only for two multiple points, and the proof for t multiple points uses exactly the same arguments.

We will prove it directly in the general case. By homotopic rotations and movements and a proper choosing of the line at infinity, we can get the following line arrangement from any line arrangement with two multiple points:

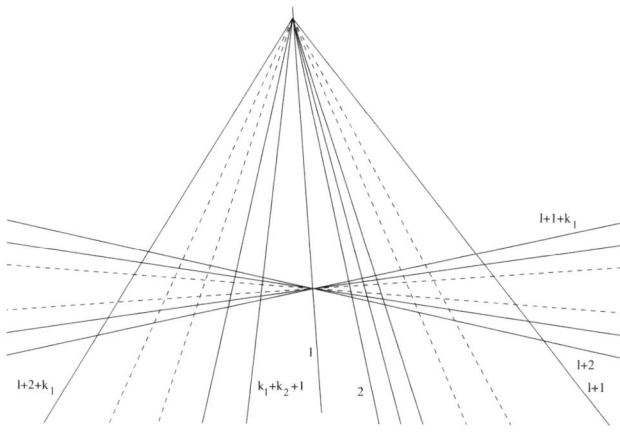

In the first Lemma (4.2), we already wrote down the set of Lefschetz pairs of this line arrangement. In order to calculate the induced relations of the multiple points, we have to compute their braid monodromy according to the Moishezon–Teicher algorithm (2.7) and then we have to use the Van Kampen theorem (2.13) to get their induced relations.

Skeleton representing the $\mathcal{L}.\mathcal{V}.\mathcal{C}.$ of $\varphi(\delta_{lk_1+1})$: The Lefschetz pair is

$$(1, k_1 + 1),$$

then the skeleton representing the local $\mathcal{L}.\mathcal{V}.\mathcal{C}.$ is:

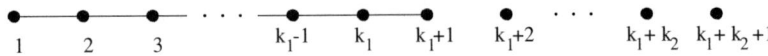

According to the algorithm, we have to apply on the skeleton the following sequence of braids:

$$\Delta \langle k_1 + 1, k_1 + 2 \rangle \Delta \langle k_1, k_1 + 1 \rangle \ldots \Delta \langle 2, 3 \rangle$$

$$\vdots$$

$$\Delta \langle k_1 + l, k_1 + l + 1 \rangle \Delta \langle k_1 + l - 1, k_1 + l \rangle \ldots \Delta \langle l + 1, l + 2 \rangle$$

The first sequence acts as follows:

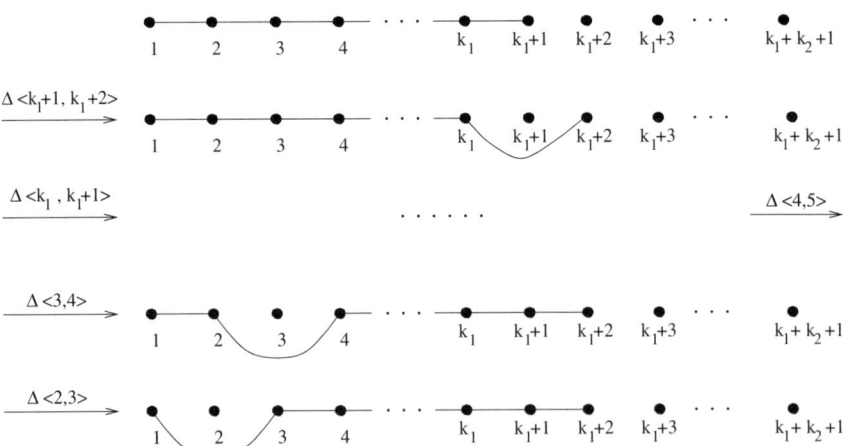

Each of the next $l - 1$ sequences moves the right side of the skeleton one step right, so we get the following:

Fundamental group's structure

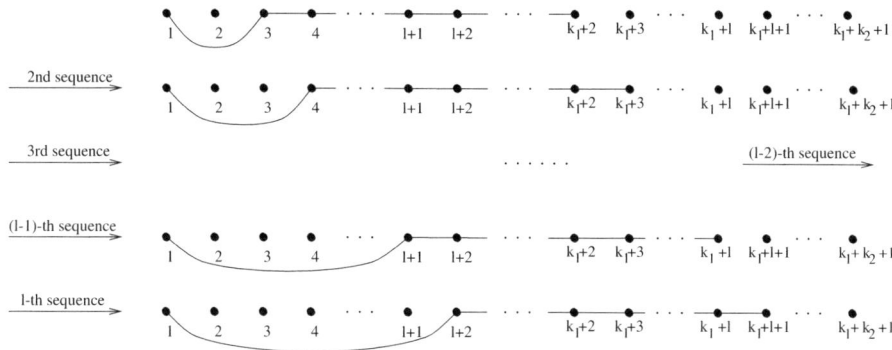

Skeleton representing the ℒ.V.C. of $\varphi(\delta_{lk_1+2})$: The Lefschetz pair is

$$(k_1 + 1, k_1 + k_2 + 1),$$

therefore the skeleton representing the local ℒ.V.C. is:

According to the algorithm, we have to apply on the skeleton the following sequence of braids:

$$\Delta\langle 1, k_1 + 1\rangle$$
$$\Delta\langle k_1 + 1, k_1 + 2\rangle\Delta\langle k_1, k_1 + 1\rangle \ldots \Delta\langle 2, 3\rangle$$
$$\vdots$$
$$\Delta\langle k_1 + l, k_1 + l + 1\rangle\Delta\langle k_1 + l - 1, k_1 + l\rangle \ldots \Delta\langle l + 1, l + 2\rangle.$$

The effect of the braid $\Delta\langle 1, k_1 + 1\rangle$ is:

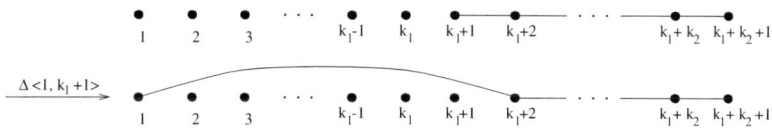

The first sequence acts as follows:

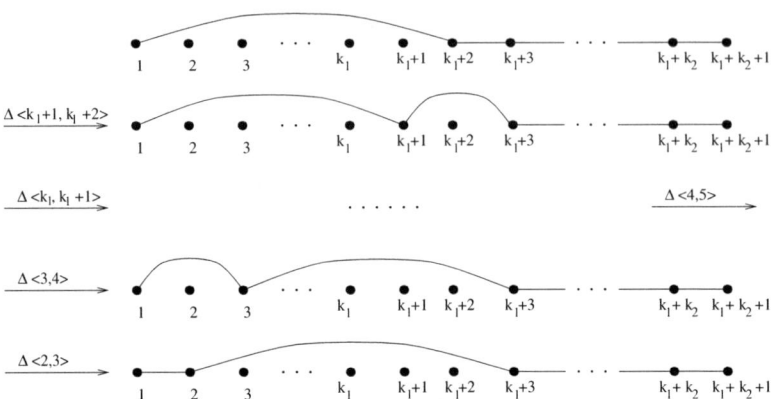

Each of the next $l-1$ sequences moves the left side of the skeleton one step left, so we get the following:

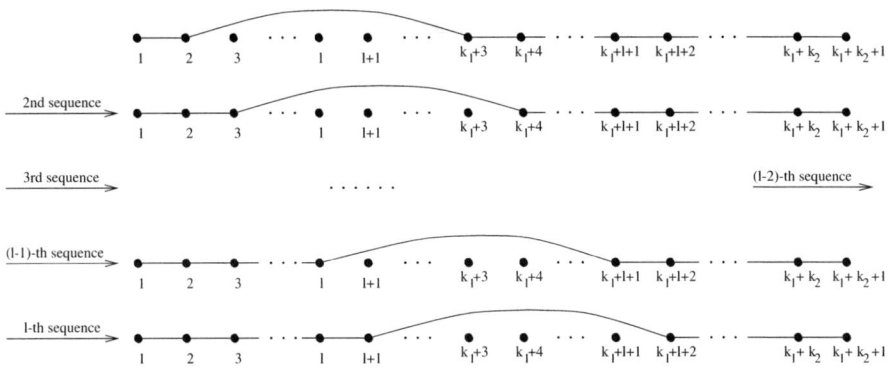

After we have calculated the skeletons representing $\mathcal{L}$.V.C.s for the braid monodromy, we can calculate the relations which they induced.

Fundamental group's structure

The relations which are induced from $\varphi(\delta_{lk_1+1})$:

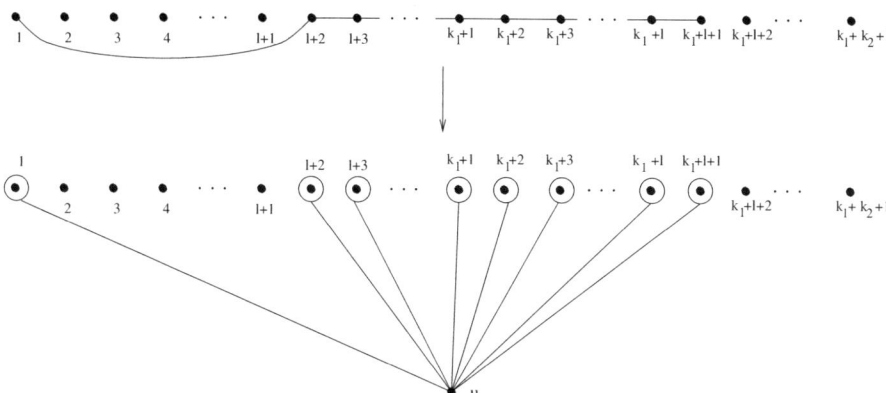

Therefore, according to Lemma 2.14, the relations are:

$$\Gamma_{k_1+l+1}\Gamma_{k_1+l}\ldots\Gamma_{l+2}\Gamma_1 = \Gamma_{k_1+l}\ldots\Gamma_{l+2}\Gamma_1\Gamma_{k_1+l+1} = \cdots = \Gamma_1\Gamma_{k_1+l+1}\ldots\Gamma_{l+2}$$

The relations which are induced from $\varphi(\delta_{lk_1+2})$:

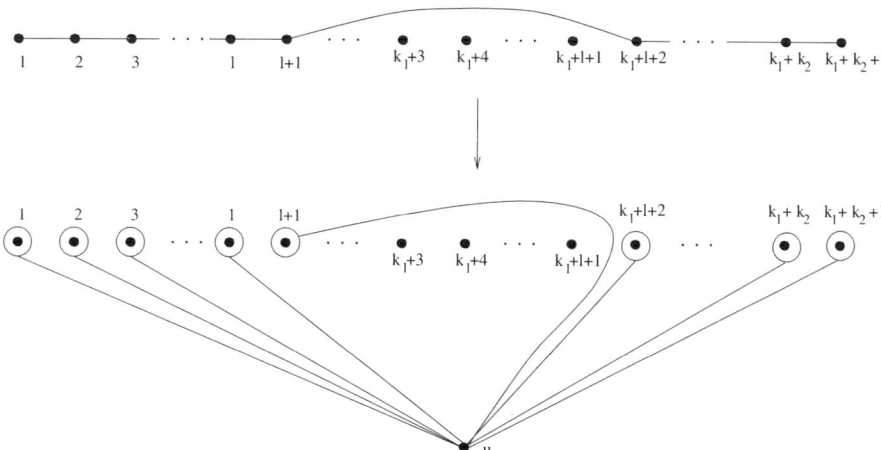

Therefore, according to Lemma 2.14, the relations are:

$$\Gamma_{k_1+k_2+1}\Gamma_{k_1+k_2}\ldots\Gamma_{k_1+l+2}(\Gamma_{k_1+l+1}\ldots\Gamma_{l+2}\Gamma_{l+1}\Gamma_{l+2}^{-1}\ldots\Gamma_{k_1+l+1}^{-1})\Gamma_l\ldots\Gamma_1$$
$$= \Gamma_{k_1+k_2}\ldots\Gamma_{k_1+l+2}(\Gamma_{k_1+l+1}\ldots\Gamma_{l+2}\Gamma_{l+1}\Gamma_{l+2}^{-1}\ldots\Gamma_{k_1+l+1}^{-1})\Gamma_l\ldots\Gamma_1\Gamma_{k_1+k_2+1} = \cdots$$
$$= \Gamma_1\Gamma_{k_1+k_2+1}\Gamma_{k_1+k_2}\ldots\Gamma_{k_1+l+2}(\Gamma_{k_1+l+1}\ldots\Gamma_{l+2}\Gamma_{l+1}\Gamma_{l+2}^{-1}\ldots\Gamma_{k_1+l+1}^{-1})\Gamma_l\ldots\Gamma_2.$$

Now, according to the first Lemma (4.2), second case, Γ_{l+1} commutes with all Γ_j, $l+2 \leq j \leq k_1 + l + 1$, therefore:

$$\Gamma_{k_1+l+1} \ldots \Gamma_{l+2} \Gamma_{l+1} \Gamma_{l+2}^{-1} \ldots \Gamma_{k_1+l+1}^{-1} = \Gamma_{l+1}.$$

Hence, the last set of relations comes to the following simplified form:

$$\Gamma_{k_1+k_2+1} \Gamma_{k_1+k_2} \ldots \Gamma_{k_1+l+2} \Gamma_{l+1} \Gamma_l \ldots \Gamma_1$$
$$= \Gamma_{k_1+k_2} \ldots \Gamma_{k_1+l+2} \Gamma_{l+1} \Gamma_l \ldots \Gamma_1 \Gamma_{k_1+k_2+1}$$
$$= \cdots = \Gamma_1 \Gamma_{k_1+k_2+1} \Gamma_{k_1+k_2} \ldots \Gamma_{k_1+l+2} \Gamma_{l+1} \Gamma_l \ldots \Gamma_2.$$

According to the proof of Proposition 2.15, these two sets of relations (of the two multiple points) are equivalent to the following two sets, respectively:

$$[\Gamma_{k_1+l+1} \Gamma_{k_1+l} \ldots \Gamma_{l+2} \Gamma_1, \Gamma_i] = 1, \ \forall i \in \{1, l+2, \ldots, k_1+l+1\}$$

$$[\Gamma_{k_1+k_2+1} \Gamma_{k_1+k_2} \ldots \Gamma_{k_1+l+2} \Gamma_{l+1} \Gamma_l \ldots \Gamma_1, \Gamma_i] = 1,$$
$$\forall i \in \{1, \ldots, l+1, k_1+l+2, \ldots, k_1+k_2+1\}.$$

And the second Lemma (4.3) is proved.

4.4. Proof of Theorem 4.1

For simplicity, we prove the theorem only for two multiple points, and the proof for t multiple points uses exactly the same arguments.

Till now, we got the following set of generators:

$$g = \{\Gamma_1, \Gamma_2, \ldots, \Gamma_{k_1+k_2+1}\}$$

and the following sets of relations:

(1) $\Gamma_i \Gamma_j = \Gamma_j \Gamma_i;\ 2 \leq i \leq l+1,\ l+2 \leq j \leq k_1+l+1$,

(2) $\Gamma_i \Gamma_j = \Gamma_j \Gamma_i;\ l+2 \leq i \leq k_1+l+1,\ k_1+l+2 \leq j \leq k_1+k_2+1$,

(3) $[\Gamma_{k_1+l+1} \Gamma_{k_1+l} \ldots \Gamma_{l+2} \Gamma_1, \Gamma_i] = 1,\ \forall i \in \{1, l+2, \ldots, k_1+l+1\}$,

(4) $[\Gamma_{k_1+k_2+1} \Gamma_{k_1+k_2} \ldots \Gamma_{k_1+l+2} \Gamma_{l+1} \Gamma_l \ldots \Gamma_1, \Gamma_i] = 1,\ \forall i \in \{1, \ldots, l+1, k_1+l+2, \ldots, k_1+k_2+1\}$.

We have to show that this finitely presented group is isomorphic to

$$\mathbb{F}^{k_1} \oplus \mathbb{F}^{k_2} \oplus \mathbb{Z}.$$

Let us modify the set of generators by replacing the generator Γ_1 by the generator

$$\Gamma' = \Gamma_{k_1+k_2+1} \Gamma_{k_1+k_2} \ldots \Gamma_{k_1+1} \Gamma_{k_1} \ldots \Gamma_2 \Gamma_1$$

Now, we have to check that after the modifications we get an equivalent set of generators, and then we have to calculate the new set of relations.

Corollary 4.4. *After replacing Γ_1 by Γ' (which was defined above) in g, we again get a set of generators. We denote this set of generators by $\tilde{g}$.*

Proof. We have to show that $\Gamma_1 \in \langle \tilde{g} \rangle$. But this is obvious, because:
$$\Gamma_1 = \Gamma_2^{-1}\Gamma_3^{-1}\ldots\Gamma_{k_1+k_2+1}^{-1}\Gamma'. \qquad \square$$

The next step is the calculation of the new set of relations for $\tilde{g}$. The sets (1) and (2) of the old sets of relations have not been changed (because these generators in the relations have not been replaced). We have to deal with the sets (3) and (4).

Corollary 4.5. $[\Gamma', \Gamma] = 1, \forall \Gamma \in \tilde{g}$.

Proof. Obviously, $\Gamma'\Gamma' = \Gamma'\Gamma'$. We will split the rest of the proof into two cases:
(a) $\Gamma \in \{\Gamma_2, \ldots, \Gamma_{l+1}, \Gamma_{k_1+l+2}, \ldots, \Gamma_{k_1+k_2+1}\}$:

$$\Gamma'\Gamma \stackrel{\text{Def}}{=} \Gamma_{k_1+k_2+1}\ldots\Gamma_{k_1+l+2}\Gamma_{k_1+l+1}\ldots\Gamma_{l+2}\Gamma_{l+1}\ldots\Gamma_1\Gamma$$

$$\stackrel{(2)}{=} \Gamma_{k_1+l+1}\ldots\Gamma_{l+2}\Gamma_{k_1+k_2+1}\ldots\Gamma_{k_1+l+2}\Gamma_{l+1}\ldots\Gamma_1\Gamma$$

$$\stackrel{(4)}{=} \Gamma_{k_1+l+1}\ldots\Gamma_{l+2}\Gamma\Gamma_{k_1+k_2+1}\ldots\Gamma_{k_1+l+2}\Gamma_{l+1}\ldots\Gamma_1$$

$$\stackrel{(1)(2)}{=} \Gamma\Gamma_{k_1+l+1}\ldots\Gamma_{l+2}\Gamma_{k_1+k_2+1}\ldots\Gamma_{k_1+l+2}\Gamma_{l+1}\ldots\Gamma_1 \stackrel{(2)+\text{Def}}{=} \Gamma\Gamma'.$$

(b) $\Gamma \in \{\Gamma_{l+2}, \ldots, \Gamma_{k_1+l+1}\}$:

$$\Gamma'\Gamma \stackrel{\text{Def}}{=} \Gamma_{k_1+k_2+1}\ldots\Gamma_{k_1+l+2}\Gamma_{k_1+l+1}\ldots\Gamma_{l+2}\Gamma_{l+1}\ldots\Gamma_1\Gamma$$

$$\stackrel{(1)}{=} \Gamma_{k_1+k_2+1}\ldots\Gamma_{k_1+l+2}\Gamma_{l+1}\ldots\Gamma_2\Gamma_{k_1+l+1}\ldots\Gamma_{l+2}\Gamma_1\Gamma$$

$$\stackrel{(3)}{=} \Gamma_{k_1+k_2+1}\ldots\Gamma_{k_1+l+2}\Gamma_{l+1}\ldots\Gamma_2\Gamma\Gamma_{k_1+l+1}\ldots\Gamma_{l+2}\Gamma_1$$

$$\stackrel{(1)+(2)}{=} \Gamma\Gamma_{k_1+k_2+1}\ldots\Gamma_{k_1+l+2}\Gamma_{l+1}\ldots\Gamma_2\Gamma_{k_1+l+1}\ldots\Gamma_{l+2}\Gamma_1 \stackrel{(1)+\text{Def}}{=} \Gamma\Gamma'. \qquad \square$$

Now, we can claim:

Claim 4.2. *The following set is a complete set of relations for $\tilde{g}$ (we denote it by $\mathcal{R}'$):*

(1') $\Gamma_i\Gamma_j = \Gamma_j\Gamma_i$; $2 \leq i \leq l+1, l+2 \leq j \leq k_1+l+1$.

(2') $\Gamma_i\Gamma_j = \Gamma_j\Gamma_i$; $l+2 \leq i \leq k_1+l+1, k_1+l+2 \leq j \leq k_1+k_2+1$.

(3') $[\Gamma', \Gamma] = 1$; $\forall \Gamma \in \tilde{g}$.

Proof. We have to show that $\{(1),(2),(3),(4)\}$ is equivalent to $\{(1'), (2'), (3')\}$ (with respect to the required replacements). In the previous claim, we proved that $\{(1),(2),(3),(4)\} \Rightarrow \{(1'), (2'), (3')\}$. We have to prove the opposite direction. Assume the set of relations $\{(1'), (2'), (3')\}$, and prove the relations $\{(1),(2),(3),(4)\}$:

(1) and (2): they are the same as $(1')$ and $(2')$, respectively.

(3): We have to prove that
$$[\Gamma_{k_1+l+1}\Gamma_{k_1+l}\ldots\Gamma_{l+2}\Gamma_1, \Gamma_i] = 1, \ \forall i \in \{1, l+2, \ldots, k_1+l+1\}.$$

From $(3')$ we know that $[\Gamma', \Gamma_i] = 1$. Therefore, we have:

$$\Gamma_{k_1+k_2+1}\Gamma_{k_1+k_2}\ldots\Gamma_1\Gamma_i = \Gamma_i\Gamma_{k_1+k_2+1}\Gamma_{k_1+k_2}\ldots\Gamma_1$$

$$\stackrel{(1')}{\Rightarrow} \Gamma_{k_1+k_2+1}\ldots\Gamma_{k_1+l+2}\Gamma_{l+1}\ldots\Gamma_2\Gamma_{k_1+l+1}\ldots\Gamma_{l+2}\Gamma_i$$

$$= \Gamma_i\Gamma_{k_1+k_2+1}\ldots\Gamma_{k_1+l+2}\Gamma_{l+1}\ldots\Gamma_2\Gamma_{k_1+l+1}\ldots\Gamma_{l+2}$$

$$\stackrel{(1')+(2')+(l+2\leq i\leq k_1+l+1)}{\Rightarrow} \Gamma_{k_1+k_2+1}\ldots\Gamma_{k_1+l+2}\Gamma_{l+1}\ldots\Gamma_2\Gamma_{k_1+l+1}\ldots\Gamma_{l+2}\Gamma_i$$

$$= \Gamma_{k_1+k_2+1}\ldots\Gamma_{k_1+l+2}\Gamma_{l+1}\ldots\Gamma_2\Gamma_i\Gamma_{k_1+l+1}\ldots\Gamma_{l+2}$$

$$\stackrel{\Gamma_2^{-1}\ldots\Gamma_{l+1}^{-1}\Gamma_{k_1+l+2}^{-1}\ldots\Gamma_{k_1+k_2+1}^{-1}}{\Longrightarrow} [\Gamma_{k_1+l+1}\Gamma_{k_1+l}\ldots\Gamma_{l+2}\Gamma_1, \Gamma_i] = 1,$$

$$\forall i \in \{l+2, \ldots, k_1+l+1\}.$$

Now, it remains to prove that:
$$[\Gamma_{k_1+l+1}\Gamma_{k_1+l}\ldots\Gamma_{l+2}\Gamma_1, \Gamma_1] = 1.$$

$$\Gamma_{k_1+l+1}\ldots\Gamma_{l+2}\Gamma_1\Gamma_1$$

$$= \Gamma_{k_1+l+1}\ldots\Gamma_{l+2}\Gamma_1(\Gamma_{l+2}^{-1}\ldots\Gamma_{k_1+l+1}^{-1}\Gamma_{k_1+l+1}\ldots\Gamma_{l+2})\Gamma_1$$

$$\stackrel{(3) + ab=ba \Rightarrow ab^{-1}=b^{-1}a}{=} (\Gamma_{l+2}^{-1}\ldots\Gamma_{k_1+l+1}^{-1})\Gamma_{k_1+l+1}\ldots\Gamma_{l+2}\Gamma_1(\Gamma_{k_1+l+1}\ldots\Gamma_{l+2})\Gamma_1$$

$$= \Gamma_1\Gamma_{k_1+l+1}\ldots\Gamma_{l+2}\Gamma_1.$$

(4) Same arguments as (3). □

We return to the proof of the theorem. Using the above claim, we can find the structure of the calculated group:

$$G = \langle g|\mathcal{R}\rangle \cong \langle \tilde{g}|\mathcal{R}'\rangle = \langle \Gamma_2, \ldots, \Gamma_{k_1+k_2+1}, \Gamma' \mid \mathcal{R}'\rangle$$
$$\cong \langle \Gamma'\rangle \oplus \langle \Gamma_2, \ldots, \Gamma_{k_1+k_2+1} \mid \mathcal{R}'\rangle$$
$$\cong \langle \Gamma'\rangle \oplus \langle \Gamma_2, \ldots, \Gamma_{l+1}, \Gamma_{k_1+l+2}, \ldots, \Gamma_{k_1+k_2+1}\rangle \oplus \langle \Gamma_{l+2}, \ldots, \Gamma_{k_1+l+1}\rangle$$
$$\cong \mathbb{Z} \oplus \mathbb{F}^{k_2} \oplus \mathbb{F}^{k_1}.$$

Hence, we finished the proof of Theorem 4.1.

4.5. The projective case

Now, we will investigate the projective case.

Theorem 4.6. *Let $\mathcal{L}$ be a real line arrangement in $\mathbb{CP}^2$ where all the t multiple points are on the same line $L \in \mathcal{L}$. Let $k_i + 1$ be the multiplicity of the multiple point P_i, $1 \leq i \leq t$. Then:*

$$\pi_1(\mathbb{CP}^2 - \mathcal{L}) \cong \bigoplus_{i=1}^{t} \mathbb{F}^{k_i}.$$

Proof. For simplicity, we will prove it for two multiple points and the proof for t multiple points uses exactly the same arguments.

From the last theorem, we get:

$$\pi_1(\mathbb{C}^2 - \mathcal{L}) \cong \mathbb{F}^{k_1} \oplus \mathbb{F}^{k_2} \oplus \mathbb{Z}.$$

According to claim 4.2, we get the following presentation for this group:
Generators: $g = \{\Gamma', \Gamma_2, \ldots, \Gamma_{k_1+k_2+1}\}$.
Relations: $\mathcal{R} = \{\Gamma_i \Gamma_j = \Gamma_j \Gamma_i,\ 2 \leq i \leq l+1,\ l+2 \leq j \leq k_1+l+1;\ \Gamma_i \Gamma_j = \Gamma_j \Gamma_i,\ l+2 \leq i \leq k_1+l+1,\ k_1+l+2 \leq j \leq k_1+k_2+1;\ [\Gamma', \Gamma] = 1,\ \forall \Gamma \in g\}$ where $\Gamma' = \Gamma_{k_1+k_2+1} \ldots \Gamma_1$.

Now, when we are going to the projective case, we add one additional relation, according to Theorem 2.9:

$$\Gamma_{k_1+k_2+1} \ldots \Gamma_1 = 1.$$

In terms of Γ', this relation gets the following form:

$$\Gamma' = 1$$

Now,

$$\pi_1(\mathbb{CP}^2 - \mathcal{L}) = \langle g \mid \mathcal{R}, \Gamma' = 1 \rangle$$
$$\cong \langle \Gamma' \mid \Gamma' = 1 \rangle \oplus \langle \Gamma_2, \ldots, \Gamma_{l+1}, \Gamma_{k_1+l+2}, \ldots, \Gamma_{k_1+k_2+1} \rangle \oplus \langle \Gamma_{l+2}, \ldots, \Gamma_{k_1+l+1} \rangle$$
$$\cong \mathbb{F}^{k_2} \oplus \mathbb{F}^{k_1}.$$
$\square$

As a consequence of the last theorem, we get:

Corollary 4.7.

$$\pi_1(\mathbb{C}^2 - \mathcal{L}) \cong \pi_1(\mathbb{CP}^2 - \mathcal{L}) \oplus \mathbb{Z}.$$

Therefore, the short exact sequence which was proved by Oka (Theorem 2.10):

$$1 \to \mathbb{Z} \to \pi_1(\mathbb{C}^2 - \mathcal{L}) \to \pi_1(\mathbb{CP}^2 - \mathcal{L}) \to 1$$

splits.

5. Arrangements with more than one equivalence class

5.1. The definition of the equivalence relation

The above results can be generalized more. Let us define the following relation on the set of multiple intersection points:

Definition 5.1. Let p_1, p_2 be two multiple intersection points. We say that $p_1 \sim p_2$ if p_1 is connected to p_2 by a "path" which its vertices are multiple intersection points.

Claim 5.1. $\sim$ *is an equivalence relation on the set of multiple intersection points.*

Proof. *Reflexive:* each point is connected to itself by the empty path.

Symmetry: if p_1 is connected to p_2 by a path P, p_2 is connected to p_1 by P^{-1} - the opposite path of P (which is also a path of multiple points).

Transitive: if p_1 is connected to p_2 by P, and p_2 is connected to p_3 by Q, p_1 is connected to p_3 by $P \cdot Q$, which is the concatenation of P and Q and therefore it is a path of multiple points, because p_2 itself is a multiple point too. □

This equivalence relation induces equivalence classes on the set of multiple intersection points. We also want to show that this equivalence relation induces a partition on the lines of the arrangement:

Claim 5.2. *Let* $C_1 = \{p_1, \ldots, p_k\}$ *be the multiple points of one equivalence class and* $C_2 = \{q_1, \ldots, q_l\}$ *be the multiple points of another equivalence class. Let* $\mathcal{L}_i$ *be the set of lines which pass through one of the multiple points in* C_i.

Then: $\mathcal{L}_1 \cap \mathcal{L}_2 = \emptyset$.

Proof. Assume, on the contrary, that there exists a line L, such that $L \in \mathcal{L}_1 \cap \mathcal{L}_2$. Therefore, $L \in \mathcal{L}_1$ and $L \in \mathcal{L}_2$. From the definitions of $\mathcal{L}_1$ and $\mathcal{L}_2$, there exist points $p \in C_1$ and $q \in C_2$ such that L passes through p and q. Therefore, $p \sim q$, and hence $C_1 = C_2$, a contradiction to the assumption that C_1 and C_2 are distinct equivalence classes. □

5.2. The affine case

Now, we can claim the following:

Theorem 5.3. *Let* $\mathcal{L}$ *be a real line arrangement in* $\mathbb{CP}^2$ *consists of n lines. We choose the line at infinity such that all the lines are intersected in* $\mathbb{C}^2$. *Assume that there are k multiple intersection points* $p_1, \ldots, p_k$ *with multiplicities* $m_1, \ldots, m_k$ *respectively. Assume also that all the multiple intersection points in every equivalence class are* collinear, i.e. every equivalence class contains a unique line which connects all the multiple points of that class. Then:

$$\pi_1(\mathbb{C}^2 - \mathcal{L}, u_0) \cong \bigoplus_{i=1}^{k} \mathbb{F}^{m_i-1} \oplus \mathbb{Z}^{n-(\sum_{i=1}^{k}(m_i-1))}.$$

The number of infinite cyclic groups is a sum of two numbers: the number of equivalence classes and the number of lines which have only simple intersection points.

Proof. Let C_i, $1 \leq i \leq t$ be the different equivalence classes of multiple points. According to the last claim, we define $\mathcal{L}_i$ to be the lines which pass through points in C_i. Let $l_1, \ldots, l_r$ be lines which are not in any $\mathcal{L}_i$ (which means that they do not pass through any multiple point, or equivalently, they intersect all the other lines at simple points only).

In every $\mathcal{L}_i$, we have a line L_i which connects all the multiple points in C_i. Therefore, according to Theorem 4.1, we have:

$$\pi_1(\mathbb{C}^2 - \mathcal{L}_i) = \left(\bigoplus_{j=1}^{n_i} \mathbb{F}^{m_{P_{i,j}}-1}\right) \oplus \mathbb{Z}$$

where $n_i = \#C_i$ and $m_{P_{i,j}}$ is the multiplicity of the j-th point in C_i, $1 \leq j \leq n_i$.
For l_i, we know:

$$\pi_1(\mathbb{C}^2 - l_i) = \mathbb{Z}.$$

Now, we use the Oka–Sakamoto theorem (see Section 2.1) to get:

$$\pi_1(\mathbb{C}^2 - \mathcal{L}) = \pi_1\left(\mathbb{C}^2 - \left(\bigcup_{i=1}^{t}\mathcal{L}_i \cup \bigcup_{i=1}^{r} l_i\right)\right)$$

$$\cong \left(\bigoplus_{i=1}^{t} \pi_1(\mathbb{C}^2 - \mathcal{L}_i)\right) \oplus \left(\bigoplus_{i=1}^{r} \pi_1(\mathbb{C}^2 - l_i)\right)$$

$$\cong \left(\bigoplus_{i=1}^{t}\left(\bigoplus_{j=1}^{n_i} \mathbb{F}^{m_{P_{i,j}}-1}\right) \oplus \mathbb{Z}\right) \oplus \left(\bigoplus_{i=1}^{r} \mathbb{Z}\right) \cong \left(\bigoplus_{i=1}^{t}\bigoplus_{j=1}^{n_i} \mathbb{F}^{m_{P_{i,j}}-1}\right) \oplus \mathbb{Z}^{t+r}.$$

It remains to show that this group is equal to the group mentioned in the formulation of the theorem. First, in the double sum, every multiple point appears exactly once, because it appears in only one equivalence class. Therefore:

$$\left(\bigoplus_{i=1}^{t}\bigoplus_{j=1}^{n_i} \mathbb{F}^{m_{P_{i,j}}-1}\right) = \bigoplus_{i=1}^{k} \mathbb{F}^{m_i-1}.$$

Now we have to show that:
$$t+r = n - \left(\sum_{i=1}^{k}(m_i - 1)\right).$$

Let o_i be the number of lines in $\mathcal{L}_i$. We know that
$$\left(\sum_{i=1}^{t} o_i\right) + r = n.$$

It is easy to see that:
$$o_i = \left(\sum_{j=1}^{n_i}(m_{P_{i,j}} - 1)\right) + 1,$$

because there is a unique line which connects all the multiple points in every equivalence class.

When we combine the last two equations, we get:
$$\sum_{i=1}^{t}\sum_{j=1}^{n_i}(m_{P_{i,j}} - 1) + t + r = n.$$

As before, due to the fact that every multiple point appears exactly in one equivalence class, we get:
$$\sum_{i=1}^{t}\sum_{j=1}^{n_i}(m_{P_{i,j}} - 1) = \sum_{i=1}^{k}(m_i - 1)$$

and therefore, we get:
$$t+r = n - \left(\sum_{i=1}^{k}(m_i - 1)\right). \qquad \square$$

5.3. The projective case

Now, we will investigate the projective case.

Theorem 5.4. *Let $\mathcal{L}$ be a real line arrangement in $\mathbb{CP}^2$ consists of n lines. We choose the line at infinity such that all the lines are intersected in $\mathbb{C}^2$. Assume that there are k multiple intersection points $p_1, \ldots, p_k$ with multiplicities $m_1, \ldots, m_k$ respectively. Assume also that all the multiple intersection points in every equivalence class are collinear, i.e. every equivalence class contains a unique line which connects all the* multiple points of that class. Then:

$$\pi_1(\mathbb{CP}^2 - \mathcal{L}, u_0) \cong \bigoplus_{i=1}^{k} \mathbb{F}^{m_i-1} \oplus \mathbb{Z}^{n-1-(\sum_{i=1}^{k}(m_i-1))}.$$

The number of infinite cyclic groups is a sum of two numbers: the number of equivalence classes minus 1 and the number of lines which have only simple intersection points.

Proof. This is the projective analogue of Theorem 5.3. We induce it using the same techniques as we induced Theorem 3.3 from Theorem 3.1. □

As a consequence of the last theorem, we get:

Corollary 5.5.

$$\pi_1(\mathbb{C}^2 - \mathcal{L}) \cong \pi_1(\mathbb{CP}^2 - \mathcal{L}) \oplus \mathbb{Z}.$$

Therefore, the short exact sequence which was proved by Oka (Theorem 2.10):

$$1 \to \mathbb{Z} \to \pi_1(\mathbb{C}^2 - \mathcal{L}) \to \pi_1(\mathbb{CP}^2 - \mathcal{L}) \to 1$$

splits.

Remark 5.2. Simultaneously and independently, Fan [Fa2] got similar results (see Section 2.1), with entirely different methods, in even more general case, when there is no equivalence class which has a cycle of multiple points in it.

6. Results concerning the bigness of the fundamental group

Definition 6.1. A group G is called *big* if $\mathbb{F}^2 \subset G$.

As a result from the general Theorems (5.3, 5.4), we can say the following:

Corollary 6.1. *Let $\mathcal{L}$ be a real line arrangement in $\mathbb{CP}^2$ consisting of n lines which satisfies the conditions of Theorem 5.3. Then, the fundamental groups of its complement, $\pi_1(\mathbb{C}^2 - \mathcal{L}, u_0)$ and $\pi_1(\mathbb{CP}^2 - \mathcal{L}, u_0)$, are big.*

Proof. According to Theorem 5.3, the fundamental group of its affine complement is of the form:

$$\pi_1(\mathbb{C}^2 - \mathcal{L}, u_0) \cong \bigoplus_{i=1}^{k} \mathbb{F}^{m_i-1} \oplus \mathbb{Z}^{n-(\sum_{i=1}^{k}(m_i-1))}.$$

Now, $m_i \geq 3$ in every multiple point, and hence $\mathbb{F}^2$ is contained in this group. Therefore, the fundamental group of its affine complement is big. The proof for the projective case is the same. □

In fact, this result has been recently proven [DOZ] for any arrangement which has at least one multiple intersection point:

Theorem 6.2 (Dethloff, Orevkov, Zaidenberg). *Let $\mathcal{L}$ be a real line arrangement in $\mathbb{CP}^2$ consisting of n lines. We choose the line at infinity such that all the lines are intersected in $\mathbb{C}^2$. Assume that there exists in $\mathcal{L}$ at least one multiple intersection point.*
Then, $\pi_1(\mathbb{CP}^2 - \mathcal{L}, u_0)$ is big.

Remark 6.2. It seems that this phenomena is not happen for branch curves of surfaces, unlike previous expectations which followed earlier results of Zariski and Moishezon. Most fundamental groups of complements of branch curves are "almost solvable", i.e. they contain a solvable subgroup of finite index and they are not "big" (see [Te2]).

Acknowledgments. We thank Prof. Leonid Makar-Limanov for suggestions which led to the crucial part of the proof of Proposition 2.15.

References

[CS] Cohen, D. C. and Suciu, A. I., The braid monodromy of plane algebraic curves and hyperplane arrangements, Comment. Math. Helv. 72 (2) (1997), 285–315.

[DOZ] Dethloff, G., Orevkov, S. and Zaidenberg, M., Plane curves with a big fundamental group of the complement, in: Voronezh Winter Mathematical Schools: Dedicated to Selim Krein (P. Kuchment, V. Lin, eds.), Amer. Math. Soc. Transl. Ser. 2 184, Amer. Math. Soc., 1998.

[Fa1] Fan, K. M., Position of singularities and fundamental group of the complement of a union of lines, Proc. Amer. Math. Soc. 124 (11) (1996), 3299–3303.

[Fa2] Fan, K. M., Direct product of free groups as the fundamental group of the complement of a union of lines, Michigan Math. J. 44 (2) (1997), 283–291.

[Ga] Garber, D., On the fundamental group of complement of real line arrangements, M.Sc. thesis, Bar-Ilan University 1997.

[Mo] Moishezon, B., Stable branch curves and braid monodromies, Lecture Notes in Math. 862, Springer-Verlag, 1981, 107–192.

[MoTe1] Moishezon, B. and Teicher, M., Braid group techniques in complex geometry I, Line arrangements in $\mathbb{CP}^2$, Contemp. Math. 78, Amer. Math. Soc., 1988, 425–555.

[MoTe2] Moishezon, B. and Teicher, M., Braid group techniques in complex geometry II, From arrangements of lines and conics to cuspidal curves, Algebraic Geometry, Lecture Notes in Math. 1479, Springer-Verlag, 1990.

[MoTe3] Moishezon, B. and Teicher, M., Braid group techniques in complex geometry V: The fundamental group of a complement of a branched curve of a Veronese generic projection, Comm. Anal. Geom. 4 (1) (1996), 1–120.

[MoTe4] Moishezon, B. and Teicher, M., Braid groups, singularities, and algebraic surfaces, Academic Press, to appear.

[O] Oka, M., On the fundamental group of a reducible curve in $\mathbb{P}^2$, J. London Math. Soc. (2) 12 (1976), 239–252.

[OS] Oka, M. and Sakamoto, K., Product theorem of the fundamental group of a reducible curve, J. Math. Soc. Japan 30 (4) (1978), 599–602.

[OT] Orlik, P. and Terao, H., Arrangements of hyperplanes, Grundlehren Math. Wiss. 300, Springer-Verlag, 1992.

[Ra] Randell, R., The fundamental group of the complement of a union of complex hyperplanes, Invent. Math. 69 (1982), 103–108; Correction, Invent. Math. 80 (1985), 467–468.

[Sa] Salvetti, M., Topology of the complement of real hyperplanes in $\mathbb{C}^N$, Invent. Math. 88 (1987), 603–618.

[Te1] Teicher, M., Braid groups, algebraic surfaces and fundamental groups of complement of branch curves, Proc. Sympos. Pure Math. 62 (1), Amer. Math. Soc, 1997, 127–150.

[Te2] Teicher, M., New invariants of surfaces, Contemp. Math. 231, Amer. Math. Soc., 1999, 271–281.

[VK] Van Kampen, E. R., On the fundamental group of an algebraic curve, Amer. J. Math. 55 (1933), 255–260.

[Z] Zariski, O., On the problem of existence of algebraic functions of two variables possessing a given branch curve, Amer. J. Math. 51 (1929), 305–328.

Kählerian structures on symplectic reductions

Peter Heinzner* and Alan Huckleberry

One purpose of this note is to outline some recent results on symplectic reduction for actions of Lie groups of holomorphic transformations on complex spaces. In particular, in Section 7 we sketch the proof of the main result of [A-H-H] on the existence of the canonical Kählerian structure on the reduction of a Kählerian space in the case of a proper action.

Since the quotient theory treated here can be applied in a variety of mathematical and physical contexts, we also hope that this article will serve as an introduction to the subject for non-specialists who might find use for the techniques and results. Therefore we have given a rather lengthy historical introduction, a number of elementary examples which we find illuminating and have attempted to explain our complex analytic point of the subject.

1. Reductive groups

Throughout this text K will denote a compact Lie group. Although one should not forget that a finite group is compact, we are primarily concerned with phenomena which are present only in the positive dimensional case. For example, if $L \times M \to M$ is a smooth action of a Lie group L on a manifold M, the vector fields induced by the canonical map $\mathfrak{l} := \mathrm{Lie}(L) \to \mathrm{Vect}(M)$ play a role in many of our considerations.

If $K \times X \to X$ is an action of a compact Lie group by holomorphic transformations on a complex manifold the associated vector fields can be regarded as holomorphic $(1, 0)$-fields, i.e., there is an induced $\mathbb{C}$-linear Lie algebra morphism $\mathfrak{k}^{\mathbb{C}} \to \Gamma_{\mathcal{O}}(X, \mathrm{T}^{1,0}(X))$. The complexification $\mathfrak{g} := \mathfrak{k}^{\mathbb{C}} = \mathfrak{k} + i\mathfrak{k}$ is the Lie algebra of a complex Lie group $G := K^{\mathbb{C}}$ which can be constructed as follows. Let $\rho: K \to \mathrm{GL}(V)$ be a faithful continuous complex (and therefore real-analytic) representation and let G be the smallest complex Lie subgroup in $\mathrm{GL}(V)$ which contains $K \cong \rho(K)$. On proves that G is in fact a complex algebraic subgroup of $\mathrm{GL}(V)$ which contains $\rho(K)$ as a maximal compact subgroup. The manifold K is totally real in G and of maximal dimension with this property, i.e., $\dim_{\mathbb{R}} K = \dim_{\mathbb{C}} G$. In fact

---

*Supported by a Heisenberg Stipendium of the Deutsche Forschungsgemeinschaft

K is a real algebraic subgroup of G which is defined by an algebraic anti-holomorphic involution.

For example, if $K = S^1$ is the unit circle, then $G = \mathbb{C}^*$ is the group of units in $\mathbb{C}$ and $\sigma(z) = \bar{z}^{-1}$ is the involution which defines S^1. Similarly, for $K = SU_n$, it follows that $G = SL_n(\mathbb{C})$ and $\sigma(Z) = {}^t\bar{Z}^{-1}$. Also, if $K = SO_n(\mathbb{R})$, then $G = SO_n(\mathbb{C})$ and $\sigma(Z) = \bar{Z}$.

If $G = K^\mathbb{C}$ is a above, then its complex (resp. complex linear algebraic) group structure is unique. More generally, the complexification of K is universal in the sense that every Lie homomorphism $h : K \to \tilde{G}$ into a complex Lie group (resp. linear algebraic group) extends to a holomorphic (resp. regular) homomorphism $h^\mathbb{C} : G \to \tilde{G}$.

The universal complexification $L^\mathbb{C}$ exists in the complex setting for any Lie group L (see [Ho]), but the canonical homomorphism $L \to L^\mathbb{C}$ may not be an embedding as in the compact case.

Groups of the form $G = K^\mathbb{C}$ where K is a compact Lie group are called complex linear reductive groups. The map $\mathfrak{k}^\mathbb{C} \to \Gamma_\mathcal{O}(X, T^{1,0}(X))$ given by a K-action of holomorphic transformations on X defines a local holomorphic action of the complexified group $K^\mathbb{C}$ in the sense of Palais ([P1], see also [H-I] for the adaption of Palais' method to the complex analytic setting). In some cases this is already global and in others it can be globalised, i.e., a universal complex $K^\mathbb{C}$-space $X^\mathbb{C}$ can be constructed with a K-equivariant open embedding $X \to X^\mathbb{C}$.

For example, if $\rho : K \to GL(V)$ is a complex representation, i.e., a linear K-action on V by $\mathbb{C}$-linear mappings is given, then the universality of the embedding $K \to K^\mathbb{C}$ shows that the local $K^\mathbb{C}$-action on V is global, i.e., the K-action is the restriction of the algebraic $K^\mathbb{C}$-action $K^\mathbb{C} \times V \to V$ given by the regular representation $\rho^c : K^\mathbb{C} \to GL(V)$.

More generally, if $K \times X \to X$ is a real-algebraic action of a compact group of algebraic morphisms on an affine variety, then, since the associated representation on the coordinate ring $\mathbb{C}[X]$ is locally finite, i.e., the orbit of every function is contained in a finite dimensional subspace, there is even an equivariant algebraic embedding $X \to V$ in a K-representation space. Applying the universally of $K \to K^\mathbb{C}$ to this representation, it again follows that the local $K^\mathbb{C}$-action on X is already global. More generally, for a Stein manifold X which has no bounded plurisubharmonic functions it is not hard to see that the local $K^\mathbb{C}$-action on X is already global (see [F]).

The situation in the general complex analytic setting is different. First of all, a bounded domain, e.g., the unit ball in $\mathbb{C}^n$, is a natural place to find an action of a compact group of holomorphic transformations. A local $K^\mathbb{C}$-action on such a manifold X is not global, because by Liouville's Theorem a complex one-parameter group would necessarily act trivially. Secondly, one has examples of Stein manifolds equipped with actions of compact groups K of holomorphic transformations, with infinitely many different K-isotropy types, i.e., isotropy subgroups which are not conjugate (see [H1]). But a linear action of a compact group has only finitely many isotropy types.

The difficulty which arises, e.g., for bounded domains can be remedied by enlarging X.

Complexification Theorem. *Let $K \times X \to X$ be an action of a compact group on a holomorphically convex complex space. Then there exists a holomorphic action $K^{\mathbb{C}} \times X^{\mathbb{C}} \to X^{\mathbb{C}}$ on another holomorphically convex space $X^{\mathbb{C}}$ and an open equivariant embedding $\iota : X \to X^{\mathbb{C}}$ which contains X as an open Runge subset. If X is Stein, then so is $X^{\mathbb{C}}$.*

Remarks.

1. The entire discussion in this paper can be carried out for complex spaces. At the points where the singularities play an important role we comment directly on the matter. Otherwise, to avoid technicalities, we stick to the smooth case.

2. The embedding $\iota : X \to X^{\mathbb{C}}$ has the desired universality property:

   If $K^{\mathbb{C}} \times Y \to Y$ is a holomorphic action of the complexified group and $\phi : X \to Y$ is a K-equivariant holomorphic map, then there exists a unique extension $\tilde{\phi} : X^{\mathbb{C}} \to Y$ of ϕ.

3. The Stein case was proved in [H2]. In the more general case of holomorphic convex spaces, a holomorphic adaption of Palais globalization theory was used (see [H-I]).

4. If $K^{\mathbb{C}} \times X \to X$ is a holomorphic action on a Stein manifold, then there is a closed equivariant embedding in a representation if and only if there are only finitely many K-isotropy types ([H1]).

5. Techniques used for these results show that $K^{\mathbb{C}}$-actions on Stein spaces locally look like algebraic actions, e.g., orbits are Zariski open in their closures.

Actions of reductive groups have been the subject of much study. One reason is that they arise in numerous applications. Another is that, due to the presence of the compact subgroup which generates the ambient group over the complex numbers or to some analogous property in the case of other fields of definition, these actions provide a convenient starting point for the study of transformation groups. This is particularly transparent in the representation theory and the theory of invariants.

2. Invariant theory

Many properties of an action of a group G on a set S can be formulated in terms of invariance. Fixed points and invariant functions are immediate examples. If a function $f : S \to \mathbb{C}$ is regarded as a map to a trivial G-space, then one is led to the fact that equivariance is also a form of invariance: Let X and Y be G-spaces and regard $\text{Map}(X, Y)$ as a $G \times G$-space by $f \to g_1 \circ f \circ g_2^{-1}$. The invariants of the diagonal G-action, $f \to g \circ f \circ g^{-1}$, are exactly the equivariant maps.

Invariants of diagonal actions are of importance in many contexts. The most obvious example is that defined on G itself be the $G \times G$-action coming from left and right multiplication, i.e., conjugation. It is quite often possible to formulate relevant invariant theoretic questions in terms of the invariants of a linear action. In concrete terms let $\rho : G \to \text{GL}(V)$ be a linear representation of G on a finite dimensional complex vector space and consider the associated action on the polynomials $\mathbb{C}[V]$, $P \mapsto P \circ g^{-1} = g \cdot P$.

Explicit knowledge about a given invariant, i.e., a polynomial P with $g \cdot P = P$ for all $g \in G$, often leads to essential information related to the problem at hand. Physical "constants" typically arise in precisely this way. There are numerous examples in mathematical contexts, e.g., characteristic classes, where the particular form of an invariant is important. Thus it is only natural that invariant theorists initially devoted their time to calculations, i.e., the attempt to compute generators of the algebra of invariants $\mathbb{C}[V]^G = \{P \in \mathbb{C}[V]; g \cdot P = P \text{ for all } g \in G\}$. Of course this may be futile, because it might not be finitely generated ([N1]). Furthermore, even in a situation where $\mathbb{C}[V]^G$ is finitely generated, there may be no method of presenting a comprehensible list of natural generators. Thus it is often more appropriate to study the structure of $\mathbb{C}[V]^G$.

The proposal to move toward a structural study (Hilbert's 14th problem is exactly the question on finite-generation) was revolutionary and was certainly one of the driving forces for the development of e.g. the foundation of commutative algebra.

After results on finite generation were proved in various special contexts, e.g., by Noether [No] for finite groups and Weyl for certain compact groups [W], Nagata [N2] formulated and proved the following final result.

Theorem 2.1. *Let $G \times X \to X$ be an algebraic action of a linear reductive group on an affine variety. Then $\mathbb{C}[V]^G$ is a finitely generated algebra.*

Remark 1. The main point here is the result for $X = V$ a linear representation. The more general result follows via equivariant embedding.

The affine variety $X/\!/G := \text{Spec}\,\mathbb{C}[X]^G$ along with the natural map $\pi : X \to X/\!/G$ is regarded as a sort of quotient of X by the G-action. It is often refered to as the Hilbert quotient. Structural questions on the ring of invariants are quite often most naturally formulated in terms of the geometry of this map.

Example 1. Let $G = \text{SO}_3(\mathbb{C})$ act by its standard linear representation on $X := \mathbb{C}^3$. Then $\mathbb{C}[X]$ is generated by the polynomial $\pi(x, y, z) = x^2 + y^2 + z^2$ and the Hilbert quotient is simply given by $\pi : \mathbb{C}^3 \to \mathbb{C}^3/\!/G \cong \mathbb{C}$.

Despite the simple nature of the above example, it does give an indication of some of the key properties of the quotient map. For example, the fiber $\pi^{-1}(0)$, i.e., the cone $\{x^2 + y^2 + z^2 = 0\}$, consists of two orbits, exactly one which is closed. The other π-fibers are in fact orbits. This is not always the case, but nevertheless $X/\!/G$ can be defined as a sort of orbit space.

Property 1. *Every fiber of $\pi : X \to X/\!/G$ contains a unique closed G-orbit.*

The set of closed orbits comes equipped with a natural stratification which is defined by the G-isotropy types, but except in very special cases, is not a subvariety. In the above example of $SO_3(\mathbb{C})$ acting on $\mathbb{C}^3$ it is $\pi^{-1}(\mathbb{C}\setminus\{0\})\cup\{0\}$. By definition the Hilbert quotient is given by the equivalence relation $x \sim y$ if and only if $f(x) = f(y)$ for all $f \in \mathbb{C}[V]^G$. It follows from Property 1 that this is the same quotient as that which is defined by the continuous invariant functions.

Property 2. *The categorical quotient is also defined by the equivalence relation $x \sim y$ if and only if $\overline{G \cdot x} \cap \overline{G \cdot y} \neq \emptyset$.*

Here the closure $\overline{G \cdot x}$ is the topological closure which in fact agrees with the Zariski closure.

Property 3. *If $\phi : X \to Z$ is a G-invariant regular map, then there exists a uniquely defined morphism $\tilde{\phi} : X/\!/G \to Z$ so that $\phi = \tilde{\phi} \circ \pi$. In particular, if X is normal, then so is $X/\!/G$.*

The behaviour with respect to restriction is also as desired.

Property 4. *If Y is a G-stable closed subvariety of X, then $\pi(Y)$ is a closed subvariety of $X/\!/G$ and $\pi|Y : Y \to \pi(Y)$ is the Hilbert quotient for Y.*

Note that if we write $G = K^{\mathbb{C}}$, that a polynomial (or more generally a holomorphic function) is G-invariant if and only if it is K-invariant (Identity Principle). Thus, $\pi : X \to X/\!/K$ is defined as the G-Hilbert quotient. Of course, the second Property is no longer valid if G is replaced by K: The K-orbit space, which is an $\mathbb{R}$-semialgebraic space defined by polynomial inequalities as well as equalities ([P-S]) for a more precise statement), is real and except in trivial cases much bigger that $X/\!/K$. Nevertheless, as will be explained later, K-quotients of special subsets of X play an important role in understanding $\pi : X \to X/\!/G$.

3. Hilbert quotient for Stein spaces

Geometric properties of the Hilbert quotient can be translated to statements about the invariants. However the existence of the quotient and its basic properties are such that, except for the input of the elements of affine algebraic geometry, the initial logic would seem to go the other way. Thus one is led to look at quotient $\pi : X \to X/\!/G$ from other perspectives.

One model for another viewpoint is the geometric slice theorem for proper Lie group actions.

A continuous action $G \times X \to X$ of a topological group on a locally compact topological space X is said to be a proper action if the associated map $G \times X \to X \times X$, $(g, x) \to (g \cdot x, x)$ is proper, i.e., inverse images of compact sets are compact.

Remark 2. Let M be a Riemannian manifold and G the group of isometries. Then $G \times M \to M$ is a smooth proper action. Here G is endowed with the compact open topology. In fact also the converse statement is essentially true (see e.g. [P2]): Let $G \times M \to M$ be a proper smooth action of a Lie group G on a smooth manifold M. Then there exists a G-invariant Riemannian metric on M, i.e., G is a closed subgroup of the group of isometries of M.

It follows immediately from the definition that the orbits of a proper action are closed and the isotropy groups are compact. Let X/G denote the quotient space, i.e., the space defined by the equivalence relation $x \sim y$ if and only if $G \cdot x = G \cdot y$. For a proper action the topology defined by the natural map $q : X \to X/G$ is Hausdorff.

If $G \times M \to M$ is a smooth proper action on a manifold, then M/G may be singular. However, local questions about the quotient can be reduced to the representation theory of compact groups. In particular these quotients are locally semi-algebraic varieties where the defining inequalities are of a rather simple nature (see [P-S]).

The reduction to questions about actions of compact groups follows from the slice theorem which proves that every point $x \in M$ possesses a G-stable open neighborhood which can be derived from the isotropy representation as follows.

If a Lie group L is acting smoothly on manifold M and $x \in M$ with $L \cdot x = x$, then the isotropy representation $\rho_x : L \to \mathrm{GL}(T_x M)$ is defined by $\rho_x(l) = d\rho(x)(l)$. This is a smooth Lie group homomorphism. Of course even for an effective action this is by no means faithful, i.g., injective, e.g., because it is possible for non zero vector fields to vanish of high order at a point. On the other hand, if L is compact, then, by using averaging over the compact group L on a neighborhood of x, one proves the

Linearization for compact isotropy. *Let $L \times M \to M$ be a smooth action of a compact Lie group and suppose that x is an L-fixed point. Then there is an L-stable neighborhood U of x and an L-equivariant isomorphism of U onto an open neighborhood of $0 \in T_x M$.*

Now let $G \times M \to M$ be a smooth action by any Lie group, i.e., compact or not, and suppose that for some $x \in M$ the isotropy group G_x is compact. The tangent space $T_x(G \cdot x)$ to the orbit is G_x-stable and, since G_x is compact and acts linearly on $T_x M$, there is a complementary G_x-invariant normal space N such that $T_x M = T_x(G \cdot x) \oplus N$. Now by the linearization statement any sufficient small G_x-stable open neighborhood S of zero in the normal space N can be identified with a G_x-stable submanifold through x, which is transversal at x to the orbit $G \cdot x$.

We refer to S as a slice at x. The appropriate model for a G-stable neighborhood of the orbit $G \cdot x$ is the G-bundle $G \times_{G_x} S$ which may be thought of as an open neighborhood of the zero section of the normal bundle $G \times_{G_x} N$ of $G \cdot x$. This neighborhood can be also defined as a G_x-quotient of $G \times S$ by the diagonal action $G_x \times G \times S \to G \times S$, $(h, g, s) \to (gh^{-1}, h \cdot s)$. Since G_x acts properly and freely, this quotient is a manifold. Furthermore, the action commutes with the left action of G on the first factor of $G \times S$ and consequently this G-action descends to a smooth action on $G \times_{G_x} S$. Finally, projection on the first factor gives a G-equivariant map $G \times_{G_x} S \to G/G_x$ which is a G-homogeneous bundle with fiber S. Pulling this back to G by the standard projection $\pi : G \to G/H$ produces the following G-diagram

$$\begin{array}{ccc} G \times S & \xrightarrow{q} & G \times_{G_x} S \\ \downarrow & & \downarrow \\ G & \longrightarrow & G/G_x \end{array}$$

The quotient map q is the defining map of $G \times_{G_x} S$ as a G_x-quotient and $G \times S \to G$ is the projection onto the first factor. The following theorem is almost an consequence of the definitions, but is never the less of central importance. It can be found at various points in the literature over the years (see [P2]) and holds in the differentiable as well as in the holomorphic category.

Slice Theorem. *Let $G \times M \to M$ be a proper smooth action, $x \in M$ and let S be a slice at x. Then, after shrinking S if necessary, the natural map $G \times_{G_x} S \to M$ defined by $[g, s] \to g \cdot s$ is a diffeomorphism onto its open image.*

Remark 3. In the holomorphic case the group G is supposed to be complex and the action to be holomorphic. But M does not need to be smooth. The slice theorem is valid for any reduced complex space.

In particular, it follows that the geometric quotient $M \to M/G$ is defined by the quotient $S \to S/G_x$ of S by the compact isotropy group G_x. Recall that in the smooth case S is G_x-equivariantly isomorphic to an open neighborhood of zero in a G_x-representation N. In fact N is a sub-representation of the isotropy representation $T_x M$. In this sense the local theory for proper G-actions can be reduced to the representation theory of compact groups.

3.1. Luna's Slice Theorem

Let G be a complex Lie group acting holomorphically on a holomorphically separable complex space X. If follows that every orbit $G \cdot x_0 = G/G_{x_0}$ is holomorphically separable. If G_{x_0} were compact, i.e., the connected component of the identity $G_{x_0}^0$

would be a compact complex torus, then using the maximum principle one sees that the G_{x_0}-action would be trivial, i.e., contained in *every isotropy group*. Since one is normally in a position to assume that the ineffectivity $I := \{g \in G; g \cdot x = x \text{ for all } x \in X\}$ is at worst discrete, this shows that such isotropy groups are compact only when they are finite. Since this situation does not appear very often, a direct application of the above geometric slice theorem is not possible in most cases of interest.

However, for actions of complex reductive groups the situation is far from hopeless. Recall that if $G \times X \to X$ is an algebraic action of a reductive group on an affine variety, then every fiber of the Hilbert quotient $\pi : X \to X/\!/G$ contains a unique closed orbit. Such orbits are affine subvarieties of X. Thus we are in a position to apply the basic

Theorem of Matsushima and Onishtchik. *Let G be a complex reductive group, H a complex closed subgroup and $Z = G/H$. Then the following are equivalent.*

i. *Z is Stein.*

ii. *Z is affine.*

iii. *H is reductive.*

For similar results over other fields of definition see [BB].

For the proof of the theorem recall that G has a unique affine algebraic structure and that the right H-action $H \times G \to G$, $(h, x) \to gh^{-1}$, is algebraic. Thus by Hilbert's theorem if H is reductive, then the quotient $G/H = \operatorname{Spec} \mathbb{C}[G]^H$ is affine algebraic. Of course this implies that it is a Stein manifold. This proves that iii. $\Rightarrow$ ii. $\Rightarrow$ i.

Assume now that $Z := G/H$ is a Stein manifold. We will now explain that H has to be reductive. Stein manifolds are characterised by the existence of a strictly plurisubharmonic smooth exhaustion function ([G]). Here an exhaustion is a function which is bounded from below and is proper.

Thus let $G = K^{\mathbb{C}}$ and let $\rho : Z \to \mathbb{R}$ be a K-invariant strictly plurisubharmonic smooth function on Z. Note that K-invariance can be achieved by averaging over the compact group K. Consider the set $Z_0 := \{z \in Z; d\rho(z) = 0\}$ of critical points of ρ.

The group G can be written in polar decomposition $G = G \cdot P$, where P is the intersection of G with the set of positive definite Hermitian operators in a suitable embedding of G in $\mathrm{GL}(V)$. This decomposition corresponds to the decomposition $\mathfrak{g} = \mathfrak{k} \oplus \mathfrak{p}$ at the Lie algebra level where $\mathfrak{p} = i\mathfrak{k}$ is the -1-eigenspace of the differential of the anti-holomorphic automorphism which defines K as its set of fixed points.

In this way the set of K-invariant plurisubharmonic functions on the group G can be identified with a subset of the set of differentiable functions on the vector space $\mathfrak{p}$

which are convex if restricted to a line through the origin. Using elementary properties of such functions, by lifting ρ from $Z = G/H$ to the group G and then restrict it to $\mathfrak{p}$ one proves the following fact.

Proposition 3.1. *Let ρ be any K-invariant strictly plurisubharmonic function on a complex homogeneous space $Z = G/H$. Then either $Z_0 := \{z \in Z; d\rho(z) = 0\}$ is empty or Z_0 consists of a single K-orbit $K \cdot z_0$. In the later case $\rho : Z \to \mathbb{R}$ is a proper exhaustion and $G_{z_0} = (K_{z_0})^{\mathbb{C}}$. In particular, since H is conjugate to G_{z_0}, the isotropy group H is a complex reductive subgroup of G.*

The fact that $G_z = (K_z)^{\mathbb{C}}$ for any $z \in Z_0$ is fundamental for the construction of the slice at z. This points are appropriate for a Luna Slice with respect to a given realization of G as the complexification of a maximal compact subgroup K of G.

We now proceed with an explanation of Luna's Slice Theorem ([L]). The context is that of a closed orbit of a reductive group G in a Stein space X. However in many cases of interest, at least locally X possess a G-equivariant embedding in a representation space which of course is a Stein space and therefore it is at first sufficient to consider that case.

Let the complex reductive group G act via a holomorphic linear representation on a complex vector space V and suppose that Z is a closed G-orbit. Now $G = K^{\mathbb{C}}$, where K is a maximal compact subgroup. By averaging any Hermitian inner product it may be assumed that K is acting by unitary transformations.

For the associated K-invariant square of the norm function $\rho : V \to \mathbb{R}$, $\rho(v) = ||v||^2$, consider the Kempf-Ness set $Z_0 = \{z \in Z; \rho(z) = \min_{\xi \in Z} \rho(\xi)\}$ of a closed G-orbit Z in V (see [K-N]). For $z_0 \in Z_0$ let N be the normal space to the orbit $G \cdot z_0$. Note that N is K_{z_0}-invariant and therefore $G_{z_0} = (K_{z_0})^{\mathbb{C}}$-invariant.

Now consider the abstract normal bundle $G \times_{G_{z_0}} N \to G/G_{z_0} = G \cdot z_0$. Of course there is an algebraic map $G \times_{G_{z_0}} N \to V$; $[g, v] \to g \cdot (z_0 + v)$ onto a constructible set containing the given orbit Z. One can not hope that this map is biholomorphic onto its image, i.e., we must shrink N to a small neighborhood S of its neutral point. Let S be a sufficiently small ball around zero in N. Then it can be shown that $S^{\mathbb{C}} := G_{z_0} \cdot S$ is an open Stein submanifold of V, i.e., a domain of holomorphy. The following result is proved by using the mentioned convexity properties of ρ.

Theorem 3.2. *Let $G = K^{\mathbb{C}}$ be a complex reductive group, Z a closed orbit in a G-representation space V and $z_0 \in Z_0 := \{z \in Z; \rho(z) = \min_{\xi \in Z} \rho(\xi)\}$.*

 i. *The natural map $G \times_{G_{z_0}} S^{\mathbb{C}} \to V$ is biholomorphic onto its open image U,*

 ii. *$G \times_{G_{z_0}} S^{\mathbb{C}}$ is saturated with with respect to the quotient map $G \times_{G_{z_0}} N \to (G \times_{G_{z_0}} N) // G = N // G_{z_0}$ and U is saturated with respect to the quotient map $V \to V // G$,*

It is possible to formulate the theorem entirely in terms of algebraic geometry as follows. Let Y and X be affine G-varieties with Hilbert quotients $\pi_Y : Y \to Y // G$ and

$\pi_X : X \to X/\!/G$. For a G-equivariant regular map $\phi : Y \to X$ let $\bar\phi : Y/\!/G \to X/\!/G$ denote the induced map of quotients.

Theorem 3.3. *If $y_0 \in Y$ is such that $G \cdot y_0$ and $G \cdot \phi(y_0)$ are closed orbits in Y resp. in X, ϕ maps $G \cdot y_0$ isomorphically onto $G \cdot \phi(y_0)$ and ϕ is étale at y_0, then $\bar\phi$ is étale at $\pi_Y(y_0)$.*

Of course the theorem in this form implies the slice theorem as formulated above. The analogous result holds in the category of complex spaces as well if we replace affine varieties with Stein spaces.

3.2. Analytic Hilbert quotients for Stein spaces

Using Luna's Slice Theorem, Snow ([Sn]) constructed the canonical complex structure on the quotient of a Stein space which is acted on holomorphically by a complex reductive group G. Our formulation here is intensional: The quotient $\pi : X \to X/\!/G$ exists in the category of topological spaces, i.e., $X/\!/G$ is defined through the equivalence relation $x \sim y$ if and only if $f(x) = f(y)$ for all $f \in \mathcal{O}(X)^G$ where $X/\!/G$ is equipped with the quotient topology. It is necessary to show that the natural sheaf of invariant holomorphic functions equips the quotient with the structure of a Stein space having the desired universal property:

> *To every invariant holomorphic map $f : X \to Y$ there exists a unique holomorphic map $\bar f : X/\!/G \to Y$ such that $\bar f = f \circ \pi$.*

In a natural way the algebraic Hilbert quotient $N \to N/\!/G_z$ provides such a structure on the restricted quotient $S^{\mathbb C} \to S^{\mathbb C}/\!/G_z$ for every $z \in Z$ where Z is a closed G-orbit in X. The base $S^{\mathbb C}/\!/G_z$ can be identified with an open neighborhood of $\pi(z) \in X/\!/G$. One must show that this gives $X/\!/G$ the desired complex structure.

Although it is not explicitly seen in his proof, Snow implicitly uses convexity properties of K-invariant plurisubharmonic functions which are related to the Hamiltonian method explained here in the following section to prove the desired results, i.e., the existence of the analytic Hilbert quotient in the category of Stein spaces.

At that time Snow's proof required that X is a normal complex space. It of course followed from universality, that $X/\!/G$ is likewise normal. In [H2], before the proof of the existence of the universal complexification which will be discussed later, the first author constructed the analytic Hilbert quotient for actions of real compact Lie groups K of holomorphic transformations on Stein spaces. The normality assumption was then removed in [H3]. Note that if the local $K^{\mathbb C}$-action induced by the K-action is itself global that $X/\!/K = X/\!/K^{\mathbb C}$ by the identity principle. After proving the existence of the universal complexification $X^{\mathbb C}$ of the K-action on a Stein space X, functoriality of the construction again leads to an isomorphism $X/\!/K \cong X/\!/K^{\mathbb C}$, i.e., the existence to the categorical quotient for compact group follows from the analogous results for holomorphic actions of reductive groups (see [H3]):

Theorem 3.4. *Let K be a compact Lie group and X a Stein K-space. Then there exists a K-equivariant open holomorphic embedding of X into a Stein space $X^{\mathbb{C}}$ endowed with a holomorphic action of $G = K^{\mathbb{C}}$ such that*

i. *The inclusion $\iota : X \to X^{\mathbb{C}}$ induces an isomorphism $X /\!/ K \cong X^{\mathbb{C}} /\!/ K^{\mathbb{C}}$.*

ii. *Every K-equivariant holomorphic map $\phi : X \to Z$ where $K^{\mathbb{C}}$ acts holomorphically on Z extends to a $K^{\mathbb{C}}$-equivariant holomorphic map $\phi^c : X^{\mathbb{C}} \to Z$.*

4. Geometric Invariant Theory

Now consider a compact projective variety X equipped with an algebraic action $G \times X \to X$ of a linear algebraic group. Unless X is essentially a product of a G-homogeneous space and a variety where G acts trivially there are non closed G-orbits. In fact, usually there will be very few closed orbits.

Example 2. Let $G \to \mathrm{GL}(V)$ be a holomorphic irreducible representation of a complex reductive group G. Then there exists exactly one closed G-orbit in $\mathbb{P}(V)$.

It is therefore not surprising that a quotient theory which is appropriate in the setting of biregular geometry requires removing G-invariant sets E from X and constructing regular invariant maps on the complement $X \setminus E$

In the setting of Geometric Invariant Theory ([M-F-K]) the sets which are removed are determined in a systematic way by the invariants of a given coordinate line bundle. In reality this is essentially the same construction as in the affine case where the ample line bundle was chosen to be the trivial one. The trivial bundle has the advantage that there is a natural action on its space of sections $\mathbb{C}[X]$. This is not the case for an arbitrary line bundle.

Let G be a complex Lie group, $G \times X \to X$ a holomorphic action on a complex space and $B \to X$ a holomorphic fiber bundle. A holomorphic action $G \times B \to B$ of holomorphic bundle maps, i.e., respecting the geometry defined be the structure group, is called a lifting of the G-action to B if the projection $B \to X$ is equivariant. In this case B is refered to as a G-bundle.

Remarks.

1. It should be emphasised that a G-bundle is a bundle together with a lifting of the action on X. If such a lifting exists, then it is in general not unique. Thus a given bundle over X may have several different structures of a G-bundle.

2. If $G \times X \to X$ is an algebraic action of a linear algebraic group on a projective variety and $L \to X$ is a holomorphic line bundle, then the G-action on X can be lifted to some power L^k of L ([Su1], [Su2]). It is often really necessary to

go to some power, e.g., consider the action of $\text{Aut}(\mathbb{P}_1(\mathbb{C}))$ on $\mathbb{P}_1(\mathbb{C})$ and the hyperplane section bundle L.

3. The assumption "linear algebraic group" in the above theorem is important. For example, the natural action of an Abelian variety T on itself can only be lifted to the trivial bundle. Even by going to its universal cover $\tilde{T} \cong \mathbb{C}^n$ one only gains the fact that the $\tilde{T}$-action can be lifted to topologically trivial bundles (see [Wi] for information on the non-Abelian case).

4. Of course lifting to a line bundle may not be unique as the case of a trivial line bundle for an action of $\mathbb{C}^*$ on X shows. This phenomenon is not restricted to trivial line bundles. For example, consider $X = \mathbb{P}_2(\mathbb{C})$ and the $\mathbb{C}^*$-action defined by $\lambda[z_0, z_1, z_2] = [z_0, \lambda z_1, \lambda^2 z_2]$. Here a lifting to the hyperplane section bundle is a linear action on $\mathbb{C}^3$ so that the natural projection $\mathbb{C}^3 \setminus \{0\} \to \mathbb{P}_2(\mathbb{C})$ is equivariant. Note that the liftings $\lambda(z_0, z_1, z_2) = (z_0, \lambda z_1, \lambda^2 z_2)$ and $\lambda(z_0, z_1, z_2) = (\lambda^{-1} z_0, z_1, \lambda z_2)$ have different dynamical properties.

Given an ample G-line bundle $L \to X$ on a projective algebraic variety equipped with an algebraic action $G \times X \to X$ of a reductive group, Geometric Invariant Theory proceeds along the same lines as the Invariant Theory in the affine case with the regular functions $\mathbb{C}[X]$ being replaced by the coordinate ring $R := \oplus_{m \geq 0} \Gamma(X, L^m)$.

Note that for any section $\sigma \in \Gamma(X, L^m)$ the set $X_\sigma := \{x \in X; \sigma(x) \neq 0\}$ is affine. Thus for an invariant holomorphic section $\sigma \in \Gamma(X, L^m)^G$ we have the Hilbert quotient $X_\sigma \to X_\sigma /\!/ G$. Let $X(L)$ denote the union of all such X_σ where σ is an invariant section of some power of the ample G-line bundle L. The set $X(L)$ is called the set of semistable points. Note that this set depends not only on L but also on the lifting of the G-action to L. The Hilbert quotients $X_\sigma \to X_\sigma /\!/ G$ fit together to yield a G-invariant regular morphism $\pi : X(L) \to X(L) /\!/ G$ of $X(L)$ onto a projective algebraic variety $X(L) /\!/ G$.

The fact that π is constructed via Hilbert quotients which fits together nicely shows that the quotient $X(L) /\!/ G$ is defined by the equivalence relation $x \sim y$ if and only if $\overline{G \cdot x} \cap \overline{G \cdot y} \neq \emptyset$. In particular, the quotient $X(L) /\!/ G$ parametrises the closed G-orbits in $X(L)$. Furthermore, the structure sheaf on $X(L) /\!/ G$ is the natural one, i.e., the invariants of the direct image of the structure sheaf, and π is an affine map.

Although it does not at all reflect the complexity and interesting nature of this subject, the following gives a small indication of Geometric Invariant Theory.

Example 3. The standard representation of $G = SO_3(\mathbb{C})$ on $\mathbb{C}^3$ can be compactified in a natural way to an action of G on $X = \mathbb{P}_3(\mathbb{C})$. This is formally defined by regarding $\mathbb{C}^4$ as a direct sum of the standard representation on $\mathbb{C}^3$ and the trivial 1-dimensional representation. Then $X := \mathbb{P}(\mathbb{C}^4) = \mathbb{P}_3(\mathbb{C})$ has the induced action.

Let $\Omega \cong \mathbb{C}^3$ be embedded as the complement of the unique G-invariant hypersurface $H = \mathbb{P}(\mathbb{C}^3 \oplus \{0\})$. The G-orbit structure in Ω hast already been discussed in Example 1. In H there are exactly two orbits: a compact quadric curve C and its complement. Note that the only fixed point in X is in the closure of exactly on orbit in Ω whose closure is the cone over C.

Since everything closes up at C, it is natural to remove it in order to obtain a reasonably defined quotient. Geometric Invariant Theory does this via the hyperplane section bundle L by identifying $X(L)$ exactly with $\mathbb{P}_3 \setminus C$. The quotient $X(L)//G$ is $\mathbb{P}_1(\mathbb{C})$ and except for the affine cone, all fibers are closed in $X(L)$.

Remarks.

1. Even in the seemingly simple cases such as the diagonal $\mathrm{SL}_2(\mathbb{C})$-action on the n-fold product $\mathbb{P}_1(\mathbb{C}) \times \ldots \mathbb{P}_1(\mathbb{C})$ leads to interesting questions which are of a combinatorial nature (see e.g. [BB-S], [Po].)

2. Geometric Invariant Theory is just one way of systematically determining G-invariant open sets with G-quotients defined by natural equivalence relations. In the following sections we will discuss approaches which utilise invariant Kähler structures. Although we shall not discuss it here, we emphasise that the more combinatorial approach of Białyniki-Birula and Sommese is in certain situations much more general. In fact the quotients which are produced may not be projective ([BB-S])

5. The Hamiltonian method

At some point in the early 1980's it became clear that methods from classical mechanics would lead to better understanding of certain types of quotients in complex geometry. Numerous people have made contributions to this aspect of the subject. From this beginning phase we were particularly influenced by works stemming from a mathematical physical and Morse theoretic point of view, e.g., [A], [A-B], [G-S1], [G-S2], [G-S3], [K1], [K2], and those coming from the direction of moduli problems in algebraic geometry [Br], [K-N], [M-F-K]. In this chapter we sketch a viewpoint which is still of invariant theoretic nature for the construction of the quotients, but which utilises the Hamiltonian method for analysing properties of the quotients, in particular, the construction of Kählerian structures. In the following we discuss the construction of the quotients via the method of symplectic reduction.

5.1. Introductory remarks and examples

Recall that a Kähler manifold can be regarded as a symplectic manifold (X, ω), i.e., ω is a non-degenerate closed two form on the differentiable manifold X, where the complex structure J is compatible with the symplectic structure and induces a Riemannian metric. In other words, up to sign conventions, $\omega(v, Jv) > 0$ for all $v \neq 0$ and $\omega(Jv, Jw) = \omega(v, w)$ for all vectors v, w.

The strength of the integrability can be seen in the original definition of a Kähler structure. The pair (X, ω) where ω is a two form on X is Kähler if and only if ω has locally a strictly plurisubharmonic potential function. In other words, for all $x \in X$ there exists an open neighborhood U of x and a strictly plurisubharmonic function $\rho : U \to \mathbb{R}$ with $\omega = 2i \partial \bar{\partial} \rho$.

If K is a compact Lie group of holomorphic transformations of X, ω is a K-invariant Kähler form and U can be chosen to be K-invariant, then, by averaging over K, ρ can be chosen to be K-invariant. Following this line of thought for a moment, for $\xi \in \text{Lie}(K) = \mathfrak{k}$, consider the associated vector field $\xi_X \in \text{Vect}(X)$. By this we mean the real field $\xi_X(x) = \left(\frac{d}{dt}\right)_{t=0} \exp t\xi \cdot x$ whose associated $(1, 0)$-field is holomorphic. For example, for the standard action of $K := S^1$ on $X = \mathbb{C}$ we have $\xi_X = \frac{\partial}{\partial \theta}$ where we have used polar coordinates $z = re^{i\theta}$.

For a K-invariant potential ρ it of course follows that $\xi_X(\rho) = 0$ for all ξ. On the other hand using the complex structure we obtain an interesting function $(J\xi_X)(\rho)$ which measures the growth of ρ along a direction transversal to the orbit of the real one-parameter group defined by ξ_X.

Let $\mu_\xi := (J\xi_X)(\rho)$ and consider the map $\mu : U \to \mathfrak{k}^*$ whose coordinate is μ_ξ, i.e., regarding ξ as a functional on the dual $\mathfrak{k}^*$ of $\mathfrak{k}$, $\mu_\xi = \xi \circ \mu$. If μ exists in this way, we write $\mu = \mu^\rho$ and refer to μ^ρ as the moment map associated to ρ. Applying the definitions in a direct way, one proves the following

Lemma 5.1. *Let (X, ω) be a Kählerian manifold and $K \times X \to X$ a smooth action of a compact Lie group by holomorphic transformations. If $\omega = 2i \partial \bar{\partial} \rho$, where ρ is a K-invariant function, then the moment map $\mu^\rho : X \to \mathfrak{k}^*$ is K-equivariant.*

Remark 4. Recall that K-equivariance means that $\mu(k \cdot x) = \text{Ad}(k)^* \cdot \mu(x)$ for all $k \in K$. It should be noted that the Lemma remains valid also if the group is not compact. But in this case, even if ω has a Kählerian potential, it can not be assumed in general that it is invariant. Thus the requirement of equivariance is not quite fulfilled in certain situations. The lack of equivariance then usually has some interesting consequences.

Examples.

1. Let K act on a finite dimensional complex vector space $V \cong \mathbb{C}^n$ via a unitary representation and let $\rho = \| \ \|^2$. Then the Lemma applies to the standard symplectic form $\omega_{\text{std}} = 2i \partial \bar{\partial} \rho = 4 \sum dx_j \wedge dy_j$.

2. The example given by the unitary group acting by conjugation on the vector space of matrices is already quite educational. In precise terms let $K = U_n$ and $X = \mathbb{C}^{n \times n}$ be the space of $n \times n$ matrices. Define the strictly plurisubharmonic

exhaustion $\rho : X \to \mathbb{R}$ by $\rho(x) = \operatorname{tr}(x \,{}^t\bar{x})$. It is an interesting exercise to explicitly compute μ^ρ, its image, the zero-fiber of μ^ρ and to analyse the dynamics of the gradient flow of the energy function $\eta := \frac{1}{2}\|\mu^\rho\|^2$.

3. In the same setting as in 1., let $\rho := \log \|\ \|^2$ and $X := V \setminus \{0\}$. Then $\omega := 2i\partial\bar\partial \rho$ is degenerate along the lines through the origin in V and defines a standard associate Kählerian structure ω_{FS} on $\mathbb{P}(V)$, i.e., the Fubini–Study form associated to the given unitary structure on V. Despite the degeneracy, the associated moment map $\mu^\rho : X \to \mathfrak{k}^*$ is equivariant. Due to the degeneracy it factors through the projection $\pi : V \setminus \{0\} \to \mathbb{P}(V)$ and defines an equivariant moment map $\tilde\mu^\rho : \mathbb{P}(V) \to \mathfrak{k}^*$.

In the examples discussed up to this point a moment map has been defined via a Kähler potential. Although many considerations can be reduced to discussion of moment maps of this type it is appropriate to give a more general definition.

Let (M, ω) be a symplectic manifold and $G \times M \to M$ an action of a Lie group G of symplectic diffeomorphisms. A smooth map $\mu : M \to \mathfrak{g}^*$ is said to be a moment map if

$$d\mu_\xi = \iota_{\xi_M}\omega$$

holds for all $\xi \in \mathfrak{g}$. Here ι_{ξ_M} denotes contraction. If μ is equivariant, then we will speak about an equivariant moment map.

Given an action $G \times M \to M$ by symplectic diffeomorphisms there may or may not exist a moment map. The first obstruction appears to be topological, because the closed form $\iota_{\xi_M}\omega$ might not be exact. However, equivariance is quite often not attainable, e.g., for $M = \mathbb{R}^2$ equipped with the standard structure $\omega = dx \wedge dy$ there is no equivariant moment map for the action of $G = \mathbb{R}^2$ on M by addition.

Lack of uniqueness, even for equivariant moment maps for Kähler forms ω which are given by potentials, present itself in interesting ways.

Example 4. Let $K = S^1$ act on $X = \mathbb{C} \times \mathbb{C}^n$ by $t \cdot (z, w) = (tz, t^{-1}w)$ and let $\rho : X \to \mathbb{R}$, $\rho(z, w) = |z|^2 + \|w\|^2$ be the standard structure. Then for every $a \in \mathbb{R}$

$$\mu^a : X \to \mathbb{R} = \operatorname{Lie}(S^1)^*, \quad \mu^a(z, w) = |z|^2 - \|w\|^2 - a,$$

defines an equivariant moment map. Of course equivariance in this case means invariance. Note that $\mu = \mu^\rho$ if and only if $a = 0$.

This example reflects the following general fact.

If an equivariant moment map μ^0 exists, then the set of G-fixed points in $\mathfrak{g}^*$ corresponds to the set of equivariant moment maps of this action, i.e., if $a \in \mathfrak{g}^*$ is a G-fixed point, then $\mu^a := \mu^0 + a$ is an equivariant moment map and via versa every equivariant moment map for the given form and action is of this form.

Existence and uniqueness of equivariant moment maps can be formulated in terms of vanishing of associated cohomology classes in the Lie algebra cohomology $H^2(\mathfrak{g})$ and $H^1(\mathfrak{g})$. In particular, for G-connected and semisimple there exists a unique moment map.

5.2. Bergman Kähler forms

One of the main advantages of the recent developments in the complex analytic invariant theory is that it applies to manifolds and group actions which are a priori far away from the algebraic geometric or compact settings.

A typical example is given by a bounded domain in $\mathbb{C}^n$ equipped with its Bergman-Kähler structure ω. The construction of this form is possible in a slightly more general situation (see [Ko]). Let X be an n-dimensional complex manifold and H the Hilbert space of holomorphic n-forms which are square integrable, i.e., such that

$$||\eta||^2 = i^n \int_X \eta \wedge \bar{\eta} < \infty.$$

Assume that H is very ample, i.e., at each point $x \in X$ the closed subspace $H(x) := \{\eta \in H;\ \eta(x) = 0\}$ is a hyperplane, i.e., $H \neq H(x)$, and there exists $\eta_0 \in H \setminus H(x)$ and an $\eta \in H(x)$ such that $d(\frac{\eta}{\eta_0})(x) \neq 0$. Here $\frac{\eta}{\eta_0}$ is considerer as a holomorphic function around x. These assumptions mean that we obtain a holomorphic map $\phi : X \to \mathbb{P}(H^*)$, $\phi(x) = H(x)$, into the projective space of the dual Hilbert space H^*. Note that the group $G := \text{Aut}_\Theta(X)$ of holomorphic automorphisms of X stabilises H and that the induced linear action on H is in fact unitary. Choose an orthonormal basis $\eta_1, \eta_2, \ldots$ of H and note that $\phi(x)$ is represented by the evaluation map $\phi(x)(\eta) = \frac{\eta}{\eta_0}(x)$. Therefore we see that ϕ maps x to the projective point represented by $\eta_1(x)\eta_1^* + \eta_2(x)_2\eta_2^* + \cdots \in H^*$, where η_j^* is dual to η_j and $(\eta_1(x), \eta_2(x), \ldots)$ is only defined up to a constant. Hence

$$\rho(x) = \log\langle\phi(x), \phi(x)\rangle = \log \sum_{j=1}^\infty \eta_j(x)\overline{\eta_j(x)}$$

can be interpretated locally as a function on X and $\omega = 2i\,\partial\bar{\partial}\rho$ is a well defined Kähler form on X. By construction ω is invariant with respect to the action of the Lie group G. Although ρ is in general not G-invariant, it follows as in the finite dimensional case, that there is a G-equivariant moment map $\mu : X \to \mathfrak{g}^*$ induced by ρ.

5.3. Adapted complex structures

Let (M, g) be a Riemannian manifold and $\rho : TM \to \mathbb{R}$ be the square of the norm function, i.e., $\rho(v) = ||v||_g^2$. Using the metric g to establish an isomorphism $TM \cong$

T^*M, let ω be the natural symplectic form on T^*M viewed as a symplectic structure on TM. Regard $dd^c\rho = \omega$ as an equation for a complex structure on TM. For a real analytic Riemannian metric g it can be shown that there exists such an integrable complex structure on a neighborhood of the zero section of TM (see [G-S]). Its maximal domain X of definition, if it exists, should be regarded as an interesting invariant of the Riemannian structure.

The adapted complex structure can be defined in another way. For this note that a geodesic $\gamma : \mathbb{R} \to M$ induces a map $\gamma_* : T\mathbb{R} \to TM$. Here we assume for simplicity that the geodesic is complete. In this context one can ask for a complex structure on a neighborhood of the zero-section in TM so that, $\gamma_* : T\mathbb{R} \to TM$ is holomorphic if we identify $T\mathbb{R}$ with $\mathbb{C}$ and restrict γ_* to an appropriate neighborhood to the zero section in $T\mathbb{R}$. Again in the real-analytic case there exists a unique such structure ([L-S]) and this structure satisfies $dd^c\rho = \omega$ as well. The group of isometries $G := \mathrm{Iso}(M, g)$ acts properly on X as a group of holomorphic automorphisms and one has the associated moment map $\mu := \mu^\rho$. It would now be interesting to apply the recently developed techniques of Kählerian reductions (see Section 7) to this setting.

5.4. Complexification

Complexification of a manifold allows the use of Stein theory in consideration of symplectic geometry. Using a compatible real analytic structure, given a paracompact differentiable manifold M, one can construct a Stein manifold X which contains M as its real points, i.e., with a real analytic closed embedding $\iota : M \to X$ and an antiholomorphic involution $\sigma : X \to X$ with $\mathrm{Fix}(\sigma) = M$. Furthermore, by choosing X smaller if necessary, M is a strong deformation retract of X (see [G], [H-W]). If $G \times M \to M$ is a proper action by diffeomorphisms, then X can be chosen so that this action extends to a proper action $G \times X \to X$ by holomorphic transformations (see [H4], [H-H-K], [Ku]). Now if (M, ω_M) is symplectic, then X can be equipped with a Kähler structure ω_X so that $\iota^*\omega_X = \omega_M$. We refer to this as the Stein–Kählerian complexification of a symplectic manifold M (see [H-H-L]). If in this situation $K \times M \to M$ is an action of a compact group of symplectic diffeomorphisms, then, by averaging ω_X, one can assume that the extended action leaves ω_X invariant. Thus, at least for actions of compact groups, the strong methods of the Stein–Kähler setting can be applied. For example, a given moment map $\mu : M \to \mathfrak{k}^*$ can be extended to a unique moment map $\mu : X \to \mathfrak{k}^*$ and methods for symplectic reduction for Stein–Kähler spaces (see next section) give information on the structure of the singular symplectic reduction (see [S-L]) of the original real manifold (see [H-H-L]). It should be noted that in a certain way the observables on the reduction of the symplectic Stein–Kähler complexificationdo not depend on the choice of the extension.

6. Invariant theory and Hamiltonian actions of compact groups

Let (X, ω) be a Kähler manifold and $K \times X \to X$ a compact group of holomorphic isometries, i.e., $k : X \to X$ is holomorphic and $k^*\omega = \omega$ for every $k \in K$. Furthermore, assume that $K^{\mathbb{C}}$ is acting holomorphically on X. Finally assume that there exists a moment map $\mu : X \to \mathfrak{k}^*$. For convenience we refer to the data $(X, \omega, \mu, K^{\mathbb{C}})$ as a Kählerian $K^{\mathbb{C}}$-space with Hamiltonian data.

For such a space let $X_0 := \mu^{-1}(0)$. There are various physical reasons for considering such level sets of isotropic K-orbits, but for now we consider this only from the point of view of quotient theory.

Example 5. Let H be a closed complex subgroup of $G := K^{\mathbb{C}}$ and $X := G/H$. If a K-invariant Kähler form ω yields Hamiltonian data as above and $X_0 \neq \emptyset$, then there exists a strictly plurisubharmonic exhaustion function $\rho : X \to \mathbb{R}$ such that (see [A-L], [H3], [H-H-L])

1. ρ is K invariant and $\omega = 2i\partial\bar{\partial}\rho$

2. $\mu = \mu^\rho$ and $X_0 = \{x \in X;\ d\rho(x) = 0)\}$ is a single K-orbit.

Note that $J\xi_X(J\xi_X(\rho)) = \omega(\xi_X, J\xi_X)$ for all $\xi \in \mathfrak{k}$, i.e., every X_0 is the set where ρ attains its minimum.

The example is one key for understanding the general theory of symplectic reduction of Kählerian G-spaces with Hamiltonian data. In that setting let

$$X(\mu) := \{x \in X;\ G \cdot X \cap X_0 \neq \emptyset\}$$

be the set of semistable points with respect to μ.

6.1. Reduction in the case of Stein $K^{\mathbb{C}}$-spaces

Let us begin with a Stein space X endowed with a holomorphic action of G where $G = K^{\mathbb{C}}$ is the complexification of a compact Lie group K and $\rho : X \to \mathbb{R}$ is a K-invariant strictly plurisubharmonic exhaustion function. The Hamiltonian data will be given by $\omega = 2i\partial\bar{\partial}\rho$ and $\mu = \mu^\rho$. Now consider the set of semistable points $X(\mu)$ associated with this choice of moment map. By the observation of Azad–Loeb ([A-L]) (see Example 5), $K \cdot x$ is the set of critical points of $\rho|G \cdot x$ and $\rho|G \cdot x$ is an exhaustion for every $x \in X_0$. As a consequence we have the following basic

Lemma 6.1. *If the strictly plurisubharmonic function $\rho : X \to \mathbb{R}$ is an exhaustion, then $G \cdot X_0 = \{x \in X;\ G \cdot x$ is closed in $X\}$ and every point is semistable, i.e., $X(\mu) = X$.*

Proof. Since ρ is an exhaustion, every closed G-orbit is contained in $G \cdot X_0$. On the other hand if $x_0 \in X_0$, then $\rho|G \cdot x_0$ is an exhaustion and therefore $G \cdot x_0$ has to be closed in X. It remains to show that $X \subset X(\mu)$. This follows since $\rho|\overline{G \cdot x}$ attains its minimum at some point $x_0 \in \overline{G \cdot x}$. Since x_0 is a minimum for $\rho|G \cdot x_0$ this implies that $G \cdot x_0$ is closed and therefore $x \in X(\mu)$. $\square$

Since X is a Stein space the analytic Hilbert quotient $X/\!/G$ can be naturally identified with the set of closed G-orbits in X, i.e., every fiber of $\pi : X \to X/\!/G$ contains a unique closed G-orbit. The above observation shows that the injection $\iota : X_0 \to X$ induces a bijection $\bar{\pi} : X_0/K \to X/\!/G$.

Example 6. Let $X := \mathbb{C}^{n \times n}$ be the space of complex $n \times n$ matrices and let $G = \mathrm{GL}_n(\mathbb{C})$ act on X by conjugation. With respect to the standard $K = \mathrm{U}_n(\mathbb{C})$-invariant Kähler structure given by $\rho : X \to \mathbb{R}$, $\rho(z) = \mathrm{tr}({}^t\bar{z} z)$, the moment map $\mu = \mu^\rho$ is given by $\mu(z) = \frac{1}{2i}[z, {}^t\bar{z}] = \frac{1}{2i}(z{}^t\bar{z} - {}^t\bar{z} z)$ where we identified $\mathfrak{u}_n(\mathbb{C})^*$ with $\mathfrak{u}_n(\mathbb{C}) = \{\xi; {}^t\bar{\xi} = -\xi\}$ using the inner product $(\xi, \eta) = -\mathrm{tr}(\xi\eta)$ on $\mathfrak{u}_n(\mathbb{C})$. Thus $\mu^{-1}(0) = \{z \in X;\ z{}^t\bar{z} = {}^t\bar{z} z\}$ is the set of normal matrices in X, i.e., those which are diagonalisable by an element in $\mathrm{U}_n(\mathbb{C})$. Since $\pi : X \to X/\!/G \cong \mathbb{C}^n$ is given by $\pi(z) = (\mathrm{tr}(z), \ldots, \det(z))$ one sees directly that the inclusion $\iota : \mu^{-1}(0) \to X$ induces a bijection $\mu^{-1}(0)/K \to X/\!/G$.

In fact it can be shown that $\pi|X_0$ is proper (see [H3]) and therefore $\bar{\pi}$ is a homeomorphism. Observe that $\rho|X_0$ is K-invariant and therefore can be regarded as a continuous function $\bar{\rho}$ on $X/\!/G$. Even if $X/\!/G$ is a smooth complex manifold and ρ is smooth the singularities of π and X_0 are reflected in the singularities, i.e., non smoothness, of $\bar{\rho}$.

Example 7. Consider $X = \mathbb{C}^{n+1} = \mathbb{C} \times \mathbb{C}^n$ and let $\rho : X \to \mathbb{R}$ be the square of the standard norm function. Define a $G = \mathbb{C}^*$-action by $t \cdot (z, w) = (tz, t^{-1}w)$. The moment map $\mu = \mu^\rho$ is given by $\mu(z, w) = |z|^2 - ||w||^2$. Since $\mu^{-1}(0) = \{(z, w) \in X;\ |z| = ||w||\}$ and $\pi : X \to X/\!/\mathbb{C}^* \cong \mathbb{C}^n$ is given by $\pi(z, w) = zw =: u$, it follows that $\bar{\rho}(u) = |z|^2 + ||w||^2 = 2|u|$. Thus, although ρ is a smooth strictly plurisubharmonic function, the associated Kähler structure on $X/\!/G$, i.e., that defined by $2i\partial\bar{\partial}\bar{\rho}$ has a singularity at 0.

The possibility of "pushing down" a Kähler structure via the Marsden–Weinstein reduction as indicated in the above example has been implemented in a number of settings (see e.g. [K1], [G-S1]). Motivated by an attempt to better understand the stratified symplectic reduction of Sjamaar–Lerman ([S-L]), this was carried out in the Stein–Kähler setting in [H-H-L]

The Kähler structure on the quotient is defined in terms of potentials. A Kählerian structure on a complex space Z consists by definition of an open covering $\{U_\alpha\}$ of Z and a collection of strictly plurisubharmonic functions $\{\rho_\alpha\}$, $\rho_\alpha : U_\alpha \to \mathbb{R}$, such that $\rho_\alpha - \rho_\beta = \mathrm{Re} f_{\alpha\beta}$ where $f_{\alpha\beta}$ are holomorphic functions on $U_{\alpha\beta} = U_\alpha \cap U_\beta$. As the above example indicates, one can not require the ρ_α's to be smooth. They are required to be continuous and smooth on a complex analytic stratification of Z. In this way one defines a notion of a stratified Kählerian space (see [H-H-L] for details and [F-N] for equivalence of various definitions of strictly plurisubharmonic functions on complex spaces). In the case at hand in this section, i.e., X Stein with the Hamiltonian data given by a smooth strictly plurisubharmonic exhaustion $\rho : X \to \mathbb{R}$ the function $\bar{\rho}$ is automatically defined and smooth on a complex stratification of $X/\!/G$.

The stratification on $X/\!/G$ is defined in terms of the orbit-type stratification of X which yields a stratification of X_0 into smooth pieces. The strict plurisubharmonicity of $\bar{\rho}$ at generic points is checked by computing the Levi-form of ρ on the complex tangent space of X_0 at a point of a generic G-orbit through $x_0 \in X_0$. This lies transversal to the map π and serves as a sort of connection for $\pi|X_0 : X_0 \to X/\!/G$.

For simplicity refer to the reduced space $X_0/K \cong X/\!/G$ as $(X_{\text{red}}, \omega_{\text{red}})$. Of course ω_{red} is only generically a smooth Kähler form, $\omega_{\text{red}} = 2i\partial\bar{\partial}\bar{\rho}$. This reduced space is the appropriate phase space for K-invariant Hamiltonians on X and, at least in certain contexts, the appropriate set of observables in the algebra of $\mathcal{C}^\infty$-functions on X_{red}

The construction leads one to define a $\mathcal{C}^\infty$-function on X_{red} to be the canonical push-down of a K-invariant $\mathcal{C}^\infty$-function on X_0, i.e., regard $f|X_0$ as a function on X_{red}. Thus the question arises as to whether of not this notion of observables depends on the exhaustion $\rho : X \to \mathbb{R}$. More precisely, given two K-invariant exhaustions ρ_j of X are the algebras $\mathcal{C}(X_{\text{red}}^1)$ and $\mathcal{C}(X_{\text{red}}^2)$ isomorphic? In fact, using the Moser techniques of time dependent vector fields it can be shown that locally there is a homeomorphism ϕ which induces such an isomorphism ([St]). It should be understand that in general ϕ can not be chosen as the identity.

6.2. Reduction of Kählerian spaces with Hamiltonian data

Even if the underlying space X is Stein the set $X(\mu)$ of semistable points may be a proper subset of X.

Example 8. View Example 7, where $\mu^a = |z|^2 - ||w||^2 + a$ dynamically: For $a < 0$ the set $X_0(a) = (\mu^a)^{-1}(0)$ is of a cylindrical nature around the z-axis with "fiber" a sphere $S(z_0)$ in the affine plane $\{(z_0, w); w \in \mathbb{C}^n\}$ as a increases to 0, $X_0(a)$ pinches down to the cone $X_0(0) = \{(z, w); |z| = ||w||\}$ and as a increases to a positive real number, $X_0(a)$ becomes a cylinder over the w-axis with circles as fibers. Now consider the sets of semistable points.

$$X(\mu^a) = \begin{cases} \mathbb{C}^{n+1} \setminus \{z = 0\} & \text{for } a < 0, \\ \mathbb{C}^{n+1} & \text{for } a = 0, \\ \mathbb{C}^{n+1} \setminus \{w = 0\} & \text{for } a > 0. \end{cases} \quad (1)$$

Note that for $a = 0$ we have $\mu = \mu^\rho$ where ρ is the square of the standard norm function.

In the above example, for $a \neq 0$ the complex group $G = \mathbb{C}^*$ acts freely and properly on $X(\mu^a)$ and, just as in the case where the moment map is defined by an potential, $X_0(a)/S^1 \cong X(\mu^a)//\mathbb{C}^*$. Of course in general one can not hope that the quotient by the complex group will be geometric, i.e., that the fibers are just orbits.

Note that in the case $a < 0$ the quotient $X(\mu^a)//\mathbb{C}^*$ is the space $\mathbb{C}^n$ with the origin blown up by a σ-process. As a increases to 0 the size of this blow up point is seen to decrease to 0 and the quotient collapses to $\mathbb{C}^n$. As a increases to positive numbers this quotient remains stable, i.e., it remains to be $\mathbb{C}^n$. The "size" of the quotient can be measured by the quotient Kähler structure. Here, in the case of a free action, this is a smooth Kähler structure which arises by pushing down the structure on $X(\mu^a)$ and is given by appropriately chosen potentials. To explain this it is sufficient to consider the example $X = \mathbb{C}^2$ equipped with its standard unitary Kähler structure $\omega = 2i\partial\bar{\partial}\rho$ and the $\mathbb{C}^*$-action $t \cdot (z_1, z_2) = (tz_1, z_2)$. Choose $\mu(z, w) = |z_1|^2 + |z_2|^2 - 1$. This corresponds to $\mathbb{C}^2$ being the w-plane and $a = -1$ in the previous example. In this case $X(\mu) = \mathbb{C}^2 \setminus \{0\}$. It follows that $\mu^{\rho_j} = \mu$ on $U_j := \{z_j \neq 0\}$ where $\rho_j(z_1, z_2) = \|(z_1, z_2)\| - \log \|z_j\|$. Now apply the reduction theory of the previous section to the open sets $U_j := \{z_j \neq 0\}$ with potentials ρ_j. For this it should be observed that the Stein reduction theory of the previous section only uses the exhaustion property of ρ to prove that $X(\mu) = X$.

Let us now look at this in a bit more detail: Let X be Stein and assume that μ and ω are given and $X = X(\mu)$. Of course, this assumption is automatically satisfied if ρ is an exhaustion, but we do not assume this here. Nevertheless we claim that also in this case we have $X_{\text{red}} = X//G$, i.e., we may equip $X_{\text{red}} = \mu^{-1}(0)/K$ with the structure of a Stein space which is defined by the analytic Hilbert quotient $X//G$.

For this first note that for every $x_0 \in X_0 = \mu^{-1}(0)$ the isotropy group $G_{x_0} = (K_{x_0})^\mathbb{C}$ is reductive. Thus using a G_{x_0}-splitting $T_{x_0}X = T_{x_0}(G \cdot x_0) \oplus N$ we find a K_{x_0} stable open ball D around zero in N and a K_{x_0}-equivariant holomorphic map $\phi : D \to X$ such that $x_0 \in \phi(D)$ and $\phi(D)$ is transversal to $G \cdot x_0$ at x_0. It follows from general principles (see [H3]) that ϕ extends to an immersion $\phi^c : G_{x_0} \cdot D \to X$ and therefore induces a locally biholomorphic map $\Phi : G \times_{G_{x_0}} S \to X$, $\Phi([g, x]) = g \cdot \phi^c(x)$, where $S := G_{x_0} \cdot D \subset N$ is an open G_{x_0}-stable Stein submanifold of the G_{x_0}-representation N. Up to now it has only been used that G_{x_0} is reductive. In the special situation at hand where $x_0 \in X_0$ it turns out that after shrinking D and therefore S that Φ is biholomorphic onto its open image $U \subset X$. This statement remains true if X is singular, but in this case S has to be replaced with a suitable locally analytic subset of N. Now observe that the restriction of ω to $K \cdot x_0$ is identical zero and that $K \cdot x_0$ is a deformation retract of U. Thus $\omega|U = 2i\partial\bar{\partial}\rho$ and after modifying ρ by an appropriate plurisubharmonic function as we did in the simple example, we may assume that $\mu|U = \mu^\rho$. Moreover a slight generalisation of the exhaustion lemma in [H-H2] shows that, after shrinking sets approximately, that ρ is in fact an exhaustion along the fibers of the analytic Hilbert quotient $\pi : U \to U /\!/ G$, i.e., ρ is bounded from below and $\rho \times \pi : U \to \mathbb{R} \times U /\!/ G$ is proper.

Now $X = X(\mu)$ implies that $G \cdot x_0 \cap X_x \neq \emptyset$ for every closed G-orbit $G \cdot x_0$ and therefore using $\rho : U \to \mathbb{R}$ one sees that U is saturated with respect to the analytic Hilbert quotient $X /\!/ G$. Moreover, $(\mu^\rho)^{-1}(0) = X_0 \cap U$ and the inclusion $(\mu^\rho)^{-1}(0) \to U$ induces a homeomorphism $(\mu^\rho)^{-1}(0)/K \to U /\!/ G \subset X /\!/ G$. Thus in fact $X_0/K \cong X /\!/ G$ is a homeomorphism. Furthermore it is also clear that the Kählerian structures $\{(\omega|U)_{\text{red}}\}$ piece together to define a stratified Kähler structure on X_{red} which is the stratified Kählerian structure of Sjamaar–Lerman ([S-L]) in this Kähler context.

6.3. Kählerian reduction in the projective setting

Let (X, ω) be a Kähler G-space with Hamiltonian data. Then it can be shown that $X(\mu)$ is open, the analytic Hilbert quotient $X(\mu) \to X(\mu) /\!/ G$ exists and $\pi | X_0$ induces a homeomorphism $X_0/K \cong X(\mu) /\!/ G =: X_{\text{red}}$ (see [H-L], [S]). The reduced symplectic structure ω_{red} provides X_{red} with a stratified Kählerian structure. The sheaf of germs of holomorphic functions $\mathcal{O}(X_{\text{red}})$ is given by the G-invariant holomorphic functions on saturated open subsets of $X(\mu)$. It is important to underline that analogous to the Geometric Invariant Theory quotients $\pi : X(\mu) \to X(\mu) /\!/ G$ is a Stein map, i.e., inverse images of arbitrary Stein subvarieties are Stein (see [H-M-P]).

One of the original motivations for introducing the techniques of symplectic reduction in the complex geometric setting was to better understand quotients which were already known to exist, e.g., Geometric Invariant Theory quotients in the projective algebraic setting.

In that case recall that the polarisation in an ample G-bundle $L \to X$, where the linear reductive group $G = K^{\mathbb{C}}$ is acting algebraically on X. To obtain the Kählerian version of this polarisation, let h be a K-invariant Hermitian metric on X and c^h the associated Chern form. Just as in the case of $X = \mathbb{P}(V)$ and K-acting on V via a unitary representation (see 5.1) let Z be the bundle space L with the zero section removed and $p : Z \to X$ the G-equivariant bundle projection. A moment map $\mu : X \to \mathfrak{k}^*$ is defined by $\mu^\rho : Z \to \mathfrak{k}^*$ where $\rho = \log \| \ \|^2$ and observing that μ^ρ is constant on the fibers of p. Apriori the moment map constructed above depends on the choice of metric. However up to shifting by an invariant constant in $\mathfrak{k}$, at least for the compact Kähler case, the set $X(\mu)$ only depends on the cohomology class of the Kähler form (see [H-H2]).

The cohomology class constructed above is very special: It is an integral point in the positive Kähler cone. Furthermore, relating the coordinate ring $\oplus_k \Gamma(X, L^k)$ to the regular functions on Z, one easily shows that $X(\mu) = X(L)$. So, just as in the Stein case where $\rho : X \to \mathbb{R}$ is an exhaustion, the Kähler reduction is the same as the invariant theoretic reduction. This of course adds tools to both sides of the picture which have been utilised by numerous authors (see [M-F-K]).

It is quite interesting to investigate the variation of $X(\mu)$ as the cohomology class $[\omega]$ varies. Furthermore, the choice of μ can vary if K has a positive dimensional center (see e.g. [D-H], [R], [T]). Note that replacing ω by $r\omega$ for $r > 0$ does not change the set $X(\mu)$. In particular, the rational points in the cone $\mathrm{NS}(X)$ in $H^{1,1}(X)$ which is generated by Chern classes of positive bundles lead only to sets which are semistable in the sense of Geometric Invariant Theory and therefore to projective quotients. A continuity argument (see [H-M]) then shows that nothing new is obtained by considering Kähler reductions associated to $\alpha \in \mathrm{NS}(X)$. Now the cone $\mathrm{NS}(X)$ is usually much smaller then the Kähler cone $K(X)$ which is the cone spaned by the cohomology classes of Kähler metrics in $H^{1,1}(X)$. However in [H-M] it has been shown that also in this case nothing new is obtained by considering $\alpha \in K(X)$:

Let $\mu : X \to \mathfrak{k}^*$ be a moment map of a Kähler form ω on a projective algebraic manifold X. Then there exists an ample line bundle $L \to X$ so that $X(\mu) = X(L)$. In particular, the analytic Hilbert quotient $X(\mu)//G = X_{\mathrm{red}}$ is projective algebraic.

Of course this formulation sounds negative, i.e., there is nothing new in the projective algebraic case. However, one could look at this more optimistically (in fact in the spirit of the proof): The Kählerian reduction of a projective algebraic variety is projective algebraic. It would be interesting to prove functorial type statement in other settings, e.g., for quasi-projective varieties with "nice" Kähler forms.

7. Kählerian reduction for proper G-actions

Let us recall the Marsden–Weinstein reduction for a free proper G-action of symplectic diffeomorphisms on a symplectic manifold (M, ω). In this case $M_0 = \mu^{-1}(0)$ is smooth with $\mathrm{codim}_{\mathbb{R}} M_0 = \dim_{\mathbb{R}} G$. Consider the principal bundle $M_0 \to M_0/G =: M_{\mathrm{red}}$: The degeneracy of $\omega|M_0$ is precisely the bundle of vertical tangent vectors. Thus, by the invariance of $\omega|M_0$ it can be pushed down to a smooth symplectic structure ω_{red} on M_{red}.

In this section we wish to sketch an approach to Kähler reduction which utilises the germ X_0 in X, i.e., X_0 and its induced partial complex structure (see [A-H-H]). This has the advantage in that the invariant theoretical nature of $X_0 \to X_{\mathrm{red}}$ has only to do with this germ as opposed to some global picture in X such as semistability in Geometric Invariant Theory. On the other hand, it is valid for proper G-actions thus opening up a whole range of examples which were not covered by the discussed case of compact groups. But even in that case, the results obtained for K-actions priori to [A-H-H] required the action of $K^{\mathbb{C}}$.

Up to this point G has denoted the complexification of K. In this section G will denote a connected Lie group acting properly on a normal Kählerian space X as a group of proper Kähler isometries. Moreover we assume that there is a G-equivariant moment map $\mu : X \to \mathfrak{g}^*$ such that $X_0 := \mu^{-1}(0) \neq \emptyset$. By a CR-function f defined on an open subset U of X_0 we mean a function $f : U \to \mathbb{C}$ which is locally the restriction of a holomorphic function defined in an open neighborhood. Let $\mathcal{O}_{X_0}$ denote the sheaf of germs of CR-functions on X_0. We want to endow X_0/G with a complex structure. Of course the natural structure sheaf on X_0/G is the sheaf of G-invariant CR-functions on X_0, i.e., we set

$$\mathcal{O}_{X_0/G}(Q) := \mathcal{O}_{X_0}(\pi^{-1}(Q))^G$$

where $\pi : X_0 \to X_0/G$ denotes the quotient map. In this section we sketch the proof of the main result of [A-H-H]:

Theorem 7.1. *For a proper Hamiltonian G-action on a normal Kähler space X the quotient X_0/G is a stratified normal Kählerian space with structure sheaf $\mathcal{O}_{X_0/G}$.*

The proof of Theorem 7.1 is by induction on dimension on X. Even if one is only interested in the smooth case, this in particular means that one has to consider singular spaces which arise in the proof as quotients of X with respect to closed normal subgroups of G. One extreme case is where G is simple. For a proof in this case we refer to [A-H-H]. The other extreme case is where G acts freely. Note that this is automatically the case if G has no compact subgroup, e.g., G is simply connected and solvable.

Here we restrict our consideration to the second extreme case, i.e., we assume that G acts freely and properly on X. Moreover, for technical convenience we will assume that G is a real form of a compact Lie group $G^\mathbb{C}$. The global G-action induces in the usual way a local holomorphic $G^\mathbb{C}$-action on X. Much of our discussing will be of a local nature around a fixed point $x_0 \in X_0 := \mu^{-1}(0)$ and therefore we begin by recalling the construction of a local normal form of the $G^\mathbb{C}$-action around x_0.

Lemma 7.2. *There is an open neighborhood N of the neutral element $e \in G^\mathbb{C}$, a locally analytic subset S of X with $x_0 \in S$ and an open neighborhood U of x_0 such that*

$$\phi : N \times S \to X, \quad \phi(g, x) = g \cdot x,$$

maps $N \times S$ biholomorphically onto $U = N \cdot S$.

Here $(g, x) \to g \cdot x$ denotes the local holomorphic $G^\mathbb{C}$-action on X where (g, x) runs over and open neighborhood of $\{e\} \times X$ in $G^\mathbb{C} \times X$ (see [H-I] for details on local actions and [H-H1] for a general version of this lemma).

Properness of the G-action on X has the following consequence.

Corollary 7.3. *After shrinking $N \times S$ we may assume that $\Phi : G \cdot N \times S \to X$, $\Phi(g, x) = g \cdot x$, is defined and biholomorphic onto its image $G \cdot U$.*

Proof. We may chose N to be connected and then it is easy to see that the map ϕ of the above lemma extends to the well defined map $\Phi : G \cdot N \times S \to X$. Since Φ is locally biholomorphic it is sufficient to show that, after shrinking $N \times S$ it is injective. Assume this not to be the case. Then there would be sequences $\{s_j\} \subset S$, $\{g_j\} \subset G \setminus N$ with $(s_j, g_j \cdot s_j)$ converging to (x_0, x_0). But then some subsequence of $\{g_j\}$ would converge to $g_0 \in G$. Hence $g_0 \cdot x_0 = x_0$ and this could imply that $g_0 = e$. Of course this contradicts $\{g_j\} \subset G \setminus N$. □

Now we may assume that $X = N \times S$ where $N = GN$ is a G-stable open neighborhood of $e \in G^{\mathbb{C}}$. In this identification x_0 corresponds to $(e, x_0) \in X_0 = \mu^{-1}(0)$. If we fix $s \in S$, then $d\mu_\xi = \iota_{\xi_X}\omega$ holds on $N \times \{s\}$ by definition of a stratified Hamiltonian symplectic structure. Therefore

$$\operatorname{rang} d\mu(x) = \dim G \cdot x = \dim \mathfrak{g}^*$$

is constant on $N \times \{s\}$. In particular, the restricted moment map $\mu(s) : N \to \mathfrak{g}^*$, $\mu(s)(y) = \mu(y, s)$, is a submersion for every $s \in S$.

Let us consider the restricted moment map $\mu(s) : N \to \mathfrak{g}^*$ more closely. The G-action on $G^{\mathbb{C}}$ is proper. Thus there is a convex neighborhood D of $0 \in \mathfrak{g}$ such that $G \cdot \exp iD$ is open in $G^{\mathbb{C}}$, $\exp iD$ is relatively compact in N and moreover

$$G \times iD \to G \cdot \exp iD, (g, i\xi) \to g \cdot \exp i\xi,$$

is a diffeomorphism. We need the following useful

Lemma 7.4. *Let $N = GN$ be an open G-stable neighborhood of $e \in G^{\mathbb{C}}$. Then N contains an open G-stable neighborhood $N_0 = GN_0$ of $e \in G^{\mathbb{C}}$ such that for every equivariant moment map $\mu : N \to \mathfrak{g}^*$ with respect to a G-invariant Kähler form on N the following holds:*

If $\mu^{-1}(0) \cap N_0 \neq \emptyset$, then $\mu^{-1}(0) \cap N_0 = G \cdot x_\mu$ for some $x_\mu \in N_0$.

Proof. Let $D_1 \subset i\mathfrak{g}$ be a convex neighborhood of zero such that $N_1 := G \exp iD_1$ satisfies $N \exp iD_1 \subset N$. Now chose a smaller convex neighborhood $D_0 \subset D_1$ of zero in $\mathfrak{g}$ such that

- $\exp iD_0(\exp iD_0)^{-1} \subset N_1$

and let $\mu : N \to \mathfrak{g}^*$ be some equivariant moment map with respect to some Kähler form ω on N. We may assume that $\mu(a) = 0$ for some $a \in \exp iD_0$. Let $b \in N_0$ with $\mu(b) = 0$ be given. Since $N_0 \subset N_1 a$ by • there exist $g_0 \in G$ and $\xi \in D_0$ such that $b = g_0 \exp i\xi \, a$. Note that $\mu(\exp i\xi \, a) = 0$ and that $\alpha_t := \exp it\xi \, a \in N$ for all $0 \leq t \leq 1$. But if $\xi \neq 0$, then $\frac{d}{dt}\mu_\xi(\exp it\xi \, a) = \omega(\xi_X(\alpha_t), J\xi_X(\alpha_t)) > 0$ contradicts $\mu_\xi(a) = \mu_\xi(\exp i\xi \, a) = 0$. Thus $\xi = 0$ and $b \in Ga$ follows. □

After replacing N with N_0 we have the following

Corollary 7.5. *If $\mu(s)^{-1}(0) \neq \emptyset$, then $\mu(s)^{-1}(0) = G \cdot x_s$ is a single G-orbit.*

Replacing N with $G \cdot \exp iD$ has another advantage. If we shrink D and S appropriately, then

$$\mu(s)|\exp iD : \exp iD \to \mathfrak{g}^*$$

is a diffeomorphism onto its open image. Now $0 \in \mu(s)(\exp iD)$ is an open condition on s. Thus after shrinking S we may assume that $\mu(s)^{-1}(0) \neq \emptyset$ for every $s \in S$. This shows one part of the next

Lemma 7.6. *After shrinking, the restriction of the projection $p : N \times S \to S$ to $X_0 = \mu^{-1}(0)$ induces a homeomorphism $\bar{p} : X_0/G \to S$.*

Proof. We already have shown that $\bar{p}$ is a continuous bijection. Moreover the construction implies that $\bar{p}$ is also proper. $\square$

Proof of the theorem in the free case. On X_0 we have already introduce the structure sheaf $\mathcal{O}_{X_0/G}$. By definition a function f defined on an open subset Q of X_0/G is holomorphic if its pull back to X_0 is locally given by the restriction of a holomorphic function. This obviously implies that the induced map $\bar{p} : X_0/G \to S$ induces a morphism of sheaves $\mathcal{O}_S \to \mathcal{O}_{X_0/G}$. This morphism is injective and is also easily shown to be surjective. $\square$

References

[A-H-H] Ammon, M., Heinzner, P., Huckleberry, A., Kählerian structures on symplectic reductions, in preparation.

[A] Atiyah, M. F., Convexity and commuting Hamiltonians, Bull. London Math. Soc. 14 (1982), 1–15.

[A-B] Atiyah, M. F, Bott, R., The Yang–Mills equations over Riemann surfaces, Phil. Trans. Roy. Soc. Lond. A 308 (1982), 523–615.

[A-L] Azad, H., Loeb, J. J., Pluri-subharmonic functions and the Kempf–Ness theorem, Bull. London Math. Soc. 25 (1993), 162–168.

[BB] Białynicki-Birula, A., On homogeneous affine spaces of linear algebraic groups, Amer. J. Math. 85 (1963), 577–582.

[BB-S] Białynicki-Birula, A., Sommese A., Quotients by $\mathbb{C}^*$ and $SL(2,\mathbb{C})$ actions, Trans. Amer. Math. Soc. 274 (1983), 773–800.

[BB-S] Białynicki-Birula, A., Święcicka J. A., On exotic orbit spaces of tori acting on projective varieties, Canadian Mathematical society Conference Proceedings 10 (1989), 25–30.

[Br]   Brion, M., Sur l'image de l'application moment, Séminaire d'algèbre Paul Dubreuil et Marie-Paule Malliavin (Paris 1986), Lecture Notes in Math. 1296, Springer-Verlag, Berlin–Heidelberg–New York 1987, 177–192.

[D-H]   Dolgachev, I., Hu, Y., Variation of Geometric Invariant Theory quotients, Inst. Hautes Études Sci. Publ. Math., 1998.

[F-N]   Fornaess, J., Narasimhan, R., The Levi problem on complex spaces with singularities, Math. Ann. 248 (1980), 47–72.

[F]   Forstneric, F., Actions of $(\mathbb{R}, +)$ and $(\mathbb{C}, +)$ on complex manifolds, Math. Z. 223 (1996), 123–152.

[G]   Grauert, H., On Levi's problem and the imbedding of real-analytic manifolds, Ann. of Math. 68 (1958), 460–473.

[G-S]   Guillemin, V., Stenzel, M., Grauert tubes and the homogenous Monge–Ampére equation, J. Differential Geom. 34 (1991), 561–570.

[G-S1]   Guillemin, V., Sternberg, S., Convexity properties of the moment mapping, Invent. Math. 67 (1982), 491–513.

[G-S2]   Guillemin, V., Sternberg, S., Convexity properties of the moment mapping II, Invent. Math. 77 (1984), 533–546.

[G-S3]   Guillemin, V., Sternberg, S., Birational equivalence in the symplectic category, Invent. Math. 97 (1989), 485–522.

[H-W]   Harvey, R. F., Wells, R. O., Holomorphic approximation and hyperfunctions theory on a C^1 totally real submanifold of a complex manifold, Math. Ann. 197 (1972), 287–318.

[H1]   Heinzner, P., Linear äquivariante Einbettungen Steinscher Räume, Math. Ann. 280 (1988), 147–160.

[H2]   Heinzner, P., Kompakte Transformationsgruppen Steinscher Räume, Math. Ann. 285 (1989), 13–28.

[H3]   Heinzner, P., Geometric invariant theory on Stein spaces, Math. Ann. 289 (1991), 631–662.

[H4]   Heinzner, P., Equivariant holomorphic extensions of real analytic manifolds, Bull. Soc. Math. France 121 (1993), 445–463.

[H-H1]   Heinzner, P., Huckleberry, A., Complex geometry of Hamiltonian actions, in preparation.

[H-H2]   Heinzner, P., Huckleberry, A., Kählerian potentials and convexity properties of the moment map, Invent. Math. 126 (1996), 65–84.

[H-H-K]   Heinzner, P., Huckleberry, A. T., Heinzner, P., Kutzschebauch, F., Abels' Theorem in the real analytic case and applications to complexifications, in: Complex Analysis and Geometry, Lecture Notes in Pure Appl. Math., Marcel Dekker, 1995, 229–273.

[H-H-L]   Heinzner, P., Huckleberry, A. T., Loose, F., Kählerian extensions of the symplectic reduction, J. Reine Angew. Math. 455 (1994), 123–140.

[H-I]   Heinzner, P., Iannuzzi, A., Integration of local actions on holomorphic fiber spaces, Nagoya Math. J. 146 (1997), 31–53.

[H-L]   Heinzner, P., Loose, F., Reduction of complex Hamiltonian G-spaces, Geometric and Functional Analysis 4 (1994), 288–297.

[H-M]   Heinzner, P., Migliorini, L., Projectivity of moment map quotients, Preprint dg-ga/9712008 Dec. 1997.

[H-M-P] Heinzner, P., Migliorini, L., Polito, M., Semistable quotients, Ann. Scuola Norm. Sup. Pisa, Cl. Sci. (4) Vol. XXVI, 2 (1998), 233–248.

[Ho]   Hochschild, G., The Structure of Lie groups, Holden-Day, San Francisco–London–Amsterdam 1965.

[K-N]   Kempf, G., Ness, L., The length of vectors in representation spaces, in: Lecture Notes in Math. 732, Springer-Verlag, Berlin–Heidelberg–New York 1979, 233–243.

[K1]   Kirwan, F., Cohomology of quotients in symplectic and algebraic geometry, Math. Notes 31, Princeton University Press, Princeton New Jersey, 1984.

[K2]   Kirwan, F., Convexity properties of the moment mapping III, Invent. Math. 77 (1984), 547–552.

[Ko]   Kobayashi, S., Geometry of bounded domains, Trans. Amer. Math. Soc. 92 (1959), 267–290.

[Ku]   Kutzschebauch, F., Eigentliche Wirkungen von Liegruppen auf reell-analytischen Mannigfaltigkeiten, Preprint Bochum 1994, Schriftenreihe des Graduiertenkollegs Geometrie und Mathematische Physik.

[L-S]   Lempert, L., Szöke, R., Global solutions of the homogeneous complex Monge-Ampére equation and complex structures on the tangent bundle of Riemannian manifolds, Math. Ann. 290 (1991), 689–712.

[L]   Luna, D., Slices étales, Bull. Soc. Math. France, Mémoire 33 (1973), 81–105.

[M-F-K] Mumford, D., Forgaty, J., Kirwan, F., Geometric invariant theory, 3rd ed., Ergeb. Math. Grenzgeb., Springer-Verlag, Berlin–Heidelberg–New York 1994.

[N1]   Nagata, M., On the 14th problem of Hilbert, Amer. J. Math. 81 (1959), 766–772.

[N2]   Nagata, M., Complete reducibility of rational representations of a matrix group, J. Math. Kyoto Univ. 1 (1961), 87–99.

[No]   Noether, E., Der Endlichkeitssatz der Invarianten endlicher Gruppen, Math. Ann. 77 (1916), 89–92.

[P1]   Palais, R. S., A global formulation of the Lie theory of transformation groups, Mem. Amer. Math. Soc. (1955).

[P2]   Palais, R.S., On the existence of slices for actions of non compact Lie groups, Ann. of Math. 73 (1961), 295–323.

[Po]   Polito, M., $SL(2, \mathbb{C})$-quotients de $(\mathbb{P}^1)^n$, C. R. Acad. Sci. Paris, 321(1) (1995), 1577–178.

[P-S]   Procesi, C., Schwarz, G., Inequalities defining orbit spaces, Invent. Math 81 (1985), 539–554.

[R] Ressayere, N., The GIT-equivalence for G-line bundles, Preprint math.AG/9811053 9. Nov. 1998.

[S-L] Sjamaar, R., Lerman, E., Stratified symplectic spaces and reduction, Ann. of Math. 134 (1991), 375–422.

[S] Sjamaar, R., Holomorphic slices, symplectic reduction and multiplicities of representations, Ann. of Math. 141 (1995), 87–129.

[Sn] Snow, D. M., Reductive group actions on Stein Spaces, Math. Ann. 259 (1982), 79–97.

[St] Stratmann, B., Differentiable structures on symplectic reductions, Transform. Groups 13 (1998), 255–267.

[Su1] Sumihiro, H., Equivariant completion, J. Math. Kyoto Univ. 14 (1974), 1–28.

[Su2] Sumihiro, H., Equivariant completion, II, J. Math. Kyoto Univ. 15 (1975), 573–605.

[T] Geometric invariant theory and flips, J. of Amer. Math. Soc. 9 (1996), 691–723.

[W] Weyl, H., Classical groups, their invariants and representations, Princeton Univ. Press, Princeton 1946.

[Wi] Winkelmann, J., Complex analytic geometry of complex parallezible manifolds, to appear in Mémoires de la SMF.

Nef divisors on moduli spaces of abelian varieties

Klaus Hulek

0. Introduction

Let $\mathcal{A}_g$ be the moduli space of principally polarized abelian varieties of dimension g. Over the complex numbers $\mathcal{A}_g = \mathbb{H}_g/\Gamma_g$ where $\mathbb{H}_g$ is the Siegel space of genus g and $\Gamma_g = \mathrm{Sp}(2g, \mathbb{Z})$. We denote the torodial compactification given by the second Voronoi decomposition by $\mathcal{A}_g^*$ and call it the *Voronoi compactification*. It was shown by Alexeev and Nakamura [A] that $\mathcal{A}_g^*$ coarsely represents the stack of principally polarized stable quasiabelian varieties. The variety $\mathcal{A}_g^*$ is projective [A] and it is known that the Picard group of $\mathcal{A}_g^*$, $g \geq 2$ is generated (modulo torsion) by two elements L and D, where L denotes the ($\mathbb{Q}$-)line bundle given by modular forms of weight 1 and D is the boundary (see [Mu2], [Fa] and [Mu1] for $g = 2, 3$ and ≥ 4). In this paper we want to prove the following

Theorem 0.1. *Let $g = 2$ or 3. A divisor $aL - bD$ on $\mathcal{A}_g^*$ is nef if and only if $b \geq 0$ and $a - 12b \geq 0$.*

The varieties $\mathcal{A}_g$ have finite quotient singularities. Adding a level-n structure one obtains spaces $\mathcal{A}_g(n) = \mathbb{H}_g/\Gamma_g(n)$ where $\Gamma_g(n)$ is the principal congruence subgroup of level n. For $n \geq 3$ these spaces are smooth. However, the Voronoi compactification $\mathcal{A}_g^*(n)$ acquires singularities on the boundary for $g \geq 5$ due to bad behaviour of the second Voronoi decomposition. There is a natural quotient map $\mathcal{A}_g^*(n) \to \mathcal{A}_g^*$. Note that this map is branched of order n along the boundary. Hence Theorem (0.1) is equivalent to

Theorem 0.2. *Let $g = 2$ or 3. A divisor $aL - bD$ on $\mathcal{A}_g^*(n)$ is nef if and only if $b \geq 0$ and $a - 12\frac{b}{n} \geq 0$.*

This theorem easily gives the following two corollaries.

Corollary 0.3. *If $g = 2$ then K is nef but not ample for $\mathcal{A}_2^*(4)$ and K is ample for $\mathcal{A}_2^*(n)$, $n \geq 5$; in particular $\mathcal{A}_2^*(n)$ is a minimal model for $n \geq 4$ and a canonical model for $n \geq 5$.*

This was first proved by Borisov [Bo].

Corollary 0.4. *If $g = 3$ then K is nef but not ample for $\mathcal{A}_3^*(3)$ and K is ample for $\mathcal{A}_3^*(n)$, $n \geq 4$; in particular $\mathcal{A}_3^*(n)$ is a minimal model for $n \geq 3$ and a canonical model for $n \geq 4$.*

In this paper we shall give two proofs of Theorem (0.1). The first and quick one reduces the problem via the Torelli map to the analogous question for $\overline{M}_2$, resp. $\overline{M}_3$. Since the Torelli map is not surjective for $g \geq 4$ this proof cannot possibly be generalized to higher genus. This is the main reason why we want to give a second proof which uses theta functions. This proof makes essential use of a result of Weissauer [We]. The method has the advantage that it extends in principle to other polarizations as well as to higher g. We will also give some partial results supporting the

Conjecture. *For any $g \geq 2$ the nef cone on $\mathcal{A}_g^*$ is given by the divisors $aL - bD$ where $b \geq 0$ and $a - 12b \geq 0$.*

Acknowledgement. It is a pleasure for me to thank RIMS and Kyoto University for their hospitality during the autumn of 1996. I am grateful to V. Alexeev and R. Salvati Manni for useful discussions. It was Salvati Manni who drew my attention to Weissauer's paper. I would also like to thank R. Weissauer for additional information on [We]. The author is partially supported by TMR grant ERBCHRXCT 940557.

1. Curves meeting the interior

We start by recalling some results about the Kodaira dimension of $\mathcal{A}_g^*(n)$. It was proved by Freitag, Tai and Mumford that $\mathcal{A}_g^*$ is of general type for $g \geq 7$. The following more general result is probably well known to some specialists.

Theorem 1.1. *$\mathcal{A}_g^*(n)$ is of general type for the following values of g and $n \geq n_0$:*

g	2	3	4	5	6	≥ 7
n_0	4	3	2	2	2	1

Proof. One can use Mumford's method from [Mu1]. First recall that away from the singularities and the closure of the branch locus of the map $\mathbb{H}_g \to \mathcal{A}_g(n)$ the canonical bundle equals

$$K \equiv (g+1)L - D. \tag{1}$$

This equality holds in particular also on an open part of the boundary. If $g \leq 4$ and $n \geq 3$ the spaces $\mathcal{A}_g^*(n)$ are smooth and hence (1) holds everywhere. If $g \geq 5$ then Tai [T] showed that there is a suitable toroidal compactification $\tilde{\mathcal{A}}_g(n)$ such that all singularities are canonical quotient singularities. By Mumford's results from [Mu1] one can use the theta-null locus to eliminate D from formula (1) and obtains

$$K \equiv \left((g+1) - \frac{2^{g-2}(2^g+1)}{n2^{2g-5}}\right) L + \frac{1}{n2^{2g-5}}[\Theta_{\text{null}}]. \tag{2}$$

We then have general type if all singularities are canonical and if the factor in front of L is positive. This gives immediately all values in the above table with the exception of $(g, n) = (4, 2)$ and $(7, 1)$. In the latter case the factor in front of L is negative. The proof that $\mathcal{A}_7$ is nevertheless of general type is the main result of [Mu1]. The difficulty in the first case is that one can possibly have non-canonical singularities. One can, however, use the following argument which I have learnt from Salvati Manni: An immediate calculation shows that for every element $\sigma \in \Gamma_g(2)$ the square $\sigma^2 \in \Gamma_g(4)$. Hence if σ has a fixed point then $\sigma^2 = 1$ since $\Gamma_g(4)$ acts freely. But for elements of order 2 one can again use Tai's extension theorem (see [T, Remark after Lemma 4.5] and [T, Remark after Lemma 5.2]). $\square$

Remark 1.2. The Kodaira dimension of $\mathcal{A}_6$ is still unknown. All other varieties $\mathcal{A}_g(n)$ which do not appear in the above list are either rational or unirational: Unirationality of $\mathcal{A}_g$ for $g = 5$ was proved by Donagi [D] and also by Mori and Mukai [MM] and Verra [V]. For $g = 4$ the same result was shown by Clemens [C]. Unirationality is easy for $g \leq 3$. Igusa [I2] showed that $\mathcal{A}_2$ is rational. Recently Katsylo [Ka] proved rationality of $\mathcal{M}_3$ and hence also of $\mathcal{A}_3$. The space $\mathcal{A}_3(2)$ is rational by work of van Geemen [vG] and Dolgachev and Ortland [DO]. $\mathcal{A}_2(3)$ is the Burkhardt quartic and hence rational. This was first proved by Todd (1936) and Baker (1942). See also the thesis of Finkelnberg [Fi]. The variety $\mathcal{A}_2(2)$ has the Segre cubic as a projective model [vdG1] and is hence also rational. Yamazaki [Ya] first showed general type for $\mathcal{A}_2(n)$, $n \geq 4$.

We denote the Satake compactification of $\mathcal{A}_g$ by $\overline{\mathcal{A}}_g$. There is a natural map $\pi : \mathcal{A}_g^* \to \overline{\mathcal{A}}_g$ which is an isomorphism on $\mathcal{A}_g$. The line bundle L is the pullback of an ample line bundle on $\overline{\mathcal{A}}_g$ which, by abuse of notation, we again denote by L. In fact the Satake compactification is defined as the closure of the image of $\mathcal{A}_g$ under the embedding given by a suitable power of L on $\mathcal{A}_g$. In particular we notice that $L.C \geq 0$ for every curve C on $\mathcal{A}_g^*$ and that $L.C > 0$ if C is not contracted to a point under the map π.

Let F be a modular form with respect to the full modular group $\text{Sp}(2g, \mathbb{Z})$. Then the *order* $o(F)$ of F is defined as the quotient of the vanishing order of F divided by the weight of F.

Theorem 1.3 (Weissauer). *For every point $\tau \in \mathbb{H}_g$ and every $\varepsilon > 0$ there exists a modular form F of order $o(F) \geq \frac{1}{12+\varepsilon}$ which does not vanish at τ.*

Proof. See [We]. $\square$

Proposition 1.4. *Let $C \subset \mathcal{A}_g^*$ be an irreducible curve which is not contained in the boundary. Then $(aL - bD).C \geq 0$ if $b \geq 0$ and $a - 12b \geq 0$.*

Proof. First note that $L.C > 0$ since $\pi(C)$ is a curve in the Satake compactification. It is enough to prove that $(aL - bD).C > 0$ if $a - 12b > 0$ and $a, b \geq 0$. This is clear for $b = 0$ and hence we can assume that $b \neq 0$. We can now choose some $\varepsilon > 0$ with $a/b > 12 + \varepsilon$. By Weissauer's theorem there exists a modular form F of say weight k and vanishing order m with $F(\tau) \neq 0$ for some point $[\tau] \in C$ and $m/k \geq 1/(12+\varepsilon)$. In terms of divisors this gives us that

$$kL = mD + D_F, \quad C \not\subset D_F$$

where D_F is the zero-divisor of F. Hence

$$\left(\frac{k}{m}L - D\right) = \frac{1}{m}D_F.C \geq 0.$$

Since $a/b > 12 + \varepsilon \geq k/m$ and $L.C > 0$ we can now conclude that

$$\left(\frac{a}{b}L - D\right).C > \left(\frac{k}{m}L - D\right).C \geq 0. \qquad \square$$

Remark 1.5. Weissauer's result is optimal, since the modular forms of order $> 1/12$ have a common base locus. To see this consider curves C in $\mathcal{A}_g^*$ of the form $X(1) \times \{A\}$ where $X(1)$ is the modular curve of level 1 parametrizing elliptic curves and A is a fixed abelian variety of dimension $g - 1$. The degree of L on $X(1)$ is $1/12$ (recall that L is a $\mathbb{Q}$-bundle) whereas it has one cusp, i.e. the degree of D on this curve is 1. Hence every modular form of order $> 1/12$ will vanish on C. This also shows that the condition $a - 12b \geq 0$ is necessary for a divisor to be nef.

2. Geometry of the boundary (I)

We first have to collect some properties of the structure of the boundary of $\mathcal{A}_g^*(n)$. Recall that the Satake compactification is set-theoretically the union of $\mathcal{A}_g(n)$ and of moduli spaces $\mathcal{A}_k(n), k < g$ of lower dimension, i.e.

$$\overline{\mathcal{A}}_g(n) = \mathcal{A}_g(n) \amalg \left(\coprod_{i_1} \mathcal{A}_{g-1}^{i_1}(n)\right) \amalg \left(\coprod_{i_2} \mathcal{A}_{g-2}^{i_2}(n)\right) \ldots \amalg \left(\coprod_{i_g} \mathcal{A}_0^{i_g}(n)\right).$$

Via the map $\pi : \mathcal{A}_g^*(n) \to \overline{\mathcal{A}}_g(n)$ this also defines a stratification of $\mathcal{A}_g^*(n)$:

$$\mathcal{A}_g^*(n) = \mathcal{A}_g(n) \amalg \left(\coprod_{i_1} D_{g-1}^{i_1}(n)\right) \amalg \left(\coprod_{i_2} D_{g-2}^{i_2}(n)\right) \ldots \amalg \left(\coprod_{i_g} D_0^{i_g}(n)\right).$$

The irreducible components of the boundary D are the closures $\overline{D}_{g-1}^{i_1}(n)$ of the codimension 1 strata $D_{g-1}^{i_1}(n)$. Whenever we talk about a *boundary component* we mean one of the divisors $\overline{D}^{i_1}_{g-1}(n)$. Then the boundary D is given by

$$D = \sum_{i_1} \overline{D}^{i_1}_{g-1}(n).$$

The fibration $\pi : D^{i_1}_{g-1}(n) \to \mathcal{A}^{i_1}_{g-1}(n) = \mathcal{A}_{g-1}(n)$ is the universal family of abelian varieties of dimension $g - 1$ with a level-n structure if $n \geq 3$ resp. the universal family of Kummer varieties for $n = 1$ or 2 (see [Mu1]). We shall also explain this in more detail later on. To be more precise we associate to a point $\tau \in \mathbb{H}_g$ the lattice $L_{\tau,1} = (\tau, \mathbf{1})\mathbb{Z}^{2g}$, resp. the principally polarized abelian variety $A_{\tau,1} = \mathbb{C}^g/L_{\tau,1}$. Given an integer $n \geq 1$ we set $L_{n\tau,n} = (n\tau, n\mathbf{1}_g)\mathbb{Z}^{2g}$, resp. $A_{n\tau,n} = \mathbb{C}^g/L_{n\tau,n}$. By $K_{n\tau,n}$ we denote the Kummer variety $A_{n,\tau n}/\{\pm 1\}$.

Lemma 2.1. *Let $n \geq 3$. Then for any point $[\tau] \in \mathcal{A}^{i_1}_{g-1}(n)$ the fibre of π equals $\pi^{-1}([\tau]) = A_{n,\tau n}$.*

Proof. Compare [Mu1]. We shall also give an independent proof below. □

This result remains true for $n = 1$ or 2, at least for points τ whose stabilizer subgroup in $\Gamma_g(n)$ is $\{\pm 1\}$, if we replace $A_{n,\tau n}$ by its associate Kummer variety $K_{n,\tau n}$.

Lemma 2.2. *Let $n \geq 3$. Then for $[\tau] \in \mathcal{A}^{i_1}_{g-1}(n)$ the restriction of $D^{i_1}_{g-1}(n)$ to the fibre $\pi^{-1}([\tau])$ is negative. More precisely*

$$D^{i_1}_{g-1}(n)|_{\pi^{-1}([\tau])} \equiv -\frac{2}{n} H$$

where H is the polarization on $A_{n\tau,n}$ given by the pull-back of the principal polarization on $A_{\tau,1}$ via the covering $A_{n,\tau n} \to A_{\tau,1}$.

Proof. Compare [Mu1, Proposition 1.8], resp. see the discussion below. □

Again the statement remains true for $n = 1$ or 2 if we replace the abelian variety by its Kummer variety.

First proof of Theorem (0.1). We have already seen (see Remark 1.5) that for every nef divisor $aL - bD$ the inequality $a - 12b \geq 0$ holds. If C is a curve in a fibre of the map $\mathcal{A}^*_g(n) \to \overline{\mathcal{A}}_g(n)$, then $L.C = 0$. Lemma (2.2) immediately implies that $b \geq 0$ for any nef divisor. It remains to show that the conditions of Theorem (0.1) are sufficient to imply nefness. For any genus the Torelli map $t : \mathcal{M}_g \to \mathcal{A}_g$ extends to a morphism $\bar{t} : \overline{\mathcal{M}}_g \to \mathcal{A}^*_g$ (see [Nam]). Here $\overline{\mathcal{M}}_g$ denotes the compactification of $\mathcal{M}_g$ by stable curves. For $g = 2$ and 3 the map $\bar{t}$ is surjective. It follows that for every curve C in $\mathcal{A}^*_g$ there is a curve C' in $\overline{\mathcal{M}}_g$ which is finite over C. Hence a divisor on $\mathcal{A}^*_g$, $g = 2, 3$ is nef if and only if this holds for its pull-back to $\overline{\mathcal{M}}_g$. In the notation of

Faber's paper [Fa] $\bar{\iota}^*L = \lambda$ where λ is the Hodge bundle and $\bar{\iota}^*D = \delta_0$ where δ_0 is the boundary ($g = 2$), resp. the closure of the locus of genus 2 curves with one node ($g = 3$) (cf also [vdG2]). The result follows since $a\lambda - b\delta_0$ is nef on $\overline{\mathcal{M}}_g$, $g = 2, 3$ for $a - 12b \geq 0$ and $b \geq 0$ (see [Fa]). □

As we have already pointed out the Torelli map is not surjective for $g \geq 4$ and hence this proof cannot possibly be generalized to higher genus. The main purpose of this paper is, therefore, to give a proof of Theorem (0.1) which does not use the reduction to the curve case. This will also allow us to prove some results for general g. At the same time we obtain an independent proof of nefness of $a\lambda - b\delta_0$ for $a - 12b \geq 0$ and $b \geq 0$ on $\overline{\mathcal{M}}_g$ for $g = 2$ and 3.

We now want to investigate the open parts $D_{g-1}^{i_1}(n)$ of the boundary components $\overline{D}_{g-1}^{i_1}(n)$ and their fibration over $\mathcal{A}_{g-1}(n)$ more closely. At the same time this gives us another argument for Lemmas (2.1) and (2.2). At this stage we have to make first use of the toroidal construction. Recall that the boundary components $D_{g-1}^{i_1}(n)$ are in $1:1$ correspondence with the maximal dimensional cusps, and these in turn are in $1:1$ correspondence with the lines $l \subset \mathbb{Q}^g$ modulo $\Gamma_g(n)$. Since all cusps are equivalent under the action of $\Gamma_g/\Gamma_g(n)$ we can restrict our attention to one of these cusps, namely the one given by $l_0 = (0, \ldots, 0, 1)$. This corresponds to $\tau_{gg} \to i\infty$. To simplify notation we shall denote the corresponding boundary stratum simply by $D_{g-1}^1(n) = D_{g-1}(n)$. The stabilizer $P(l_0)$ of l_0 in Γ_g is generated by elements of the following form(cf. [HKW, Proposition I.3.87]):

$$g_1 = \begin{pmatrix} A & 0 & B & 0 \\ 0 & 1 & 0 & 0 \\ C & 0 & D & 0 \\ 0 & 0 & 0 & 1 \end{pmatrix}, \quad \begin{pmatrix} A & B \\ C & D \end{pmatrix} \in \Gamma_{g-1},$$

$$g_2 = \begin{pmatrix} \mathbf{1}_{g-1} & 0 & 0 & 0 \\ 0 & \pm 1 & 0 & 0 \\ 0 & 0 & \mathbf{1}_{g-1} & 0 \\ 0 & 0 & 0 & \pm 1 \end{pmatrix},$$

$$g_3 = \begin{pmatrix} \mathbf{1}_{g-1} & 0 & 0 & {}^tN \\ M & 1 & N & 0 \\ 0 & 0 & \mathbf{1}_{g-1} & -{}^tM \\ 0 & 0 & 0 & 1 \end{pmatrix}, \quad M, N \in \mathbb{Z}^{g-1},$$

$$g_4 = \begin{pmatrix} \mathbf{1}_{g-1} & 0 & 0 & 0 \\ 0 & 1 & 0 & S \\ 0 & 0 & \mathbf{1}_{g-1} & 0 \\ 0 & 0 & 0 & 1 \end{pmatrix}, \quad S \in \mathbb{Z}.$$

We write $\tau = (\tau_{ij})_{1 \leq i, j \leq g}$ in the form

$$\begin{pmatrix} \tau_{11} & \cdots & \tau_{1,g-1} & \tau_{1g} \\ \vdots & & \vdots & \vdots \\ \tau_{1,g-1} & \cdots & \tau_{g-1,g-1} & \tau_{g-1,g} \\ \tau_{1g} & \cdots & \tau_{g-1,g} & \tau_{gg} \end{pmatrix} = \begin{pmatrix} \tau_1 & {}^t\tau_2 \\ \hline \tau_2 & \tau_3 \end{pmatrix}.$$

Then the action of $P(l_0)$ on $\mathbb{H}_g$ is given by (cf. [HKW, I.3.91]):

$$g_1(\tau) = \begin{pmatrix} (A\tau_1 + B)(C\tau_1 + D)^{-1} & * \\ \tau_2(C\tau_1 + D)^{-1} & \tau_3 - \tau_2(C\tau_1 + D)^{-1} C\, {}^t\tau_2 \end{pmatrix},$$

$$g_2(\tau) = \begin{pmatrix} \tau_1 & * \\ \pm\tau_2 & \tau_3 \end{pmatrix},$$

$$g_3(\tau) = \begin{pmatrix} \tau_1 & * \\ \tau_2 + M\tau_1 + N & \tau_3' \end{pmatrix}$$

where $\tau_3' = \tau_3 + M\tau_1\,{}^tM + M\,{}^t\tau_2 + {}^t(M\,{}^t\tau_2) + N\,{}^tM$,

$$g_4(\tau) = \begin{pmatrix} \tau_1 & \tau_2 \\ \tau_2 & \tau_3 + S \end{pmatrix}.$$

The parabolic subgroup $P(l_0)$ is an extension

$$1 \longrightarrow P'(l_0) \longrightarrow P(l_0) \longrightarrow P''(l_0) \longrightarrow 1$$

where $P'(l_0)$ is the rank 1 lattice generated by g_4. To obtain the same result for $\Gamma_g(n)$ we just have to intersect $P(l_0)$ with $\Gamma_g(n)$. Note that g_2 is in $\Gamma_g(n)$ only for $n = 1$ or 2. The first step in the construction of the toroidal compactification of $\mathcal{A}_g^*(n)$ is to divide $\mathbb{H}_g$ by $P'(l_0) \cap \Gamma(n)$ which gives a map

$$\begin{array}{rcl} \mathbb{H}_g & \longrightarrow & \mathbb{H}_{g-1} \times \mathbb{C}^{g-1} \times \mathbb{C}^* \\ \begin{pmatrix} \tau_1 & {}^t\tau_2 \\ \tau_2 & \tau_3 \end{pmatrix} & \longmapsto & (\tau_1, \tau_2, e^{2\pi i \tau_3/n}). \end{array}$$

Partial compactification in the direction of l_0 then consists of adding the set $\mathbb{H}_{g-1} \times \mathbb{C}^{g-1} \times \{0\}$. It now follows immediately from the above formulae for the action of $P(l_0)$ on $\mathbb{H}_g$ that the action of the quotient group $P''(l_0)$ on $\mathbb{H}_{g-1} \times \mathbb{C}^{g-1} \times \mathbb{C}^*$ extends to $\mathbb{H}_{g-1} \times \mathbb{C}^{g-1} \times \{0\}$. Then $D_{g-1}(n) = (\mathbb{H}_{g-1} \times \mathbb{C}^{g-1})/P''(l_0)$ and the map to $\mathcal{A}_{g-1}(n)$ is induced by the projection from $\mathbb{H}_{g-1} \times \mathbb{C}^{g-1}$ to $\mathbb{H}_{g-1}$. This also shows that $D_{g-1}(n) \to \mathcal{A}_{g-1}(n)$ is the universal family for $n \geq 3$ and that the general fibre is a Kummer variety for $n = 1$ and 2.

Whenever $n_1 | n_2$ we have a Galois covering

$$\pi(n_1, n_2) : \mathcal{A}_g^*(n_2) \longrightarrow \mathcal{A}_g^*(n_1)$$

whose Galois group is $\Gamma_g(n_1)/\Gamma_g(n_2)$. This induces coverings $\overline{D}_{g-1}(n_2) \to \overline{D}_{g-1}(n_1)$, resp. $D_{g-1}(n_2) \to D_{g-1}(n_1)$. In order to avoid technical difficulties we assume for the moment that $\mathcal{A}_g^*(n)$ is smooth (this is the case if $g \leq 4$ and $n \geq 3$). In what follows we will always be able to assume that we are in this situation. Then we denote the normal bundle of $\overline{D}_{g-1}(n)$ in $\mathcal{A}_g^*(n)$ by $N_{\overline{D}_{g-1}(n)}$, resp. its restriction to $D_{g-1}(n)$ by $N_{D_{g-1}(n)}$. Since the covering map $\pi(n_1, n_2)$ is branched of order n_2/n_1 along the boundary, it follows that

$$\pi^*(n_1, n_2) n_1 N_{\overline{D}_{g-1}}(n_1) = n_2 N_{\overline{D}_{g-1}}(n_2).$$

We now define the bundle

$$\overline{M}(n) := -n N_{\overline{D}_{g-1}(n)} + L.$$

This is a line bundle on the boundary component $\overline{D}_{g-1}(n)$. We denote the restriction of $\overline{M}(n)$ to $D_{g-1}(n)$ by $M(n)$. We find immediately that

$$\pi^*(n_1, n_2) \overline{M}(n_1) = \overline{M}(n_2).$$

The advantage of working with the bundle $\overline{M}(n)$ is that we can explicitly describe sections of this bundle. For this purpose it is useful to review some basic facts about theta functions. For every element $m = (m', m'')$ of $\mathbb{R}^{2g}$ one can define the theta-function

$$\Theta_{m'm''}(\tau, z) = \sum_{q \in \mathbb{Z}^g} e^{2\pi i [(q+m')\tau^t(q+m')/2 + (q+m')^t(z+m'')]}.$$

The transformation behaviour of $\Theta_{m'm''}(\tau, z)$ with respect to $z \mapsto z + u\tau + u'$ is described by the formulae ($\Theta 1$)–($\Theta 5$) of [I1, pp. 49, 50]. The behaviour of $\Theta_{m'm''}(\tau, z)$ with respect to the action of $\Gamma_g(1)$ on $\mathbb{H}_g \times \mathbb{C}^g$ is given by the theta transformation formula [I1, Theorem II.5.6] resp. the corollary following this theorem [I1, p. 85].

Proposition 2.3. *Let $n \equiv 0 \mod 4p^2$. If $m', m'', \overline{m}', \overline{m}'' \in \frac{1}{2p}\mathbb{Z}^{g-1}$, then the functions $\Theta_{m'm''}(\tau, z) \Theta_{\overline{m}'\overline{m}''}(\tau, z)$ define sections of the line bundle $M(n)$ on $D_{g-1}(n)$.*

Proof. It follows from ($\Theta 3$) and ($\Theta 1$) that for $k, k' \in n\mathbb{Z}^{g-1}$ the following holds:

$$\Theta_{m',m''}(\tau, z + k\tau + k') = e^{2\pi i [-\frac{1}{2}k\tau^t k - k^t(z+k')]} \Theta_{m',m''}(\tau, z).$$

Similarly, of course, for $\Theta_{\overline{m}', \overline{m}''}(\tau, z)$. Moreover the theta transformation formula together with formula ($\Theta 2$) gives

$$\Theta_{m',m''}(\tau^\#, z^\#) = e^{2\pi i [\frac{1}{2}z(C\tau+D)^{-1}C^t z]} \det(C\tau + D)^{1/2} u \Theta_{m',m''}(\tau, z)$$

for every element $\gamma = \begin{pmatrix} A & B \\ C & D \end{pmatrix} \in \Gamma_{g-1}(n)$ and

$$\tau^\# = \gamma(\tau), \quad z^\# = z(C\tau + D)^{-1}.$$

Here u^2 is a character of $\Gamma_{g-1}(1, 2)$ with $u^2|_{\Gamma_{g-1}(4)} \equiv 1$.

On the other hand the boundary component $D_{g-1}(n)$ is defined by $t_3 = 0$ with $t_3 = e^{2\pi i \tau_3/n}$. We have already described the action of $P''(l_0)$ on $\mathbb{H}_{g-1} \times \mathbb{C}^{g-1}$. The result then follows by comparing the transformation behaviour of $(t_3/t_3^2)^n$ with respect to g_1 and g_3 with the above formulae together with the fact that the line bundle L is defined by the automorphy factor $\det(C\tau + D)$. □

This also gives an independent proof of Lemma (2.2).

3. Geometry of the boundary (II)

So far we have described the stratum $D_{g-1}(n)$ of the boundary component $\overline{D}_{g-1}(n)$ and we have seen that there is a natural map $D_{g-1}(n) \to \mathcal{A}_{g-1}(n)$ which identifies $D_{g-1}(n)$ with the universal family over $\mathcal{A}_{g-1}(n)$ if $n \geq 3$. We now want to describe the closure $\overline{D}_{g-1}(n)$ in some detail. In order to do this we have to restrict ourselves to $g = 2$ and 3. First assume $g = 2$. Then the projection $D_1(n) \to \mathcal{A}_1(n) = X^0(n)$ extends to a projection $\overline{D}_1(n) \to X(n)$ onto the modular curve of level n and in this way $\overline{D}_1(n)$ is identified with Shioda's modular surface $S(n) \to X(n)$. The fibres are either elliptic curves or n-gons of rational curves (if $n \geq 3$). Similarly the fibration $D_2(n) \to \mathcal{A}_2(n)$ extends to a fibration $\overline{D}_2(n) \to \mathcal{A}_2^*$ whose fibres over the boundary of $\mathcal{A}_2^*(n)$ are degenerate abelian surfaces. This was first observed by Nakamura [Nak] and was described in detail by Tsushima [Ts] whose paper is essential for what follows.

We shall now explain the toroidal construction which allows us to describe the fibration $\overline{D}_2(n) \to \mathcal{A}_2^*(n)$ explicitly. Here we shall concentrate on a description of this map in the most difficult situation, namely in the neighbourhood of a cusp of maximal corank.

The toroidal compactification $\mathcal{A}_g^*(n)$ is given by the second Voronoi decomposition Σ_g. This is a rational polyhedral decomposition of the convex hull in $\operatorname{Sym}_g^{\geq 0}(\mathbb{R})$ of the set $\operatorname{Sym}_g^{\geq 0}(\mathbb{Z})$ of integer semi-positive $(g \times g)$-matrices. For $g = 2$ and 3 it can be described as follows. First note that $\operatorname{Gl}(g, \mathbb{Z})$ acts on $\operatorname{Sym}_g^{\geq 0}(\mathbb{R})$ by $\gamma \mapsto {}^t M \gamma M$. For $g = 2$ we define the standard cone

$$\sigma_2 = \mathbb{R}_{\geq 0}\gamma_1 + \mathbb{R}_{\geq 0}\gamma_2 + \mathbb{R}_{\geq 0}\gamma_3$$

with

$$\gamma_1 = \begin{pmatrix} 1 & 0 \\ 0 & 0 \end{pmatrix}, \quad \gamma_2 = \begin{pmatrix} 0 & 0 \\ 0 & 1 \end{pmatrix}, \quad \gamma_3 = \begin{pmatrix} 1 & -1 \\ -1 & 1 \end{pmatrix}.$$

Then

$$\Sigma_2 = \{M(\sigma_2);\ M \in \operatorname{Gl}(2, \mathbb{Z})\}.$$

Similarly for $g = 3$ we consider the standard cone

$$\sigma_3 = \mathbb{R}_{\geq 0}\alpha_1 + \mathbb{R}_{\geq 0}\alpha_2 + \mathbb{R}_{\geq 0}\alpha_3 + \mathbb{R}_{\geq 0}\beta_1 + \mathbb{R}_{\geq 0}\beta_2 + \mathbb{R}_{\geq 0}\beta_3$$

with

$$\alpha_1 = \begin{pmatrix} 1 & 0 & 0 \\ 0 & 0 & 0 \\ 0 & 0 & 0 \end{pmatrix}, \quad \alpha_2 = \begin{pmatrix} 0 & 0 & 0 \\ 0 & 1 & 0 \\ 0 & 0 & 0 \end{pmatrix}, \quad \alpha_3 = \begin{pmatrix} 0 & 0 & 0 \\ 0 & 0 & 0 \\ 0 & 0 & 1 \end{pmatrix},$$

$$\beta_1 = \begin{pmatrix} 0 & 0 & 0 \\ 0 & 1 & -1 \\ 0 & -1 & 1 \end{pmatrix}, \quad \beta_2 = \begin{pmatrix} 1 & 0 & -1 \\ 0 & 0 & 0 \\ -1 & 0 & 1 \end{pmatrix}, \quad \beta_3 = \begin{pmatrix} 1 & -1 & 0 \\ -1 & 1 & 0 \\ 0 & 0 & 0 \end{pmatrix}.$$

Then

$$\Sigma_3 = \{M(\sigma_3); \ M \in \mathrm{Gl}(3, \mathbb{Z})\}.$$

We consider the lattices

$$N_3 = \mathbb{Z}\gamma_1 + \mathbb{Z}\gamma_2 + \mathbb{Z}\gamma_3$$

$$N_6 = \mathbb{Z}\alpha_1 + \mathbb{Z}\alpha_2 + \mathbb{Z}\alpha_3 + \mathbb{Z}\beta_1 + \mathbb{Z}\beta_2 + \mathbb{Z}\beta_3.$$

The fans Σ_2 resp. Σ_3 define torus embeddings $T^3 \subset X(\Sigma_2)$ and $T^6 \subset X(\Sigma_3)$. We denote the divisors of $X(\Sigma_3)$ which correspond to the 1-dimensional simplices of Σ_3 by $\mathcal{D}^i$. Let $\mathcal{D} = \mathcal{D}^1$ be the divisor corresponding to $\mathbb{R}_{\geq 0}\alpha_3$. An open part of $\mathcal{D}$ (in the $\mathbb{C}$-topology) is mapped to the boundary component $\overline{D}_2(n)$. In order to understand the structure of $\mathcal{D}$ we also consider the rank 5 lattice

$$N_5 = \mathbb{Z}\alpha_1 + \mathbb{Z}\alpha_2 + \mathbb{Z}\beta_1 + \mathbb{Z}\beta_2 + \mathbb{Z}\beta_3 \cong N_6/\mathbb{Z}\alpha_3.$$

The natural projection $\rho : N_{6,\mathbb{R}} \to N_{5,\mathbb{R}}$ maps the cones of the fan Σ_3 to the cones of a fan $\Sigma_3' \subset N_{5,\mathbb{R}}$. This fan defines a torus embedding $T^5 = (\mathcal{D} \setminus \bigcup_{i \neq 1} \mathcal{D}^i) \subset X(\Sigma_3') = \mathcal{D}$.

The projection

$$\lambda : N_{6,\mathbb{R}} \cong \mathrm{Sym}_3(\mathbb{R}) \longrightarrow N_{3,\mathbb{R}} \cong \mathrm{Sym}_2(\mathbb{R})$$

$$\begin{pmatrix} a & b & d \\ b & c & e \\ d & e & f \end{pmatrix} \longmapsto \begin{pmatrix} a & b \\ b & c \end{pmatrix}$$

maps Σ_3 to Σ_2 and factors through $N_{5,\mathbb{R}}$. In this way we obtain an induced map

$$\begin{array}{ccc} \mathcal{D} = X(\Sigma_3') & \longrightarrow & X(\Sigma_2) \\ \cup & & \cup \\ T^5 & \longrightarrow & T^3. \end{array}$$

In order to describe this map we first consider the standard simplices $\sigma_3 \subset N_{6,\mathbb{R}}$ and $\sigma_2 \subset N_{3,\mathbb{R}}$, resp. $\sigma_3' = \rho(\sigma_3) \subset N_{5,\mathbb{R}}$. On the torus T^6 (and similarly on T^5 and T^3)

we introduce coordinates by

$$t_{ij} = e^{2\pi i \tau_{ij}/n} \qquad (1 \leq i, j \leq 3).$$

These coordinates correspond to the dual basis of the basis U_{ij}^* of $\mathrm{Sym}(3, \mathbb{Z})$ where the entries of U_{ij}^* are 1 in positions (i, j) and (j, i) and 0 otherwise. One easily checks that $T_{\sigma_3} \cong \mathbb{C}^6 \subset X(\Sigma_3)$ and as coordinates on T_{σ_3} one can take the coordinates which correspond to the dual basis of the generators $\alpha_1, \ldots, \beta_3$. Let us denote these coordinates by $T_1, \ldots, T_6$. A straightforward calculation shows that the inclusion $T^6 \subset T_{\sigma_3}$ is given by

$$\begin{aligned}
T_1 &= t_{11}t_{13}t_{12}, & T_2 &= t_{22}t_{23}t_{12}, & T_3 &= t_{33}t_{13}t_{23}, \\
T_4 &= t_{23}^{-1}, & T_5 &= t_{13}^{-1}, & T_6 &= t_{12}^{-1}.
\end{aligned} \qquad (1)$$

Then $\mathcal{D} \cap T_{\sigma_3} = \{T_3 = 0\}$. For genus 2 the corresponding embedding $T^3 \subset T_{\sigma_2}$ is given by

$$T_1 = t_{11}t_{12}, \qquad T_2 = t_{22}t_{12}, \qquad T_3 = t_{12}^{-1}.$$

Finally we consider $T_{\sigma_3'} \cong \mathbb{C}^5 \subset X(\Sigma_3')$. The projection $\mathcal{D} = X(\Sigma_3') \to X(\Sigma_2)$ map $T_{\sigma_3'}$ to T_{σ_2}. We can use T_1, T_2, T_4, T_5, T_6 as coordinates on $T_{\sigma_3'}$. Since $\lambda(\alpha_1) = \lambda(\beta_2) = \gamma_1$, $\lambda(\alpha_2) = \lambda(\beta_1) = \gamma_2$ and $\lambda(\alpha_3) = \gamma_3$ we find that

$$\begin{aligned}
T_{\sigma_3'} &\cong \mathbb{C}^5 & &\longrightarrow & T_{\sigma_2} &\cong \mathbb{C}^3 \\
(T_1, T_2, T_4, T_5, T_6) & & &\longmapsto & (T_1 T_5, T_2 T_4, T_6).
\end{aligned} \qquad (2)$$

Given any (maximal dimensional) cone $\sigma' = \rho(\sigma)$ in Σ_3' we can describe the map $T_{\sigma'} \to T_{\lambda(\sigma)}$ in terms of coordinates by the method described above. In this way we obtain a complete description of the map $\mathcal{D} \to X(\Sigma_2)$.

Let us now return to the toroidal compactification $\mathcal{A}_3^*(n)$ of $\mathcal{A}_3(n)$. Let $u_0 \subset \mathbb{Q}^6$ be a maximal isotropic subspace. Then we obtain the compactification of $\mathcal{A}_3(n)$ in the direction of the cusp corresponding to u_0 as follows: The parabolic subgroup $P(u_0) \subset \Gamma_3(n)$ is an extension

$$1 \longrightarrow P'(u_0) \longrightarrow P(u_0) \longrightarrow P''(u_0) \longrightarrow 1$$

where $P'(u_0)$ is a lattice of rank 6. We have an inclusion $\mathbb{H}_g/P'(u_0) \subset T^6 \subset X(\Sigma_3)$ and we denote the interior of the closure of $\mathbb{H}_g/P'(u_0)$ in $X(\Sigma_3)$ by $X(u_0)$. Then $P''(u_0)$ acts on $X(u_0)$ and we obtain a neighbourhood of the cusp corresponding to u_0 by $X(u_0)/P''(u_0)$. We have already described the partial compactification in the direction of a line (in our case l_0). Similarly we can define a partial compactification in the direction of an isotropic plane h_0. The space $\mathcal{A}_3^*(n)$ is then obtained by glueing all these partial compactifications.

The result of Nakamura and Tsushima can then be stated as follows: The restriction of the map $\pi : \mathcal{A}_3^*(n) \to \overline{\mathcal{A}}_3(n)$ to the boundary component $\overline{D}_2(n)$ admits a factorisation

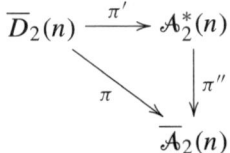

where $\pi'' : \mathcal{A}_2^*(n) \to \overline{\mathcal{A}}_2(n)$ is the natural map of the Voronoi compactification $\mathcal{A}_2^*(n)$ of $\mathcal{A}_2(n)$ to the Satake compactification $\overline{\mathcal{A}}_2(n)$. The map $\pi' : \overline{D}_2(n) \to \mathcal{A}_2^*(n)$ is a flat family of surfaces extending the universal family over $\mathcal{A}_2(n)$. In order to describe the fibres over the boundary points of $\mathcal{A}_2^*(n)$ recall that every boundary component of $\mathcal{A}_2^*(n)$ is isomorphic to the Shioda modular surface $S(n)$. We explain the *type* of a point P in $\mathcal{A}_2^*(n)$ as follows:

P has type I $\iff P \in \mathcal{A}_2(n)$

P has type II $\iff P$ lies on a smooth fibre of a boundary component $S(n)$

P has type IIIa $\iff P$ is a smooth point on a singular fibre of $S(n)$

P has type IIIb $\iff P$ is a singular point of an n-gon in $S(n)$.

Points of type IIIb are also often called *deepest points*.

Proposition 3.1 (Nakamura, Tsushima). *Assume $n \geq 3$. Let P be a point in $\mathcal{A}_2^*(n)$ and denote the fibre of the map $\pi' : \overline{D}_2(n) \to \mathcal{A}_2^*(n)$ over P by A_P. Then the following holds:*

(i) *If $P = [\tau] \in \mathcal{A}_2(n)$ is of type I then A_P is a smooth abelian surface, more precisely $A_P \cong A_{n,\tau n}$.*

(ii) *if P is of type II, then A_P is a cycle of n elliptic ruled surfaces.*

(iii) *If P is of type IIIa, then A_P consists on n^2 copies of $\mathbb{P}^1 \times \mathbb{P}^1$.*

(iv) *If P is of type IIIb, then A_P consists of $3n^2$ components. These are $2n^2$ copies of the projective plane $\mathbb{P}^2$ and n^2 copies of $\tilde{\mathbb{P}}^2$, i.e. $\mathbb{P}^2$ blown up in 3 points in general position.*

Proof. The proof consists of a careful analysis of the map $\overline{D}_2(n) \to \mathcal{A}_2^*(n)$ using the description of the map $\mathcal{D} \to X(\Sigma_2)$. For details see [Ts, section 4]. □

Remarks. (i) The degenerations of type IIIa and IIIb are usually depicted by the diagrams

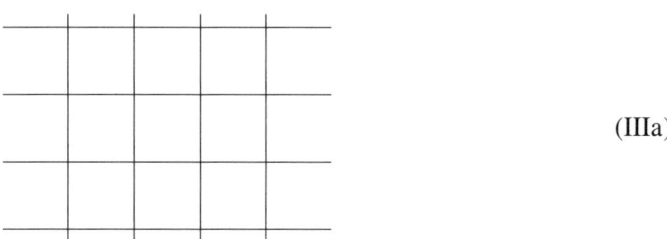
(IIIa)

where each square stands for a $\mathbb{P}^1 \times \mathbb{P}^1$, resp.

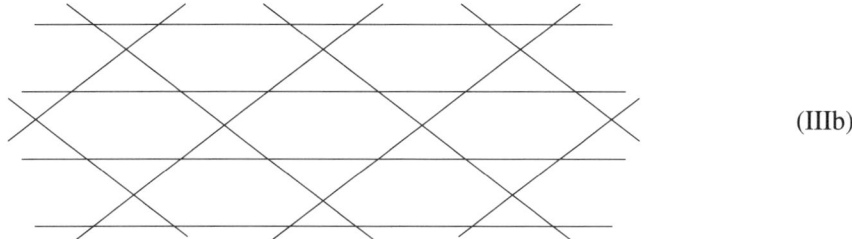
(IIIb)

where the triangles stand for projective planes $\mathbb{P}^2$ and the hexagons for blown-up planes $\tilde{\mathbb{P}}^2$.

(ii) The singular fibres are degenerate abelian surfaces (cf. [Nak], [HKW]).

(iii) This description must be modified for $n = 1$ or 2. Then the general fibre is a Kummer surface $K_{n,\tau n}$ and the fibres of type (IIIb) consist of 8 ($n = 2$), resp. 2 copies of $\mathbb{P}^2$.

The following is a crucial technical step:

Proposition 3.2. *Let* $n \equiv 0 \bmod 8p^2$. *If* $m', m'', \overline{m}', \overline{m}'' \in \frac{1}{2p}\mathbb{Z}^2$ *then the sections* $\Theta_{m'm''}(\tau, z)\Theta_{\overline{m}'\overline{m}''}(\tau, z)$ *of the line bundle* $M(n)$ *on* $D_2(n)$ *extend to sections of the line bundle* $\overline{M}(n)$ *on* $\overline{D}_2(n)$.

Proof. We have to prove that the sections in question extend to the part of $D_2(n)$ which lies over the boundary of $\mathcal{A}_2^*(n)$. This is a local statement. Moreover it is enough to prove extension in codimension 1. Due to symmetry considerations we can restrict ourselves to one boundary component in $\mathcal{A}_2^*(n)$. We shall use the above description of the toroidal compactifications $\mathcal{A}_2^*(n)$ and $\mathcal{A}_3^*(n)$ and of the map $\overline{D}_2(n) \to \mathcal{A}_2^*(n)$. We consider the boundary component of $\mathcal{A}_2^*(n)$ given by $\{T_2 = 0\} \subset T_{\sigma_2} \subset X(\Sigma_2)$. Recall the theta functions

$$\Theta_{m'm''}(\tau, z) = \sum_{q \in \mathbb{Z}^2} e^{2\pi i [\frac{1}{2}(q+m')\tau^t(q+m')+(q+m')^t(z+m'')]}$$

In our situation

$$\tau = \begin{pmatrix} \tau_{11} & \tau_{12} \\ \tau_{12} & \tau_{22} \end{pmatrix}, \quad z = (z_1, z_2) = (\tau_{13}, \tau_{23}).$$

In level n we have the coordinates

$$t_{ij} = e^{2\pi i \tau_{ij}/n}$$

and $\Theta_{m'm''}(\tau, z)$ becomes

$$\Theta_{m'm''}(\tau, z) = \sum_{q=(q_1,q_2)\in\mathbb{Z}^2} t_{11}^{\frac{1}{2}(q_1+m'_1)^2 n} t_{12}^{(q_1+m'_1)(q_2+m'_2)n} t_{22}^{\frac{1}{2}(q_2+m'_2)^2 n}$$

$$t_{13}^{(q_1+m'_1)n} t_{23}^{(q_2+m'_2)n} e^{2\pi i (q+m')^t m''}.$$

We use the coordinates T_1, T_2, T_4, T_5, T_6 on $T_{\sigma'_3}$. It follows from (1) that

$$t_{11} = T_1 T_5 T_6, \qquad t_{22} = T_2 T_4 T_6,$$
$$t_{23} = T_4^{-1}, \qquad t_{13} = T_5^{-1}, \qquad (3)$$
$$t_{12} = T_6^{-1}.$$

This leads to the following expression for the theta-functions

$$\Theta_{m'm''}(\tau, z) = \sum_{q\in\mathbb{Z}^2} T_1^{\frac{1}{2}(q_1+m'_1)^2 n} T_2^{\frac{1}{2}(q_2+m'_2)^2 n} T_4^{\frac{1}{2}(q_2+m'_2)(q_2+m'_2-2)n}$$

$$T_5^{\frac{1}{2}(q_1+m'_1)(q_1+m'_1-2)n} T_6^{\frac{1}{2}((q_1+m'_1)-(q_2+m'_2))^2 n} e^{2\pi i (q+m')^t m''}.$$

By (2) the locus over $T_2 = 0 \subset T_{\sigma_2}$ in $T_{\sigma'_3}$ is given by $T_2 T_4 = 0$. The equation for the boundary component $\overline{D}_2(n)$ is given by $t_{33} = 0$. Since by (1) we have $t_{33} = T_3 T_4 T_5$ we can assume that the normal bundle and hence $\overline{M}(n)$ (more precisely its pullback to $X(\Sigma'_3)$) is trivial outside $T_4 T_5 = 0$. Since the exponent of T_2 is a non-negative integer (here we use $n \equiv 0 \mod 8p^2$) this shows that the sections extend over $T_2 = 0$, $T_4 \neq 0$. To deal with the other components of $T_{\Sigma'_3}$ which lie over $\{T_2 = 0\}$ in T_{σ_2} we use the matrices

$$v_{nm} = \begin{pmatrix} 1 & 0 & m \\ 0 & 1 & n \\ 0 & 0 & 1 \end{pmatrix} \quad (n, m \in \mathbb{Z})$$

(cf. [Ts]) which act on $\operatorname{Sym}_3^{\geq 0}(\mathbb{Z})$ by

$$\gamma \mapsto {}^t v_{nm} \gamma v_{nm}.$$

Via λ this action lies over the trivial action on $\operatorname{Sym}_2^{\geq 0}(\mathbb{Z})$. This action also factors through ρ. Let $(\sigma'_3)_{nm} = \rho({}^t v_{nm} \sigma_3 v_{nm})$. We can then either argue with the symmetries induced by this operation or repeat directly the above calculation for $T_{(\sigma'_3)_{nm}}$.

Acting with ν_{0m}, $m \in \mathbb{Z}$, we can thus treat all components in $X(\Sigma_3')$ lying over $\{T_2 = 0\}$ in $X(\Sigma_2)$. □

4. Curves in the boundary

We can now treat curves contained in a boundary component. The following technical lemma will be crucial. Its proof uses the ideas of [We, Abschnitt 4] in an essential way and it can be generalized in a suitable form to arbitrary g. We consider the boundary component $\overline{D}_2(n)$ which belongs to the line $l_0 = (0, \ldots, 0, 1) \subset \mathbb{Q}^6$. Recall that the open part $D_2(n)$ of $\overline{D}_2(n)$ is of the form $D_2(n) = \mathbb{C}^2 \times \mathbb{H}_2/(P''(l_0) \cap \Gamma(n))$ and that the group $P''(l_0)/(P''(l_0) \cap \Gamma(n))$ acts on $\overline{D}_2(n)$. Recall also the fibration $\pi' : \overline{D}_2(n) \to \mathcal{A}_2^*(n)$. We shall denote the boundary of $\mathcal{A}_2^*(n)$ by B.

Proposition 4.1. *Let $(z, \tau) \in \mathbb{C}^2 \times \mathbb{H}_2$. For every $\varepsilon > 0$ there exist integers n, k and a section $s \in H^0(\overline{M}(n)^k)$ such that*
(i) $s([z, \tau]) \neq 0$ *where* $[z, \tau] \in D_2(n) = \mathbb{C}^2 \times \mathbb{H}_2/(P''(l_0) \cap \Gamma(n))$,
(ii) s *vanishes on* $\pi^* B$ *of order* λ *with* $\frac{\lambda}{k} \geq \frac{n}{12+\varepsilon}$.

Proof. Let $p \geq 3$ be a prime number (which will be chosen later). For $l = 2p$ we consider the set of characteristics $\mathcal{M}$ in $(\frac{1}{l}\mathbb{Z}/\mathbb{Z})^6$ of the form $m = (m_p, m_2)$ in $(\frac{1}{p}\mathbb{Z}/\mathbb{Z})^6 \oplus (\frac{1}{2}\mathbb{Z}/\mathbb{Z})^6$ with $m_p \notin \mathbb{Z}^6$. The group $\Gamma_3(1)$ acts on $\mathcal{M}$ with 2 orbits. Assume $\varepsilon > 0$ is given and that $\tilde{\mathcal{M}}$ is a subset of $\mathcal{M}$ with

$$\#\tilde{\mathcal{M}} < \varepsilon \# \mathcal{M}.$$

Then set

$$\Theta_{\mathcal{M},\tilde{\mathcal{M}}}(\tau, z) = \prod_{m \in \mathcal{M} \setminus \tilde{\mathcal{M}}} \Theta_m^l(\tau, z).$$

Let $n = 8p^2$. By Proposition (3.2) the functions $\Theta_m^l(\tau, z)$ define sections in $\overline{M}(n)^p$. Let $M_1, \ldots, M_N \in \Gamma_2(1)$ be a set of generators of $\Gamma_2(1)/\Gamma_2(n) \cong \mathrm{Sp}(4, \mathbb{Z}/n\mathbb{Z})$. Then $M_1, \ldots, M_N$, considered as elements in $P(l_0)$, act on the line bundle $\overline{M}(n)$. We set

$$F_r(\tau, z) = \sum_{i=1}^{N} M_i^* \Theta_{\mathcal{M},\tilde{\mathcal{M}}}^r.$$

This is a $\Gamma_2/\Gamma_2(n)$-invariant section of $\overline{M}(n)^{pr}$.

Now consider the abelian surface $A = A_{\tau,1} = \mathbb{C}^2/(\mathbb{Z}^2\tau + \mathbb{Z}^2)$. Then $A_{n\tau,n} = \mathbb{C}^2/((n\mathbb{Z})^2\tau + (n\mathbb{Z})^2)$ is the fibre of π over the point $[\tau] \in \mathcal{A}_2(n)$. Let

$$\tilde{\mathcal{M}} = \{m \in \mathcal{M}; \Theta_m(\tau, z) = 0\}.$$

The argument of Weissauer shows that

$$\#\tilde{M} < \varepsilon \# M$$

for p sufficiently large. For some r the section $F_r(\tau, z)$ does not vanish at $[z, \tau] \in D_2(n)$. Let B' be a boundary boundary component of $\mathcal{A}_2^*(n)$. The inverse image D' of B' under π' consists of several components. Using the matrices v_{nm} which were introduced in the proof of Proposition (3.2) one can, however, show that the vanishing order of the sections $\Theta_m^l(\tau, z)$ on the components of D' only depends on B'. Hence one can argue as in [We] and finds that the vanishing order along $\pi^* B$ goes to $\frac{prn}{12}$ as p goes to infinity. Setting $k = pr$ this gives (ii). □

We can now start giving the proof of Theorem (0.1). Let

$$H = aL - bD \qquad b > 0, \quad 12a - \frac{b}{n} > 0$$

be a divisor on $\mathcal{A}_g^*(n)$. In view of Proposition (1.4) it remains to consider curves C which are contained in the boundary. To simplify notation we write the decomposition of the boundary D as

$$D = \sum_{i=1}^{N} \overline{D}_{g-1}^i(n)$$

where $N = N(n, g)$ can be computed explicitly. Then

$$H|_{\overline{D}_{g-1}^1(n)} = \left(aL - b\sum_{i \neq 1} \overline{D}_{g-1}^i(n)\right)\Big|_{\overline{D}_{g-1}^1(n)} - b\overline{D}_{g-1}^1(n)|_{\overline{D}_{g-1}^1(n)}. \qquad (4)$$

Now let $g = 2$ or 3 where we have the fibration

$$\pi' : \overline{D}_{g-1}^1(n) \longrightarrow \mathcal{A}_{g-1}^*(n).$$

We shall denote the boundary of $\mathcal{A}_{g-1}^*(n)$ by B. Also note that the restriction of L to the boundary equals $\pi'^* L_{\mathcal{A}_{g-1}^*(n)}$ where we use the notation L for both the line bundle on $\mathcal{A}_g^*(n)$ and $\mathcal{A}_{g-1}^*(n)$. Thus we find that

$$H|_{\overline{D}_{g-1}^1(n)} = \pi'^*(aL - bB) - b\overline{D}_{g-1}^1(n)|_{\overline{D}_{g-1}^1(n)}. \qquad (5)$$

In view of the definition of the line bundle $\overline{M}(n)$ this gives

$$H|_{\overline{D}_{g-1}^1(n)} = \pi'^*\left(\left(a - \frac{b}{n}\right)L - bB\right) + \frac{b}{n}\overline{M}(n). \qquad (6)$$

Proof of Theorem (0.1) for $g = 2$. In this case the boundary components $\overline{D}_1^i(n)$ are isomorphic to Shioda's modular surface $S(n)$ and the projection π' is just projection to the modular curve $X(n)$. The degree of L on $X(1)$ is $\frac{1}{12}$ and we have one cusp.

Hence
$$\deg_{X(n)}(aL - bB) = \mu(n)\left(\frac{a}{12} - \frac{b}{n}\right)$$

where $\mu(n)$ is the degree of the Galois covering $X(n) \to X(1)$, i.e. $\mu(n) = |\mathrm{PSL}(2, \mathbb{Z}/n\mathbb{Z})|$. This is non-negative if and only if $a - 12\frac{b}{n} \geq 0$. The normal bundle of $\overline{D}_1^i(n)$ can also be computed explicitly. This can be done as follows: Using the degree 10 cusp form which vanishes on the reducible locus one finds the equality $10L = 2H_1 + D$ on $\mathcal{A}_2^*$ where H_1 is the Humbert surface parametrizing polarized abelian surfaces which are products. Hence we conclude for the canonical bundle on $\mathcal{A}_2^*(n)$ that $K = (3 - \frac{10}{n})L + \frac{2}{n}H_1$. The restriction of the divisor H_1 to a boundary component $\overline{D}_1^i(n) \cong S(n)$ is the sum of the n^2 sections L_{ij} of $S(n)$. The canonical bundle of the surfaces $S(n)$ is equal to the pull-back via π' of $3L$ minus the divisor of the cusps on the modular curve $X(n)$ (see also [BH]). Hence adjunction together with an easy calculation gives

$$-n\overline{D}_1^i(n)|_{\overline{D}_1^i(n)} = 2{\pi'}^*L_{X(n)} + 2\sum L_{ij}$$

Since $L_{ij}|_{L_{ij}} = -L_{X(n)}$ one sees immediately that this line bundle is nef and positive on the fibres of $\pi' : S(n) \to X(n)$. The result now follows directly from (5). $\square$

We shall now turn to the case $g = 3$. As we have remarked before it remains to consider curves which are contained in the boundary of $\mathcal{A}_3^*(n)$. Among those curves we shall first deal with curves whose image under the map π' meets the interior of $\mathcal{A}_2(n)$.

Proposition 4.2. *Let $H = aL - bD$ be a divisor on $\mathcal{A}_3^*(n)$ with $a - 12\frac{b}{n} > 0$, $b > 0$. For every curve C in a boundary component $\overline{D}_2(n)$ with $\pi'(C) \cap \mathcal{A}_2(n) \neq \emptyset$ the intersection number $H.C > 0$.*

Proof. We shall use (6) and Proposition (4.1). If we replace n by some multiple and consider the pull-back of H the coefficient b/n is not changed. The inverse image of C may have several components. All of these are, however, equivalent under some finite sympectic group and it is sufficient to prove that the degree of H is positive on one (and hence on every) component lying over C. After this reduction we can again assume that C is irreducible and by Proposition (4.1) we can find for every $\varepsilon > 0$ a divisor $\mathcal{C}$ not containing C with

$$\overline{M}(n) = \mathcal{C} + \frac{\lambda}{k}\pi^*B, \qquad \frac{\lambda}{k} \geq \frac{n}{12 + \varepsilon}.$$

By (6)
$$H|_{\overline{D}_2(n)} = \pi^*\left(\left(a - \frac{b}{n}\right)L - b\left(1 - \frac{\lambda}{nk}\right)B\right) + \frac{b}{n}\mathcal{C}.$$

The assertion follows from the corresponding result for $g = 2$ provided

$$\left(a - \frac{b}{n}\right) - 12\frac{b}{n}\left(1 - \frac{\lambda}{nk}\right) \geq \left(a - 12\frac{b}{n}\right) - \frac{b}{n}\left(1 - \frac{12}{12 + \varepsilon}\right) > 0.$$

Since $a - 12b/n > 0$ this is certainly the case for ε sufficiently small. □

We are now left with curves in the boundary of $\mathcal{A}_3^*(n)$ whose image under π' is contained in the boundary of $\mathcal{A}_2^*(n)$. These are exactly the curves which are contained in more than 1 boundary component of $\mathcal{A}_3^*(n)$. Before we conclude the proof, we have to analyze the situation once more. First of all we can assume by symmetry arguments that C is contained in $\overline{D}_2(n) = \overline{D}_2^1(n)$. Let B' be a component of the boundary B of $\mathcal{A}_2^*(n)$ which contains $\pi'(C)$. Let $D' = (\pi')^{-1}(B')$. Then D' consists of n irreducible components and we have the following commutative diagram ($n \geq 3$):

$$\begin{array}{ccc} \overline{D}_2^1(n) & \xrightarrow{\pi'} & \mathcal{A}_2^*(n) \\ \cup & & \cup \\ D' & \xrightarrow{\pi'} & B' \cong S(n) \\ & \searrow_{\pi} & \downarrow_{\pi''} \\ & & X(n). \end{array}$$

Altogether there are three possibilities:
(1) $\pi'(C) = pt$, i.e. C' is contained in a fibre of π'.
(2) $\pi(C) = pt, \pi'(C) \neq pt$. Then $\pi'(C)$ is either a smooth fibre of $S(n)$ or a component of a singular n–gon.
(3) $\pi(C) = X(n)$.
The final step in the proof of Theorem (0.1) is the following:

Proposition 4.3. *Let $C \subset \overline{D}_2(n)$ be a curve whose image $\pi'(C)$ is contained in the boundary of $\mathcal{A}_2^*(n)$. If $H = aL - bD$ is a divisor with $b > 0, a - 12\frac{b}{n} > 0$ then $H.C > 0$.*

Proof. By induction on g and formula (5) it is enough to prove that there is some $\overline{D}_2^j(n)$ with $C.\overline{D}_2^j(n) \leq 0$. Consider the inverse image D' of B' under π'. Then D' consists of n irreducible components each of which is of the form $\overline{D}_2^i(n) \cap \overline{D}_2^1(n)$ for some $i \neq 1$. We already know that $-B'|_{B'}$ is nef. Hence

$$\left(\sum_{i \in I} \overline{D}_2^i(n) \cap \overline{D}_2^1(n)\right).C \leq 0$$

where I is a suitable set of indices consisting of n elements. In particular $\overline{D}_2^j(n).C \leq 0$ for some index j. □

Remarks. (i) If $\pi'(C) = pt$, then one can give an alternative proof of $\overline{D}_2(n).C > 0$ by computing the normal bundle of $\overline{D}_2(n)$ restricted to the singular fibres of π'. The conormal bundle is ample as in the smooth case (cf. Lemma (2.2)).

(ii) If $\pi'(C) \neq pt$ one can also use the theta functions $\Theta_{m'm''}$ with $m', m'' \in \frac{1}{2}\mathbb{Z}^2$ to construct sections of $\overline{M}(n)$ which, after subtracting suitable components of the form $\overline{D}_2^i(n) \cap \overline{D}_2^1(n)$, do not vanish identically on C. In this way one can compute similarly to the proof of Proposition (4.2) that $H.C > 0$.

Proof of Theorem (0.1) ($g = 3$). This follows now immediately from Proposition (1.4), Proposition (4.2) and Proposition (4.3). □

Proof of the corollaries. These follow immediately from Theorem (0.1) since the moduli spaces are smooth and since

$$K \equiv (g+1)L - D.$$

Obviously

$$(g+1) - \frac{12}{n} \geq 0 \Leftrightarrow \begin{cases} n \geq 4 & \text{if } g = 2 \\ n \geq 3 & \text{if } g = 3. \end{cases}$$

Hence K is nef if $g = 2, n \geq 4$ and $g = 3, n \geq 3$, resp. numerically positive if $g = 2, n \geq 5$ and $g = 3, n \geq 4$. It follows from general results of classification theory that K is ample in the latter case. □

References

[A] V. Alexeev, Complete moduli of (co) abelian varieties, Preprint 1996.

[Bo] L. Borisov, A finiteness theorem for Sp(4, $\mathbb{Z}$), Preprint 1995.

[BH] W. Barth, K. Hulek, Projective models on Shioda modular surfaces, Manuscripta Math. 50 (1985), 73–132.

[C] H. Clemens, Double solids, Adv. Math. 47 (1983), 107–230.

[DO] I. Dolgachev, D. Ortland, Point sets in projective spaces and theta functions. Astérisque 165 (1988).

[D] R. Donagi, The unirationality of $\mathcal{A}_5$, Ann. of Math. 119 (1984), 269–307.

[Fa] C. Faber, Chow rings of moduli spaces of curves I: The Chow ring of $\overline{\mathcal{M}}_3$, Ann. of Math. 132 (1990), 331–419.

[Fi] H. Finkelnberg, On the geometry of the Burkhardt quartic, Thesis. Leiden (1989).

[F] E. Freitag, Siegelsche Modulfunktionen, Springer-Verlag, 1983.

[HKW]   K. Hulek, C. Kahn, S. Weintraub, Moduli spaces of abelian surfaces: Compactification, Degenerations and Theta Functions, de Gruyter Exp. Math. 12, Walter de Gruyter, Berlin–New York 1993.

[I1]   J. I. Igusa, Theta functions, Springer-Verlag, 1972.

[I2]   J. I. Igusa, Arithmetic theory of moduli for genus 2, Ann. of Math. (2) 72 (1960), 612–649.

[Ka]   P. Katsylo, Rationality of the moduli variety of curves of genus 3, Comment. Math. Helvetici 71 (1996), 507–524.

[MM]   S. Mori, S. Mukai, The uniruledness of the moduli space of curves of genus 11, in: Algebraic Geometry (Tokyo/Kyoto 1982), 334–353. Lecture Notes in Math. 1016, Springer-Verlag, 1983.

[Mu1]   D. Mumford, On the Kodaira dimension of the Siegel modular variety, in: Algebraic geometry – open problems, Proceedings, Ravello 1982, SLN 997, 348–375.

[Mu2]   D. Mumford, Towards an enumerative geometry on the moduli space of curves, in: Arithmethic and Geometry, vol. II, Progr. Math. 36, Birkhäuser, 1983, 271–328.

[Nak]   I. Nakamura, On moduli of stable quasi abelian varieties, Nagoya Math. J. 58 (1975), 149–214.

[Nam]   Y. Namikawa, A new compactification of the Siegel space and degeneration of abelian varieties I, II, Math. Ann. 221 (1977), 97–141 and 201–241.

[T]   Y.-S. Tai, On the Kodaira dimension of the moduli space of abelian varieties. Invent. Math. 68 (1982), 425–439.

[Ts]   R. Tsushima, A formula for the dimension of spaces of Siegel modular cusp forms, Amer. J. Math. 102 (1980), 937–977.

[vdG1]   G. van der Geer, On the geometry of a Siegel modular threefold, Math. Ann. 260 (1982), 317–350.

[vdG2]   G. van der Geer, The Chow ring of the moduli space of abelian threefolds, J. Algebraic Geom. 7 (1998), 753–770.

[vG]   B. van Geemen, The moduli space of curves of genus 3 with level 2 structure is rational, Preprint.

[V]   A. Verra, A short proof of the unirationality of $\mathcal{A}_5$, Indagationes Math. 46 (1984), 339–355.

[We]   R. Weissauer, Untervarietäten der Siegelschen Modulmannigfaltigkeiten von allgemeinem Typ, Math. Ann. 275 (1986), 207–220.

[Ya]   T. Yamazaki, On Siegel modular forms of degree two. Amer. J. Math. 98 (1976), 39–53.

Abelian surfaces with two plane cubic curve fibrations and Calabi–Yau threefolds

*Klaus Hulek and Kristian Ranestad**

1. Introduction 275
2. Numerical possibilities 277
3. Elliptic scrolls and degenerate Calabi–Yau 3-folds in $\mathbf{P}^5$ 279
4. Non-normal Del Pezzo 3-folds in $\mathbf{P}^5$ 284
5. Non-normal Calabi–Yau 3-folds in $\mathbf{P}^5$ 295
6. Equations of elliptic scrolls in $\mathbf{P}^5$ 298
7. Heisenberg symmetry of elliptic scrolls in $\mathbf{P}^5$ 301
8. Conclusion 313

1. Introduction

Abelian surfaces of small degree are contained in nodal Calabi–Yau 3-folds, similarly many Calabi Yau 3 folds of small degree specialize to nodal Calabi–Yau 3-folds with abelian surfaces on them. The first assertion is intimately connected with the fact that the moduli space of abelian surfaces of small degree is uniruled: An abelian surface on a Calabi–Yau 3-fold moves in a linear pencil, and therefore gives rise to a $\mathbf{P}^1$ in the moduli space of abelian surfaces. This idea was taken up and explored by Gross and Popescu [GP1, GP2] starting with a very singular Calabi–Yau variety, the secant variety of an elliptic normal curve. The translation scrolls inside the secant variety are degenerate abelian surfaces and form a $\mathbf{P}^1$ on the boundary of the moduli of abelian surfaces. They show that the secant variety deforms to nodal Calabi–Yau 3-folds with only isolated singularities and with a pencil of abelian surfaces as long as the degree of the elliptic curve is less than 11. This limit is related to the Del Pezzo bound for the possible smoothing of minimal elliptic surface singularities.

---

* Both authors were partially supported by the HCM contract AGE (Algebraic Geometry in Europe), no ERBCHRXCT940557. The first author would also like to thank MSRI for its hospitality.

We explore a similar setting. In a $\mathbf{P}^2$-scroll over an elliptic curve, any anticanonical divisor, if there is one, is a, possibly degenerate, abelian surface. If one can glue two $\mathbf{P}^2$-scrolls over elliptic curves along an anticanonical divisor, the union is a singular Calabi–Yau 3-fold. Furthermore, if the anticanonical divisor moves in a pencil on at least one of the two scrolls, then we are in a position like above.

We start in Section 2 by asking for smooth abelian surfaces with two pencils of plane cubic curves on it. The two pencils would then define $\mathbf{P}^2$-scrolls whose union is Calabi–Yau. It turns out that purely numerical considerations bound the degree of these abelian surfaces by 18. This bound is obtained by abelian surfaces which form the complete intersection $((0, 3), (3, 0))$ in the $\mathbf{P}^2 \times \mathbf{P}^2$ with its Segre embedding. For each even degree $10 \le d \le 18$ there are numerical possibilities which are realized. In this paper we study the associated elliptic scrolls and Calabi–Yau 3-folds in the case $d = 12$, i.e. the case of abelian surfaces embedded linearly normally in $\mathbf{P}^5$.

In Section 3 we find and describe the abelian surfaces of degree 12 and the two scrolls defined by their pencils of plane cubic curves. The union of the two scrolls is a non-normal Calabi–Yau 3-fold of degree 12.

In Sections 4 and 5 a separate approach leads to constructions via projected Del Pezzo 3-folds of non-normal Calabi–Yau 3-folds in degrees 10,11, 12 and 13. The projected Del Pezzo 3-folds are bilinked to the non-normal Calabi–Yau 3-folds. In the last three sections we prepare the argument that the reducible Calabi–Yau 3-folds described in Section 3 may also be obtained via bilinkage from projected Del Pezzo 3-folds.

After finding equations for elliptic scrolls in Section 6, we devote Section 7 to the Heisenberg symmetry of elliptic scrolls and provide a description of the family of Calabi–Yau 3-folds that are unions of two scrolls.

More precisely, let H_6 be the Heisenberg group of level 6 and let N_6 be its normalizer in $GL(6, \mathbf{C})$. In N_6 there is a natural involution ι which restricts to the abelian surfaces as multiplication by -1. Let $G_6 = \langle H_6, \iota \rangle$. Then the space of cubics in $\mathbf{P}^5$ contains a 4-dimensional vector space of G_6-invariant pencils.

We let $\mathbf{P} = \mathbf{P}^3$ be the parameter space for these G_6-invariant pencils of cubics. H_6 contains a subgroup isomorphic to H_2 and four subgroups $H(K_1), \ldots, H(K_4)$ of index 3 containing this subgroup. For every subgroup $H(K_i)$ there is a set of three lines in $\mathbf{P}^5$ containing $H(K_i)$ in its stabilizer and left invariant by the action of H_6. Similarly there exists for each subgroup $H(K_i)$ a line l_i in $\mathbf{P}$ parametrizing pencils of cubics which contain the corresponding three lines.

A general point on any of these four lines l_i in $\mathbf{P}$ corresponds to a pencil of cubics which defines an elliptic scroll singular along the corresponding three lines in $\mathbf{P}^5$. The scroll is residual to three $\mathbf{P}^3$'s in the complete intersection of the two cubics.

Between each pair of lines l_i, l_j, there is a 1 : 1 correspondence defined by the pairs of points which correspond to elliptic scrolls that intersect along a possibly degenerate abelian surface. The lines spanned by corresponding points form a conic section in the Grassmannian of lines in $\mathbf{P}$. Altogether there are 6 disjoint conic sections in the

Grassmannian of lines in **P** which parametrize abelian surfaces with two plane cubic curve fibrations.

In the final section we state and prove the main theorem of the paper: Let X_E and X_F be elliptic scrolls in $\mathbf{P}^5$ that intersect precisely along a $(1, 6)$-polarized abelian surface.

Theorem 8.3. *The reducible 3-fold $Y = X_E \cup X_F$ is a degeneration of irreducible non-normal Calabi–Yau 3-folds of degree 12 in $\mathbf{P}^5$. The general such 3-fold is singular precisely along 6 disjoint lines.*

Chang has described smooth 3-folds of degree 12 that are birational to Calabi–Yau 3-folds (cf. [Ch], [DP]). Similar to the ones described in this paper they are bilinked to Fano 3-folds of degree 7, but they differ by their sectional genus. This difference is manifested in the appearance of non-normal singularities. The Calabi–Yau 3-folds of Theorem 8.3 do not deform to smooth ones: The double point formulas for 3-folds in $\mathbf{P}^5$ give the class of the non-normal singular locus in terms of the coefficients of the Hilbert polynomial, so any deformation also has non-normal singularities.

We find three open problems related to the topics of this paper particularly interesting:

Problem 1.1. Find H_6-invariant non-normal but irreducible Calabi–Yau 3-folds, with a pencil of $(1, 6)$-polarized abelian surfaces, degenerating to the union of two elliptic scrolls.

Problem 1.2. Consider the normalization of the Calabi–Yau 3-folds of degrees $10, \ldots, 13$ of Section 5. Find the invariants, the Betti and Hodge numbers and and describe the Kähler cone of these Calabi–Yau 3-folds.

Problem 1.3. Describe the elliptic scrolls and the reducible Calabi–Yau 3-folds in the cases $d = 14, 16, 18$.

We work over the complex numbers.

2. Numerical possibilities

Assume that an abelian surface $A \subset \mathbf{P}^n$ has two fibrations

$$p : A \to E \quad \text{and} \quad q : A \to F$$

in plane cubic curves. These fibrations define two $\mathbf{P}^2$-scrolls. If the fibers of the maps p and q have intersection number ≥ 2, then the planes in the two fibrations intersect in at least a line and the two $\mathbf{P}^2$-scrolls coincide. So for our purposes we can assume that F and E are sections of p and q with $E \cdot F = 1$. In particular $A = E \times F$.

When there is no isogeny between E and F then A is the product of $E \times F$ in the Segre embedding of $\mathbf{P}^2 \times \mathbf{P}^2$ in $\mathbf{P}^8$. We shall now assume that E is general in the sense that E has $\operatorname{End} E \cong \mathbf{Z}$ and that $\gamma : E \to F$ is a primitive isogeny of degree l. Then $NS(A)$ is generated by E, F and Γ where Γ is the graph of γ. The numerical equivalence of a hyperplane divisor on A may therefore be expressed as

$$H \equiv aE + bF + c\Gamma, \qquad a, b, c \subset \mathbf{Z}.$$

Notice that these surfaces and these divisors really exist for any general elliptic curve E. We investigate for which numerical data they give us two plane cubic curve fibrations. The intersection numbers are given by the table:

	E	F	Γ
E	0	1	l
F	1	0	1
Γ	l	1	0

With the requirements

$$H \cdot E = b + lc = 3, \quad H \cdot F = a + c = 3 \quad \text{and} \quad H \cdot \Gamma = al + b \geq 2,$$

we get

$$d/2 := H^2/2 = 9 - lc^2.$$

This means that $d/2 \leq 9$ and that there are the following possibilities for H:

Table 1.

d	l	c	a	b	$H \cdot \Gamma$
10	1	2	1	1	2
10	4	1	2	-1	7
12	3	1	2	0	6
14	2	1	2	1	5
16	1	1	2	2	4
18	*	0	3	3	*

The $*$ in Table 1 means that there is no isogeny between E and F involved.

Proposition 2.1. *There exist abelian surfaces with two plane curve fibrations of degree d in $\mathbf{P}^{\frac{d}{2}-1}$, when $d = 10, 12, 14, 16, 18$.*

Proof. To give examples it is now enough to check that H is very ample. For this we use Reider's criterion [Re], which in these cases reduces to check that there are no elliptic curves C on A with $H \cdot C \leq 2$. Any such curve C, not equivalent to E, F or Γ, must intersect each of these strictly positively, i.e.

$$C \cdot E > 0, \quad C \cdot F > 0, \quad \text{and} \quad C \cdot \Gamma > 0.$$

But in each case except the second $H = aE + bF + c\Gamma$ with $a + b + c \geq 3$ so $H \cdot C \geq 3$. In the second case $H = 2E - F + \Gamma$, so if $C = \alpha E + \beta F + \gamma \Gamma$, then $H \cdot C < 3$ implies that
$$3\alpha + 3\beta + 7\gamma \leq 2.$$

The other inequalities above yield
$$\beta + 4\gamma \geq 1, \quad \alpha + \gamma \geq 1 \quad \text{and} \quad 4\alpha + \beta \geq 1,$$
while $C^2 = 0$, since C is elliptic, yields
$$\alpha\beta + 4\alpha\gamma + \beta\gamma = 0.$$

From the first four inequalities we get $\beta \leq -(\alpha + \gamma)$ which combines with the last equality to yield
$$0 = 4\alpha\gamma + \beta(\alpha + \gamma) \leq 4\alpha\gamma - (\alpha + \gamma)^2 = -(\alpha - \gamma)^2.$$

This only occurs when $\alpha = \gamma$, i.e. from the relation $C^2 = 0$, when $\alpha = 0$ or $2\alpha + \beta = 0$. The former is impossible since $\alpha + \gamma > 0$, while the latter is impossible since then $C \cdot H = 10\alpha - 6\alpha = 4\alpha \geq 4$.

In the first case $H \cdot \Gamma = 2$, so H is not very ample. In fact $|H|$ maps A two to one to a quintic elliptic scroll in $\mathbf{P}^4$. In this case $E = F$ and the scroll is the symmetric product of E. Thus the two fibrations coincide in the image. In each of the other cases $|H|$ defines an embedding. □

Remark 2.2. When $d = 18$ there is the simple example of
$$E \times F = E \times \mathbf{P}^2 \cap \mathbf{P}^2 \times F \subset \mathbf{P}^2 \times \mathbf{P}^2$$
in its Segre embedding in $\mathbf{P}^8$. Clearly the union of the two scrolls deform in this case to Calabi–Yau 3-folds.

In this paper we shall concentrate on the case $d = 12$.

3. Elliptic scrolls and degenerate Calabi–Yau 3-folds in $\mathbf{P}^5$

From now on we consider abelian surfaces $A \subset \mathbf{P}^5$ of degree 12, i.e. with a $(1, 6)$-polarization, and with two fibrations
$$p : A \to E \quad \text{and} \quad q : A \to F$$
in plane cubic curves. Furthermore we assume that E is general. We denote by X_E the scroll of planes of the fibration p, and by X_F the scroll of planes of the fibration q. The corresponding $\mathbf{P}^2$-bundles are denoted V_E and V_F, respectively.

In the notation of the previous section and Table 1 there is an isogeny

$$\gamma : E \to F$$

of degree 3. Furthermore, if Γ is the graph of the isogeny, then the hyperplane divisor is

$$H = 2E + \Gamma.$$

We fix an origin $o \in E$ and let $s_1 = \gamma(t_1)$, for some 3-torsion point t_1 on E, not in the kernel of γ. Let $h : F \to F$ be translation by s_1, and let $\gamma^* : \text{Pic}^0 F \to \text{Pic}^0 E$ be the isogeny dual to γ. If $\mathcal{O}_F(o)$ is the line bundle of degree 1 whose unique section vanishes at o, then $h^*(\mathcal{O}_F(o)) \otimes \mathcal{O}_F(-o)$ generates the kernel of γ^*, i.e.

$$\ker \gamma^* = \{\mathcal{O}_F, h^*(\mathcal{O}_F(o)) \otimes \mathcal{O}_F(-o), (h^2)^*(\mathcal{O}_F(o)) \otimes \mathcal{O}_F(-o)\} \quad (h^3 = \text{id}).$$

Proposition 3.1. *The rank 3 vector bundle associated to the $\mathbf{P}^2$-bundle V_F decomposes into $\mathcal{E} = \mathcal{L}_0 \oplus h^* \mathcal{L}_0 \oplus (h^2)^* \mathcal{L}_0$, where $\mathcal{L}_0$ is a line bundle of degree 2 on F and h is as above. Furthermore the scroll X_F is singular precisely along three lines, which span $\mathbf{P}^5$.*

Proof. First, notice that Γ is contained in a pencil of hyperplanes defined by the pencil $|H - \Gamma| = |2E|$ on A. This pencil is the pullback by q of a divisor Δ_0 of degree 2 on F. Since any divisor in this pencil is a pair of plane cubic curves contained in a hyperplane, their planes intersect in a point. Thus the linear system $|\Delta_0|$ defines a morphism of F of degree 2 into the double locus of X_F.

Next we consider the translates Γ_t of Γ on A by a point on t on E and find which Γ_t are contained in a pencil of hyperplanes. This happens precisely when $H - \Gamma_t$ is the pullback of a divisor of degree 2 from F, or equivalently when the restriction of $\Gamma - \Gamma_t$ to a fiber $E_f = q^{-1}(f)$ of q is trivial. But the translate Γ_t can be represented as

$$\{(x + t, \gamma(x)) | x, t \in E\}$$

so the intersection

$$(\Gamma - \Gamma_t) \cap E_f = \gamma^{-1}(f) - \gamma^{-1}(f) + 3t$$

which is trivial precisely when $3t = o$. Since γ is already an isogeny of degree 3, the 3-torsion points in the kernel of γ leave Γ invariant under translation, so in fact we may choose $t = t_1$ as above and we get 3 distinct translates of Γ which are contained in a pencil of hyperplanes in $\mathbf{P}^5$. We denote them by $\Gamma_0 (= \Gamma), \Gamma_1, \Gamma_2$. They each span a $\mathbf{P}^3$ and determine a linear system of degree 2 on F, which we denote by $|\Delta_0|, |\Delta_1|, |\Delta_2|$. Each linear system $|\Delta_i|$ defines a morphism of degree 2 of F to the double locus of X_F. The images of these three maps are disjoint lines, and clearly the planes of X_F are spanned by the respective images by these three maps. Therefore V_F is defined by a decomposable rank 3 bundle $\mathcal{E}$ of degree 6. If we denote the line bundle associated to the divisor Δ_i by $\mathcal{L}_i$, then

$$\mathcal{E} = \mathcal{L}_0 \oplus \mathcal{L}_1 \oplus \mathcal{L}_2.$$

To check the differences between the line bundles $\mathcal{L}_i$, we consider the intersection

$$(\Gamma - \Gamma_t) \cap F_e = \gamma(e) - \gamma(e-t) = \gamma(t).$$

Since $3t_1 = o$, translation by $s_1 = \gamma(t_1)$ on F is a 3-torsion element that generates the kernel of the dual isogeny, $\gamma^* : \text{Pic}^0 F \to \text{Pic}^0 E$, i.e.

$$\mathcal{L}_i = (h^i)^* \mathcal{L}_0,$$

where h is the translation on F by s_1.

It remains only to check the singularities of X_F. Any singular point of X_F is the intersection of two, possibly infinitely close, planes of X_F. But two planes intersect only if they span at most a hyperplane. The two planes are defined by a section of a line bundle $q^*\mathcal{L}$ for a line bundle $\mathcal{L}$ of degree 2 on F. They intersect precisely when

$$\mathcal{E} \otimes \mathcal{L}^{-1}$$

has a section. But this is the case precisely when $\mathcal{L} = \mathcal{L}_i$ for $i \in \{0, 1, 2\}$. Furthermore, for each of these three cases $\mathcal{E} \otimes \mathcal{L}_i^{-1}$ has precisely one section, so the corresponding planes span a hyperplane and the intersection of the two planes is only a point. $\square$

The abelian surface A is an anticanonical divisor on V_F. We compute the sections of $-K_{V_F}$.

Lemma 3.2. $h^0(\mathcal{O}_{V_F}(-K_{V_F})) = 4$.

Proof. The natural isomorphism

$$H^0(\mathcal{O}_{V_F}(-K_{V_F})) \cong H^0(F, \text{Sym}^3 \mathcal{E} \otimes \mathcal{L}_0^{-1} \otimes \mathcal{L}_1^{-1} \otimes \mathcal{L}_2^{-1}),$$

reduces the computation to counting trivial summands of the rank 10 vector bundle

$$\text{Sym}^3 \mathcal{E} \otimes \mathcal{L}_0^{-1} \otimes \mathcal{L}_1^{-1} \otimes \mathcal{L}_2^{-1}.$$

Since $\mathcal{L}_0^{-1} \otimes \mathcal{L}_1^{-1} \otimes \mathcal{L}_2^{-1} = \mathcal{L}_0^{-3}$ this count is the number of summands in

$$\text{Sym}^3 \mathcal{E} = \text{Sym}^3(\mathcal{L}_0 \oplus h^*\mathcal{L}_0 \oplus (h^2)^*\mathcal{L}_0)$$

which equal $\mathcal{L}_0^3$. As $h^3 = \text{id}$ this number is 4. $\square$

Lemma 3.3. *The $\mathbf{P}^2$-bundle V_F is the quotient of a trivial bundle $\mathbf{P}^2 \times E$ by a cyclic group of order 3.*

Proof. Consider again the isogeny $\gamma : E \to F$. Since $h^*(\mathcal{O}_F(o))\otimes \mathcal{O}_F(-o)$ generates the kernel of γ^* the isogeny dual to γ,

$$\gamma^*\mathcal{L}_0 \cong \gamma^*h^*\mathcal{L}_0 \cong \gamma^*(h^2)^*\mathcal{L}_0.$$

Therefore the pullback of $V_F = \mathbf{P}(\mathcal{E})$ over F via γ trivializes the bundle. The kernel of γ is a cyclic group of order three which acts on the pullback $\gamma^*\mathcal{E}$. □

Thus we may construct V_F by starting with $\mathbf{P}^2 \times E$, and dividing by a suitable diagonal action of the cyclic group of order 3. For this we consider a vector space $V = \langle x_0, x_1, x_2 \rangle$ with the action

$$\tau : x_i \mapsto \epsilon^i x_i, \quad i \in \mathbf{Z}_3.$$

As above, let $o \in E$ be the origin of E, and consider the linear system $|3o|$ on E. It embeds E as a plane cubic curve in $\mathbf{P}^2$. We may choose $\langle e_0, e_1, e_2 \rangle$ as a basis for the underlying vector space of $\mathbf{P}^2$ such that

$$\tau : e_i \mapsto \epsilon^{-i} e_i$$

induces the action of translation by a 3-torsion point t_0 on E. The diagonal action defined by $\tau \in \mathbf{Z}_3$ on $V \times E$:

$$\tau : v \times e \mapsto \tau(v) \times e + t_0$$

acts without fixed points, so the quotient is a rank 3 vector bundle on $E/\langle t_0 \rangle = F$. The action of τ on V decomposes into the characters

$$V = \langle x_0 \rangle \oplus \langle x_1 \rangle \oplus \langle x_2 \rangle.$$

The anticanonical divisors on V_F pull back to anticanonical divisors on $\mathbf{P}^2 \times E$ which are invariant under the action of τ. But the anticanonical divisors on $\mathbf{P}^2 \times E$ are just the pullbacks of the cubic curves on the plane. The action of τ on the plane has the following basis of invariant cubics:

$$\langle x_0^3, x_1^3, x_2^3, x_0 x_1 x_2 \rangle.$$

Since these have no basepoints, there are no basepoints for the system of anticanonical divisors on V_F, and the general one is smooth. Notice furthermore that this linear system of invariant cubics contains the Hesse pencil

$$\langle x_0^3 + x_1^3 + x_2^3, x_0 x_1 x_2 \rangle,$$

and recall that the singular curves in this pencil are 4 triangles. In fact it is easy to check that these four triangles are the only triangles in the linear system of invariant cubics. The vertices of the triangle $x_0 x_1 x_2 = 0$ are mapped to the singular lines of X_F. The vertices of the three other triangles sweep out elliptic normal curves of degree 6 as we shall see next.

Proposition 3.4. *The scroll X_F is the 3-torsion translation scroll of an elliptic normal curve in $\mathbf{P}^5$.*

Proof. Consider an elliptic normal curve C of degree 6 in $\mathbf{P}^5$, embedded by the linear system $|6o|$. For any $P \in C$ consider the translation scroll
$$V_P = \cup_{y \in C} \langle y, y + P, y + 2P \rangle.$$
This is, for general P, a $\mathbf{P}^2$-scroll of degree 18. When $3P = o$ the points y, $y + P$ and $y + 2P$ generate the same plane, so then the translation scroll has degree 6. In this case
$$\langle y, y + P, y + 2P \rangle \cap \langle z, z + P, z + 2P \rangle \neq \emptyset$$
precisely when $3y + 3z = o$. But then the pencil $|y + z|$ defines a map from C to the double locus of V_P. Now $3(y + z) = 3(y + z + P) = 3(y + z + 2P)$, so this map factors through the isogeny $C \to C/\langle P \rangle$. Thus the 9 linear systems $|y + z|$ with $3(y + z) = 0$ define three pencils of pairs of planes each defining a double line for the translation scroll. The translation scroll is clearly a scroll over $C/\langle P \rangle$ and its associated vector bundle decomposes into the sum of three line bundles of degree 2 which define the three double lines. The differences between any two of these line bundles are the powers of some 3-torsion line bundle. If $C \cong E$ and $C/\langle P \rangle = F$, we have $V_P \cong X_F$. □

Finally, we consider the union of the two elliptic scrolls $X_1 = X_E$ and $X_2 = X_F$ in $\mathbf{P}^5$ which both contain the $(1, 6)$-polarized abelian surface A. Let $V_1 = V_E$ and $V_2 = V_F$ be the corresponding $\mathbf{P}^2$-bundles. Then V_1 and V_2 are the normalizations of X_1 and X_2 and $A \in |-K_{V_1}|$, resp. $A \in |-K_{V_2}|$. Let $Y = X_1 \cup X_2$.

Proposition 3.5. *Y has a partial desingularization $Y_0 = V_1 \cup V_2$, which is a Calabi–Yau 3-fold, i.e. $K_{Y_0} = \mathcal{O}_{Y_0}$ and $q = h^1(\mathcal{O}_{Y_0}) = 0$. In particular $V_1 \cap V_2 = A$.*

Proof. First, we may use the previous notation and let $A = E \times F$ with hyperplane divisor $H = 2E + \Gamma$.

Notice that the abelian surface A does not intersect the singular lines of X_F. In fact $|H - E_f|$, for a plane cubic curve E_f passing through a singular point, would then have a basepoint. This is impossible by Reider's criterion [Re].

We show next that X_1 and X_2 intersect transversally along A: Near A the 3-fold Y is the intersection of two smooth 3-folds. If the intersection is not transversal the tangent spaces of the two scrolls at some point coincide. This is possible only if the planes of the two scrolls at the given point intersect along a line. Those two planes intersect A in elliptic cubic curves E_f and F_e which meet in a point. Any curve in $|H - E_f - F_e|$ has degree 6 and arithmetic genus 2. Since there is a pencil of hyperplanes through the two planes, this curve moves in a pencil on A. This is impossible by Riemann Roch, so transversality follows. If $X_1 \cap X_2$ contains a point disjoint from A, then a plane in X_1 meets a plane in X_2 along a line and the argument above applies again. Thus X_1 and X_2 meet transversally along A.

Now, look at the partial desingularization Y_0 of Y obtained by normalizing X_1 and X_2. Thus we may write $Y_0 = V_1 \cup V_2$ and $V_1 \cap V_2 = A$. To show that $q(Y_0) = 0$ we consider $\text{Pic}^0 Y_0$. An element in $\text{Pic}^0 Y_0$ is a pair $(\mathcal{M}_1, \mathcal{M}_2)$ with $\mathcal{M}_i \in \text{Pic}^0 V_i$ which glue along A. Consider the fibrations

$$p_1 : V_1 \to F \quad \text{and} \quad p_2 : V_2 \to E.$$

Now, $\mathcal{M}_i = p_i^*(\mathcal{N}_i)$, where $\mathcal{N}_1 \in \text{Pic}^0(F)$ resp. $\mathcal{N}_2 \in \text{Pic}^0(E)$. On A we have sections of each p_i which are fibres of the opposite map. We identify these sections with F and E respectively. Then $\mathcal{M}_1|_E = \mathcal{O}_E$ and $\mathcal{M}_2|_F = \mathcal{O}_F$. So if we want to glue $\mathcal{M}_1$ and $\mathcal{M}_2$ we must have that

$$\mathcal{O}_E = \mathcal{M}_1|_E = \mathcal{N}_2$$

and

$$\mathcal{O}_F = \mathcal{M}_2|_F = \mathcal{N}_1,$$

and hence $(\mathcal{M}_1, \mathcal{M}_2) = (\mathcal{O}_{X_1}, \mathcal{O}_{X_2})$. This shows that $\text{Pic}^0 Y_0 = \{\mathcal{O}_{Y_0}\}$. Since $\text{Pic}^0 Y_0$ is a reduced group scheme in characteristic 0 and since $\text{H}^1(Y_0, \mathcal{O}_{Y_0})$ is the tangent space at the origin, it follows that $q(Y_0) = \text{h}^1(Y_0, \mathcal{O}_{Y_0}) = 0$.

Since $A \in |-K_{V_1}|$, resp. $A \in |-K_{V_2}|$ it is clear that the restriction of K_{Y_0} to V_1 and to V_2 is trivial. But $\text{Pic}^0 Y_0 = \{\mathcal{O}_{Y_0}\}$, so $K_{Y_0} = \mathcal{O}_{Y_0}$. □

The 3-fold $Y = X_E \cup X_F$ is non-normal, singular along six lines three on each scroll in addition to the surface $A = X_E \cap X_F$. In the next section we describe a series of non-normal Del Pezzo 3-folds. In Section 5 we show that these are bilinked to non-normal Calabi–Yau 3-folds. After some further analysis of the equations of Y in Sections 6 and 7 we show in Section 8 that Y is a degeneration of these non-normal Calabi–Yau 3-folds.

4. Non-normal Del Pezzo 3-folds in $\mathbf{P}^5$

Del Pezzo 3-folds are 3-folds W for which $K_W \cong -2H$, where H is ample. Accordingly any smooth surface on W in $|H|$ is a Del Pezzo surface.

Let $V_t \subset \mathbf{P}^{t+1}$ with $t = 3, \ldots, 8$ be the image of $\mathbf{P}^3$ by the map defined by all quadrics through $8 - t$ general points in $\mathbf{P}^3$. Then V_t is a Del Pezzo 3-fold. We describe the image $W_t \subset \mathbf{P}^5$ of the general projection of V_t.

In particular we want to describe the singular non-normal locus. Thus we are interested in the cases $t = 5, 6, 7, 8$ and will prove

Theorem 4.1. W_t is non-normal along $\binom{t-3}{2}$ skew lines and has $\binom{8-t}{2}$ additional ordinary double points when $t = 5, 6, 7, 8$.

To prove this we will use a result of Reye on linear systems of quadrics. To explain Reye's result we need the notion of apolarity applied to quadrics in $\mathbf{P}^3$. Thus let $S = k[x_0, \ldots, x_3]$ and $T = k[y_0, \ldots, y_3]$, and define a pairing $S_2 \times T_2 \to k$ by letting S operate as differential operators on T and vice versa.

We say that quadrics in S and T are apolar if they are orthogonal with respect to this pairing. In our situation we think of S as the coordinate ring of $\mathbf{P}^3$ and T as the coordinate ring of $\check{\mathbf{P}}^3$. Starting with a 6-dimensional subspace of quadrics $V \subset S_2$, there is a 4-dimensional subspace i.e. a web of quadrics $V^\perp \subset T_2$.

Since any quadric in V is apolar to any member of $V^\perp$, we say that V and $V^\perp$ are apolar sets of quadrics. Consider the discriminant D of the space $V^\perp$ of quadrics. This is a quartic surface defined by the determinant of a symmetric 4×4 matrix with linear entries. The quadrics in $V^\perp$ of rank 1 and 2 are respectively triple and double points on D. The possible numbers of rank 1 and rank 2 quadrics are given in the following table:

Table 2.

rank 1 quadrics	rank 2 quadrics
4	$0/\infty$
3	1
2	3
1	6
0	10

This follows from a few lemmas which have independent interest for us.

Lemma 4.2. *Each rank 1 quadric in $V^\perp$ determines a basepoint for the quadrics in V and vice versa. Each rank 2 quadric in $V^\perp$ determines a line contained in 4 quadrics in V and vice versa. Alternatively, if ρ is the map defined by the linear system V of quadrics, then each rank 2 quadric in $V^\perp$ determines a line in the source double point locus of ρ and vice versa.*

Proof. Note that if $a = (a_0, a_1, a_2, a_3)$, $L = \sum_{i=0}^{3} a_i y_i$ and $q \in S_2$, then

$$L^2(q) = 2q(a).$$

Now each rank 1 quadric in $V^\perp$ has the form L^2 for some point a, and apolarity says that $L^2(q) = 2q(a) = 0$ for every quadric q in V so a is a basepoint for V. Conversely, if $a = (a_0, a_1, a_2, a_3)$ is a basepoint for V and $L = \sum_{i=0}^{3} a_i y_i$, then $L^2(q) = 2q(a) = 0$ for every $q \in V$ so $L^2 \in V^\perp$.

Each rank 2 quadric in $V^\perp$ has the form $L_1^2 + L_2^2$ for some linear forms $L_1 = \sum_{i=0}^{3} a_i y_i$ and $L_2 = \sum_{i=0}^{3} b_i y_i$. Let $l \subset \mathbf{P}^3$ be the line spanned by the points $a = (a_0, a_1, a_2, a_3)$ and $b = (b_0, b_1, b_2, b_3)$. Let $V_l \subset V$ be the subspace of quadrics which vanish on l. Then

$$V_l = \{q \in V | L_1^2(q) = 2q(a) = L_2^2(q) = 2q(b) = L_1 L_2(q) = 0\}.$$

But $(L_1^2 + L_2^2)(q) = 0$ for every $q \in V$, so V_l has codimension 2 in V. Conversely if V_l has codimension 2 in V, then some linear combination of L_1^2, L_2^2 and $L_1 L_2$ is contained in $V^\perp$. But any such linear combination is a rank 2 (or rank 1) quadric, and the lemma follows. □

Porteous' formula [cf. Fu 14.4.11] computes the number of rank 2 quadrics in a general web of rank 4 quadrics. This number is 10. Reye found a geometric interpretation of these 10 rank 2 quadrics considering apolar twisted cubic curves to the web of quadrics, i.e. twisted cubic curves whose defining net of quadrics is apolar to $V^\perp$.

By a determinental net of quadrics we mean a net (i.e. a 3-dimensional space) of quadrics which is generated by the 2×2 minors of a 2×3 matrix of linear forms. The general determinental net of quadrics generates the ideal of a twisted cubic curve.

Lemma 4.3 (Reye). *The general 6-dimensional subspace $V \subset S_2$ contains precisely two determinental nets of quadrics, which together span V. If C_1 and C_2 are the twisted cubic curves defined by these two nets, then every rank 1 quadric in $V^\perp$ corresponds to a point of intersection between C_1 and C_2 and vice versa. Furthermore every rank 2 quadric in $V^\perp$ corresponds to a common secant line for C_1 and C_2 and vice versa.*

Proof. The number of determinental nets of quadrics in a general 6-dimensional space V of quadrics is nowadays computable by quantum cohomology [Kre] (compute the number of twisted cubic surface scrolls through nine points in $\mathbf{P}^4$ in the quantum cohomology of the Grassmannian of lines and intersect with a general $\mathbf{P}^3$), a few years ago by modern intersection theory [ES] and in the ancient times by direct geometric arguments [Rey]. We leave the choice of reference to the reader.

Given two determinental nets which span V, the correspondence between points of intersection and the rank 1 quadrics follows from Lemma 4.2.

For the second correspondence consider first a common secant line to C_1 and C_2. This line and any one of the two curves form a complete intersection $(2, 2)$. Therefore the line lies in a pencil of quadrics from each of the two determinental nets. Together the two pencils form a web of quadrics in V which by Lemma 4.2 corresponds to a rank 2 quadric in $V^\perp$. On the other hand the secant lines to a twisted cubic curve form a congruence of bidegree $(1, 3)$ in the Grassmannian of lines in $\mathbf{P}^3$. Thus two general twisted cubic curves have

$$(1, 3) \cdot (1, 3) = 1 + 9 = 10$$

common secant lines. This is exactly the number of rank 2 quadrics in the web $V^\perp$, so the second correspondence follows. □

To fill in the remainder of Table 2 we want to compute how much the number of rank 2 quadrics decreases when the web acquires a rank 1 quadric. We give an argument using Reye's geometric interpretation. Our arguments will depend on a genericity assumption, i.e. the space V is general with a given number of basepoints. The argument would go through without this assumption also, but then the numbers in Table 2 would have different interpretations. Since we will only use general systems V we do not consider the degenerate cases.

When the space V of quadrics has basepoints, then there will be infinitely many apolar twisted cubic curves to V, but taking two of them will always suffice for our argument. First note that as long as the two twisted cubic curves have less than 4 common points, the corresponding nets of quadrics do not intersect.

When the web contains one rank 1 quadric, then the two twisted cubic curves have one common point. The number of common secant lines passing through this point is, with our genericity assumption, easily computed by projection from the point, it is 4, the number of intersection points between two conics in the plane. So the web has 6 rank 2 quadrics in addition to the rank 1 quadric. This is the second row in the table.

When the web contains 2 rank 1 quadrics, then the two twisted cubic curves have two common points. In this case there are 4 common secant lines through each of the two intersection points, and one of these is the line passing through the two points, so there are exactly 3 common secant lines which do not pass through any of the two common points. Thus the web has 3 rank 2 quadrics in addition to the two rank 1 quadrics. This is the third row in the table.

When the web contains 3 rank 1 quadrics, the two twisted cubic curves have 3 points in common. There are 4 common secant lines through each intersection point, and adding up three are counted twice, so we get only one common secant line which does not pass through any of the intersection points. Thus the web has one rank 2 quadric in addition to the three rank 1 quadrics. This is the fourth row in the table.

When the web has more than 3 rank 1 quadrics, the twisted cubics have at least 4 points of intersection. In this case the number of common secant lines that does not pass through the intersection points is infinite or zero depending on whether the two determinental nets intersect or not. This covers the remaining row in the table.

Lemma 4.4. *Consider the map ρ defined by the linear system V of quadrics, and consider a subscheme Z of length 2 which does not intersect the baselocus of V. Then Z is mapped to a point by ρ if and only if either the restriction of ρ to the unique line passing through Z is $2:1$ onto a line, or this line contains two base points for V.*

Proof. The linear system V restricted to the line through Z has degree 2. If this linear system has no basepoints, then the image of the line by ρ is a line or a conic section. It is a line if and only if some subscheme of length 2 is mapped to a point. If the line through Z intersects the baselocus, the intersection must contain two points, such that the line is contracted by V. □

Proof of Theorem 4.1. Combining Lemmas 4.2, 4.3 and 4.4 we find that a general 6-dimensional linear system of quadrics with $8 - t$ basepoints and $t = 5, 6, 7, 8$ defines a rational map from $\mathbf{P}^3$ to $\mathbf{P}^5$, whose image has degree t and whose non-normal double locus consists of $\binom{t-3}{2}$ disjoint lines. Any line between basepoints is contracted to an isolated singularity. The number of isolated singular points is therefore $\binom{8-t}{2}$. It is easily checked by restriction to general hyperplane sections that these singular points are ordinary double points. □

By abuse of notation we call the varieties W_t Del Pezzo 3-folds. Their normalizations have only isolated double points from the contracted lines between basepoints, these occur when $t \leq 6$. For each t we want to describe the ideal of W_t and understand their linkage class.

Proposition 4.5. *A 3-fold W_5 is bilinked to the union of two $\mathbf{P}^3$'s which span $\mathbf{P}^5$. A 3-fold W_6 is bilinked to the union of three $\mathbf{P}^3$'s. The ideals of general Del Pezzo 3-folds W_7 and W_8 are generated by quartics and quintics, and quintics and sextics respectively.*

Proof. To understand the ideal of W_t we first investigate the ideal of the singular lines. For $t = 5$ there is one line so this case is trivial. The case $t = 6$ is also easy since there are three lines; they span $\mathbf{P}^5$ as soon as the projection is general, as is easily verified. For $t = 7$ and $t = 8$ the situation is a bit more involved.

Lemma 4.6. *The 6 singular lines of a general Del Pezzo 3-fold W_7 lie in a determinental net of quadrics in $\mathbf{P}^5$, in fact in the Segre embedding of $\mathbf{P}^1 \times \mathbf{P}^2$.*

Proof. Let V be the linear system of quadrics which define the rational map of $\mathbf{P}^3$ onto W_7. By Lemma 4.3 the linear system V contains and is spanned by at least two determinental nets. Of course, the corresponding twisted cubic curves pass through the basepoint, so the union of any two is rational and is therefore contained in a cubic surface. By genericity we may assume that this surface is smooth. On this cubic surface there is a pencil of twisted cubic curves through the basepoint with six common secant lines. This pencil of curves corresponds to a linear pencil of nets of quadrics, so since two nets are contained in V they all are. The images of these twisted cubic curves on W_7 are plane curves. Since the twisted cubic curves pass through the basepoint of V these plane curves have degree 5. The 6 common secant lines are mapped to the 6 singular lines on W_7 and they account for the 6 singular points on each of these plane rational quintic curves.

Consider the space of quadrics passing through the 6 singular lines. If any of these quadrics intersects a plane of the plane quintic curves properly, its intersection would be a conic section passing through the 6 singular points of the quintic curve. But this is impossible by Bezout. Therefore any quadric which passes through the 6 lines must contain the planes of these curves. The intersection of these quadrics is therefore at least a threefold. Now, 6 lines impose at most 18 conditions on quadrics, so there are at least 3 such quadrics and the intersection is a threefold. A codimension 2 variety contained in 3 quadrics is a rational cubic scroll. If this scroll is singular, it is a cone, and any two planes meet. In this case two planes span a hyperplane. Pulled back to the cubic surface of the twisted cubic curves, this hyperplane corresponds to a quadric which contains the two curves. But the union of the two curves has arithmetic genus 0, so it is not contained in a quadric and a cubic surface. Therefore the scroll is smooth, isomorphic to the Segre 3-fold scroll. □

Proposition 4.7. W_7 *is contained in five quartic hypersurfaces, they define an arithmetically Cohen Macaulay 3-fold, the union of W_7 and the rational cubic scroll R which contains the singular lines of W_7. The quartic hypersurfaces are defined by the maximal minors of a 4×5 matrix with linear entries.*

Proof. Again let V be the linear system of quadrics which define the rational map of $\mathbf{P}^3$ onto W_7. In the proof of Lemma 4.6 we saw that there is a pencil of twisted cubic curves through the basepoint p of V whose defining nets of quadrics are contained in V. These twisted cubic curves sweep out a cubic surface S_3 whose image contains the singular lines on W_7 and is contained in the rational cubic scroll R. The linear system of quadrics V restricts to S_3 with one base point, so the image has degree 11. Consider the quartic surfaces through S_3 and singular in p; they consist of the union of S_3 and planes through p, so these quartics form a net. On the other hand, the image of S_3 in $\mathbf{P}^5$ is contained in the net of quadrics through R, and these quadrics pulled back to $\mathbf{P}^3$, correspond to quartic surfaces through S_3 singular at p, i.e. to the above net of quartics. Since the net of quartics has no unassigned basepoints, the quadrics through R define precisely the image of S_3 on W_7, i.e. the intersection $W_7 \cap R$ is precisely the image of the surface S_3. The union $W_7 \cup R$ has degree 10 and genus 11. To conclude that the union is arithmetically Cohen Macaulay we give an example. Let V be the space of quadrics

$$\langle x_0^2 + x_1x_2, x_1^2 + x_2x_3, x_2^2 + x_3x_0, x_0x_1, x_0x_1 + x_2x_3, x_0x_2 + x_1x_3 \rangle.$$

A straightforward computation in [MAC] shows that the Del Pezzo 3-fold W_7 defined by V lies in precisely 5 quartics, the 4×4 minors of a 4×5 matrix with linear entries, i.e. these quartics define an arithmetically Cohen Macaulay scheme of degree 10 and genus 11. Since this is an open condition in the Hilbert scheme [Ell], the same is true for the general projection W_7. □

For W_8 we get somewhat less.

Lemma 4.8. W_8 *contains two plane rational sextic curves, and each of the singular lines is spanned by a pair of nodes of these two sextic curves. In particular there are sextic generators in the ideal of W_8.*

Proof. In this case there are two apolar twisted cubic curves, these are mapped to plane sextic curves with 10 double points at the intersection of these planes with the 10 singular lines of W_8. So we need sextic generators in the ideal of W_8. □

We are now ready to give some numerical results for the ideals of general Del Pezzo 3-folds W_t of degree $5 \leq t \leq 8$.

Lemma 4.9. *Table 3 gives the degrees d and the number of generators in the ideal of W_t, for $t = 5, 6, 7, 8$.*

Table 3.

	$d=2$	$d=3$	$d=4$	$d=5$	$d=6$
$t=5$	0	5	0	0	0
$t=6$	0	1	7	0	0
$t=7$	0	0	5	5	0
$t=8$	0	0	1	10	≥ 1

Proof. The following spaces V of quadrics

$$\langle x_1x_2 + x_0x_3, x_3^2 + x_2x_0, x_3^2 + x_1x_0, x_3^2 + x_2x_1, x_0x_1 + x_2x_3, x_0x_2 + x_1x_3 \rangle,$$

$$\langle x_0^2 + x_1x_2, x_1^2 + x_2x_3, x_3x_0, x_0x_1, x_0x_1 + x_2x_3, x_0x_2 + x_1x_3 \rangle,$$

$$\langle x_0^2 + x_1x_2, x_1^2 + x_2x_3, x_2^2 + x_3x_0, x_0x_1, x_0x_1 + x_2x_3, x_0x_2 + x_1x_3 \rangle,$$

and

$$\langle x_0^2 + x_1x_2, x_1^2 + x_2x_3, x_2^2 + x_3x_0, x_3^2 + x_0x_1, x_0x_1 + x_2x_3, x_0x_2 + x_1x_3 \rangle,$$

have respectively 3,2,1 and no basepoints. The ideals of the corresponding Del Pezzo 3-folds in $\mathbf{P}^5$ are easily computed in [MAC] and have Betti numbers as in the table.

For the proof of the lemma we show that the table represents a lower bound on the number of generators in each of the given degrees. The lemma then follows by semicontinuity.

In each case we define Σ to be the union of the singular lines L_i on W_t, i.e. precisely the non-normal double point locus of W_t. Let V_t be the normalization of W_t, i.e. V_t is isomorphic to $\mathbf{P}^3$ blown up in $8-t$ points, and with $\binom{8-t}{2}$ lines contrated to ordinary double points. The map $\varphi : V_t \to W_t$ is double precisely along Σ, thus we get an exact sequence of sheaves

$$0 \to \mathcal{O}_{W_t} \to \varphi_* \mathcal{O}_{V_t} \to \mathcal{O}_\Sigma(-1) \to 0,$$

where $\mathcal{O}_\Sigma = \oplus_{i=1}^l \mathcal{O}_{L_i}$ and $l = \binom{t-3}{2}$. Since V_t is a Del Pezzo 3-fold, the cohomology of this sequence gives $h^3(\mathcal{O}_{W_t}(k)) = h^3(\varphi_* \mathcal{O}_{V_t}(k)) = 0$, when $k \geq 3$. Furthermore the Euler characteristic of the relevant twists of the ideal sheaf of W_t is easily computed from this exact sequence together with the exact sequence

$$0 \to \mathcal{I}_{W_t}(k) \to \mathcal{O}_{\mathbf{P}^5}(k) \to \mathcal{O}_{W_t}(k) \to 0.$$

We collect the results in the following table.

Table 4.

	k	$\chi(\mathcal{O}_{W_t}(k))$	$h^0(\varphi_*\mathcal{O}_{V_t}(k))$	$h^0(\mathcal{O}_\Sigma(k-1))$	$h^0(\mathcal{O}_{\mathbf{P}^5}(k))$	$\chi(\mathcal{I}_{W_t}k)$
$t=5$						
	3	51	54	3	56	5
$t=6$						
	3	55	64	9	56	1
	4	113	125	12	126	13
$t=7$						
	4	121	145	24	126	5
	5	221	251	30	252	31
$t=8$						
	4	125	165	40	126	1
	5	236	286	50	252	16

By restriction to general $\mathbf{P}^3$ sections of W_t it is easy to check that $h^1(\mathcal{O}_{W_t}(k)) = 0$ for the values of k in the table. Since additionally $h^3(\mathcal{O}_{W_t}(k)) = 0$ when $k \geq 3$, the Euler characteristic of the twisted ideal is a lower bound for $h^0(\mathcal{I}_{W_t}(k))$. Thus the Betti numbers of the ideal follow except in the case of quintics and W_7 and sextics and W_8. These cases are accounted for in Proposition 4.7 and Lemma 4.8. In fact, from Lemma 4.7 we get 4 linear syzygies among the 5 quartics in the ideal of W_7. Therefore there are also 4 extra quintic generators in the ideal, 5 altogether. Lemma 4.8 says that there are sextic generators in the ideal of W_8. □

To finish the proof of Proposition 4.5 it remains to consider the linkage classes of W_5 and W_6. The 3-fold W_5 is linked (3, 3) to a rational 3-fold scroll of degree 4. This lies in a quadric and is linked (2, 3) to two $\mathbf{P}^3$'s, which clearly is minimal in its even biliaison class.

W_6 is linked (3, 4) to a 3-fold U with sectional genus 1 which lies on two cubics. Consider a $\mathbf{P}^3$ spanned by two singular lines in W_6. It intersects W_6 in the two lines and in two additional skew lines each intersecting both the singular lines. Clearly every cubic through W_6 contain this $\mathbf{P}^3$, so U must intersect it in a quartic surface singular along the two singular lines. U is therefore an elliptic scroll linked (3, 3) to the union of three $\mathbf{P}^3$'s (cf. also Section 6). This concludes the proof of Proposition 4.5. □

In the last section we need a converse to Proposition 4.7.

Proposition 4.10. *Let W be a 3-fold with sectional genus 1, with non-normal double points along 6 skew lines, no 3 in a $\mathbf{P}^3$ and not all 6 on a rational normal quartic scroll. Assume that there is a Segre cubic scroll R in $\mathbf{P}^5$ containing the 6 skew lines transverse to its planes and intersecting W in a surface of degree 11, linked to 4 planes in the intersection of R with a quintic hypersurface. Assume furthermore that $W \cup R$ is scheme-theoretically defined by the 4×4 minors of a 4×5 matrix with linear entries, and that the only common singularities of these quartics are the 6 lines. Then W is a Del Pezzo 3-fold W_7.*

Proof. Consider the 4×5 matrix M whose maximal minors define $R \cup W$. Let $\langle z_0, \ldots, z_3 \rangle$ be the coordinates of $\mathbf{P}^3$, then the 5 bilinear equations

$$(z_0, z_1, z_2, z_3) \cdot M = (0, 0, 0, 0, 0)$$

define a 3-fold T in $\mathbf{P}^3 \times \mathbf{P}^5$. The projection to $\mathbf{P}^5$ is onto $W \cup R$, while the other projection is onto $\mathbf{P}^3$. The fiber in T over any point in $\mathbf{P}^3$ is linear, defined by the 5 linear equations in the 6 coordinate functions of $\mathbf{P}^5$, so for the general point in $\mathbf{P}^3$ the fiber is a single point. The fiber over any point in $W \cup R$ is also linear, with dimension equal to 3 minus the rank of M at the target point. The points where M has rank at most 2, are singular on $W \cup R$, in fact they are singular on any quartic minor of M, so by our assumption, these points all lie on the 6 singular lines. Outside the 6 singular lines M has rank 3 and therefore there is a rational map

$$\varphi: W \cup R ----> \mathbf{P}^3$$

which is a morphism outside these 6 lines. This map is birational on one component of $W \cup R$ and contracts every other component to a surface, a curve or a point. The inverse map

$$\psi : \mathbf{P}^3 ----> W \cup R$$

is defined by the maximal minors of a 5×6-matrix M' with linear entries. M' is obtained from M by interchanging rows and linear forms. Surfaces that are images of components of $W \cup R$ are fixed components for the linear system of quintic minors of M'.

Our first aim is to show that W has to be irreducible. We analyse carefully possible reducible components of the surface $S = W \cap R$. For this we start with the intersection of S with a general plane in R.

On R the surface $S \equiv 5h - 4f$, where h is the class of a hyperplane section, while f is the class of a plane. Since W is non-normal along 6 lines transverse to the planes of R, the intersection of S with a general plane is a curve C of degree 5 singular in 6 points, the points of intersection between the plane and the singular lines. Furthermore, no three singular lines span a $\mathbf{P}^3$, so no three of the singular points of C are collinear. More generally, any effective divisor of type $ah - af$ is the product of a plane curve of degree a and $\mathbf{P}^1$. Geometrically each $\mathbf{P}^1$ is a line transverse to all planes in R. Conversely, any irreducible divisor of type $h + bf$ that contains more than one line transverse to all planes has $b = -1$ and is a quadric surface, while any irreducible divisor of type $2h + bf$ that contains more than 4 lines transverse to all planes has $b = -2$ and is a rational normal quartic scroll. In our situation, this means that three singular lines intersect no plane in collinear points. Also, since not all 6 singular lines are on a rational normal quartic scroll, the 6 singular points cannot lie on a conic section. Therefore, if C is reducible it is either the union of 2 conics and a line, or the union of a conic and a cubic. In the former case the singular points are the points of intersection between the two conics and one point of intersection between the line and each conic, while in the latter case 5 singular points are points of intersection between the conic and the cubic, while the last point is singular on the cubic.

Next, the surface S intersects each general line in R transverse to the planes in one point. Since the intersection of S with a general plane has at most 3 components, S itself has at most 4 components: Only the planes in R do not intersect the general plane in a curve, so since exactly one component intersects every line transverse to the planes, there are at most 4 components of S.

To describe the possible components of S we first note that on R any effective divisor is of type $ah - bf$, with $a \geq 0$ and $b \leq a$. In our two cases of possible reducible curves in general planes we have the following possible decompositions of S into irreducible components:

$$S \equiv 5h - 4f = b_0 f + (h - b_1 f) + (2h - b_2 f) + (2h - b'_2 f)$$

where $b_1 + b_2 + b'_2 - b_0 = 4$ and

$$S \equiv 5h - 4f = b_0 f + (2h - b_2 f) + (3h - b_3 f)$$

where $b_2 + b_3 - b_0 = 4$. With the above restriction on the b_i, the former case occurs only when $(b_0, b_1, b_2, b'_2) = (1, 1, 2, 2)$, $(b_0, b_1, b_2, b'_2) = (0, 0, 2, 2)$ or $(b_0, b_1, b_2, b'_2) = (0, 1, 2, 1)$. The latter case occurs only when $(b_0, b_2, b_3) = (1, 2, 3)$, $(b_0, b_2, b_3) = (0, 2, 2)$ or $(b_0, b_2, b_3) = (0, 1, 3)$.

We are now ready to analyse possible components of W. Clearly any two components of W intersect each other along a surface or a curve (or both). If two components of $W \cup R$ intersect each other along a curve, then every hypersurface through $W \cup R$ has to be singular along this curve, so by our assumption this happens only along the 6 singular lines. The sectional genus of W is 1, thus for the general $\mathbf{P}^3$ section of W the arithmetic genus is 1 plus the number of intersection points with the singular lines. This means that for a general $\mathbf{P}^3$ passing through a plane of R, the contribution to the arithmetic genus of the plane curve C is 0. The degree of W is 7, so residual to C in this intersection is a curve of degree 2. It must intersect C in 2 points, and it must itself have arithmetic genus 0, so it is a conic section, two connected lines or a double line. This very much restricts the possiblities for the irreducible components of W.

In our analysis of the components W_d of W, we shall index them by their degree, unless the degree is not specified, in which case we use index 0. This notation applies only in this proof and should not be confused with Del Pezzo 3-folds W_t. A component W_0 of W must intersect R in a curve or a surface. If it intersects R only in a curve, then this curve meets the general plane in R only in points, thus this component intersects a general $\mathbf{P}^3$ through a general plane in R in a curve not contained in the plane. Therefore this component has degree 1 or 2. Similarly if W_0 intersects the general plane in R in a curve of degree a, then the degree of the component is a, $a+1$ or $a+2$. In fact, if W_0 spans $\mathbf{P}^5$, i.e. when $a \geq 2$, then the degree is strictly greater than a. Of course, while W_0 is irreducible, $W_0 \cap R$ may still be reducible. Anyway, we may enumerate the different cases in a table, where the columns are ordered by the degree of the curve of intersection between a component and a general plane, while the entries are the degrees of the components themselves:

Table 5.

0	1	2	3	4	5
1	0	0	0	0	6
1	1	0	0	5	0
0	1	0	0	0	6
0	1	3, 3	0	0	0
0	0	3	4	0	0
0	0	0	0	0	7

In the first two cases of Table 5 the first linear component W_1 may intersect R in a plane and a line, or in a connected curve of degree 3. The latter may be excluded since no connected cubic curve is supported on the 6 singular lines. In the former case the line must be one of the 6 singular lines. The remaining part of W is a component of degree 6 or two components of degree 1 and 5, respectively. These remaining components must intersect R in a surface of type $5h - 5f$, i.e. they do not intersect the general line in R transverse to the planes. In the first case the linear component intersects the component W_6 of degree 6 along a quadric surface, since the sectional genus is 1. The linear component W_1 intersects R in a plane, but W_6 intersects the general plane in R in a curve of degree 5, so this plane must also be contained in W_6. This contradicts the fact that W_6 does not intersect the general line in R transverse to the planes. The second case is impossible for the same reason.

In the third case of the table there is a linear component W_1 that intersects R along a quadric surface and a component W_6 that intersects R in a surface of type $4h - 3f$. Again the two components have also to intersect along a quadric surface. The intersection of the two components on R is a curve of type $(h - f)(4h - 3f) = 4h^2 - 7hf$, i.e. a curve of degree 5. On the other hand on W_1 each of the two other components intersect along a quadric surface and two singular lines. This adds up to a curve of degree 6 in the intersection of all three components. This is a contradiction. The fourth case is entirely similar.

In the fifth case there are two components W_3 and W_4 of degree 3 and 4 respectively. They intersect each other in a surface of degree 2. Furthermore they intersect on R in a curve of type $(2h - f)(3h - 2f) = 6h^2 - 7hf$, i.e. a curve of degree 11. The union of the 5 singular lines in this intersection is a curve of type $5h^2 - 10hf$, so the remaining part is of type $h^2 + 3hf$. This is a curve of degree 6. Again the surface $W_3 \cap W_4$ has degree 2. If it is irreducible, then the lines of one of the rulings are at least 3-secants to R, impossible, so the quadric surface is contained in R, again contradicting our assumption. If the surface $W_3 \cap W_4$ is the union of two planes, then at least one of the planes intersects R in a curve of degree at least 3, again contradicting our assumption, so this case is also impossible. This concludes the proof that W is irreducible.

It remains to show that R is contracted by φ. If not, W must be contracted. First of all this is possible only if W is swept out by lines, i.e. has a line through every point. For this consider a general plane in R. It intersects W in a plane quintic curve with 6 double points. W has degree 7, so in a general $\mathbf{P}^3$ through the plane W has a curve of degree 2 residual to the plane curve. Since the arithmetic genus of W is 1 and the plane curve is rational, this residual curve is a conic or two connected lines. Now, W can only have a line through each point if this residual curve is always two lines. On the other hand if these two lines are contracted by φ, they are contracted to the same point, since the matrix M has rank 3 even in the point of intersection. But all fibers of φ are linear, so W must contain a plane through the two lines, this is absurd. Thus W is mapped birationally to $\mathbf{P}^3$ by φ.

It follows that the restriction of φ to R is a contraction onto a surface. Clearly it must contract the lines transverse to the planes. From the matrix M we see that the map is defined by cubics on each plane, and it is birational, so the image has degree at least 3. Therefore, the inverse map is defined by hyperplanes or quadrics. Since W has degree 7 the inverse map must be defined by quadrics, and in fact with one basepoint, i.e. W is a Del Pezzo 3-fold W_7. □

5. Non-normal Calabi–Yau 3-folds in $\mathbf{P}^5$

We shall show that W_t for $t = 5, 6, 7$ is bilinked on a quintic hypersurface to a non-normal Calabi–Yau 3-fold. W_t is assumed to be general in the sense of Section 4, including in the case of W_7 any Del Pezzo 3-fold characterized as in Proposition 4.10. Most of the argument goes through also for W_8, but since we shall not need this later we do not conclude in this case (see Remark 5.3).

Lemma 5.1. *The general Del Pezzo 3-fold W_t for $t = 5, 6, 7, 8$ is contained in an irreducible quintic hypersurface. The general such quintic hypersurface is normal, it has double points along the singular lines of W_t and has only canonical singularities.*

Proof. We start with a general W_t and a general quintic hypersurface Q through W_t. Since W_t has non-normal double points along the singular lines and quintics generate its ideal, Q has multiplicity 2 at a general point on a singular line. Let

$$p' : \mathrm{Bl}_s(\mathbf{P}^5) \to \mathbf{P}^5$$

be the blowup of $\mathbf{P}^5$ along the singular lines of W_t, and denote by W'_t and Q' the strict transforms of W_t and Q. Then W'_t has at most ordinary double points as singularities.

Next we consider the blowup

$$p'' : \mathrm{Bl}_{s,W}(\mathbf{P}^5) \to \mathrm{Bl}_s(\mathbf{P}^5)$$

of $\mathrm{Bl}_s(\mathbf{P}^5)$ along W'_t. Over the singular points of W'_t, the blowup $\mathrm{Bl}_{s,W}(\mathbf{P}^5)$ will have isolated double points. Denote by Q'' the strict transform of Q' and let W''_t be the strict transform of W'_t in Q''. By the fundamental property of blowup, W''_t is a Cartier divisor on Q''.

We now analyse the situation, assuming that we have chosen Q general. For $t \leq 7$, the ideal of W_t is generated by quintics by Proposition 4.5, so in this case Q'' is smooth outside W''_t. We claim that Q'' is smooth and that p'' restricted to Q'' is a small resolution of Q'. Since the quintics generate the ideal of W_t, the conormal sheaf of W'_t in $\mathrm{Bl}_s(\mathbf{P}^5)$ twisted by the class of Q' is generated by its global sections. In particular, any divisor equivalent to Q' induces a section of this sheaf, which is a rank 2 bundle outside the singular points. Its zero locus is the subscheme of W'_t defined by the Jacobian ideal of Q'. For general Q this is a smooth curve which does not pass through the singular points of W'_t. Thus Q' is singular only along some smooth curve on W'_t not passing through the singularities of W'_t. The map p'' defines an isomorphism between W''_t and W'_t. Moreover since W''_t is a Cartier divisor on Q'' it follows that Q'' is smooth and that p'' defines a small resolution of Q' showing that Q' has only canonical singularities. For W_8 one may show with [MAC] in the example of Lemma 4.9 that the base locus of the quintics through W_8 is precisely the singular lines and the two planes. Therefore also in this case Q'' is smooth and we can argue as before.

Finally Q is smooth in codimension 1 and since it is a hypersurface, it is normal. It remains to prove that Q has canonical singularities along the singular lines. First recall that Q has multiplicity 2 at the generic point of the singular lines. Moreover the blowup p' defines a resolution of Q over the points on the singular lines outside the singular curve of Q'. Hence Q has transversal ordinary double points outside the finitely many points on the singular lines which are the intersection with the image under p' of the singular curve of Q'. In particular Q has canonical singularities outside these finitely many points. Combining this with the argument that p'' defines a small resolution of Q' gives the claim of the lemma. □

Proposition 5.2. *The general Del Pezzo 3-fold W_t for $t = 5, 6, 7$ is bilinked $(5, 4)$ and $(5, 5)$ on a quintic hypersurface to a variety Y of degree $t + 5$ which has non-normal singularities along the singular lines of W_t. The normalization of Y is a smooth Calabi–Yau 3-fold.*

Proof. We start with a general quintic hypersurface Q which contains W_t as in Lemma 5.1, and keep the same setup and notation as in the above proof. In addition we let $H_{Q'}$ and $H_{Q''}$ be the pullback by p' and p'' of a hyperplane H restricted to Q. A 3-fold Y

bilinked in hypersurfaces of degree 4 and 5 to W_t on Q is nothing but a Weil divisor equivalent to $W_t + H_Q$, where H_Q is the restriction of H to Q. The strict transform Y'' of Y on Q'' is a Cartier divisor linearly equivalent to $W_t'' + H_{Q''}$. To analyze the singularities and the canonical sheaf of Y we perform adjunction on the smooth 4-fold Q''.

Let E'' be the pullback of the exceptional divisor of $\mathrm{Bl}_s(\mathbf{P}^5)$ to $\mathrm{Bl}_{s,W}(\mathbf{P}^5)$. Then the canonical line bundle on Q'' is $\mathcal{O}_{Q''}(E'' - H_{Q''})$. By adjunction on Q'', the canonical line bundle on W_t'' is $\mathcal{O}_{W_t''}(W_t'' + E'' - H_{Q''})$. Consider now the normalization $\tilde{W}_t$ of W_t. The map $W_t'' \to W_t$ factors through this normalization. On $\tilde{W}_t$ there are no exceptional divisors since the singular locus has codimension 2. Therefore, the map $W_t'' \to \tilde{W}_t$ is the blowup along a curve with exceptional line bundle $\mathcal{O}_{W_t''}(E'')$. But $\tilde{W}_t$ is a Del Pezzo 3-fold with only canonical Gorenstein singularities, so the canonical divisor of W_t'' is also $\mathcal{O}_{W_t''}(E'' - 2H_{Q''})$. Therefore $\mathcal{O}_{W_t''}(W_t'' - H_{Q''}) = \mathcal{O}_{W_t''}(-2H_{Q''})$ and we obtain the equality

$$\mathcal{O}_{W_t''}(W_t'' + H_{Q''}) = \mathcal{O}_{W_t''}.$$

Now, consider the exact sequences of sheaves on Q''

$$0 \longrightarrow \mathcal{O}_{Q''}(H_{Q''}) \longrightarrow \mathcal{O}_{Q''}(W_t'' + H_{Q''}) \longrightarrow \mathcal{O}_{W_t''}(W_t'' + H_{Q''}) \longrightarrow 0,$$

$$0 \longrightarrow \mathcal{O}_{Q''} \longrightarrow \mathcal{O}_{Q''}(H_{Q''}) \longrightarrow \mathcal{O}_{H_{Q''}}(H_{Q''}) \longrightarrow 0$$

and

$$0 \longrightarrow \mathcal{O}_{H_{Q''}} \longrightarrow \mathcal{O}_{H_{Q''}}(H_{Q''}) \longrightarrow \mathcal{O}_{H_{Q''} \cap H'_{Q''}}(H_{Q''}) \longrightarrow 0$$

for general hyperplanes H and H'. Since Q and $Q \cap H$ have only canonical singularities, Q'' and $H_{Q''}$ are regular. Thus the second and third sequence remain exact on global sections. Furthermore $H \cap H' \cap Q$ is a smooth quintic surface, so $h^1(\mathcal{O}_{H \cap H' \cap Q}(H)) = h^1(\mathcal{O}_{H_{Q''} \cap H'_{Q''}}(H_{Q''})) = 0$. Therefore $h^1(\mathcal{O}_{H_{Q''}}(H_{Q''})) = h^1(\mathcal{O}_{Q''}(H_{Q''})) = 0$ and also the first sequence remains exact after taking global sections. Thus the line bundle in the middle is generated by global sections, i.e. the linear system $|W_t'' + H_{Q''}|$ of divisors on Q'' has no basepoints. By Bertini, we may conclude that a general member Y'' of this linear system is smooth.

The exceptional locus of p'' on Q'' is a surface scroll, and W_t'' intersects this scroll in a section over the smooth base curve. The restriction of the linear system $|W_t'' + H_{Q''}|$ to this scroll has no basepoints so the general member is again a section over the base curve. Therefore the restriction of p'' to Y'' is also an isomorphism, and the image Y' on Q' is smooth, while Y is singular only along the singular lines. Since W_t has non-normal double points along the singular lines, the same is the case for Y.

The canonical line bundle on Y'' is easily computed by adjunction on Q''. In fact it is

$$\mathcal{O}_{Y''}(Y'' + E'' - H_{Q''}) = \mathcal{O}_{Y''}(W_t'' + E'').$$

Since $\mathcal{O}_{W_t''}(Y'') = \mathcal{O}_{W_t''}(W_t'' + H_{Q''}) = \mathcal{O}_{W_t''}$, we have $\mathcal{O}_{Y''}(W_t'') = \mathcal{O}_{Y''}$. Therefore the canonical line bundle on Y'' is $\mathcal{O}_{Y''}(E'')$.

We turn to the singular variety Y. It is singular precisely along the singular lines of W_t where it has non-normal double points. Consider the normalization $\tilde{Y} \to Y$. As in the case of W_t, the resolution of singularities $Y'' \to Y$ factors through this normalization. We will show that $\tilde{Y}$ is already smooth. For this we first specialize the quintic hypersurface Q to a hypersurface containing W_t and a 3-fold Z that is smooth along the singular lines of W_t. In case $t = 7$, we take Z to be the Segre cubic scroll R, while in the cases $t = 5$ and $t = 6$ we may take Z to be one and two $\mathbf{P}^3$'s respectively. We make the computation explicit in the case when $t = 7$, which we will use later. The other cases are similar. Assume that Q is a quintic which contains W_t and the rational cubic scroll R. Let $Q_0 \to Q$ be the blow up of Q along R. This defines a small resolution of the singularities of Q along the lines. Over each line there is an exceptional scroll. This scroll is isomorphic to some Hirzebruch surface F_a for some $a \geq 0$. The strict transform of W_t meets this scroll in a rational curve which is a bisection on the scroll, since W_t is double along the line. Since Y as a Weil divisor is equivalent to $W_t + H_Q$ on Q, and the strict transform of H_Q intersects the scroll in a ruling, we get that the strict transform of Y meets the exceptional scroll in a bisection which is an elliptic curve. The normalization of Y factors through this small resolution and, therefore, has elliptic curves lying over the singular lines. In particular the normalization is smooth over the singular lines. By deformation to the general quintic Q, the normalization of Y is smooth.

Now, the normalization $\tilde{Y}$ of Y is isomorphic to Y in codimension 1, therefore the canonical line bundle $\mathcal{O}_{Y''}(E'')$ on Y'' is the exceptional line bundle of the map $Y'' \to \tilde{Y}$ and the canonical line bundle of $\tilde{Y}$ is trivial.

The irregularity of $\tilde{Y}$ equals the irregularity of Y'', since they are birational and both are smooth. But on Y'' we get

$$h^1(\mathcal{O}_{Y''}) = h^2(\mathcal{O}_{Q''}(-W_t'' - H_{Q''})) = h^1(\mathcal{O}_{W_t''}(-H_{Q''})) = 0$$

so the normalization of Y is regular. Thus the normalization of Y is a smooth Calabi–Yau 3-fold and the proposition follows. □

Remark 5.3. The above proof goes through also for W_8, except for the question of a smooth normalization.

6. Equations of elliptic scrolls in $\mathbf{P}^5$

In the remaining sections we prove that the union of two elliptic scrolls which intersect along an abelian surface is bilinked to a Del Pezzo 3-fold W_7. On the way we describe the family of these reducible Calabi–Yau 3-folds using Heisenberg symmetry. But first we study the ideal of elliptic scrolls without using this symmetry.

Recall from Section 3 that two elliptic scrolls whose intersection is a $(1,6)$-polarized abelian surface in $\mathbf{P}^5$, are each singular along three lines. Therefore we restrict our attention to this kind of elliptic scrolls. We start with a lemma which gives a quick construction of such scrolls.

Proposition 6.1. *The union of three $\mathbf{P}^3$'s in $\mathbf{P}^5$, which meet pairwise in lines, is linked $(3, 3)$ to an elliptic scroll. Furthermore any elliptic 3-fold scroll of degree 6, singular along three lines which span $\mathbf{P}^5$, is linked $(3, 3)$ to three $\mathbf{P}^3$'s.*

Proof. First, consider three $\mathbf{P}^3$'s which meet pairwise in lines and two general cubic hypersurfaces containing them. The linked variety X has degree 6 and sectional genus 1 (cf. [PS]). Since the complete intersection has trivial canonical bundle, each component intersects the rest along an anticanonical divisor. Therefore X meets each of the $\mathbf{P}^3$'s in a quartic surface, singular along the two lines of intersection with the other two $\mathbf{P}^3$'s. These quartic surfaces clearly are elliptic scrolls: Through every point on the surface outside the two singular lines there is a unique line in the surface through the point intersecting the two lines, by Bezout, so the quartic surface is a scroll of lines. Through every point on the singular lines there are two rulings of the scroll. In the Grassmannian of lines the curve parametrizing the rulings has a double cover to each of the two singular lines, so the curve is of type $(2, 2)$ on $\mathbf{P}^1 \times \mathbf{P}^1$, i.e. it is elliptic. The normalization of X therefore contains elliptic scrolls in parts of hyperplane sections. On the other hand, residual to each quartic surface in hyperplane sections, there is a pencil of surfaces of degree two on X. Residual to the $\mathbf{P}^3$ they form part of a complete intersection of two quadrics in the hyperplane. The other two $\mathbf{P}^3$'s intersect this hyperplane in two planes passing through a point, so the surface of degree two on X must be two planes residual to these in the complete intersection. This displays the scroll structure of X. Clearly the scroll is elliptic.

Next, consider an elliptic scroll X_F of degree 6 and singular along three lines that span $\mathbf{P}^5$. Let

$$\varphi : \mathbf{P}(\mathcal{E}) \to X_F$$

be the normalization map. Like in Section 3 the three lines correspond to a decomposition

$$\mathcal{E} = \mathcal{L}_0 \oplus \mathcal{L}_1 \oplus \mathcal{L}_2,$$

where the $\mathcal{L}_i$ are line bundles of degree 2. We have

$$H^0(F, \mathcal{E}) \cong H^0(X_F, \mathcal{O}_{X_F}(1)) \cong H^0(\mathbf{P}^5, \mathcal{O}_{\mathbf{P}^5}(1))$$

since the linear system that maps $\mathbf{P}(\mathcal{E})$ into $\mathbf{P}^5$ is complete. There is an isomorphism

$$H^0(\mathcal{O}_{\mathbf{P}(\mathcal{E})}(n)) \cong H^0(F, S^n \mathcal{E}) \cong H^0(F, \mathcal{L}_0^n \oplus \mathcal{L}_0^{n-1} \otimes \mathcal{L}_1 \oplus \cdots \oplus \mathcal{L}_2^n)$$

where the number of summands is

$$\binom{n+2}{2} = \frac{(n+1)(n+2)}{2}$$

and the degree of each summand is $2n$. This shows that
$$h^0(\mathcal{O}_{\mathbf{P}(\mathcal{E})}(n)) = h^0(F, S^n\mathcal{E}) = n(n+1)(n+2).$$
We now compare this to the situation on the singular scroll X_F. Recall that there are three sections $F_i \subset \mathbf{P}(\mathcal{E})$ such that the map φ restricted to F_i induces a double cover $\varphi : F_i \to L_i$ onto a line L_i. This is branched over 4 points and hence
$$\varphi_* \mathcal{O}_{F_i} = \mathcal{O}_{L_i} \oplus \mathcal{O}_{L_i}(-2),$$
and we have the exact sequence
$$0 \to \mathcal{O}_{L_i} \to \varphi_*\varphi^* \mathcal{O}_{L_i} \to \mathcal{O}_{L_i}(-2) \to 0$$
where $\varphi_*\varphi^*\mathcal{O}_{L_i} = \varphi_*\mathcal{O}_{F_i}$.

Lemma 6.2. *There is an exact sequence of sheaves*
$$0 \to \mathcal{O}_{X_F} \to \varphi_*\mathcal{O}_{\mathbf{P}(\mathcal{E})} \to \oplus_{i=0}^2 \mathcal{O}_{L_i}(-2) \to 0.$$

Proof. The cokernel clearly has support on the locus on X_F where the map is not an isomorphism, i.e. on the three lines L_i. Restricted to these lines the exact sequence reduces to the one above. Since the lines are disjoint, the cokernel is a direct sum. □

Lemma 6.3. *After tensoring with $\mathcal{O}_{X_F}(n)$ for any nonnegative n the sequence of Lemma 6.2 is exact on global sections.*

Proof. For $n = 0, 1$ the statement is immediate, since the third term has no sections. Furthermore $h^1(\mathcal{O}_{X_F}(1)) = h^1(\mathcal{O}_{\mathbf{P}(\mathcal{E})}(1)) = 0$ and $h^2(\mathcal{O}_{X_F}(n)) = h^2(\mathcal{O}_{\mathbf{P}(\mathcal{E})}(n)) = 0$ for $n \geq 1$. Let H be a general hyperplane, and $P = \mathbf{P}^3$ be a general 3-space inside H. Then $X_F \cap P$ is a smooth elliptic curve, so $h^1(\mathcal{O}_{X_F \cap P}(n)) = 0$ for $n \geq 1$. Furthermore $h^1(\mathcal{O}_{X_F \cap H}(1)) = 0$. Inductively $h^1(\mathcal{O}_{X_F \cap H}(n)) \leq h^1(\mathcal{O}_{X_F \cap H}(n-1)) = 0$ for $n \geq 2$, and similarly
$$h^1(\mathcal{O}_{X_F}(n)) \leq h^1(\mathcal{O}_{X_F}(n-1)) = 0$$
for $n \geq 2$ and the lemma follows. □

Lemma 6.3 allows us to compute
$$h^0(\mathcal{O}_{X_F}(n)) = n(n+1)(n+2) - 3(n-1), \quad n \geq 1.$$
The natural maps $H^0(\mathcal{O}_{\mathbf{P}^5}(n)) \to H^0(\mathcal{O}_{X_F}(n))$ and the dimension
$$h^0(\mathcal{O}_{\mathbf{P}^5}(n)) = \dim S^n H^0(\mathcal{E}) = \binom{n+5}{5},$$

gives us

$$h^0(\mathcal{I}_{X_F}(n)) - h^1(\mathcal{I}_{X_F}(n)) = \binom{n+5}{5} + 3(n-1) - n(n+1)(n+2).$$

In particular $h^0(\mathcal{I}_{X_F}(3)) \geq 2$. Clearly the complete intersection of two cubics through X_F contains the $\mathbf{P}^3$'s spanned by the pairs of singular lines, therefore $h^0(\mathcal{I}_{X_F}(3)) = 2$ and Proposition 6.1 follows. □

Corollary 6.4. $h^0(\mathcal{I}_{X_F}(5)) = 54$.

Proof.
$$h^0(\mathcal{I}_{X_F}(5)) - h^1(\mathcal{I}_{X_F}(5)) = 54,$$

as follows from the computation in the proof of Proposition 6.1. But $X_F \cap \mathbf{P}^2$ for a general $\mathbf{P}^2$ is 4-regular in the sense of Castelnuovo–Mumford, so $h^1(\mathcal{I}_{X_F}(n)) = 0$ for $n \geq 3$. □

7. Heisenberg symmetry of elliptic scrolls in $\mathbf{P}^5$

First we collect some basic observations. Recall the Heisenberg group of level 6:

$$1 \to \mu_6 \to H_6 \to \mathbf{Z}/6 \times \mathbf{Z}/6 \to 0.$$

Given any subgroup $G \subset \mathbf{Z}/6 \times \mathbf{Z}/6$ we can consider the preimage

$$H(G) := \pi^{-1}(G) \subset H_6.$$

Let $\langle x_0, \ldots, x_5 \rangle$ be a basis for a 6-dimensional complex vector space V and let $\langle e_0, \ldots, e_5 \rangle$ be a basis for V^*. The Schrödinger representation $\rho : H \to GL(V, \mathbf{C})$ is defined by

$$\sigma : x_i \mapsto x_{i+1}, \quad \tau : x_i \mapsto \rho^i x_i$$

where $\rho = \rho = e^{2\pi i/6}$ and indices are taken modulo 6. It defines, by restriction, representations

$$\rho_G : H(G) \to GL(V, \mathbf{C}).$$

The group $\mathbf{Z}/6 \times \mathbf{Z}/6$ contains unique subgroups isomorphic to $\mathbf{Z}/2 \times \mathbf{Z}/2$, resp. $\mathbf{Z}/3 \times \mathbf{Z}/3$. The preimages of these groups in H_6 are isomorphic to the Heisenberg groups H_2 and H_3 of level 2 and 3. Note, however, that in the case of H_3 the induced representation differs from the Schrödinger respresentation by the non-trivial automorphism of the Galois group.

Lemma 7.1. (i) *There are 4 subgroups isomorphic to $\mathbf{Z}/3$ in $\mathbf{Z}/3 \times \mathbf{Z}/3$.*

(ii) *There are 4 subgroups isomorphic to $\mathbf{Z}/2 \times \mathbf{Z}/6$ in $\mathbf{Z}/6 \times \mathbf{Z}/6$.*

Proof. Claim (i) is trivial. Every subgroup isomorphic to $\mathbf{Z}/2 \times \mathbf{Z}/6$ is generated by the elements of order 2 and one element of order 3 in $\mathbf{Z}/6 \times \mathbf{Z}/6$, i.e. by the subgroup $\mathbf{Z}/2 \times \mathbf{Z}/2 \subset \mathbf{Z}/6 \times \mathbf{Z}/6$ and a subgroup of order 3 of the group $\mathbf{Z}/3 \times \mathbf{Z}/3 \subset \mathbf{Z}/6 \times \mathbf{Z}/6$. This shows (ii). □

Remark 7.2. At the same time this gives us a natural 1 : 1 correspondence between the 4 groups of part (i) of Lemma 7.1 and the 4 groups of part (ii). We shall use this frequently in what follows.

We shall denote the 4 subgroups of $\mathbf{Z}/6 \times \mathbf{Z}/6$ which are isomorphic to $\mathbf{Z}/2 \times \mathbf{Z}/6$ by $K_1, \ldots, K_4$. For every group G in $\mathbf{Z}/6 \times \mathbf{Z}/6$ we set

$$H(G)_\iota = \langle H(G), \iota \rangle \subset G_6.$$

We denote by G_2 resp. G_3 the groups $\langle H_2, \iota \rangle$, resp. $G_3 = \langle H_3, \iota \rangle$. Next we need some elementary representation theory. We denote by U the Schrödinger representation of H_2.

Lemma 7.3. (i) *As an H_2-module $V \cong 3U$,*

(ii) *as a G_2-module $V \cong 2U_+ \oplus U_-$, where U_+, resp. U_- means that ι acts by $+1$, resp. -1 on U,*

(iii) *as K_i-modules $V \cong U \oplus U' \oplus \bar{U}'$. Here the subgroup of K_i which is isomorphic to $\mathbf{Z}/3$, acts on U' by a non-trivial character and on $\bar{U}'$, by the inverse of this character. The involution ι leaves U fixed and interchanges U' and $\bar{U}'$.*

Proof. (i), (ii). We can decompose V as an H_2-module as follows:

$$\langle x_0, x_3 \rangle, \langle x_2, x_5 \rangle, \langle x_4, x_1 \rangle.$$

From this the claim is obvious.

(iii). It is enough to consider one of the groups K_i. The others can be done in the same way, or one can use the normalizer N_6 of H_6 in $GL(6, \mathbf{C})$. Here we shall consider the subgroup K given by τ^2. One immediately checks that τ^2 acts by $1, \rho^4, \rho^2 (\rho = e^{2\pi i/6})$ on the above submodules of V, hence giving the claim. □

We next want to associate basic geometric objects in $\mathbf{P}^5$ to the subgroups K_i, $i = 1, 2, 3, 4$.

Lemma 7.4. (i) *To every subgroup $H(K_i), i = 1, 2, 3, 4$ of H_6 one can associate a unique set of 3 lines in $\mathbf{P}^5$ which is an H_6-orbit such that $H(K_i)$ is the stabilizer of each of these lines. The distinguished subgroup of order 3 in K_i fixes these lines pointwise.*

(ii) *Every H_6-orbit of lines in $\mathbf{P}^5$ consisting of 3 lines is one of the above.*

Proof. Let $\{L_1, L_2, L_3\}$ be an H_6-orbit of lines in $\mathbf{P}^5$. Then every line L_j has a stabilizer in $\mathbf{Z}/6 \times \mathbf{Z}/6$ of order 12. This must then be one of the groups K_i, $i = 1, 2, 3, 4$. Without loss of generality it suffices to consider the group generated by $\langle \tau, \sigma^3 \rangle$. Then the action of τ on the vector space associated to such a line L_j splits into a sum of two different characters. Therefore L_j is spanned by two basis vectors e_k, e_l. To obtain invariance under σ^3 the only possibilities are $\langle e_0, e_3 \rangle$, $\langle e_1, e_4 \rangle$ and $\langle e_2, e_5 \rangle$. Furthermore the distinguished subgroup order 3 generated by τ^2 fixes these three lines pointwise. $\square$

Lemma 7.5. (i) *To every subgroup $H(K_i)$, $i = 1, 2, 3, 4$ of H_6 one can associate a unique set of three 3-spaces in $\mathbf{P}^5$ which is an H_6-orbit such that $H(K_i)$ is the stabilizer of each of these $\mathbf{P}^{3'}$s.*

(ii) *Every H_6-orbit of 3-spaces in $\mathbf{P}^5$ consisting of three $\mathbf{P}^{3'}$s is one of the above.*

Proof. This is the dual statement to Lemma 7.4. Given three lines L_1, L_2, L_3 which form an H_6-orbit, the three $\mathbf{P}^{3'}$s are the spaces spanned by two of these lines. $\square$

Next we turn to the space of cubic forms $H^0(\mathcal{O}_{\mathbf{P}^5}(3))$ which we want to study as an H_6-, resp. G_6-module.

Lemma 7.6. *The G_6-module $H^0(\mathcal{O}_{\mathbf{P}^5}(3))$ is a sum of four 2-dimensional and twelve 4-dimensional representations. As H_3-representation it is a sum of characters. The trivial character corresponds to the four pencils. The other 8 come in pairs (given by the involution ι) and each pair determines three 4-dimensional irreducible G_6-representations. The subspace of 2-dimensional representations is spanned by*

$$\langle x_0^3 + x_2^3 + x_4^3, x_1^3 + x_3^3 + x_5^3 \rangle$$

$$\langle x_0 x_2 x_4, x_1 x_3 x_5 \rangle$$

$$\langle x_3^2 x_0 + x_5^2 x_2 + x_1^2 x_4, x_4^2 x_1 + x_0^2 x_3 + x_2^2 x_5 \rangle$$

$$\langle x_1 x_2 x_3 + x_3 x_4 x_5 + x_5 x_0 x_1, x_2 x_3 x_4 + x_4 x_5 x_0 + x_0 x_1 x_2 \rangle.$$

Proof. This is a straightforward computation. $\square$

Since all the 2-dimensional representations are mutually isomorphic this defines a $\mathbf{P}^3$ of pencils of cubics.

Proposition 7.7. *For every G_6-orbit of 3 lines in $\mathbf{P}^5$, there is a unique pencil of G_6-invariant pencils of cubics containing these lines.*

Proof. We can assume that the 3 lines in question are $\langle e_0, e_3 \rangle$, $\langle e_1, e_4 \rangle$ and $\langle e_2, e_5 \rangle$. A general pencil of G_6-invariant pencils of cubics is of the form

$$a\langle x_0^3 + x_2^3 + x_4^3, x_1^3 + x_3^3 + x_5^3 \rangle$$
$$+ b\langle x_0 x_2 x_4, x_1 x_3 x_5 \rangle$$
$$+ c\langle x_3^2 x_0 + x_5^2 x_2 + x_1^2 x_4, x_4^2 x_1 + x_0^2 x_3 + x_2^2 x_5 \rangle$$
$$+ d\langle x_1 x_2 x_3 + x_3 x_4 x_5 + x_5 x_0 x_1, x_2 x_3 x_4 + x_4 x_5 x_0 + x_0 x_1 x_2 \rangle.$$

Such a pencil contains the above lines if and only if $a = c = 0$. □

Remark 7.8. (i) It is also easy to determine the pencil of pencils containing the other minimal H_6-orbits of lines. E.g. the three lines fixed by the elements σ and τ^3 are the line $\langle (1, 1, 1, 1, 1, 1), (1, -1, 1, -1, 1, -1) \rangle$ and its τ-translates. The corresponding pencil of pencils is given by $3a + b = c + d = 0$.

(ii) Every pencil of G_6-invariant pencils of cubics containing a minimal orbit $\{L_1, L_2, L_3\}$ also contains the three $\mathbf{P}^{3'}$s spanned by two of these lines. (This can be seen by direct inspection). Hence every such pencil has a base locus consisting of three $\mathbf{P}^{3'}$s and a residual 3-fold X of degree 6. By Proposition 6.1, X is an elliptic scroll. Furthermore, every elliptic 3-fold scroll X_F, as in Section 3, is G_6-invariant, so it is contained in a G_6-invariant pencil of cubics. Hence for a general pencil of cubics the residual X must be of the form X_E for a suitable elliptic curve E.

Our next aim is to study the Heisenberg action on the embedded abelian surfaces $E \times F$. Recall that every abelian surface with a very ample $(1, 6)$-polarization can be embedded G_6-equivariantly into $\mathbf{P}^5$ and that the choice of such an embedding is equivalent to the choice of a level-6 structure on $E \times F$ (i.e. a canonical level struture associated to the polarization H by which we mean a symplectic basis of the kernel of the map $\lambda_H : A \to \hat{A}$). We denote the family of all G_6-invariant abelian surfaces in $\mathbf{P}^5$ with two plane elliptic fibrations by $\mathcal{A}$. As in Section 1 we start with curves E and F and a $3:1$ morphism $\gamma : E \to F$. We can assume that $E = \mathbf{C}/(\mathbf{Z}\tau + \mathbf{Z})$, $F = \mathbf{C}/(\mathbf{Z}3\tau + \mathbf{Z})$ and that the map γ is induced by $\gamma(z) = 3z$. We denote the generators τ and 1 of the lattice $\mathbf{Z}\tau + \mathbf{Z}$ by e_1 and e_3 and set $s_6 = e_1/6$ and $t_6 = e_3/6$. Moreover we denote the generators 3τ and 1 of the lattice $\mathbf{Z}3\tau + \mathbf{Z}$ by e_2 and e_4. Then $\gamma(e_1) = e_2$, $\gamma(e_3) = 3e_4$. The point $u_6 = \gamma(s_6)$ is represented by $e_2/6$ and $\gamma(t_6) = 3v_6$ is represented by $e_4/2$. We choose v_6 as the point represented by $e_4/6$. In this set-up the product $A = E \times F$ is given by the period matrix

$$\begin{pmatrix} \tau & 0 & 1 & 0 \\ 0 & 3\tau & 0 & 1 \end{pmatrix}.$$

The first step is to understand the polarization $H = 2E + \Gamma$ in terms of a Riemann form. This is necessary to understand the level-6 structures on A. First we look at the semi-positive line bundle defined by E. The corresponding Riemann form with respect to the lattice $L = \mathbf{Z}e_1 + \mathbf{Z}e_2 + \mathbf{Z}e_3 + \mathbf{Z}e_4$ is clearly given by

$$H_E = \begin{pmatrix} 0 & 0 & 0 & 0 \\ 0 & 0 & 0 & 1 \\ 0 & 0 & 0 & 0 \\ 0 & -1 & 0 & 0 \end{pmatrix}.$$

Next we want to identify the form H_Γ with respect to the chosen basis $e_1, \ldots, e_4$. Since $\Gamma \cdot F = 1$ we can write $A = \Gamma \times F$. By abuse of notation, let $\gamma = (\mathrm{id}, \gamma) : E \to E \times F$ be the embedding of E into A. Then $\gamma(E) = \Gamma$. We have $\gamma(e_1) = e_1 + e_2 =: f_1$, $\gamma(e_3) = e_3 + 3e_4 =: f_2$. We can also choose f_1, f_2, e_2, e_4 as a basis for L. With respect to this basis the semi-positive form H_Γ is given by $(e_2, e_4) = 1$ and all other products 0. A straightforward calculation then shows that in terms of the basis $e_1, \ldots, e_4$ the form H_Γ is given by:

$$H_\Gamma = \begin{pmatrix} 0 & 0 & 3 & -1 \\ 0 & 0 & -3 & 1 \\ -3 & 3 & 0 & 0 \\ 1 & -1 & 0 & 0 \end{pmatrix}.$$

Since $H = 2E + \Gamma$ it follows that the corresponding form with respect to the basis $e_1, \ldots, e_4$ is given by

$$H = \begin{pmatrix} 0 & 0 & 3 & -1 \\ 0 & 0 & -3 & 3 \\ -3 & 3 & 0 & 0 \\ 1 & -3 & 0 & 0 \end{pmatrix}.$$

Note that this is indeed the form associated to a $(1, 6)$-polarization, since

$$\det \begin{pmatrix} 3 & -1 \\ -3 & 3 \end{pmatrix} = 6.$$

Our next aim is to identify the group $\Theta(H) = \ker(\lambda_H : A \to \hat{A})$ as a subgroup of $A^{(6)} = E^{(6)} \times F^{(6)}$. What we have to do is to find a basis of $L^\vee/L$ where $L^\vee$ is the dual lattice with respect to the form H. General theory tells us that $L^\vee/L \cong \mathbf{Z}/6 \times \mathbf{Z}/6$. It is a straightforward calculation to check that $e_1/2 + e_2/6, e_3/6 + e_4/2 \in L$. We can take these elements as generators of $L^\vee/L$. Note that as points in $A = E \times F$ these are just the points $(3s_6, u_6)$ and $(t_6, 3v_6)$.

Let $w_1 = e_1/2 + e_2/6$, $w_2 = e_3/6 + e_4/2$. Then a straightforward calculation shows that for the Weil pairing with respect to H we have

$$(w_1, w_2) = \left(e^{2\pi i/6} \right),$$

i.e. these points define a level-6 structure. Note that $2w_1 \in F = \{0\} \times F$ and $2w_2 \in E = E \times \{0\}$. In particular the 2 groups of order 3 generated by $2w_1$ and $2w_2$ each respect one of the two plane elliptic fibrations of $A = E \times F$. Since the embedded abelian surface $A = E \times F$ is G_6-invariant the same holds for the scrolls X_E and X_F

defined by the 2 plane cubic fibrations. Since the groups of order 3 generated by $2w_1$, resp. $2w_2$ each respect one of these fibrations it follows that they act trivially on the 3 singular lines of the scroll X_E, resp. X_F. In particular this gives 2 groups $H(K_i)$ and $H(K_j)$ which each has the 3 singular lines of the scrolls X_E, resp. X_F as one of its orbits. Any other choice of a level-6 structure on A gives an analogous picture. We shall return to this in a moment.

We now want to understand the variety $\mathcal{A}$ parametrizing G_6-invariant abelian surfaces with 2 plane cubic fibrations. We have already observed that each of these two fibrations determines a singular scroll X_E, resp. X_F and a group $H(K_i)$, resp. $H(K_j)$. The abelian surface A is the intersection of the scrolls X_E and X_F (cf. Proposition 3.5.) This defines a decomposition of the family $\mathcal{A}$ into six families $\mathcal{A}_{ij}$ where $\{i, j\} \subset \{1, 2, 3, 4\}$. We want to exhibit a concrete parametrization of the families $\mathcal{A}_{ij}$, thereby also showing that the $\mathcal{A}_{ij}$ form six irreducible components. To do this we go back to an elliptic curve E as before and the level-6 structure on E given by (s_6, t_6). We can perform the above construction and associate to these data the surface $A = E \times F$, the polarization $H = 2E + \Gamma$ and the level-6 structure (w_1, w_2). This gives us a morphism

$$\psi_{ij} : X^0(6) \to \mathcal{A}_{ij}$$

from the (open) elliptic modular curve $X^0(6)$ parametrizing elliptic curves with a level-6 strucure to $\mathcal{A}_{ij}$. Note that the (compact) modular curve $X(6)$ is an elliptic curve.

Lemma 7.9. *The map $\psi_{ij} : X^0(6) \to \mathcal{A}_{ij}$ is surjective onto the component $\mathcal{A}_{ij}$ and has degree 3.*

Proof. Here we shall treat the case where K_i and K_j are the groups determined by $\langle \sigma^2 \rangle$ and $\langle \tau^2 \rangle$. This is no loss of generality. Going back to the abelian surface $A = E \times F$ we want to study the possible embeddings of A into $\mathbf{P}^5$ such that $A \in \mathcal{A}_{ij}$. Since G_6-invariant embeddings of A correspond to the choice of a level-6 structure (w_1', w_2') we have to look for those level-6 structures (w_1', w_2') such that $2w_1' \in F = \{0\} \times F$ and $2w_2' \in E = E \times \{0\}$ or $2w_1' \in E$ and $2w_2' \in F$. These two cases correspond to changing the role of E and F and it is, therefore, enough to look at the first possibility, namely $2w_1' \in F$ and $2w_2' \in E$. Since (w_1, w_2) is a basis of $\Theta(H)$ we can write $w_1' = \alpha w_1 + \beta w_2$, $w_2' = \gamma w_1 + \delta w_2$. Moreover $2w_1 \in F$ and hence $2w_1' \in F$ if and only if $2\beta w_2 \in F$ which is only the case for $\beta = 0$ or 3. Moreover w_1' has to be an element of order 6. This gives us three possibilities for $\pm w_1'$, namely $w_1, w_1 + 3w_2, 2w_1 + 3w_2$. Since (w_1', w_2') and $(-w_1', -w_2')$ define the same level-6 structure we can assume that w_1' is one of the 3 points above. A similar argument can be applied to w_2' and altogether we find the following 6 possibilities for level-6 structure (w'_1, w'_2) with $2w'_1 \in F$ and $2w'_2 \in E$:

$$(w_1, w_2), (w_1, w_2 + 3w_1), (w_1 + 3w_2, w_2),$$
$$(-2w_1 + 3w_2, w_2 + 3w_1), (-2w_1 + 3w_2, -2w_2 + 3w_1), (w_1 + 3w_2, -2w_2 + 3w_1).$$

For every pair (w'_1, w'_2) as above the pair $(-w'_2, w'_1)$ is a level-6 structure with $-2w'_2 \in E$, $2w'_1 \in F$. In this way we obtain 12 level-6 structures which belong to the pair $\{i, j\}$. This fits in with the number of level-6 structures which is given by

$$\frac{1}{2} 6^3 \left(1 - \frac{1}{4}\right) \left(1 - \frac{1}{9}\right) = 72 = 12 \cdot 6,$$

where 6 corresponds to the number of pairs $\{i, j\} \subset \{1, 2, 3, 4\}$.

The map $\psi_{ij} : X^0(6) \to \mathcal{A}_{ij}$ was defined by associating to an elliptic curve E with level-6 structure (s_6, t_6) the abelian surface $A = E \times F$ with level-6 structure (w_1, w_2). We want to prove that this map is surjective which implies in particular that $\mathcal{A}_{ij}$ is irreducible. So we have to show that we can obtain all abelian surfaces $A = E \times F$ and all level-6 structures (w'_1, w'_2) as above by varying E and the level-6 structure (s_6, t_6). Let (s'_6, t'_6) be any level-6 structure on E. To this we associate a surface $A' = E \times F'$ with $F' = E/\langle 2t'_6 \rangle$. If we want that $F' = F$ we must (at least for general F) have that $\langle 2t'_6 \rangle = \langle 2t_6 \rangle$. Moreover t'_6 must be a point of order 6. Up to sign this leaves us with the possibilities $t'_6 = t_6, 3s_6 + t_6, 3s_6 + 2t_6$. Altogether we obtain 18 possible level-6 structures, namely

$$(s_6 + 2it_6, t_6), (s_6 + (2i+1)t_6, t_6)$$
$$(s_6 + 2it_6, 3s_6 + t_6), (-2s_6 + (2i+1)t_6, 3s_6 + t_6)$$
$$(-2s_6 + (2i+1)t_6, 3s_6 - 2t_6), (s_6 + (2i+1)t_6, 3s_6 - 2t_6)$$

where in each case $i = 0, 1, 2$. Each of the values $i = 0, 1, 2$ gives the same level-6 structure on A. Hence under the morphism $\psi_{ij} : X^0(6) \to \mathcal{A}_{ij}$ the pairs $(E, (s'_6, t'_6))$ are mapped 3 : 1 to $(A = E \times F, (w'_1, w'_2))$ where (w'_1, w'_2) runs through all 6 possible level-6 structures on A with $2w'_1 \in F$ and $2w'_2 \in E$. In particular the map $X^0(6) \to \mathcal{A}_{ij}$ is 3 : 1 and surjective and $\mathcal{A}_{ij}$ is irreducible. $\square$

Next we consider the family $\mathcal{V}$ of G_6-invariant scrolls which arise as $\mathbf{P}^2$-scrolls defined by a plane cubic fibration of a G_6-invariant abelian surface A. Such a scroll is singular along 3 lines which form an orbit of one of the groups $H(K_i)$ and hence there is a natural decomposition $\mathcal{V} = \mathcal{V}^1 \cup \mathcal{V}^2 \cup \mathcal{V}^3 \cup \mathcal{V}^4$ where $\mathcal{V}^i$ is the set of those scrolls which are invariant under $H(K_i)$. These scrolls are in 1 : 1 correspondence with points in an open set of the pencil of pencils of cubics associated to the group $H(K_i)$ (cf. Remark 7.8. ii). In particular the varieties $\mathcal{V}^i$ are irreducible and rational.

The G_6-action on the elliptic scrolls restricts to pencils of G_6-invariant abelian surfaces. We describe these before we return to the surfaces which lie on two scrolls.

Recall that the normalization $\varphi : \mathbf{P}_F(\mathcal{E}) \to X_F \subset \mathbf{P}^5$ of the scroll X_F is a $\mathbf{P}^2$-bundle over the elliptic curve F associated to the rank 3 vector bundle

$$\mathcal{E} = \mathcal{L}_0 \oplus h^*\mathcal{L}_0 \oplus (h^2)^*\mathcal{L}_0, \qquad (h^3 = \mathrm{id}).$$

X_F is singular along 3 lines L_i, and

$$(h^i)^*\mathcal{L}_0 = \varphi^*\mathcal{O}_{L_i}(1), \qquad i = 0, 1, 2$$

are line bundles of degree 2 on F. If we assume that say

$$L_0 = \langle e_0, e_3 \rangle, \quad L_1 = \langle e_1, e_4 \rangle, \quad \text{and} \quad L_2 = \langle e_2, e_5 \rangle,$$

then

$$\mathrm{Stab}(L_i) = \langle \tau, \sigma^3 \rangle.$$

Recall from Section 3 that we may find $\mathbf{P}_F(\mathcal{E})$ as a quotient of $\mathbf{P}^2 \times E$ by a subgroup $\mathbf{Z}/3 \subset H_3$. This subgroup leaves three sections $E \to \mathbf{P}^2 \times E$ invariant, and these three sections are mapped in the quotient to the three sections $F \to \mathbf{P}_F(\mathcal{E})$ which again are mapped 2 : 1 to the singular lines in X_F. In this set-up we may describe the G_6-action on the abelian surfaces on X_F. These surfaces are all pulled back to anticanonical divisors on $\mathbf{P}_F(\mathcal{E})$. Recall that $h^0(\mathcal{O}_{\mathbf{P}(\mathcal{E})}(-K)) = 4$. Elements in $|-K|$ pull back to products

$$E' \times E \subset \mathbf{P}^2 \times E$$

where E' is a cubic curve invariant under the subgroup of order 3; in suitable coordinates it is defined by a form in the web

$$\langle x_0^3, x_1^3, x_2^3, x_0 x_1 x_2 \rangle.$$

These forms are precisely the invariants of degree 3 of a subgroup of order 3 in H_3. In fact the elements of order 3 of H_6 which fix the lines L_i pointwise, leave each plane in X_F invariant and they lift to an action on $\mathbf{P}_F(\mathcal{E})$ and $\mathbf{P}^2 \times E$ which leaves each plane invariant and fixes the three special sections pointwise. These sections, in suitable coordinates, meet each plane in the points $(1, 0, 0)$, $(0, 1, 0)$, $(0, 0, 1)$ and the action is given by $\tau \in H_3$ in the plane.

The involution ι leaves L_0 fixed while L_1 and L_2 are interchanged. This also lifts to $\mathbf{P}^2 \times E$. The ι-invariant plane cubics in the above web form the net

$$\langle x_0^3, x_1^3 + x_2^3, x_0 x_1 x_2 \rangle.$$

G_6 finally permutes the three lines cyclically, so the corresponding plane cubics are defined by invariant forms in the variables permuted cyclically. This action is the one defined by $\sigma \in H_3$. The G_6-invariant anticanonical divisors therefore correspond precisely to the Hesse-pencil

$$\langle x_0^3 + x_1^3 + x_2^3, x_0 x_1 x_2 \rangle.$$

As an H_3-module $H^0(\mathcal{O}_{\mathbf{P}^2}(3))$ splits as the Hesse-pencil plus 8 characters. Two of these give rise to anticanonical divisors on $\mathbf{P}_F(\mathcal{E})$. The remaining 6 characters of H_3 give rise to bielliptic surfaces in $|-K+T|$ where T is some torsion divisor.

At this point we can also understand the different G_6-embeddings of a scroll X_F into $\mathbf{P}^5$. For this we start with a G_6-invariant scroll X_F. Its desingularisation is $\mathbf{P}_F(\mathcal{E})$. First note that the Heisenberg group H_2 acts on the planes of X_F and that this induces actions of H_2 on the base curve F of $\mathbf{P}_F(\mathcal{E})$ and on the 3 sections which are mapped to the singular lines of X_F. To define a non-degenerate map from $\mathbf{P}_F(\mathcal{E})$ to $\mathbf{P}^5$ is the same as defining an isomorphism from $H^0(\mathcal{O}_{\mathbf{P}_F(\mathcal{E})}(1)) = H^0(\mathcal{E})$ to V. The decomposition $\mathcal{E} = \mathcal{L}_0 \oplus h^*\mathcal{L}_0 \oplus (h^2)^*\mathcal{L}_0$ defines a decomposition $H^0(\mathcal{E}) = U_1 \oplus U_2 \oplus U_3$ where each of the spaces U_i is the space of sections of a degree 2 line bundle on F and hence has dimension 2. The level 2 structure on F which comes from the action of G_6 on X_F gives us an identification (unique up to a scalar) of each of the U_i with the H_2-module U. Now we pick one of the 4 subgroups K_j. Recall that as an H_2-module $V = 3U$, whereas as an $H(K_j)$-module $V = U \oplus U' \oplus \bar{U}'$ (cf. Lemma 7.3). In order to map the 3 decomposing sections of $\mathbf{P}_F(\mathcal{E})$ to the singular lines associated to the group K_j we must map each of the spaces U_i to one of the spaces U, U' and $\bar{U}'$. The group G_6 acts transitively on the lines associated to K_j and hence we can (up to an element in G_6) assume that U_1, U_2, U_3 map to $U, U', \bar{U}'$. This defines the isomorphism from $H^0(\mathcal{E})$ to V up to an element in $(\mathbf{C}^*)^3$. On the other hand the Heisenberg group H_3 acts irreducibly on the 3-dimensional space given by the decomposition $V = 3U$. Hence, by Schur's lemma, the isomorphism from $H^0(\mathcal{E})$ to V is uniquely defined (up to a scalar). This shows that given a G_6-invariant embedding of X_F in $\mathbf{P}^5$ we can find four such embeddings, one for each of the subgroups K_j.

We can now consider the incidence correspondence

$$\mathcal{I}(A, X) = \{(A, X) | A \in \mathcal{A}, X \in \mathcal{V}, A \subset X\} \subset \mathcal{A} \times \mathcal{V}.$$

Proposition 7.10. *This is a* 2 : 3 *correspondence.*

Proof. Clearly $A \in \mathcal{A}$ is the intersection of two scrolls. On the other hand, the number of such surfaces in a scroll is the answer to the question: how many abelian surfaces are there in the pencil $H \subset |-K_{\mathbf{P}_F(\mathcal{E})}|$ coming from the Hesse pencil which are isomorphic to the product $E \times F$? For this we want to find an embedding $E \subset \mathbf{P}^2 \times E$ which after projection to $\mathbf{P}^2$ maps E to an element in the Hesse pencil which is $\mathbf{Z}/3$-equivariant (here $\mathbf{Z}/3$ is the group which acts on $\mathbf{P}^2 \times E$ with quotient $\mathbf{P}_F(\mathcal{E})$). To embed E as an element in the Hesse pencil is the same as choosing a level 3 structure on E. Say $\mathbf{Z}/3$ acts on E by translation with an element σ' of order 3. So we have to ask in how may ways we can extend σ' to a level 3 structure. If τ' is another 3-torsion point with $(\sigma', \tau') = 1$ (here $(,)$ is the Weil pairing), we have the possibilities (σ', τ'), $(\sigma', \tau'\sigma')$, $(\sigma', \tau'\sigma'^2)$ and no others. This gives us the three possibilities. □

Remark 7.11. Notice that the three choices of $\mathbf{Z}/3$-subgroups of $\mathbf{Z}/3 \times \mathbf{Z}/3$ correspond precisely to the three subgroups K_i distinct from the subgroup K_j which stabilizes the three singular lines of the scroll X_F.

Corollary 7.12. *Given two distinct subgroups K_i and K_j, the incidence correspondence $\mathcal{I}(A, X)$ defines a $1 : 1$ correspondence between elliptic scrolls X whose singular lines are invariant under these two subgroups.*

Proof. A scroll singular along one of the triples of lines contains exactly one abelian surface which forms the intersection with a scroll singular along the other triple of lines. □

Corollary 7.13. *The abelian surfaces with two plane cubic curve fibrations are contained in precisely a pencil of G_6-invariant pencils of cubic hypersurfaces.*

Proof. The space of G_6-invariant pencils of cubics is a $\mathbf{P}^3$, and the G_6-invariant scrolls singular along a triple of lines are defined by points on four lines in $\mathbf{P}^3$ corresponding to the four subgroups K_i. An abelian surface in the intersection of two scrolls is contained in the pencils of cubics corresponding to a line joining two of these lines. If there were more than a pencil of invariant pencils of cubics through the surface, then it would be contained in four scrolls. This is impossible (see Proposition 7.10). □

Consider the Grassmannian of lines in the space $\mathbf{P}^3$ of G_6-invariant pencils of cubics. The four lines of pencils defining G_6-invariant scrolls are pairwise disjoint. The lines corresponding to abelian surfaces $A \in \mathcal{A}_{ij}$ define a one to one correspondence between two skew lines, so they form a conic section in the Grassmannian. Summing up we have 6 disjoint conic sections in the Grassmannian parametrizing $\mathcal{A}$.

We can now sum up our discussion as follows:

Proposition 7.14. *The variety $\mathcal{A}$ consists of 6 irreducible components $\mathcal{A}_{ij}$ indexed by the pairs $\{i, j\} \subset \{1, 2, 3, 4\}$. The incidence variety $\mathcal{I}(A, X)$ consists of 12 components $\mathcal{I}_{ij}^k$ indexed by pairs $\{i, j\}$ and an element $k \in \{i, j\}$. For every component $\mathcal{I}_{ij}^k$ there is a diagram*

$$X^0(6) \xrightarrow{\varphi} \mathcal{I}_{ij}^k \xrightarrow{q} \mathcal{V}^k$$
$$p \downarrow$$
$$\mathcal{A}_{ij}$$

where φ is $3 : 1$ and p and q are $1 : 1$.

Proof. The map φ is the map which associates to each pair $(E, (s_6, t_6))$ the abelian surface $A = E \times F$, the level 6 structure (w_1, w_2) and the scroll X which is the

$\mathbf{P}^2$-scroll attached to A which is K_k-invariant. This map factors through $\mathcal{A}_{ij}$ (giving the map ψ_{ij} of Lemma 7.9) and in particular the projection p has an inverse.

The map q is $1:1$ by Corollary 7.12. □

Remark 7.15. The components $\mathcal{A}_{ij}$, $\mathcal{I}_{ij}^k$ and $\mathcal{V}^k$ are all rational. We have already observed this for the varieties $\mathcal{V}^k$ which are isomorphic to an open set of a pencil of cubics. Since the maps p and q are birational, this is also true for the other varieties.

For our conclusion on 3-folds $Y = X_E \cup X_F$ we study a G_6-invariant rational cubic scroll. It plays a crucial role when we later bilink Y to a 3-fold of degree 7 (cf. Propositions 4.7 and 5.2).

The subgroups $H(K_i)$ have a nontrivial intersection, namely the subgroup $\langle \tau^3, \sigma^3 \rangle$. This subgroup therefore fixes all 4 triples of lines. We shall find small G_6-orbits of planes intersecting all these lines. Consider the subgroup

$$G_3 = \langle \sigma^2, \tau^2, \iota \rangle \subset G_6.$$

We look for G_3-invariant planes.

Now the action of τ^2 is defined by

$$\tau^2 = \mathrm{diag}(1, \eta, \eta^2, 1, \eta, \eta^2),$$

where $\eta = e^{2\pi i/3}$ while σ^2 sends $x_i \mapsto x_{i+2}$ and ι sends $x_i \mapsto x_{-i}$. This implies easily that the 4 planes in the G_6-orbit of any such plane must be of the form

$$\begin{array}{ll}
\alpha x_0 + \beta x_3 = \alpha x_2 + \beta x_5 = \alpha x_4 + \beta x_1 = 0 & P_0 \\
\alpha x_0 - \beta x_3 = \alpha x_2 - \beta x_5 = \alpha x_4 - \beta x_1 = 0 & P_1 \\
\beta x_0 + \alpha x_3 = \beta x_2 + \alpha x_5 = \beta x_4 + \alpha x_1 = 0 & Q_0 \\
\beta x_0 - \alpha x_3 = \beta x_2 - \alpha x_5 = \beta x_4 - \alpha x_1 = 0 & Q_1.
\end{array}$$

In particular we notice that there is a 1-parameter family of such planes. The union of these planes forms a rational cubic scroll R: In fact the union of the planes is defined by

$$\mathrm{rank} \begin{pmatrix} x_0 & x_2 & x_4 \\ x_3 & x_5 & x_1 \end{pmatrix} \leq 1,$$

so the scroll R is the Segre embedding of $\mathbf{P}^1 \times \mathbf{P}^2$ in $\mathbf{P}^5$.

Remark 7.16. The orbits of planes in R do not all have length 4. In fact

$$P_0 = P_1 \quad \text{and} \quad Q_0 = Q_1 \leftrightarrow \alpha\beta = 0$$

$$P_0 = Q_0 \quad \text{and} \quad P_1 = Q_1 \leftrightarrow \alpha^2 - \beta^2 = 0$$

and

$$P_0 = Q_1 \quad \text{and} \quad P_1 = Q_0 \leftrightarrow \alpha^2 + \beta^2 = 0.$$

Thus the orbits are of length 2 precisely when

$$(\alpha, \beta) = (1, 0), (0, 1), (1, 1), (1, -1), (1, i), (1, -i).$$

Remark 7.17. The cubic scroll R contains all four sets of G_6-invariant triples of lines. These lines are all transverse to the planes of R. Thus the six singular lines of two scrolls X_E and X_F which intersect along an abelian surface $E \times F$ are all contained in a rational cubic scroll.

In Lemma 8.1 we shall show that the union $Y = X_E \cup X_F$ is contained in 6 quintics. These are all singular along two triples of lines in the scroll R. We analyze these quintics more closely. With bihomogeneous coordinates s, t and y_0, y_1, y_2 on $R \cong \mathbf{P}^1 \times \mathbf{P}^2$, the restriction of quintics singular along the two triples of lines generated by τ and σ have the form

$$y_0^2 y_1^3 - y_0^3 y_1 y_2 - y_0 y_1^2 y_2^2 + y_0^2 y_2^3$$
$$y_0^3 y_1^2 - y_0 y_1^3 y_2 - y_0^2 y_1 y_2^2 + y_1^2 y_2^3$$
$$y_0^2 y_1^2 y_2 - y_0^3 y_2^2 - y_1^3 y_2^2 + y_0 y_1 y_2^3,$$

multiplied by any quintic in s, t. The 6 quintics in the ideal of two scrolls are determined by a pencil of quintics in s, t. This pencil is G_3-invariant, so it is an element in the net

$$\langle s^5, t^5 \rangle \oplus \langle s^4 t, st^4 \rangle \oplus \langle s^3 t^2, s^2 t^3 \rangle$$

of pencils. The basepoints of this pencil define precisely the planes of R common to all quintics through W. If the pencil has no basepoints, then the two scrolls would intersect R in only the singular lines. We shall see in Proposition 7.18 that this is not the case. On the other hand every point of intersection of the two scrolls with R outside the 6 singular lines lies in a plane which must be defined by a basepoint of the pencil, so there are at most 4 planes with such an intersection. These planes clearly form orbits under G_6 so there are 2 or 4 planes as explained above.

We carry this analysis a bit further in order to show that the union of two scrolls is bilinked to a Del Pezzo 3-fold W_7. First we consider the intersection of the cubic scroll R and an elliptic scroll X. We use the fact that X is G_6-invariant and that each plane in R is invariant under the subgroup $G_3 \subset G_6$ generated by $\langle \sigma^2, \tau^2 \rangle$. Thus any point in a plane has an orbit by this subgroup of order divisible by 3, and any invariant curve has degree divisible by 3.

Proposition 7.18. *The intersection $X \cap R$ is a curve of degree 18. It has the following decomposition into irreducible components:*

$$C = 2L_1 + 2L_2 + 2L_3 + l_1 + \cdots + l_{12}$$

where L_i are the singular lines of X and as such meet every plane in R. The lines l_i form 4 triangles in 4 planes of the $\mathbf{P}^2$-bundle R.

Proof. First we prove that the intersection is a curve. If not, it contains an irreducible surface, call it T. If T contains a plane, this is common to R and X. But no plane in X is stabilized by G_3, in fact X intersects only one of the 4 triples of lines, so this is impossible. Thus T intersects each plane in R in some curve. This curve has degree 3 since X is contained in two cubics and the curve is invariant under G_3.

As noted above, the three singular lines of the scroll X all lie in R, and are transverse to the planes in R. Since every cubic through X is singular along the three lines, the intersection of a cubic with R is a triangle in each plane. In fact T in $R = \mathbf{P}^2 \times \mathbf{P}^1$ must equal $T = T_0 \times \mathbf{P}^1$, where T_0 is this triangle, i.e. T is the union of three quadric surfaces. But X does not contain any quadric surface, so the intersection $X \cap R$ cannot contain a surface. It is therefore a curve, call it Γ. Since the intersection is proper, this curve has, by Bezout, degree 18.

We have seen already that the curve Γ contains the three singular lines. In fact on $\tilde{X}$, the normalization of X, the preimage $\tilde{\Gamma}$ of Γ contains the curves F_i which lie 2 : 1 over the singular lines L_i. Thus on $\tilde{X}$ we have

$$\tilde{\Gamma} = F_1 + F_2 + F_3 + ah^2 + bhf,$$

in notation as in the proof of Proposition 4.10, with $6a + b = 12$. Since any plane which intersects R properly intersects in a scheme of length 3, the general plane of X must intersect R in three points, the points of intersection between the plane and the lines L_i. Therefore $a = 0$, and $\tilde{\Gamma} = F_1 + F_2 + F_3 + l_1 + \cdots + l_{12}$ where the l_i are lines in the planes of $\tilde{X}$.

Finally if a plane in R intersects X along a curve, this curve has degree divisible by three, in fact equal three, since X lies in two cubics. Therefore the twelve lines l_i form triangles in four planes of R. $\square$

8. Conclusion

We shall conclude by showing that the union of two scrolls X_E and X_F which intersect along an abelian surface $E \times F$ is bilinked to a Del Pezzo W_7.

Let $Y = X_E \cup X_F$.

Lemma 8.1. $h^0(\mathcal{I}_Y(5)) = 6$ and Y lies on irreducible quintic hypersurfaces.

Proof. Consider the exact sequence

$$0 \to \mathcal{I}_Y(5) \to \mathcal{I}_{X_E}(5) \oplus \mathcal{I}_{X_F}(5) \to \mathcal{I}_{E \times F}(5) \to 0.$$

First, the intersection $E \times F \cap \mathbf{P}^3$ of $E \times F$ with a general $\mathbf{P}^3$ is at least 5-regular in the sense of of Castelnuovo–Mumford, so the same is true for $E \times F$, i.e. $h^1(\mathcal{I}_{E \times F}(k)) = 0$, when $k \geq 4$. In particular, $h^0(\mathcal{I}_{E \times F}(5)) = 102$. Similarly, the intersection $Y \cap \mathbf{P}^2$ of Y with a general plane is 5-regular in the sense of Castelnuovo–Mumford so $h^1(\mathcal{I}_Y(k)) = 0$ for $k \geq 4$. Furthermore $h^0(\mathcal{I}_{X_E}(5)) = h^0(\mathcal{I}_{X_F}(5)) = 54$ by Corollary 6.4. Therefore $h^0(\mathcal{I}_Y(5)) = 6$. If every quintic in the ideal of Y is reducible, then they all have a fixed quartic hypersurface as a component. But Y is G_6-invariant, and there are no G_6-invariant quartic hypersurfaces, so the lemma follows. $\square$

Now, according to Proposition 7.18 the two elliptic scrolls X_E and X_F intersect the rational scroll R in the six singular lines and in four triangles each. Since $X_E \cap X_F$ is a surface which does not meet the six singular lines, the two sets of four triangles must lie pairwise in the same four planes of R. Thus the six quintics through the two scrolls $X_E \cup X_F$ contain these four planes. In particular, if Z is linked (5, 5) to the union $X_E \cup X_F$, then Z has degree 13 and contains four planes of R. Furthermore since each quintic intersects the planes in R in curves singular in the six points of intersection with the singular lines, Z will intersect each plane in one point in addition to the six singular points. The general quintic in the pencil intersects a plane in an irreducible curve, so, by Bezout, there are no quartic curves singular in the six singular points. Therefore any quartic hypersurface containing Z must also contain R.

Now, the partial normalization Y' of $Y = X_E \cup X_F$ along the 6 singular lines is Calabi–Yau (cf. Proposition 3.5). In particular the dualizing sheaf $\omega_{Y'}$ is trivial and has one global section. It follows from the next lemma that the dualizing sheaf ω_Y also has a section.

Let $f : Y' \to Y$ be a finite morphism of projective schemes and denote by ω_Y the dualizing sheaf of Y. Then $f^! \omega_Y$ is a dualizing sheaf of Y' (See [Ha, Ex III, 7.2] and for the definition of $f^! \omega_Y$ see [Ha, Ex III, 6.10]). Hence we can put $\omega_{Y'} = f^! \omega_Y$.

Lemma 8.2. *If* $H^0(Y', \omega_{Y'}) \neq 0$, *then also* $H^0(Y, \omega_Y) \neq 0$.

Proof. Using [Ha, Ex III,6.10(b)] we have that

$$\begin{aligned} H^0(Y', \omega_{Y'}) &= \mathrm{Hom}_{\mathcal{O}_{Y'}}(\mathcal{O}_{Y'}, \omega_{Y'}) = \mathrm{Hom}_{\mathcal{O}_{Y'}}(\mathcal{O}_{Y'}, f^! \omega_Y) \\ &= H^0(\mathcal{H}om_{\mathcal{O}_{Y'}}(\mathcal{O}_{Y'}, f^! \omega_Y)) \\ &= H^0(f_* \mathcal{H}om_{\mathcal{O}_{Y'}}(\mathcal{O}_{Y'}, f^! \omega_Y)) \\ &= H^0(\mathcal{H}om_{\mathcal{O}_Y}(f_* \mathcal{O}_{Y'}, \omega_Y)) \quad ([\text{Ha., Ex. III. 6.10(b)}]) \\ &= \mathrm{Hom}_{\mathcal{O}_Y}(f_* \mathcal{O}_{Y'}, \omega_Y). \end{aligned}$$

Hence a section s of $\omega_{Y'}$ gives rise to a morphism $\varphi_s : f_* \mathcal{O}_{Y'} \to \omega_Y$. Combining this with the natural morphism $\mathcal{O}_Y \to f_* \mathcal{O}_{Y'}$ we obtain a morphism $\mathcal{O}_Y \to \omega_Y$ and hence a section of ω_Y. $\square$

In the cohomology of the liaison exact sequence (cf. [PS])

$$0 \to \omega_Y \to \mathcal{O}_{Y \cup Z}(4) \to \mathcal{O}_Z(4) \to 0,$$

a section of ω_Y corresponds to a section of $h^0(\mathcal{I}_Z(4))$, i.e. to a quartic hypersurface containing Z. This quartic must contain R.

Let W be linked to Z in this quartic and a general quintic through Z. Then W must intersect R in a surface, in fact in a surface linked to 4 planes in the intersection of R with a quintic hypersurface. This is clearly a surface of degree 11. Furthermore the arithmetic genus of W is 1 by linkage, so $W \cup R$ has arithmetic genus 11. From the liaison exact sequences

$$0 \to \omega_Z(1) \to \mathcal{O}_{Y \cup Z}(5) \to \mathcal{O}_Y(5) \to 0,$$

$$0 \to \omega_Z(1) \to \mathcal{O}_{W \cup Z}(4) \to \mathcal{O}_W(4) \to 0$$

we get that $h^0(\mathcal{I}_W(4)) = h^0(\mathcal{I}_Y(5)) - 1 = 5$.

Clearly all the quartics through W have to contain also R. Thus $W \cup R$ is contained in 5 quartics.

For later we need that the 5 quartics are minors of a 4×5 matrix with linear entries. If we can show this in a special case, we may conclude by semicontinuity that so is $W \cup R$ in general (cf. [Ell]).

We do this by considering, in the notation of Remark 7.8, the points $(1, -3, 0, 0)$ and $(0, 0, 0, 1)$ in the $\mathbf{P}^3$ of G_6-invariant pencils of cubics. They are points on two distinct lines corresponding to two distinct subgroups $H(K_i)$. The corresponding pencils of cubics each define an elliptic scroll (in fact a reducible scroll) residual to three $\mathbf{P}^3$'s. The union of these scrolls lies on 6 quintics and is bilinked $(5, 5)$ and $(5, 4)$ to a 3-fold of degree 7 which lies in 5 quartic hypersurfaces. These 5 quartics define a determinantal 3-fold of degree 10. These claims are easily checked with i.e. [MAC].

Recall from Proposition 4.10 that W is a non-normal Del Pezzo 3-fold W_7 as soon as we have checked that no three of the singular lines lie in a $\mathbf{P}^3$, and that not all six lie in a rational normal quartic scroll, and finally that the common singular locus of the quartic hypersurfaces through W is precisely the 6 singular lines. While the former two requirements follows easily from our analysis of the singular lines in the previous section, the latter requirement is easily checked in the above example. Therefore we may, by Proposition 5.2, bilink W in complete intersections $(5, 4)$ and $(5, 5)$ to a non-normal Calabi–Yau 3-fold Y_t, non-normal only along 6 lines. Clearly, we may perform the bilinkage in a family, so the reducible 3-fold Y is a degeneration of Y_t. We have shown

Theorem 8.3. *The reducible 3-fold $Y = X_E \cup X_F$ is a degeneration of irreducible non-normal Calabi–Yau 3-folds of degree 12 in $\mathbf{P}^5$. The general such 3-fold is singular precisely along 6 disjoint lines.*

Remark 8.4. Computing the normalizer N_3 of $G_3 \subset GL(6, \mathbf{C})$ and its representations, one may show that there is an N_3-invariant linear complex in the Grassmannian

G of lines in the space of G_6-invariant pencils of cubics. This defines a one to one correspondence between any two of the four lines defined by the subgroups K_i. Therefore it is natural to guess that this complex defines the correspondences $\mathcal{A}_{ij}$, in fact that it parametrizes the set of all G_6-invariant abelian surfaces. This is proved by Gross and Popescu [GP2].

References

[Ch] Chang, M.-C., Classification of Buchsbaum subvarieties of codimension 2 in projective space, J. Reine Angew. Math. 401 (1989), 101–112.

[DP] Decker, W., Popescu, S., On surfaces in $\mathbf{P}^4$ and 3-folds in $\mathbf{P}^5$, in: Vector bundles in algebraic geometry, eds. Hitchin, N. J., Newstead, P. E., Oxbury, W. M., London Math. Soc. Lecture Notes Ser. 208, Cambridge University Press, 1995, 69–100.

[Ell] Ellingsrud, G., Sur le schema de Hilbert des variétés de codimension 2 dans $\mathbf{P}^e$ à cone de Cohen Macaulay, Ann. Sci. École Norm. Sup (4) 8 (1975), 423–432.

[ES] Ellingsrud, G., Strømme, S. A., The number of twisted cubic curves on the general quintic threefold, Math. Scand. 76 (1995), 5–34.

[Fu] Fulton, W., Intersection Theory, Ergeb. Math. Grenzgeb. (3) 2, Springer-Verlag, New York–Berlin–Heidelberg 1984.

[GP1] Gross, M., Popescu, S., Equations of $(1, d)$-polarized abelian surfaces, Math. Ann. 310 (1998), 333–377.

[GP2] Gross, M., Popescu, S., Calabi–Yau 3-folds and moduli of abelian surfaces, I and II, to appear.

[Ha] Hartshorne, R., Algebraic Geometry. Grad. Texts in Math. 52, Springer-Verlag, New York 1977.

[Kre] Kresch, A., FARSTA, a computer program for quantum cohomology, Appendix to: Quantum Cohomology at the Mittag-Leffler Institute, ed. by P. Aluffi, Scuola Normale Superiore, Pisa 1997.

[MAC] Bayer, D., Stillman, M., MACAULAY: A system for computation in algebraic geometry and commutative algebra, source and object code available for Unix and Macintosh computers, contact the authors, or download from zariski.harvard.edu via anonymous ftp.

[PS] Peskine, C., Szpiro, L., Liaison des variétés algébriques I, Invent. Math. 26 (1974), 271–302.

[Re] Reider, I., Vector bundles of rank 2 and linear systems on algebraic surfaces, Ann. of Math. 127 (1988), 309–316.

[Rey] Reye, T., Ueber lineare Systeme und Gewebe von Flächen zweiten Grades, J. Reine Angew. Math. 82 (1877), 54–83.

Real algebraic threefolds IV. Del Pezzo fibrations

János Kollár

Contents

1	Introduction	317
2	Real surfaces with Du Val singularities	322
3	The basic set-up	324
4	The classification of S^r	329
5	The real points of singular fibers	337
6	Proof of the main theorem	340
7	Examples	342

1. Introduction

This paper continues the study of the topology of real algebraic threefolds begun in [Kollár97b, Kollár97c, Kollár98a], but the current work is independent of the previous ones in its methodology.

The present aim is to understand the topology of the set of real points of threefolds which admit a morphism to a curve whose general fiber is a rational surface. This class of threefolds also appears as one of the 4 possible outcomes of the minimal model program (cf. [Kollár-Mori98, Sec. 3.7]). Our main theorem gives a nearly complete description of the possible topological types of the set of real points of such a threefold in the orientable case.

Theorem 1.1. *Let X be a smooth projective real algebraic threefold such that the set of real points $X(\mathbb{R})$ is orientable. Assume that there is a morphism $f : X \to C$ onto a real algebraic curve C whose general fibers are rational surfaces. Let $M \subset X(\mathbb{R})$*

be any connected component. Then

$$M \sim N \mathbin{\#} a\mathbb{RP}^3 \mathbin{\#} b(S^1 \times S^2) \quad \text{for some } a, b \geq 0,$$

where one of the following holds.

1. *N is a connected sum of lens spaces (cf. (1.8)).*

2. *N is Seifert fibered over a topological surface (cf. (1.9)).*

3. *N is either an $S^1 \times S^1$-bundle over S^1 or is doubly covered by such a bundle.*

1.2 (First reduction step). As with many results for 3-folds, the proof starts with a suitable minimal model program. The minimal model program over $\mathbb{R}$ was studied in greater generality in [Kollár97c] so I just outline the main conclusions. The end result is that (1.1) follows from (1.3) and for the rest of the paper we do not have to know anything about the 3-dimensional minimal model program.

Let us run the relative minimal model program for $X \to C$ over $\mathbb{R}$ (cf. [Kollár-Mori98, Sec. 3.7] or [Kollár97c, Sec. 3]). The change in the topology of real points in the course of the program is described in [Kollár97c, 1.2]. There are two possibilities for the final step of the program.

First, we may get a conic bundle over a surface. These are described in [Kollár98a] and we obtain cases 1 or 2 of (1.1).

Otherwise the program ends with a morphism $f^m : X^m \to C$ which has the following properties:

1. X^m has only isolated singularities which are $\mathbb{Q}$-factorial over $\mathbb{R}$ (that is, if D is a Weil divisor defined over $\mathbb{R}$ then mD is Cartier for some $m > 0$).

2. $-K_{X^m}$ is f^m-ample.

3. K_{X^m} is Cartier at all real points.

4. Every fiber of f is irreducible (over $\mathbb{R}$).

5. The topological normalization $\overline{X^m(\mathbb{R})}$ of $X^m(\mathbb{R})$ is a 3-manifold.

In view of [Kollár97c, 1.2] and [Kollár98a, 1.1], (1.1) is reduced to proving the following:

Theorem 1.3. *Let X be a projective real algebraic threefold and $f : X \to C$ a morphism onto a real algebraic curve C. Assume that X satisfies the conditions (1.2.1–5) and that $\overline{X(\mathbb{R})}$ is orientable. Let $M \subset \overline{X(\mathbb{R})}$ be any connected component. Then*

$$M \sim N \mathbin{\#} a\mathbb{RP}^3 \mathbin{\#} b(S^1 \times S^2) \quad \text{for some } a, b \geq 0,$$

where one of the following holds.

1. N is a connected sum of lens spaces.

2. N is either an $S^1 \times S^1$-bundle over S^1 or is doubly covered by such a bundle.

Remark 1.4. The conclusion of the theorem can probably be considererably improved.

In case (1.3.1) I do not have any examples where N is not a single lens space. A more detailed analysis of the present methods may prove that this is always the case.

$(S^1 \times S^1)$-bundles over S^1 are classified by the monodromy map on $H_1(S^1 \times S^1, \mathbb{Z}) \cong \mathbb{Z}^2$. If this map is hyperbolic then N has a geometry modelled on *Sol* (cf. [Scott83, p. 470]). I believe that N can never be of this type but the methods of my proof do not say anything about the monodromy.

1.5 (Second reduction step). In this step we reduce the proof of (1.3) to the study of the singular fibers of f.

Let $A \sim S^1$ be a connected component of $C(\mathbb{R})$ and $p_1, \ldots, p_s \in A$ the points (in cyclic order) over which f is not smooth. For each i pick a point $q_i \in (p_i, p_{i+1})$. Then $X_{q_i} := f^{-1}(q_i)$ is a smooth Del Pezzo surface and $X_{q_i}(\mathbb{R})$ is orientable. Thus $X_{q_i}(\mathbb{R})$ is either $S^1 \times S^1$ or a disjoint union of copies of S^2 by a result of [Comessatti14]. (See also [Silhol89, V.3.4, VI.4.6 and VI.6.3].) Gluing 3-manifolds along such surfaces is a relatively simple operation, thus one can expect to get a good description of $X(\mathbb{R})$ by describing the pieces $Z_i := (f^{-1}[q_{i-1}, q_i])(\mathbb{R})$ for every i; see (6.3).

$f : Z_i \to [q_{i-1}, q_i]$ is a function whose only critical value is p_i. Thus Z_i can be viewed as a regular neighborhood of the critical level set $(f^{-1}(p_i))(\mathbb{R})$. In fact, once we have a topological description of $(f^{-1}(p_i))(\mathbb{R})$, it is easy to figure out what Z_i is, at least up to finite ambiguity. This is done in (6.1).

The complex projective surface $S_i := f^{-1}(p_i)$ is a "singular Del Pezzo" surface which appears as a degeneration of smooth Del Pezzo surfaces. Quite a lot is known about such surfaces. A structural theory of normal Del Pezzo surfaces is developed in [Keel-McKernan98]. Degenerations of $\mathbb{P}^2$ and some other Del Pezzo surfaces are studied in [Manetti91, Manetti93]. Unfortunately, the conclusion of these works is that a complete classification of such singular Del Pezzo surfaces is not feasible because of the combinatorial complexity of the problem. On the other hand, the methods developed by these and other authors can be used to answer many questions about singular Del Pezzo surfaces.

The main part of this work is devoted to understanding the topology of the set of real points of certain singular Del Pezzo surfaces. It should be emphasized that, as opposed to almost all previous studies, we have to consider nonnormal surfaces as well. In fact, as far as the topology of the real points is concerned, normal surfaces present no difficulties.

Nonnormal Del Pezzo surfaces whose canonical divisor is Cartier have been enumerated in [Reid94]. Many of the irreducible ones appear as singular fibers in our case. For all such examples one can perform a small perturbation of $f : Z_i \to [q_{i-1}, q_i]$ such that we obtain a Morse function whose fibers stay *real algebraic* Del Pezzo surfaces. These are again very easy to understand topologically.

I have been unable to find any other nonnormal irreducible examples, thus it is possible that the work of Sections 2–4 is entirely superfluous. On the other hand, the methods of these sections lead to a description of the singular fibers in many cases and it may be possible to develop them further to obtain a complete list of the irreducible fibers occurring in (1.3).

Some interesting examples are given in Section 7.

1.6 (The fundamental group of a real Del Pezzo surface). Let $S := f^{-1}(p_i)$ be one of the fibers and $Z := (f^{-1}[q_{i-1}, q_i])(\mathbb{R})$ the corresponding piece of the above decomposition of $X(\mathbb{R})$.

Here I would like to outline a simple argument giving some information about the fundamental group of Z. Since $S(\mathbb{R})$ is a retract of Z, we see that $\pi_1(Z) \cong \pi_1(S(\mathbb{R}))$.

Let $B \subset S$ be the singular locus and U_j the connected components of $S(\mathbb{R}) \setminus B(\mathbb{R})$. $\pi_1(B(\mathbb{R}))$ is a free group and each U_j gives one relation. Hence we see that the number of relations of $\pi_1(S(\mathbb{R}))$ is bounded by the number of connected components of $S(\mathbb{R}) \setminus B(\mathbb{R})$.

Since S is a limit of smooth Del Pezzo surfaces, $|-K_S|$ has a member; call it D. (This is a bit nebulous since S may not even be normal; see (3.3) for its precise meaning.) Let $h : S' \to S$ be the minimal resolution of S. It is easy to see that $|-K_{S'}|$ has a unique member D' such that $h_*(D') = B + D$. Let $f : S' \to S''$ be a minimal model and set $D'' := f_* D'$.

f is a composite of blow ups and one checks that

$$|\pi_0(S''(\mathbb{R}) \setminus D''(\mathbb{R}))| \geq |\pi_0(S'(\mathbb{R}) \setminus D'(\mathbb{R}))|.$$

Using [Kollár-Mori98, 3.39] one easily checks that every fiber of h is either contained in Supp D' or is disjoint from it. This gives that

$$|\pi_0(S'(\mathbb{R}) \setminus D'(\mathbb{R}))| \geq |\pi_0(S(\mathbb{R}) \setminus (B+D)(\mathbb{R}))|.$$

Thus we are led to counting the number of connected components of $S''(\mathbb{R}) \setminus D''(\mathbb{R})$. A typical case is when $S'' = \mathbb{P}^2$ and D'' is a cubic curve. We see that there are at most 4 connected components, and 4 is achieved when D'' consists of 3 lines. If we take into account that we want to use only $S(\mathbb{R}) \setminus B(\mathbb{R})$ and not $S(\mathbb{R}) \setminus (B+D)(\mathbb{R})$, we obtain that $S(\mathbb{R}) \setminus B(\mathbb{R})$ has at most 2 connected components in this case.

The other possible minimal models S'' can be similarly treated. (Ad hoc arguments are needed to exclude the case when S'' is a minimal ruled surface with negative section E and $D'' = 2E +$ (many fibers).) At the end we obtain that

(∗) $\pi_1(Z)$ is the free product of groups with 1 relation.

On the one hand (∗) is quite strong since most 3-manifold groups do not have this property. On the other hand, (∗) is not strong enough to exclude all hyperbolic 3-manifolds.

The proof of (1.3) given in Sections 2–6 is essentially an elaboration of this approach. The steps of going from S to S' and S'' are studied in more detail. At the end we obtain a rather complete geometric description of $S(\mathbb{R})$.

Remark 1.7 (PL three manifolds). In this paper I usually work with *piecewise linear manifolds* ([Rourke-Sanderson82] is a good introduction). Every real algebraic variety carries a natural PL structure (cf. [BCR87, Sec.9.2]).

In dimension 3 every compact topological 3-manifold carries a unique PL-manifold structure (cf. [Moise77, Sec. 36]) and a PL-structure behaves very much like a differentiable structure. For instance, let M^3 be a PL 3-manifold, N a compact PL-manifold of dimension 1 or 2 and $g : N \hookrightarrow M$ a PL-embedding. Then a suitable open neighborhood of $g(N)$ is PL-homeomorphic to a real vector bundle over N (cf. [Moise77, Secs. 24 and 26]). (The technical definition of such neighborhoods is given by the notion of *regular neighborhood*, see [Rourke-Sanderson82, Chap. 3]). If $f : M \to N$ is a PL-map and $X \subset N$ a compact subcomplex then there is a regular neigborhood $X \subset U \subset N$ such that $f^{-1}(U)$ is a regular neigborhood $f^{-1}(X) \subset M$ (cf. [Rourke-Sanderson82, 2.14]).

Definition 1.8 (Lens spaces). For relatively prime $0 < q < p$ consider the action $(x, y) \mapsto (e^{2\pi i/p}x, e^{2\pi i q/p}y)$ on the unit sphere $S^3 \sim (|x^2| + |y^2| = 1) \subset \mathbb{C}^2$. The quotient is a 3-manifold called the *lens space* $L_{p,q}$.

Another way to obtain lens spaces is to glue two solid tori together. The result is a lens space, S^3 or $S^1 \times S^2$. Sometimes one writes $L_{1,0} = S^3$ and $L_{0,1} = S^1 \times S^2$, though these are usually not considered lens spaces. (See, for instance, [Hempel76, p. 20].)

Definition 1.9 (Seifert fiber spaces). A 3-manifold M is called *Seifert fibered* if there is a morphism $f : M \to F$ to a topological surface such that every $P \in F$ has a neighborhood $P \in U \subset F$ such that $f : f^{-1}(U) \to U$ is fiber preserving homeomorphic to one of the normal forms $f_{c,d}$ defined below.

Let $S^1 \subset \mathbb{C}$ be the unit circle with coordinate u and $D^2 \subset \mathbb{C}$ the closed unit disc with coordinate z. For a pair of integers c, d satisfying $0 \leq c < d$ and $(c, d) = 1$, define

$$f_{c,d} : S^1 \times D^2 \to D^2 \quad \text{by} \quad f_{c,d}(u, z) = u^c z^d.$$

$f_{c,d}$ restricts to a fiber bundle $S^1 \times (D^2 \setminus \{0\}) \to D^2 \setminus \{0\}$. The fiber of $f_{c,d}$ over the origin is still S^1, but $f_{c,d}^{-1}(0)$ has multiplicity d.

(This is the classical definition of Seifert fibered spaces, which is slightly more restrictive than the one in [Scott83].)

Acknowledgments. I thank M. Kapovich and S. Kahrlamov for answering my numerous questions about 3-manifold topology and real algebraic geometry. Partial financial support was provided by the NSF under grant number DMS-9622394.

2. Real surfaces with Du Val singularities

The minimal model theory of real surfaces has been studied in detail in the papers [Comessatti14, Silhol89, Kollár97a]. It is not difficult to generalize these results to the case when we allow Du Val singularities (2.1). These results were explained in [Kollár98a].

Definition 2.1. Let $(0 \in S)$ be a normal surface singularity over $\mathbb{R}$ with minimal resolution $g : S' \to S$. $(0 \in S)$ is called a *Du Val* singularity (or rational double point) iff $K_{S'} \cong g^* K_S$. Equivalently, $(0 \in S)$ is Du Val iff every g-exceptional curve is a smooth rational curve with selfintersection -2. (See [Reid85] or [Kollár-Mori98, 4.2] for the relevant background on Du Val singularities over $\mathbb{C}$.)

It is not hard to see that every real Du Val singularity is real analytically equivalent to one of the following normal forms (cf. [AGV85, I.17.1] or [Kollár-Mori98, Sec. 4.2]).

A_n^+ $(x^2 + y^2 - z^{n+1} = 0)$ for $n \geq 2$,

A_n^- $(x^2 - y^2 - z^{n+1} = 0)$ for $n \geq 0$,

A_n^{++} $(x^2 + y^2 + z^{n+1} = 0)$ for n odd,

D_n^+ $(x^2 + y^2 z + z^{n-1} = 0)$ for $n \geq 4$,

D_n^- $(x^2 + y^2 z - z^{n-1} = 0)$ for $n \geq 4$,

E_6^+ $(x^2 + y^3 + z^4 = 0)$,

E_6^- $(x^2 + y^3 - z^4 = 0)$,

E_7 $(x^2 + y^3 + yz^3 = 0)$,

E_8 $(x^2 + y^3 + z^5 = 0)$.

Definition 2.2. Let S be a surface with Du Val singularities. A curve $C \subset S$ is called a (-1)-*curve* if its birational transform on the minimal resolution is a smooth rational curve with selfintersection -1. Thus $(C \cdot K_S) = -1$.

Definition 2.3. Let $P \in S(\mathbb{R})$ be a smooth real point and x, y local coordinates at P. The surface $S' \subset S \times \mathbb{P}^1_{(u:v)}$ given by equation $ux - vy^m = 0$ is called a $(1, m)$-*blow up* of P on S. For $m = 1$ this is the ordinary blow up. A $(1, m)$-blow up has a unique singular point of type A_{m-1}^- at $(0, 0, 0, 1)$.

It should be noted that for $m \geq 2$ the $(1, m)$-blow up does depend on the choice of the local coordinates.

It is frequently better to think of a $(1, m)$-blow up as follows. First blow up $0 \in S$ to get $S_1 \to S$. Then blow up a point on the exceptional divisor of $S_1 \to S$ to obtain $S_2 \to S_1$. Then blow up a point on the exceptional divisor of $S_2 \to S_1$ which is not on (the birational transform of) any previous exceptional divisor to obtain $S_3 \to S_2$. After m-times we have m exceptional curves in the following configuration:

$$\overset{-1}{\circ} - \overset{-2}{\circ} - \cdots - \overset{-2}{\circ}.$$

We can now contract all the (-2)-curves to get a $(1,m)$-blow up. This shows that the exceptional curve of a $(1,m)$-blow up is a (-1)-curve.

If $P, \bar{P} \in S(\mathbb{C})$ are smooth and conjugate complex points, then we can choose conjugate coordinate systems to do a $(1,m)$-blow up at both points. The result is again a real algebraic surface with a conjugate pair of A_{m-1}-points (for nonreal points the signs in the equations do not matter).

Definition 2.4. Let F_1, F_2 be real algebraic surfaces with Du Val singularities and assume that $g : F_1 \to F_2$ is a composite of $(1,m)$-blow ups. A $(1,m)$-blow up of a conjugate point pair is an isomorphism in the neighborhood of the real points, so if a $(1,m)$-blow up of a conjugate point pair is followed by a $(1,m)$-blow up of a real point then their order can be reversed. Repeating if necessary, g can be factored uniquely as

$$g : F_1 \xrightarrow{g^c} F^r \xrightarrow{g^r} F_2$$

where g^c is a composite of $(1,m)$-blow ups of conjugate point pairs and g^r is a composite of $(1,m)$-blow ups of real points.

Definition 2.5. Let F be a normal projective surface such that K_F is $\mathbb{Q}$-Cartier. F is called a (singular) *Del Pezzo* surface if $-K_F$ is ample. F is called a *weak Del Pezzo* surface if $-K_F$ is nef and big. A morphism $g : F \to C$ is called a *conic bundle* if every fiber is isomorphic to a plane conic (which can be smooth, a pair of intersecting lines or a double line). Every conic bundle can be embedded into a $\mathbb{P}^2$-bundle over C such that the fibers become conics.

Combining the results of [Kollár98a, Sec. 9] with (2.4) we obtain the following.

Theorem 2.6. *Let F be a projective surface over $\mathbb{R}$ with Du Val singularities. Then there are surfaces and morphisms*

$$g : F \xrightarrow{g^c} F^r \xrightarrow{g^r} F^*$$

with the following properties.

1. *F^r and F^* are projective surfaces over $\mathbb{R}$ with Du Val singularities.*

2. *g^c is a composite of $(1,m)$-blow ups of conjugate point pairs. In particular $F(\mathbb{R}) \cong F^r(\mathbb{R})$.*

3. *g^r is a composite of $(1,m)$-blow ups of real points.*

4. *F^* falls in one of the following 3 cases:*

(C) *(Conic bundle)* $\rho(F^*) = 2$ and F^* is a conic bundle over a smooth curve A.

(D) *(Del Pezzo surface)* $\rho(F^*) = 1$ and $-K_{F^*}$ is ample.

(N) *(Nef canonical class)* K_{F^*} is nef. □

Next we collect some auxiliary results that are needed elsewhere.

Lemma 2.7. *Let K be a field and F a Del Pezzo surface over K with Du Val singularities. Assume that $\rho(F) = 1$ and that there is an effective Cartier divisor $0 \neq B \subset F$ such that $-(K + B)$ is ample.*
Then F is isomorphic to $\mathbb{P}^2$ or to a quadric hypersurface in $\mathbb{P}^3$.

Proof. Let H be a generator of $\mathrm{Pic}(F)/(\mathrm{torsion})$ and write $B \equiv bH$, $K_F \equiv -aH$. This implies that $(K_F^2) = a^2(H^2)$. By assumption $a > b > 0$, so $a \geq 2$. Since $(K_F^2) \leq 9$, there are three possibilities:

1. $a = 3$ and $(H^2) = 1$. Then $F_{\bar{K}} \cong \mathbb{P}^2$ and F contains a line defined over K, thus $F \cong \mathbb{P}^2$.

2. $a = 2$ and $(H^2) = 2$. Then $F_{\bar{K}}$ is a quadric and $\mathcal{O}(1)$ is defined over K. Thus F itself is a quadric.

3. $a = 2$ and $(H^2) = 1$. Then $2p_a(B) - 2 = B \cdot (B + K_F) = H \cdot (-H) = -1$, which is impossible. □

Lemma 2.8. *Let F be a real surface with Du Val singularities and $\rho(F) = 2$. Let $F \to A$ be a conic bundle with a section H. Then H intersects every singular fiber at a singular point of F.*

Proof. A section can never intersect a multiple fiber at a smooth point of F. If $f^{-1}(a)$ is a pair of conjugate lines then their intersection point P is the only real point, hence H passes through P. If P is a smooth point of F then the intersection number of H and $f^{-1}(a)$ is at least 2, which is impossible. □

3. The basic set-up

Notation 3.1. Let X be a real algebraic threefold and $f : X \to C$ a proper morphism to a smooth real algebraic curve. As a generalization of the conditions of (1.3) we assume the following:

1. X has isolated $\mathbb{Q}$-factorial singularities,
2. $-K_X$ is $\mathbb{Q}$-Cartier and f-ample, and

3. every fiber of f is irreducible (over $\mathbb{R}$).

Let $0 \in C(\mathbb{R})$ be a point and $X_0 := f^{-1}(0)$ the fiber over 0.

Lemma 3.2. *Notation as above. There are 3 possibilities for X_0.*

1. *X_0 is reduced and geometrically irreducible.*
2. *$X_0 = m Z_0$ for some $m \geq 2$ where Z_0 is reduced and geometrically irreducible.*
3. *$X_0 = m(Z_0 + \bar{Z}_0)$ for some $m \geq 1$ where Z_0 and $\bar{Z}_0$ are conjugate, reduced and irreducible.*

Proof. Write $(X_0)_\mathbb{C} = \sum a_i Z_i$ as the sum of its irreducible and reduced components. For any Z_i let $Z'_i := Z_i$ if Z_i is defined over $\mathbb{R}$ and $Z'_i := Z_i + \bar{Z}_i$ otherwise. Z'_i is defined over $\mathbb{R}$ hence $(X_0)_\mathbb{C} = m Z'_i$ for some $m \geq 1$ since X_0 is irreducible over $\mathbb{R}$. $\square$

3.3. Let X be a real algebraic threefold and $f : X \to C$ a proper morphism to a smooth real algebraic curve, $0 \in C(\mathbb{R})$ a point and $Z := f^{-1}(0)$ the fiber over 0. Assume that X has isolated singularities, $-K_X$ is $\mathbb{Q}$-Cartier and f-ample, and Z is reduced and geometrically irreducible. We would like to explain what it means that Z is a (possibly nonnormal) Del Pezzo surface.

Let Σ be the finite set of points of X where K_X is not Cartier. Set $X^0 := X \setminus \Sigma$ and $Z^0 := Z \setminus \Sigma$. (To be completely precise, I should let Σ be the set of points where X is not Gorenstein. At the end, however, this does not make any difference.) Z^0 is a Cartier divisor on the variety X^0, thus

$$\omega_{Z^0} \cong \omega_{X^0}(Z^0) \otimes \mathcal{O}_{Z^0} \cong \omega_{X^0} \otimes \mathcal{O}_{Z^0}.$$

By assumption $\omega_{X^0}^{-m}$ is f-very ample for $m \gg 1$, so we obtain that $\omega_{Z^0}^{-m}$ is very ample for $m \gg 1$.

Let $p : \bar{Z} \to Z$ be the normalization and set $\bar{Z}^0 := \pi^{-1}(Z^0)$. There is an adjunction map $\omega_{\bar{Z}^0} \to \pi^* \omega_{Z^0}$ (cf. [Hartshorne77, Ex.III.7.2.a]). (The adjunction map is defined even over $\bar{Z}$ but $\pi^* \omega_Z$ may be messy since ω_Z need not be locally free.) $\bar{Z}^0$ is a normal surface, so there is an effective divisor $B^0 \subset \bar{Z}^0$ such that

$$\mathcal{O}_{\bar{Z}^0}(K_{\bar{Z}^0} + B^0) \cong \omega_{\bar{Z}^0}(B^0) \cong \pi^* \omega_{Z^0}.$$

(Further information about B^0 is given in (3.7).) Here both sides are locally free (since the right hand side is), so we can raise them to any power to get isomorphisms

$$\mathcal{O}_{\bar{Z}^0}(m(K_{\bar{Z}^0} + B^0)) \cong \pi^* \omega_{Z^0}^m.$$

Now choose $m > 0$ such that $m K_X$ is Cartier. The reflexive sheaves

$$\mathcal{O}_{\bar{Z}}(-m(K_{\bar{Z}} + B)) \quad \text{and} \quad \pi^* \mathcal{O}_X(-m K_X)$$

are isomorphic on $\bar{Z}^0$, thus outside finitely many points. Hence they are isomorphic. Since $\mathcal{O}_X(-m K_X)$ is locally free we conclude that

1. $K_{\bar{Z}} + B$ is $\mathbb{Q}$-Cartier and $-(K_{\bar{Z}} + B)$ is ample, and

2. $(K_{\bar{Z}} + B)^2 = K_{X_t}^2$ where X_t is any smooth fiber of X/C.

We can also get some information about the global sections. One has to be a little careful since K_X may not be Cartier everywhere. As usual, let $\omega_X^{[r]}$ denote the double dual of $\omega_X^{\otimes r}$. $\omega_X^{[-n]}$ is reflexive, so $\omega_X^{[-n]} \otimes \mathcal{O}_Z$ has no embedded points. Thus

$$H^0(Z^0, \omega_{Z^0}^{-n}) = H^0(Z^0, \omega_X^{[-n]} \otimes \mathcal{O}_{Z^0}) \supset H^0(Z, \omega_X^{[-n]} \otimes \mathcal{O}_Z)$$
$$\geq H^0(X_t, \omega_{X_t}^{[-n]}) = \binom{n+1}{2}(K_{X_t}^2) + 1,$$

where the inequality holds by semicontinuity of H^0 and the last equality is a fact about smooth Del Pezzo surfaces. We can pull back these sections to $\bar{Z}^0$ and extend them across the finitely many points. Thus we conclude that

3. $H^0(\bar{Z}, \mathcal{O}_{\bar{Z}}(-(K_{\bar{Z}} + B))) \geq (K_{X_t}^2) + 1 \geq 2$.

Choose a pencil in $|-(K_{\bar{Z}} + B)|$ and write it as $D + M$ where D is the fixed part and M the moving part. If K_X is Cartier at all points of $X(\mathbb{R})$ then $Z(\mathbb{R}) \subset Z^0$ and also $\bar{Z}(\mathbb{R}) \subset \bar{Z}^0$. This shows that $K_{\bar{Z}} + B$ is Cartier at all real points. Setting $S := \bar{Z}$ we obtain a quadruplet (S, B, D, M) and we have proved the following.

Proposition 3.4. *Let $f : X \to C$ be a family of Del Pezzo surfaces satisfying the conditions (3.1.1–3). Let $Z := f^{-1}(0)$ be a reduced and geometrically irreducible fiber. Then the quadruplet (S, B, D, M) constructed above satisfies the conditions (3.5).* □

Condition 3.5. For a quadruplet (S, B, D, M) consider the following properties.

1. S is a normal projective surface over $\mathbb{R}$,

2. B and D are Weil divisors on S,

3. M is a pencil on S without fixed components,

4. $K_S + B + D + M \sim 0$.

5. $K_S + B$ is Cartier at all real points of S, and

6. $-(K_S + B)$ is $\mathbb{Q}$-Cartier and ample.

Remark 3.6. (1) Most quadruplets satisfying the above conditions do not arise as special fibers of families of Del Pezzo surfaces. An obvious numerical conditions is that $(K_S + B)^2$ be an integer beween 1 and 9. Even if this holds, there is no reason to assume that the surface Z is smoothable.

(2) Even if we know (S, B, D, M), it is not always easy to reconstruct the fiber Z. First of all, we need to know the map $B \to$ (singular curve of Z); a set theoretic information. If B is reduced then this determines the scheme structure of Z at least

when Z is S_2 (which holds if X has terminal singularities). If, however, B is not reduced, additional scheme theoretic information is needed and this seems rather complicated. See [Reid94] for a closely related case.

3.7. Let X be a smooth 3-fold and $Z \subset X$ a reduced surface which is not normal along a curve $C \subset Z$. Let $\pi : \bar{Z} \to Z$ be the normalization and set $\text{red}(\pi^{-1}(C)) = \sum_i \bar{C}_i$. As in (3.3), we see that $\omega_{\bar{Z}}(\sum b_i \bar{C}_i) \cong \pi^* \omega_Z$ for some $b_i \geq 0$. We would like to establish a relationship between the coefficients b_i and the local structure of Z along C.

We can cut everything by a general hyperplane, and our problem is reduced to a curve question which has been classically studied in detail, since $(\sum b_i)/2$ is exactly the contribution of the singular point to the arithmetic genus of a curve. We need the classification of singularities with small b_i:

1. $b_1 = 0$ iff Z is smooth along C.

2. $b_1 = 1$ iff Z has 2 branches meeting transversely along C.

3. $b_i \leq 2$ for every i iff either Z has one branch with a cusp along C, or 2 branches which are simply tangent along C or 3 branches meeting pairwise transversely along C or $b_i \leq 1$ for every i.

3.8. In general S is singular and so we try to study it through a suitable resolution. The most natural choice would be to take its minimal resolution, but the following partial resolution turns out to be more convenient. (The main reason for this choice is explained in (5.2.3).)

There is a unique morphism $f : S^m \to S$ such that

1. f is an isomorphism above $P \in S$ if S is smooth at P or if S has a Du Val singularity at P and $P \notin \text{Supp } B$

2. f is the minimal resolution over $P \in S$ otherwise.

(Note that this is not the same as the so called "minimal Du Val resolution".)

f is a birational map between normal surfaces, thus we can pull back any divisor by f if we allow rational coefficients. Hence we can define B^m by the formula $K_{S^m} + B^m \equiv f^*(K_S + B)$. $-B^m \equiv_f K_{S^m}$ has nonnegative intersection number with any exceptional curve, hence B^m is effective (cf. [Kollár-Mori98, 3.39]). Moreover, every exceptional curve appears in B^m with positive coefficient (cf. [Kollár-Mori98, 4.3–5]). (Here we use that we did not resolve Du Val points.) Write $f^*M = D' + M^m$ where D' is the fixed part (which may have rational coefficients) and M^m is a pencil without fixed components. Set $D^m := f^*D + D'$.

(∗) In this paper, the quadruplet (S^m, B^m, D^m, M^m) always denotes the one constructed above starting with (S, B, D, M).

Condition 3.9. For a quadruplet (S^m, B^m, D^m, M^m) consider the following properties.

1. S^m is a projective surface over $\mathbb{R}$ with Du Val singularities.

2. B^m and D^m are effective $\mathbb{Q}$-divisors on S such that $B^m + D^m$ is an integral divisor.

3. M^m is a pencil on S^m without fixed components.

4. $K_{S^m} + B^m + D^m + M^m \sim 0$.

5. If $C \subset S^m$ is a geometrically irreducible real curve then C appears in B^m with integer coefficient.

6. $-(K_{S^m} + B^m)$ is nef and big.

7. $-(K_{S^m} + B^m)$ has positive intersection number with every (-1)-curve (2.2).

8. S is smooth along Supp B.

9. $-(K_{S^m} + B^m)$ is nef and has positive intersection number with every curve not in Supp B.

Proposition 3.10. *If (S, B, D, M) satisfies the conditions (3.5) then (S^m, B^m, D^m, M^m) satisfies the conditions (3.9).*

Proof. (1,3,4,6,8) are clear from the construction. $B^m + D^m \sim -K_{S^m} - M^m$ now implies (2). If $C \subset S$ is a geometrically irreducible real curve and $f(C)$ is a curve then C appears in B and in B^m with the same integer coefficient. If $f(C) = P$ is a point then P is real, so $K_S + B$ is Cartier there and every curve above P appears with integral coefficient in B^m.

(9) holds since every f-exceptional curve is in B. Since we took minimal resolutions, there are no f-exceptional (-1)-curves, and this shows (7). □

Lemma 3.11. *Let (F, B, D, M) be a quadruplet. Let $g : F \to F'$ be a birational morphism and set $B' := g_*B$, $D' := g_*D$ and $M' := g_*M$. Then:*

1. *If (F, B, D, M) satisfies one of the conditions (3.9.2–6 or 9) then (F', B', D', M') also satisfies the same condition.*

2. *If g^{-1} is a composite of $(1, m)$-blow ups and (F, B, D, M) satisfies one of the conditions (3.9.1–9) then (F', B', D', M') also satisfies the same condition.*

Proof. This is clear for (3.9.2,3 and 5) and $K_{F'} = g_* K_F$ implies it for (3.9.4).

Let E_i be the connected components of the exceptional set of g. We can write $K_F + B \equiv g^*(K_{F'} + B') + E$ where E is g-exceptional. $-(K_F + B)$ is nef, so

E is effective and $E_i \subset \mathrm{Supp}\, E$ unless $K_F + B$ is numerically trivial along E_i (cf. [Kollár-Mori98, 3.39]). If $C \subset F$ is an irreducible curve which is not g-exceptional then

$$g_*(C) \cdot (K_{F'} + B') = C \cdot g^*(K_{F'} + B') = C \cdot (K_F + B) - C \cdot E \leq C \cdot (K_F + B).$$

This shows that $-(K_{F'} + B')$ is nef if $-(K_F + B)$ is, settling the case (3.9.9). If (3.9.6) holds on F then bigness of $-(K_{F'} + B')$ follows from

$$(K_{F'} + B')^2 = (K_F + B) \cdot g^*(K_{F'} + B') = (K_F + B)^2 - (K_F + B) \cdot E \geq (K_F + B)^2.$$

Assume now that g^{-1} is a single $(1, m)$-blow up. The points $g(E_i)$ are smooth on F' by (2.3), which implies the claim for (3.9.8). Every connected component of the exceptional set is a (-1)-curve by (2.3) and $K_F + B$ has negative intersection number with it. So $\mathrm{Supp}\, E$ coincides with the exceptional set and if C intersects the exceptional set then

$$g(C) \cdot (K_{F'} + B') \leq -C \cdot E < 0.$$

This shows the claim for (3.9.7). By induction this gives (3.11.2). $\square$

4. The classification of S^r

Notation 4.1. Throughout this section (S^m, B^m, D^m, M^m) denotes a quadruplet which satisfies the conditions (3.9.1–8). Let $g : S^m \to S^*$ be a minimal model of S^m and set $B^* := g_*(B^m)$. For simplicity we assume S^* is not obtained by $(1, m)$-blow ups from another surface. (This could happen in only a handful of cases. For instance, the blow up of $\mathbb{P}^2$ at a point is also a $\mathbb{P}^1$-bundle so it could be S^* as in (2.6.4.C).)

4.2 (How to determine S^m?). We use the following method to get information about S^m.

(1) First we determine the possible surfaces S^*. This was in fact done using the MMP for real surfaces with Du Val singularities.

(2) Then we get a list of all possible quadruplets (S^*, B^*, D^*, M^*). This involves finding all possible ways of writing $-K_{S^*} \sim (B^* + D^*) + M^*$, so this is equivalent to classifying all pencils in $|-K_{S^*}|$. This is easy to do if $|-K_{S^*}|$ is small. Unfortunately, $|-K_{S^*}|$ gets arbitrarily large for minimal ruled surfaces and I do not know of any useful classification of all anticanonical pencils.

(3) Given a quadruplet (S^*, B^*, D^*, M^*) we can try to find all possible ways it came from an (S^m, B^m, D^m, M^m). We factor $S^m \to S^*$ into $(1, m)$-blow ups

$$S^m \to \cdots \to S_2 \to S_1 \to S^*.$$

For any of the intermediate stages $g : S^m \to S_i$ set $B_i := g_* B^m$, $D_i := g_* D^m$, $M_i := g_* M^m$. A key point to observe is that by (3.11) all the (S_i, B_i, D_i, M_i) satisfy the conditions (3.9.1–8). Thus we can work our way backwards one blow up at a time starting with S^*.

(4) Assume that we already have (S_i, B_i, D_i, M_i) and that $\pi : S_{i+1} \to S_i$ is the ordinary blow up of a point $P \in S_i$ with exceptional curve $E \subset S_{i+1}$. By (3.9.4)

$$K_{S_{i+1}} + B_{i+1} + D_{i+1} + M_{i+1} = \pi^*(K_{S_i} + B_i + D_i + M_i),$$

so we conclude that the

$$\text{coefficient of } E \text{ in } (B_{i+1} + D_{i+1}) = \text{mult}_P(B_i + D_i) + \text{mult}_P M_i - 1.$$

In particular, we can blow up only points in $\text{Supp}(B_i + D_i)$ and the base points of M_i.

(5) Given (S^*, B^*, D^*, M^*) and the sequence of blow ups leading to S^m, we have determined S^m, $B^m + D^m$ and M^m. Conditions (3.9.5–8) give further restrictions on B^m. In many cases these are impossible to satisfy.

(6) The role of (3.9.5) turns out to be crucial for us. Frequently there are many possibilities for B^m such that (S^m, B^m, D^m, M^m) satisfies all the conditions (3.9) except (3.9.5), but there are no choices of B^m where (3.9.5) also holds.

(3.9.5) is especially useful if there are many geometrically irreducible curves in $\text{Supp}(B^m + D^m)$. This is the reason for studying S^r since in $S^r \to S^*$ all exceptional curves are geometrically irreducible. It turns out that S^r is obtained from S^* by at most 1 blow up and this allows us to understand all possible (S^r, B^r, D^r, M^r) quite well.

(7) In (4.3) we subdivide the possible pairs (S^*, B^*) into 5 cases and then do a separate classification in each case.

Lemma 4.3. *With the above notation the pair (S^*, B^*) satisfies one of the following conditions.*

1. *S^* is a Del Pezzo surface with $\rho(S^*) = 1$ which is neither $\mathbb{P}^2$ nor a quadric in $\mathbb{P}^3$.*

2. *S^* is a weak Del Pezzo surface with $\rho(S^*) = 2$ and $B^* = 0$.*

3. *There is a $\mathbb{P}^1$- bundle structure $S^* \to A$ such that $B_\mathbb{C}^*$ has a unique irreducible component which dominates A. This component is a section.*

4. *There is a conic bundle structure $S^* \to A$ such that $B_\mathbb{C}^*$ has two irreducible components which dominate A. These components are conjugate sections.*

5. *S^* is $\mathbb{P}^2$ or a quadric in $\mathbb{P}^3$.*

Proof. If S^* is a Del Pezzo surface with $\rho(S^*) = 1$ then we have either (1) or (5). In all other cases there is a conic bundle structure $S^* \to A$. Let $F \subset S^*$ be a general fiber. Then

$$2 = -K_{S^*} \cdot F = (B^* + D^*) \cdot F + M^* \cdot F \geq (B^* + D^*) \cdot F.$$

$B^* + D^*$ is a sum of curves with integral coefficients, so it intersects F in at most 2 points. Thus $B_{\mathbb{C}}^*$ has at most 2 horizontal components.

$-(K_{S^*} + B^*) \cdot F > 0$ since $-(K_{S^*} + B^*)$ is big, thus $B^* \cdot F < 2$.

If $B_{\mathbb{C}}^*$ has 1 horizontal component H, then H is defined over $\mathbb{R}$ and hence H appears in B^* with integer coefficient. Together with $B^* \cdot F < 2$ this shows that H is a section. Then S^* is a $\mathbb{P}^1$-bundle by (2.8) and (3.9.8), so we are in case (3).

If $B_{\mathbb{C}}^*$ has 2 horizontal components and they are both defined over $\mathbb{R}$ then arguing as above leads to a contradiction.

If $B_{\mathbb{C}}^*$ has 2 horizontal components which are conjugates $H, \bar{H}$ then they are both sections and we are in case (4).

If $B^* = 0$ then we are in case (2).

Finally it may happen that $B_{\mathbb{C}}^* \neq 0$ has no horizontal components. $\rho(S^*) = 2$ and so $\rho(S^*/A) = 1$. S^* is $\mathbb{Q}$-factorial hence every curve in S^* is $\mathbb{Q}$-Cartier. This implies that any vertical curve is numerically equaivalent to a (rational) multiple of a general fiber, in particular B^* is nef. This implies that $-K_{S^*}$ is ample. By the Cone Theorem (cf. [Kollár-Mori98, 3.7]) S^* has another extremal ray with contraction $S^* \to W$. This is not a birational contraction by assumption (4.1). So $S^* \to W$ is another conic bundle structure and the vertical $B_{\mathbb{C}}^*$ becomes horizontal for $S^* \to W$. These cases were treated already. $\square$

Theorem 4.4. *Assume that S^* is a weak Del Pezzo surface with Du Val singularities, not isomorphic to $\mathbb{P}^2$ or to a quadric in $\mathbb{P}^3$. Assume furthermore that either $\rho(S^*) = 1$ or $B^* = 0$. Then S^r is a weak Del Pezzo surface with Du Val singularities and $B^r = 0$.*

Proof. Assume first that $B^* \neq 0$. Write $B^* + D^* = C + C'$ where $\operatorname{Supp} C = \operatorname{Supp} B^*$. $-K \equiv C + C' + M^*$ and C is a Cartier divisor by (3.9.8) which is not empty if $B^* \neq 0$. On S^* every effective Cartier divisor is ample since $\rho(S^*) = 1$, so $C' + M^*$ is ample. This is impossible by (2.7).

Thus $B^* = 0$ and $S^* \not\cong \mathbb{P}^2$. If $B^r = 0$ then $-(K_{S^r} + B^r) = -K_{S^r}$ is nef and big, so S^r is a weak Del Pezzo surface with Du Val singularities.

We need to exclude the case when $B^r \neq 0$. Let

$$\cdots \to S_2 \to S_1 \to S^*$$

be the series of blow ups leading to S^r and $S_{i+1} \to S_i$ the last blow up with exceptional curve $E \subset S_{i+1}$ such that $B_i = 0$. Then $B_{i+1} = aE$ for some $a > 0$ and a is an integer since E is geometrically irreducible. S_{i+1} is smooth along B_{i+1}, so $S_{i+1} \to S_i$ is an ordinary blow up. This is impossible by (4.5). $\square$

Proposition 4.5. *Let F be a projective surface with Du Val singularities over $\mathbb{C}$ and $p : G \to F$ the blow up of a smooth point with exceptional curve $E \subset G$. Assume that $-(K_G + aE)$ is nef and has positive intersection number with every (-1)-curve for some $a \geq 1$.*

Then $F \cong \mathbb{P}^2$ and $a < 2$.

Proof. Let $F' \to F$ and $G' \to G$ be the minimal resolutions. Then F' is the blow up of G' at a point and $-(K_{G'} + aE')$ is the pull back of $-(K_F + aE)$. Thus it is sufficient to consider the case when F itself is smooth.

If $F \not\cong \mathbb{P}^2$ then there is a morphism $g : F \to \mathbb{P}^1$ whose general fiber is $\mathbb{P}^1$. $-K_F$ is g-nef, so it is easy to see that every fiber of g has dual graph

$$\overset{0}{\circ} \quad \text{or} \quad \overset{-1}{\circ} - \overset{-2}{\circ} - \overset{-2}{\circ} - \cdots - \overset{-2}{\circ} - \overset{-1}{\circ}.$$

If P is on a fiber of the second type then E intersects a rational curve with self intersection -2 or -3 and $-(K_G + aE)$ is not nef for $a > 0$. In the first case, the birational transform of the fiber becomes a (-1)-curve F intersecting E. Thus $-(K_G + aE) \cdot F = 1 - a \leq 0$. $\square$

Theorem 4.6. *Assume that $S^* \to A$ is a $\mathbb{P}^1$-bundle and $H \subset B^*$ is a (real) section. Then $S^r = S^*$.*

Proof. If $S^r \neq S^*$ then there is a last contraction $S_1 \to S^*$. The inverse of this is a $(1, m)$-blow up of a real point $P \in S^*$ and we need to show that this can not happen. Let F' be the fiber of $S^* \to A$ containing P.

If $S_1 \to S^*$ is an ordinary blow up then the exceptional curve $E \subset S_1$ and the birational transform F of F' are both (-1)-curves intersecting in a point. By (3.9.7) this implies that $E \cdot (K_{S_1} + B_1) < 0$ and $F \cdot (K_{S_1} + B_1) < 0$. We can write $B_1 = b_e E + b_f F + B'$ where B' does not have E, F as components. H intersects either E or F, so $E \cdot B' \geq 1$ or $F \cdot B' \geq 1$. We get a contradiction by (4.8).

The $(1, m)$-blow ups are excluded by (4.7). $\square$

4.7 (Excluding $(1, m)$ blow ups). Given a pair (S^*, B^*) assume that $S_1 \to S^*$ is a $(1, m)$-blow up for some $m \geq 2$ with exceptional curve E. Then S_1 is singular at a point $P \in E$, so E is not in B_1. Let $\pi : \tilde{S}_1 \to S_1$ be the minimal resolution of P. Write $\pi^*(D_1 + M_1) = \tilde{D}_1 + \tilde{M}_1$ where $\tilde{M}_1$ has no fixed components. Then $(\tilde{S}_1, \pi^* B_1, \tilde{D}_1, \tilde{M}_1)$ satisfies the conditions (3.9.1–8) and $\tilde{S}_1$ is obtained from S^* by m ordinary blow ups (2.3). So once we prove that there can be at most one ordinary blow up, this implies that there are no $(1, m)$-blow ups at all for $m \geq 2$.

Lemma 4.8. *Let S be a surface, $E, F \subset S$ effective curves and B' an effective $\mathbb{Q}$-divisor on S whose support does not contain E and F. Assume that $(E^2) = (F^2) = (E \cdot K_S) = (F \cdot K_S) = -1$ and $(E \cdot F) = 1$. Let b_e, b_f be integers and assume that*

$$E \cdot (K_S + b_e E + b_f F + B') < 0 \quad \text{and} \quad F \cdot (K_S + b_e E + b_f F + B') < 0.$$

Then $b_e = b_f$, $E \cdot B' < 1$ and $F \cdot B' < 1$. $\square$

Theorem 4.9. *Assume that $S^* \to A$ is a conic bundle and $B_{\mathbb{C}}^*$ contains a pair of conjugate sections. Then A is rational and either $S^r = S^*$ or S^r is obtained from S^* by one ordinary blow up.*

Proof. Let $H, \bar{H} \subset \operatorname{Supp} B^*$ be the conjugate sections. First we determine $B^* + D^*$. We can write $-K_{S^*} \equiv H + \bar{H} + aF$ for some a where F is a general fiber. $B^* + D^*$ is a Weil divisor containing $H + \bar{H}$, so $(B^* + D^* - H - \bar{H}) + M \equiv aF$ is effective and moves in a pencil, hence $a \geq 1$. By the adjunction formula

$$2g(H) - 2 = H \cdot (K_{S^*} + H) = -(H \cdot \bar{H}) - a.$$

Thus $g(H) = 0$ and either $H \cdot \bar{H} = 1$ and $B^* + D^* = H + \bar{H}$ or $H \cdot \bar{H} = 0$ and $B^* + D^* = H + \bar{H} + \epsilon C$ where $\epsilon \in \{0, 1\}$ and C is a fiber. In the first case let C denote the fiber passing through the unique point of $H \cap \bar{H}$. In both cases $M = |F|$ is base point free. The real points of $H + \bar{H} + \epsilon C$ are in $C(\mathbb{R})$. So all real blow ups take place over C by (4.2.4). We distinguish three cases according to the type of C.

(1): C is a double fiber. Any section intersects a double fiber at a singular point of S^*. Since $H \subset B^*$ and B^* is disjoint from the singular points of S^*, we see that $S^* \to A$ does not have any double fibers.

(2): C is a smooth fiber. Consider the case of 2 ordinary real blow ups $S_2 \to S_1 \to S^*$. There are 3 cases and after blow up we get one of the following curve configurations where • denotes the birational transform of C.

$$\overset{-1}{\circ} - \overset{-2}{\bullet} - \overset{-1}{\circ} \quad \text{or} \quad \overset{-1}{\circ} - \overset{-2}{\circ} - \overset{-1}{\bullet} \quad \text{or} \quad \overset{-2}{\circ} - \overset{-1}{\circ} - \overset{-2}{\bullet}.$$

Let E_1, E_2, E_3 be these curves from left to right. $B_2 = e_1 E_1 + e_2 E_2 + e_3 E_3 + H_2 + \bar{H}_2$, the e_i are integers, $-(K_{S_2} + B_2)$ is positive on the (-1)-curves and nonnegative on the (-2)-curves. By solving these inequalities we obtain that $H_2 + \bar{H}_2$ does not intersect the (-2)-curves, $e_1 = e_2 = e_3$ in the first 2 cases and $e_1 = e_2/2 = e_3$ in the third case.

If $C \cap (H + \bar{H})$ is a conjugate point pair, then $H_2 + \bar{H}_2$ intersects the curve • and this leads to a contradiction in the first and third cases. In the second case we use (4.2.4) to conclude that $B^* + D^* = H + \bar{H} + C$ and $c_2 = 0$, giving a contradiction.

If $C \cap (H + \bar{H})$ is a real point P, then C has coefficient zero in $B^* + D^*$ hence we get that $e_1 = e_2 = e_3 = 0$ and the only point that we can blow up on S^* is P. Thus the first case can not happen and $H_2 + \bar{H}_2$ intersects E_2 in the second case and E_1 in the third case. Both are impossible.

$(1, m)$-blows up are excluded by (4.7).

(3): C is a reducible fiber. The only real point of C is its singular point, hence we have to start by blowing it up. There is only 1 case and after 2 blow ups we get the following curve configuration.

$$\overset{-1}{\circ} - \overset{-2}{\circ} - \bullet,$$

where • denotes the birational transform of C (thus it is geometrically reducible). Let E_1, E_2, E_3 be these curves from left to right. $B_2 = e_1 E_1 + e_2 E_2 + e_3 E_3 + H_2 + \bar{H}_2$, e_1, e_2 are integers, $-(K_{S_2} + B_2)$ is positive on the (-1)-curve and nonnegative on the (-2)-curves. By solving these inequalities we obtain that $e_1 = e_2 = e_3$ and $H_2 + \bar{H}_2$

does not intersect •. This is impossible since H, a section, does not pass through the singular point of the fiber C.

$(1, m)$-blow ups are excluded as above. $\square$

Notation 4.10. There are 5 normal quadrics over $\mathbb{R}$ up to isomorphism. The smooth ones are $Q^{4,0} := (x^2 + y^2 + z^2 + t^2 = 0)$, $Q^{3,1} := (x^2 + y^2 + z^2 - t^2 = 0)$ and $Q^{2,2} := (x^2 + y^2 - z^2 - t^2 = 0)$. The quadric cones are $Q^{3,0} := (x^2 + y^2 + z^2 = 0)$ and $Q^{2,1} := (x^2 + y^2 - z^2 = 0)$.

$Q^{4,0}(\mathbb{R}) = \emptyset$ and $Q^{3,0}(\mathbb{R})$ is a single point, so they are not very interesting for us.

Theorem 4.11. *Assume that S^* is $\mathbb{P}^2$ or a quadric in $\mathbb{P}^3$. Then one of the following holds.*

1. *S^r is $\mathbb{P}^2$ or a quadric in $\mathbb{P}^3$.*

2. *S^r is weak Del Pezzo surface and $B^r = 0$.*

3. *S^r is obtained from $\mathbb{P}^2$ by one ordinary blow up.*

4. *S^r is $\mathbb{P}^2$ blown up in 2 points and $B^r(\mathbb{R}) = \emptyset$.*

5. *S^r is one of the above cases blown up at conjugate pairs of points.*

(The last case occurs since blowing up the quadric $Q^{3,1}$ at a real point is the same as blowing up $\mathbb{P}^2$ at a conjugate pair of points. A more economical choice of S^* would have eliminated this case.)

Proof. Assume that we blow up 2 (possibly infinitely near) real points in $\mathbb{P}^2$. There is a unique line passing through the center of both blow ups. This line and the two exceptional curves form a configuration

$$\overset{-1}{\circ} - \overset{-1}{\circ} - \overset{-1}{\circ} \quad \text{or} \quad \overset{-1}{\circ} - \overset{-1}{\circ} - \overset{-2}{\circ}.$$

Let E_1, E_2, E_3 be these curves from left to right. $B_2 = e_1 E_1 + e_2 E_2 + e_3 E_3 + B'_2$, the e_i are integers, $-(K_{S_2} + B_2)$ is positive on the (-1)-curves and nonnegative on the (-2)-curve. By solving these inequalities we obtain that $e_1 = e_2 = e_3 = 0$ and $E_i \cdot B'_2 < 1$ for every i.

If $B_2 = 0$ then we end up in case (2) as in (4.4). If B^* contains a curve with positive integer coefficient then its birational transform intersects $E_1 + E_2 + E_3$ and one of the inequalities $E_i \cdot B'_2 < 1$ is violated.

A similar argument excludes the possibility of even one blow up when S^* is a quadric and B^* contains a curve with positive integer coefficient.

We are left with three cases: $S^* = \mathbb{P}^2$, $B^* = c(L + \bar{L})$ for a conjugate pair of lines, or $S^* = Q^{3,1}$, $B^* = c(L + \bar{L})$ for a conjugate pair of intersecting lines or $S^* = Q^{2,2}$, $B^* = c(L + \bar{L})$ for a conjugate pair of nonintersecting lines. In the latter case $S^* \cong \mathbb{P}^1 \times \mathbb{P}^1$ and one of the projections lands us in case (4.9), so this is already treated.

Assume that $S^* = \mathbb{P}^2$ and $B^* = c(L + \bar{L})$ for a conjugate pair of lines intersecting at a (necessarily real) point P. Then $B^* + D^* = L + \bar{L}$ and M is a pencil of lines with a base point Q. By (4.2.4) the only possibilities to blow up are P and Q. If $Q \neq P$ and we blow up Q then projection from Q becomes a $\mathbb{P}^1$-bundle and $L, \bar{L}$ become conjugate sections. This case was treated in (4.9) and we get that blowing up P is the only possible further blow up.

If we blow up P (possibly $P = Q$) and the exceptional curve E appears in B with coefficient ≥ 1 then we are in case (4.6) and no more blow ups are possible. So E appears in B with coefficient 0. Another blow up on E would create a (-2)-curve with coefficient 0 in B_2 intersecting the birational transform of L, a contradiction. Thus we can blow up only $Q \neq P$ and then we are in the already discussed case.

As we showed, we can never blow up infinitely near points, so there are no $(1, m)$-blow ups by (4.7).

If $S^* = Q^{3,1}$ then a one point blow up of S^* is also a blow up of $\mathbb{P}^2$ at a conjugate pair of points. So we are reduced to considering the blow ups of $\mathbb{P}^2$. These cases can also be treated directly. □

We can summarize the above results as follows.

Proposition 4.12. *Assume that the quadruplet* (S^m, B^m, D^m, M^m) *satisfies the conditions* (3.9). *Then* (S^r, B^r) *satisfies one of the following conditions.*

1. S^r *is a weak Del Pezzo surface and* $B^r = 0$.

2. S^r *is a* $\mathbb{P}^1$*-bundle,* B^r *consists of a section, at most 2 real fibers and some conjugate pairs of fibers.*

3. S^r *is a conic bundle,* B^r *consists of a conjugate pair of sections and at most one fiber.*

4. S^r *is a conic bundle with one point blown up,* B^r *consists of a conjugate pair of sections and possibly one fiber which is the union of a* $\mathbb{P}^1$ *and of a conjugate pair of rational curves.*

5. $S^r = \mathbb{P}^2$ *and* B^r *is either a line or a pair of lines or a conic.*

6. $S^r = Q^{2,2} \cong \mathbb{P}^1 \times \mathbb{P}^1$ *and* B^r *is either a line or a pair of intersecting lines or a conic.*

7. $S^r = Q^{2,1}$ *and* B^r *is a hyperplane section not through the vertex.*

8. $S^r(\mathbb{R})$ *is a finite set.*

Proof. There are only a few points which have not been settled earlier.

In the $\mathbb{P}^1$-bundle case write $B^r = H + B'$. From the adjunction formula we get that

$$-2 \leq 2g(H) - 2 = H \cdot (K + H) = -H \cdot (B' + D^r + M^r) \leq -(H \cdot B'),$$

so B^r contains at most 2 fibers.

In the conic bundle case we either have no blow ups and get (3) or let C denote the fiber containing the center of the blow up. There are 2 cases to consider. C can be singular giving case (4). If C is smooth then after one blow up we get 2 intersecting (-1)-curves and both appear with coefficient 0 in B^r. □

The following result could have been proved much earlier.

Lemma 4.13. *Let (S, B, D, M) be a quadruplet satisfying the conditions (3.5). Then one of the following holds.*

1. *S is a cone over an elliptic curve and $B = 0$.*

2. *$S_{\mathbb{C}}$ is a rational surface with rational singularities and every irreducible component of B^m is a smooth rational curve.*

Proof. In the previous classification, a nonrational surface can occur only in (4.6). So S^* is a $\mathbb{P}^1$-bundle and $H \subset B^*$ is a section. From the adjunction formula

$$2g(H) - 2 = H(K + H) = -H(B^* - H + D^* + M^*) \leq 0.$$

So H is elliptic, $B^* = H$ and $H(K + B^*) = 0$. We have proved that there are no real blow ups and a similar computation shows that there are no complex blow ups either. So $S^m = S^*$ and $|-r(K + B^*)|$ contracts H for $r \gg 1$.

Let $h : S' \to S$ be any resolution. The Leray spectral sequence gives an exact sequence

$$H^1(S', \mathcal{O}_{S'}) \to R^1 h_* \mathcal{O}_{S'} \to H^2(S, \mathcal{O}_S).$$

$H^1(S', \mathcal{O}_{S'}) = 0$ if S' is rational. $h^2(S, \mathcal{O}_S) = h^0(S, \mathcal{O}_S(K_S))$ and the latter is zero since $\mathcal{O}_S(-rK_S)$ has sections for $r \gg 1$. Thus $R^1 h_* \mathcal{O}_{S'} = 0$ and so S has rational singularities.

Irreducible components of B^* are smooth and rational by the classification and all other curves in B^m appear as exceptional curves of $(1, m)$-blow ups, so they are smooth and rational. □

Lemma 4.14. *Let S be a real algebraic surface with rational singularities and $h : S' \to S$ a proper birational morphism from a normal surface S'. Then $S'(\mathbb{R}) \to S(\mathbb{R})$ has connected fibers.*

Proof. For any $P \in S(\mathbb{R})$, $H^1(h^{-1}(P), \mathcal{O}) = 0$ since S has rational singularities. Thus $C := \text{red}(h^{-1}(P))$ is a tree of smooth rational curves. If $A, B \in C(\mathbb{R})$ are two points then C has a unique chain of rational curves $C_{A,B}$ connecting A and B. Complex conjugation fixes the two ends of the chain, so it fixes every complex irreducible component and every singular point. Thus $C_{A,B}(\mathbb{R})$ is a connected chain of circles. □

5. The real points of singular fibers

In (3.8) B^m was defined by the formula $K_{S^m}+B^m \equiv f^*(K_S+B)$. This shows that S is obtained from S^m by contracting all the curves $A \subset S^m$ such that $(A \cdot (K_{S^m}+B^m)) = 0$. By (3.9) all such curves are in Supp B^m. Using (4.14), a version of this also holds for real points.

Proposition 5.1. *From* $(S^m(\mathbb{R}), B^m(\mathbb{R})) \cong (S^r(\mathbb{R}), B^r(\mathbb{R}))$ *we obtain* $(S(\mathbb{R}), B(\mathbb{R}))$ *by contracting certain connected subcurves of* $B^r(\mathbb{R})$ *to points and by adding isolated points.* □

Remark 5.2. (1) A real singular point of S may be isolated in $S(\mathbb{R})$ and so invisible in the real part of the resolution.

(2) In the applications I will be able to compute only a Zariski neighborhood of $S(\mathbb{R})$ in $S(\mathbb{C})$. (5.1) allows us to compute the pair $(S(\mathbb{R}), B(\mathbb{R}))$ up to finite ambiguity.

(3) The somewhat complicated choice of the partial resolution $f : S^m \to S$ is important for (5.1). If we take the minimal resolution then the exceptional curves over Du Val points appear in B with zero coefficient. Thus $S^m \to S$ would also involve exceptional curves which are not controled by B^m. It is probably possible to analyze this but some complications definitely do appear.

We have a list of all possible pairs $(S^r(\mathbb{R}), B^r(\mathbb{R}))$, so using (5.1) we can get a list of all possible pairs $(S(\mathbb{R}), B(\mathbb{R}))$. From these pairs the real points of the fibers are obtained by gluing B to itself. It should be kept in mind that if a point $P \in B$ is glued to its conjugate $\bar{P}$ then we obtain a real point. Thus some points of $X_0(\mathbb{R})$ may not come from a point of $S(\mathbb{R})$.

The gluing process is easy if $B(\mathbb{R})$ is not complicated, say empty or a single circle but it may be subtle in general. Some of the worst cases can be avoided if we assume that $X(\mathbb{R})$ is orientable though this is not essential.

Next we define a certain class of 2-complexes. The reason for this rather unnatural definition is that this is what I can prove about the real points of singular fibers.

Definition 5.3. Let C be a circle, $(B_i, \partial B_i)$ 2-discs and $\phi_i : \partial B_i \to C$ PL-maps. We use the ϕ_i to glue the discs to C. The resulting 2-complex $C \cup_{\phi_i} B_i$ is called a *circle with discs attached*. A disc B_i is called *inessential* if ϕ_i has degree 0. $\pi_1(C \cup_{\phi_i} B_i) = \mathbb{Z}/m$ where m is the gcd of the degrees of the ϕ_i.

Condition 5.4. We consider compact 2-complexes K which satisfy the following conditions:

1. There is a 2-complex K' and a surjective PL-map $h : K' \to K$.

2. K' is the disjoint union of points, intervals, spheres, real projective planes, tori, Klein bottles and circles with discs attached.

3. There are finite subsets $A \subset K$ and $A' \subset K'$ such that $h : K' \setminus A' \to K \setminus A$ is a homeomorphism. Moreover, A' does not contain any interior point of an interval.

4. If K' contains a torus or a Klein bottle then this is its only 2-dimensional connected component.

Observe that the 2-dimensional part of K' is uniquely determined by K. In the applications the lower dimensional parts will play only a minor role.

It is easy to see that if K satisfies the above conditions and G is a finite group acting on K without fixed points then K/G also satisfies the above conditions.

Theorem 5.5. *Let X be a real algebraic threefold and $f : X \to C$ a morphism to a smooth real algebraic curve. Assume that X has isolated singularities which are $\mathbb{Q}$-factorial over $\mathbb{R}$, $-K_X$ is f-ample and the smooth part of $X(\mathbb{R})$ is orientable. Let $0 \in C(\mathbb{R})$ be a point such that $X_0 := f^{-1}(0)$ is irreducible (over $\mathbb{R}$).*

Then $X_0(\mathbb{R})$ is a 2-complex satisfying the conditions (5.4).

Proof. Assume first that X_0 is geometrically reducible. Then red $X_0 = Z_0 + \bar{Z}_0$ and all real points of X_0 are in $Z_0 \cap \bar{Z}_0$. Thus $X_0(\mathbb{R})$ is a 1-complex and we are done.

Next consider the case when X_0 is geometrically irreducible and reduced. The list of all $(S^r(\mathbb{R}), B^r(\mathbb{R}))$ can be read off from (4.12) and (5.1) then gives a longer list for the pairs $(S(\mathbb{R}), B(\mathbb{R}))$. The orientability assumption can be used through the following two consequences:

1. $S^r(\mathbb{R}) \setminus B^r(\mathbb{R})$ is orientable, and

2. If $\{C_t\}$ is a base point free pencil on S^r such that $(C_t \cdot (K_{S^r} + B^r))$ is odd then the general C_t intersects at least one irreducible component of B^r which gets contracted in S.

The first of these holds since $S^r(\mathbb{R}) \setminus B^r(\mathbb{R})$ injects into $S(\mathbb{R}) \setminus B(\mathbb{R})$ by (5.1) and $X_0(\mathbb{R})$ is 2-sided in $X(\mathbb{R})$, hence its smooth part is orientable. The second assertion holds since otherwise $\{C_t\}$ would give a base point free family of curves in X such that $(C_t \cdot K_X)$ is odd. This is impossible since $X(\mathbb{R})$ is orientable (cf. [Kollár97a, 2.8]).

A real point of $X_0(\mathbb{R})$ is either in the image of $S(\mathbb{R})$ or is in $\text{Sing}(X_0)$. The latter is a 1-complex, and adding a 1-complex does not change the conditions (5.4).

Let us now go through the list of (4.12).

If S^r is a weak Del Pezzo surface such that the smooth part of $S^r(\mathbb{R})$ is orientable then $S^r(\mathbb{R})$ is obtained from spheres and tori by identifying some points. In the smooth case this was proved by [Comessatti14], see also [Silhol89]. The singular case is discussed in [Kollár98a, 9.9] or it can be read off the list of real plane quartics in [GUT66]. See also [Wall95].

Next let S^r be a conic bundle. Then $S^r(\mathbb{R})$ is a union of spheres with possibly some points identified. $B^r(\mathbb{R})$ is either empty or a circle in a sphere. The first case goes as before. If $B^r(\mathbb{R})$ is contracted, we get one more sphere. Otherwise as we go from S to X_0 we have to glue $B^r(\mathbb{R})$ to itself. Since $B^r(\mathbb{R})$ cuts the sphere into 2 discs, we obtain a circle with two discs attached. If S^r is a conic bundle with one point blown up then we have spheres and one $\mathbb{RP}^2$ and $B^r(\mathbb{R})$ is a line in $\mathbb{RP}^2$. If it is contracted then we get a sphere, otherwise we obtain a circle with one disc attached.

Similar arguments apply every time $B(\mathbb{R})$ is empty or a circle. Some of these cases can not happen, for instance ($\mathbb{RP}^2$, smooth conic) does not occur since the complement of a smooth conic is not orientable.

In case ($\mathbb{RP}^2$, two lines) a general line in $\mathbb{P}^2$ has intersection number -1 with $(K_{S^r} + B^r)$. Thus at least one of the lines gets contracted in S. Similarly, if $S^r = \mathbb{P}^1 \times \mathbb{P}^1$ then all of B^r has to be contracted.

We are left with the case when S^r is a $\mathbb{P}^1$-bundle. A general fiber in S^r has intersection number -1 with $(K_{S^r} + B^r)$, and this shows that the section H gets contracted. $S^r(\mathbb{R})$ is a torus or a Klein bottle. If we contract $H(\mathbb{R})$ it becomes a sphere with a pair of points pinched together. If B^r contains at most one real fiber then we are in one of the previously studied cases.

When B^r contains two real fibers F_1^r, F_2^r, further discussions are needed. Let $F_i \subset S$ be the birational transform of F_i^r. Since H is contracted and B^r does not contain any other sections, F_1 and F_2 intersect at a single point $P \in S$.

If F_1 and F_2 are mapped to the same curve in X_0 then we obtain a circle with 2 discs attached. This is the typical case and it occurs when X_0 is a cone over a nodal rational curve.

We are left with the cases when the normalization $\pi : S \to X_0$ maps $F_i \to C_i$ and $C_1 \neq C_2$. Each F_i appears in B with coefficient 1, so $F_i \to C_i$ is a degree 2 map and the two branches of X intersect transversely along C_i by (3.7.2).

There are two different degree 2 maps $\mathbb{RP}^1 \to \mathbb{RP}^1$. The first is $(z : 1) \mapsto (z^2 : 1)$. Under this map the circle $\mathbb{P}^1(\mathbb{R})$ maps to an interval. If $F_1 \to C_1$ is such a map then $\pi(S(\mathbb{R}))$ is the circle $C_2(\mathbb{R})$ with a disc attached and we have a 1-complex $C_2(\mathbb{R}) \setminus \pi(F_1(\mathbb{R}))$ added to it.

Finally it may happen that both $F_i(\mathbb{R}) \to C_i(\mathbb{R})$ are given as $(z : 1) \mapsto (z : z^2-1)$. I claim that near $\pi(P)$ this leads to a configuration which can not be realized in $\mathbb{R}^3$.

If (H^2) is odd then locally near P

$$(S(\mathbb{R}), F_1(\mathbb{R}), F_2(\mathbb{R}), P) \sim_s (\mathbb{R}^2, (x = 0), (y = 0), (0, 0)),$$

where $\sim_s$ denotes a PL homeomorphism of stratified spaces. There are 2 more branches of $X_0(\mathbb{R})$ passing through $\pi(P)$. One branch B_1 passes through $F_1(\mathbb{R})$ the other branch B_2 passes through $F_2(\mathbb{R})$) and these two branches do not intersect each other. Moreover, all intersections are generically transverse. Intersect everything with a small sphere S_ϵ^2 around the origin. The upper hemi sphere $(z \geq 0)$ is a topological disc and B_1 and B_2 intersect it in 2 curves. One of these curves connects the points $(0, 1)$, $(0, -1)$ and the other the points $(1, 0)$, $(-1, 0)$. Hence $B_1 \cap B_2 \cap S_\epsilon^2 \neq \emptyset$, a contradiction.

If (H^2) is even then locally near P

$$(S(\mathbb{R}), F_1(\mathbb{R}), F_2(\mathbb{R}), P)$$
$$\sim_s ((xy - z^2 = 0), (x = z = 0), (y = z = 0), (0, 0, 0)) \subset \mathbb{R}^3.$$

There are 2 more branches of $X_0(\mathbb{R})$ passing through $\pi(P)$. One passes through $F_1(\mathbb{R})$ the other through $F_2(\mathbb{R})$ and these two branches do not intersect each other. Moreover, all intersections are generically transverse. This is again impossible in $\mathbb{R}^3$.

We are left with the case when X_0 is a multiple fiber. Let $m > 0$ be the smallest integer such that mX_0 is Cartier. Then $\mathcal{O}_X(mX_0)$ is a locally free sheaf isomorphic to $\mathcal{O}_X$ in a neighborhood of X_0. The section $1 \in H^0(X, \mathcal{O}_X) \cong H^0(X, \mathcal{O}_X(mX_0))$ determines the corresponding m-sheeted cyclic cover $Y \to X$ (cf. [Kollár-Mori98, Sec. 2.4]). Since X_0 is Cartier at real points, $Y \to X$ is unramified at real points. Thus $Y(\mathbb{R}) \to X(\mathbb{R})$ is a homeomorphism if m is odd. If m is even then over each connected component we get either a 2-sheeted cover or the empty set. In the latter case we switch to $-1 \in H^0(X, \mathcal{O}_X)$.

We have already proved that $Y_0(\mathbb{R})$ satisfies the properties (5.4), thus the same holds for $X_0(\mathbb{R})$, as remarked at the end of (5.4). □

6. Proof of the main theorem

In this section I use (5.5) and a purely topological argument to complete the proof of (1.3). First, following (1.5), we identify the neighborhoods of the singular fibers. Then we show how these pieces are assembled to form $\overline{X(\mathbb{R})}$.

Lemma 6.1. *Let M be an orientable 3-manifold and $K \subset M$ a 2-complex satisfying the conditions (5.4). Let $K \subset U \subset M$ be a regular neighborhood of K and assume that ∂U is a union of spheres and tori. Then every connected component of U is one of the following:*

1. *connected sum of lens spaces and $S^1 \times S^2$ minus balls,*

2. *connected sum of lens spaces, $S^1 \times S^2$ and of a solid torus minus balls,*

3. *interval bundle over a torus or a Klein bottle.*

Proof. As a first step we replace $h : K' \to K$ with another map $\tilde{h} : \tilde{K}' \to \tilde{K}$ such that

4. $\tilde{h} : \tilde{K}' \to \tilde{K}$ satisfies the conditions (5.4),

5. U is a regular neighborhood of $\tilde{K}$,

6. $\tilde{h}$ is injective on the set of 2-dimensional points of $\tilde{K}'$.

To achieve this, note that there are only finitely many points $p \in K$ over which h is not one-to-one. We will get rid of these one at a time. Let $p \in V$ be a regular neighborhood. $K \cap \partial V$ is a union of connected 1-complexes A_j and $K \cap V$ is the cone over $K \cap \partial V$. If all the A_j have dimension 0 then we do not need to do anything.

Otherwise, there is a 1-dimensional component (say A_1) and a PL-homeomorphism $\psi : (V, \partial V) \to (B^3, S^2)$ such that $\psi(A_1)$ is in the northern hemisphere and all the other 1-dimensional A_j map to the southern hemisphere. Now cut the 3-ball B^3 along the equator and move the two halves apart a little. The center of the ball sweeps out a small interval. We add this interval to obtain K_1. K'_1 is K' union an interval.

Repeating this procedure if necessary, at the end we obtain $\tilde{h} : \tilde{K}' \to \tilde{K}$.

To simplify notation let us assume that $h : K' \to K$ already satisfies the above conditions 4–6. Next we use induction on the number of 2-dimensional components of K'.

If there is an $S^2 \sim L \subset K'$ then $h(L)$ is an embedded S^2. Cut U along $h(L)$ and glue 3-balls to the resulting two spheres to get U_1. K' is replaced by $K'_1 := K' \setminus L$. U is obtained from U_1 by attaching a 1-handle which is either taking connected sum of two components or taking connected sum with $S^1 \times S^2$. If there is an $\mathbb{RP}^2 \sim L \subset K'$ then $h(L)$ is an embedded $\mathbb{RP}^2$ and the boundary of its regular neighborhood is S^2 (since M is orientable). Cutting along this S^2 corresponds to connected sum with $\mathbb{RP}^3$ (which is the lens space $L_{2,1}$).

Assume now that there is an $L \subset K'$ which is a circle with discs attached. Let V be a regular neighborhood of $h(C)$ which we may assume to be an embedded circle. V is a solid torus. Consider the case when one of the discs (say B_1) is inessential. Then $\partial V \cap h(N_1)$ bounds a disc $D \subset \partial V$. We may asume that no other B_i intersects this disc. We change $h(B_1)$ by replacing $V \cap B_1$ with D and adding an interval connecting D with $h(C)$ to K. This way we have not changed U, we have removed one inessential disc and we created a new embedded sphere. By repeating this procedure we may assume that there are no inessential discs. Then each $h(B_i)$ intersects ∂V in a simple closed curve which is not null homotopic. Since these curves are all disjoint, they are in the same homotopy class γ. Thus $V \cup_i h(B_i)$ is a solid torus with discs attached along parallel curves in ∂V. The boundary of a regular neighborhood W of $V \cup_i h(B_i)$ is a union of spheres. We can again cut U along these spheres. W can also be obtained as attaching first a solid torus V' to V such that the meridian of V' maps to γ and then removing some balls from V'. Gluing two solid tori gives a lens space (1.8).

We are left to deal with the 1-dimensional part of K. Every connected component can be collapsed to a bouquet of circles. If there are ≥ 2 circles then the boundary of its regular neighborhood has genus ≥ 2, a contradiction. Thus we get either a ball (if there are no cicrcles) or a solid torus (if there is one circle).

If there is an $L \subset K'$ which is a torus or a Klein bottle then the boundary of a regular neighborhood of $h(L)$ is either 2 or 1 tori. There are no other 2-dimensional components in K' by (5.4.4). If K has a 1-dimensional subcomplex which does not collapse into $h(L)$ then it leads to a genus ≥ 2 component in ∂U, a contradiction. □

Remark 6.2. Using (3.7) we obtain that the only lens spaces that can appear in (6.1) are $\mathbb{RP}^3 = S^3/\mathbb{Z}_2$ and $S^3/\mathbb{Z}_3$. I have no example for the latter.

6.3 (Proof of (1.3)). We have established that M is glued together from the following pieces:

1. $S^1 \times S^2$ minus open balls,

2. lens space minus open balls,

3. solid torus minus open balls,

4. interval bundle over a torus or a Klein bottle.

An orientable interval bundle over a torus is a torus times an interval and the boundary of an orientable interval bundle over a Klein bottle is a torus (cf. [Kollár97c, 1.6]).

If the gluing involves a torus times an interval then this piece can be thrown away, except when the two boundary components are glued to each other. This gives a torus bundle over a circle.

Two copies of an interval bundle over a Klein bottle glued together map to an interval such that the fibers are tori over interior porints and Klein bottles over the two boundary points. This 3-manifold is doubly covered by a torus bundle over a circle.

An interval bundle over a Klein bottle glued to a solid torus is homeomorphic to $\mathbb{RP}^3 \# \mathbb{RP}^3$ (cf. [Kollár97c, 12.7]), so in this case we can change the decomposition to one that does not involve any interval bundles over a Klein bottle.

We are left with the case when M is glued together from lens spaces minus open balls. We do one gluing at a time. If a new lens space is glued in, that is connected sum. If two boundary components of a connected component are glued together then that is the same as taking connected sum with $S^1 \times S^2$. □

7. Examples

Example 7.1. Consider $\mathbb{P}^4(1, 1, 1, 2, 2)$ with coordinates $(x : y : z : u : v)$ and the affine line $\mathbb{A}^1$ with coordinate t. Let X be the complete intersection

$$X \subset \mathbb{P}^4(1, 1, 1, 2, 2) \times \mathbb{A}^1 \quad \text{given by equations}$$
$$u^2 + v^2 = f_4(x, y, z) \quad \text{and} \quad tv = q_2(x, y, z).$$

Let $\pi : X \to \mathbb{A}^1$ be the second projection. For $t \neq 0$ we can eliminate v to obtain a degree 2 Del Pezzo surface

$$X_t \cong (u^2 = f_4(x, y, z) - t^{-2} q_2(x, y, z)^2) \subset \mathbb{P}^3(1, 1, 1, 2).$$

For $t = 0$ the equations become $u^2 + v^2 = f_4(x, y, z)$ and $q_2(x, y, z) = 0$. This has two points of index 2 at $(0 : 0 : 0 : 1 : \pm\sqrt{-1})$, both are analytically isomorphic to $\mathbb{C}^2/\mathbb{Z}_4(1, 1)$. The projection $(x : y : z : u : v) \mapsto (x : y : z)$ is defined outside these two points and the minimal resolution of X_0 becomes a conic bundle over the conic $(q_2(x, y, z) = 0)$. The singular fibers correspond to the solutions of the equations $q_2 = f_4 = 0$. We get various cases depending on how the curves $(q_2 = 0)$ and $(f_4 = 0)$ intersect.

Another way to obtain this model is as follows. Consider the family

$$Y \subset \mathbb{P}^3(1, 1, 1, 2) \times \mathbb{A}^1 \quad \text{given by} \quad u^2 = t^2 f_4(x, y, z) - q_2(x, y, z)^2.$$

Outside the origin Y is isomorphic to X via the transformation

$$(x, y, z, u, t) \mapsto (x, y, z, ut, t).$$

The central fiber Y_0 consists of a conjugate pair of planes intersecting along the conic $(q_2 = 0)$. Y is singular along $(u = t = q_2 = 0)$. If we blow up the singular curve, the two planes in the central fiber become disjoint. Contracting them gives the 3-fold X.

Example 7.2. Consider $\mathbb{P}^4(1, 1, 2, 3, 3)$ with coordinates $(x : y : z : u : v)$ and the affine line $\mathbb{A}^1$ with coordinate t. Let X be the complete intersection

$$X \subset \mathbb{P}^4(1, 1, 2, 3, 3) \times \mathbb{A}^1 \quad \text{given by equations}$$
$$u^2 + v^2 = f_6(x, y, z) \quad \text{and} \quad tv = c_3(x, y, z).$$

Let $\pi : X \to \mathbb{A}^1$ be the second projection. For $t \neq 0$ we can eliminate v to see that the fiber is the degree 1 Del Pezzo surface

$$X_t \cong (u^2 = f_6(x, y, z) - t^{-2} c_3(x, y, z)^2) \subset \mathbb{P}^3(1, 1, 2, 3).$$

For $t = 0$ the equations become $u^2 + v^2 = f_6(x, y, z)$ and $c_3(x, y, z) = 0$. This has two points of index 3 at $(0 : 0 : 0 : 1 : \pm\sqrt{-1})$, both are analytically isomorphic to $\mathbb{C}^2/\mathbb{Z}_9(1, 2)$. The projection $(x : y : z : u : v) \mapsto (x : y : z)$ is defined outside these two points and the minimal resolution of X_0 becomes a conic bundle over the curve $(c_3(x, y, z) = 0)$ blown up in one point. Although c_3 has degree 3, this curve is birational to $\mathbb{P}^1$. Indeed, since z has degree 2, it appears in c_3 only linearly, so z can be rationally expressed in terms of x, y.

One can also obtain this model from the degeneration

$$Y \subset \mathbb{P}^3(1, 1, 2, 3) \times \mathbb{A}^1 \quad \text{given by} \quad u^2 = t^2 f_6(x, y, z) - c_3(x, y, z)^2,$$

but the birational transformation between them is more complicated.

Example 7.3. Start with the trivial family of quadrics

$$\mathbb{P}^1 \times \mathbb{P}^1 \times \mathbb{A}^1 \quad \text{with coordinates} \quad (x_1 : x_2), (y_1 : y_2), t.$$

Consider the $\mathbb{Z}_2$-action

$$\tau_1 : (x_1, x_2, y_1, y_2, t) \mapsto (x_2, -x_1, y_2, -y_1, -t).$$

This has 4 fixed points at $(1, \pm\sqrt{-1}, 1, \pm\sqrt{-1}, 0)$. Set $X_1 := (\mathbb{P}^1 \times \mathbb{P}^1 \times \mathbb{A}^1)/\mathbb{Z}_2(\tau_1)$. X_1 has 4 singularities of analytic type $\mathbb{C}^3/\mathbb{Z}_2(1, 1, 1)$. The central fiber is double and the reduced central fiber has 4 A_1-type points. It is easy to see that $X_1(\mathbb{R})$ is not orientable and its central fiber is $S^1 \times S^1$.

Another $\mathbb{Z}_2$-action is given by

$$\tau_2 : (x_1, x_2, y_1, y_2, t) \mapsto (x_2, -x_1, y_1, -y_2, -t).$$

This also has 4 fixed points at $(1, \pm\sqrt{-1}, 1, 0, 0)$ and $(1, \pm\sqrt{-1}, 0, 1, 0)$. Set $X_2 := (\mathbb{P}^1 \times \mathbb{P}^1 \times \mathbb{A}^1)/\mathbb{Z}_2(\tau_2)$. X_2 has 4 singularities of analytic type $\mathbb{C}^3/\mathbb{Z}_2(1, 1, 1)$, the central fiber is double and the reduced central fiber has 4 A_1-type points. One can see that $X_2(\mathbb{R})$ is orientable and its central fiber is a Klein bottle.

Example 7.4. Let $(x : y : z)$ and t be coordinates on $\mathbb{P}^2 \times \mathbb{A}^1$ and consider the surface $S := (x^2 + y^2 - tz^2 = 0)$. Let $\mathbb{Z}_n$ act on (x, y, z, t) as a rotation of order n on x, y and identity on z, t. This induces an action on $\mathcal{O}_S + \mathcal{O}_S$. Let $F \subset \mathcal{O}_S + \mathcal{O}_S$ be the locally free rank 2 subsheaf which on the $z \neq 0$ affine chart is generated by the sections (x, y) and $(-y, x)$. Outside $(x^2 + y^2 = 0)$ the subsheaf F is the same as $\mathcal{O}_S + \mathcal{O}_S$. Let $X := \mathbb{P}_S F$ be the corresponding $\mathbb{P}^1$-bundle over S. $\mathbb{Z}_n$ acts on X. This action has two isolated conjugate fixed points over $(0, 0, 1, 0) \in S$ and fixes the conjugate surfaces over the curves $(1, \pm\sqrt{-1}, 0, t) \subset S$. Set $X_n := X/\mathbb{Z}_n$. The projection $\pi : X_n \to \mathbb{A}^1$ exhibits X_n as a degeneration of quadrics and $-K_{X_n}$ is π-ample. X_n has only terminal singularities, it has 2 points of index n and it is smooth at all real points. The central fiber is geometrically reducible.

The only slight problem with this example is that $\rho(X_n/\mathbb{A}^1) = 2$ since π can be factored as $X_n \to S/\mathbb{Z}_n \to \mathbb{A}^1$. Probably one can globalize this example to get relative Picard number 1.

Example 7.5. Let $S^* \subset \mathbb{P}^3$ be a smooth quadric and $B^* = 0$. Let $P, \bar{P} \in S^*$ be a pair of conjugate points not on a line. Let $S_1 \to S^*$ be the blow up of $P + \bar{P}$ with exceptional curve $E + \bar{E}$. Set $B_1 := (1/2)(E + \bar{E})$. The pencil of planes through $P + \bar{P}$ gives S_1 a conic bundle structure with $E + \bar{E}$ as conjugate sections. Blow up 3 more pairs of conjugate points on $E + \bar{E}$ to get S^m. The birational transforms of $E, \bar{E}$ have self intersection -4, so they can be contracted $S^m \to S$. This S is among the types described in (7.1).

Example 7.6. Set $Q := \mathbb{P}^1_{(x_0:x_1)} \times \mathbb{P}^1_{(y_0:y_1)}$ and let $\sigma : Q \to Q$ be given by $(x_0 : x_1, y_0 : y_1) \mapsto (y_1 : y_0, x_0 : x_1)$. Then σ has order 4 on Q and on $H_1(Q(\mathbb{R}), \mathbb{Z})$. Take $Q \times \mathbb{P}^1_{(z_0:z_1)}$. Let τ be the action which is σ on Q and $(z_0 : z_1) \mapsto (z_0 + z_1 : -z_0 + z_1)$ on the second factor. Set $X := (Q \times \mathbb{P}^1_{(z_0:z_1)})/(\tau)$. Then $X \to \mathbb{P}^1_{(z_0:z_1)}/(\tau)$ is a quadric bundle (except over a conjugate pair of complex points). Its real part gives an $S^1 \times S^1$-bundle over S^1 with order 4 monodromy.

Example 7.7. Let Q be the degree 6 Del Pezzo surface given in $\mathbb{P}^1_{(x_0:x_1)} \times \mathbb{P}^1_{(y_0:y_1)} \times \mathbb{P}^1_{(z_0:z_1)}$ by the equation

$$x_0 y_0 z_1 + x_0 y_1 z_0 + x_1 y_0 z_0 = x_0 y_1 z_1 + x_1 y_0 z_1 + x_1 y_1 z_0.$$

Let $\sigma : Q \to Q$ be given by

$$(x_0 : x_1, y_0 : y_1, z_0 : z_1) \mapsto (z_1 : z_0, x_1 : x_0, y_1 : y_0).$$

σ has order 6 on Q and on $H_1(Q(\mathbb{R}), \mathbb{Z})$. Take $Q \times \mathbb{P}^1_{(t_0:t_1)}$. Let τ be the action which is σ on Q and

$$(t_0 : t_1) \mapsto (\cos(\pi/6)t_0 + \sin(\pi/6)t_1 : -\sin(\pi/6)t_0 + \cos(\pi/6)t_1)$$

on the second factor. Set $X := (Q \times \mathbb{P}^1_{(t_0:t_1)})/(\tau)$. Then $X \to \mathbb{P}^1_{(t_0:t_1)}/(\tau)$ is a degree 6 Del Pezzo surface bundle (except over a conjugate pair of complex points). Its real part gives an $S^1 \times S^1$-bundle over S^1 with order 6 monodromy.

Example 7.8. In $\mathbb{P}^3_{(x_0:x_1:x_2:x_3)} \times \mathbb{P}^1_{(y_0:y_1)}$ consider the 3-fold

$$X := (y_0(x_0^2 + x_1^2) = y_1(x_2^2 + x_3^2)).$$

For p odd and $(p, q) = 1$ let $\sigma : X \to X$ be the action which is rotation by $2\pi/p$ on (x_0, x_1) and rotation by $2q\pi/p$ on (x_2, x_3). Then $X(\mathbb{R}) \sim S^3$ and $X(\mathbb{R})/(\sigma) \sim L_{p,q}$.

The complex 3-fold $X/(\sigma)$ is smooth at its real points but it has nonterminal singularities at complex points. At least for $q = 1$ these are easy to resolve and one obtains a quadric bundle X' such that $X'(\mathbb{R}) \sim L_{p,1}$. The other cases seem more complicated.

References

[AGV85]   V. I. Arnold, S. M. Gusein-Zade and A. N. Varchenko, Singularities of Differentiable Maps I–II, Birkhäuser, 1985.

[BCR87]   J. Bochnak, M. Coste and M.-F. Roy, Géométrie algébrique réelle, Springer-Verlag, 1987.

[CKM88]   H. Clemens, J. Kollár and S. Mori, Higher Dimensional Complex Geometry, Astérisque 166, 1988.

[Comessatti14]   A. Comessatti, Sulla connessione delle superfizie razionali reali, Annali di Math. 23 (3) (1914), 215–283.

[Fulton95]   W. Fulton, Algebraic topology, Springer-Verlag, 1995.

[Furushima86]   M. Furushima, Singular Del Pezzo surfaces, Nagoya Math. J. 104 (1986), 1–28.

[GUT66]   D. A. Gudkov, G. A. Utkin and M. L. Taj, The complete classification of irreducible curves of the 4th order (in Russian), Math. Sb. 69 (1966), 222–256.

[Hartshorne77]   R. Hartshorne, Algebraic Geometry, Springer-Verlag, 1977.

[Hempel76]   J. Hempel, 3-manifolds, Princeton Univ. Press, 1976.

[Keel-McKernan98]   S. Keel and J. McKernan, Rational curves on quasi-projective varieties, Mem. Amer. Math. Soc., to appear.

[Kharlamov76]   V. Kharlamov, The topological type of non-singular surfaces in RP^3 of degree four, Funct. Anal. Appl. 10 (1976), 295–305.

[Kollár87]   J. Kollár, The structure of algebraic threefolds - an introduction to Mori's program, Bull. Amer. Math. Soc. 17 (1987), 211–273.

[Kollár96]   J. Kollár, Rational Curves on Algebraic Varieties, Ergeb. Math. Grenzgeb. 32, Springer-Verlag, 1996.

[Kollár97a] J. Kollár, Real Algebraic Surfaces, Notes of the 1997 Trento summer school lectures, preprint.

[Kollár97b] J. Kollár, Real Algebraic Threefolds I. Terminal Singularities, Collect. Math. 49 (1998), 335–360.

[Kollár97c] J. Kollár, Real Algebraic Threefolds II. Minimal Model Program, J. Amer. Math. Soc. 12 (1999), 33–83.

[Kollár98a] J. Kollár, Real Algebraic Threefolds III. Conic Bundles, preprint.

[Kollár98b] J. Kollár, The Nash Conjecture for threefolds, Electron. Res. Announc. Amer. Math. Soc. 4 (1998), 63–73; http://www.ams.org/era.

[Kollár-Mori98] J. Kollár - S. Mori, Birational geometry of algebraic varieties, Cambridge Univ. Press, 1998.

[Manetti91] M. Manetti, Normal degenerations of the complex projective plane, J. Reine Angew. Math. 419 (1991), 89–118.

[Manetti93] M. Manetti, Normal projective surfaces with $\rho = 1$, $P_{-1} \geq 5$, Rend. Sem. Mat. Univ. Padova 89 (1993), 195–205.

[Moise77]   E. Moise, Geometric topology in dimensions 2 and 3, Grad. Texts in Math. 47, Springer-Verlag, 1977.

[Nash52]   J. Nash, Real algebraic manifolds, Ann. of Math. 56 (1952), 405–421.

[Reid85]   M. Reid, Young person's guide to canonical singularities, in: Algebraic Geometry, Proc. Sympos. Pure Math. 46, Amer. Math. Soc., 1987, pp. 345–414.

[Reid94]   M. Reid, Nonnormal Del Pezzo surfaces, Publ. RIMS Kyoto Univ. 30 (1994), 695–728.

[Rolfsen76] D. Rolfsen, Knots and links, Publish or Perish, 1976.

[Rourke-Sanderson82] C. Rourke and B. Sanderson, Introduction to piecewise linear topology, Springer-Verlag, 1982.

[Scott83]   P. Scott, The geometries of 3-manifolds, Bull. London Math. Soc. 15 (1983), 401–487.

[Shafarevich72] R. I. Shafarevich, Basic Algebraic Geometry (in Russian), Nauka, 1972; English translation: Springer-Verlag, 1977, Second Expanded Edition, 1994.

[Silhol89]   R. Silhol, Real algebraic surfaces, Lecture Notes in Math. 1392, Springer-Verlag, 1989.

[Wall95]   C. T. C. Wall, Real rational quartic curves, in: W. L. Marar, ed., Real and complex singularities, Pitman Res. Notes in Math. 333 (1995), 1–32.

Seiberg–Witten invariants for 4-manifolds with $b_+ = 0$

Christian Okonek* and Andrei Teleman*

Abstract. We extend Seiberg–Witten theory to 4-manifolds with $b_+ = 0$. The moduli space of irreducible monopoles depends in this case on a parameter varying in an infinite dimensional space which is divided into a countable family of chambers by a codimension-1 wall. The Seiberg–Witten invariant associated with a class of $Spin^c$-structures is a functorial map which assigns an integer to every chamber and satisfies a universal wall-crossing formula for transversal wall-crossing; it should be regarded as a distinguished element in the $\mathbb{Z}$-torsor of functions satisfying the wall-crossing formula. The Seiberg–Witten invariant for a 3-manifold with $b_1 = 0$ is completely determined by the 4-dimensional invariant of the product with a circle. When the base manifold is a complex surface, a Kobayashi–Hitchin-type correspondence allows us to compute the invariants in the so-called complex geometric chambers by counting complex curves.

1991 Mathematics Subject Classification: 53 C, 32 C10, 32 L05, 32 L10, 57R57

1. Introduction

In this paper we introduce Seiberg–Witten invariants for compact oriented 4-manifolds with $b_+ = 0$.

For such manifolds the moduli space of monopoles always contains reducible solutions. The main idea of our approach is to consider $Spin^c$-structures for which the expected dimension of the Seiberg–Witten moduli space vanishes, and to count only the irreducible solutions, precisely as in 3-dimensional Floer theory. The main difficulty with this approach is that in this way one gets parameter-dependent invariants, because varying the parameters of the equations can make several irreducible solutions degenerate into reducible ones, or viceversa.

However, this phenomenon can be controlled in the following way: we show that, under certain topological assumptions, such jumps can only occur for parameters

---

*Partially supported by: AGE-Algebraic Geometry in Europe, contract No ERBCHRXCT940557 (BBW 93.0187), and by SNF, nr. 21-36111.92

belonging to a codimension-1 locus in the parameter space, which we call wall. The invariant is well defined in every connected component of the complement of the wall in the parameter space; such a connected component will be called chamber. Therefore one gets an invariant which is a map assigning an integer to every chamber. This assignment is functorial with respect to orientation preserving diffeomeorphisms. One should note that assigning any topological invariant to the chambers leads to a functorial map, which can in principle be regarded as a (very difficult to compute) differentiable invariant.

Next we show that the Seiberg–Witten invariant of a 3-manifold with $b_1(M) = 0$ is determined by the 4-dimensional invariant of the product $S^1 \times M$ (which has $b_+ = 0$).

When the base manifold is a complex surface, our invariant can be explicitely computed for a distinguished class of parameters, called complex-geometric parameters, for which the moduli space of solutions can be interpreted in terms of effective divisors via a Kobayashi–Hitchin correspondence. A chamber containing a set of complex-geometric parameters will be called complex-geometric. We make several explicit computations for surfaces of class VII_0.

Recall that the existence of curves on an arbitrary surface of class VII_0 is still an open problem. One can reformulate this problem in terms of our Seiberg–Witten invariant. This approach requires the understanding of the geometry of the wall and of the locus of complex-geometric parameters within the space of all parameters.

2. Construction of the invariant

Let X be a closed, connected, oriented 4-manifold. Recall [OT1] that a $Spin^c$-structure on X can be defined as a triple $(\Sigma^\pm, \iota, \gamma)$, consisting of a pair of Hermitian bundles $\Sigma^\pm$, called the spinor bundles of the structure, a unitary isomorphism $\iota : \det \Sigma^+ \xrightarrow{\simeq} \det \Sigma^-$, and an orientation preserving linear isomorphism

$$\gamma : \Lambda^1_X \to \mathbb{R}SU(\Sigma^+, \Sigma^-),$$

called the Clifford map of the structure. The line bundle $\det \Sigma^\pm$ is called the determinant line bundle of the structure, and its Chern class is called the Chern class of the structure. A class $c \in H^2(X, \mathbb{Z})$ is the Chern class of a $Spin^c$-structure iff it is an integral lift of $w_2(X)$.

Any Clifford map γ defines a Riemannian metric g_γ on X, defined by the condition that $\frac{1}{\sqrt{2}} \gamma$ becomes an isometry.

The set $\mathfrak{Spin}^c(X)$ of isomorphy classes of $Spin^c$-structures on X is a trivial cover of the space Met_X of metrics on X via the map $\gamma \mapsto g_\gamma$. The fibre can be identified with $\pi_0(\mathfrak{Spin}^c(X))$ and is denoted by $Spin^c(X)$. The set $Spin^c(X)$ is always nonempty and has the structure of a $H^2(X, \mathbb{Z})$-torsor.

The Seiberg–Witten equations (SW^γ) associated with a Spin^c-structure $(\Sigma^\pm, \iota, \gamma)$ are

$$\begin{cases} \slashed{D}_A \Psi = 0 \\ \gamma(F_A^+) = (\Psi\bar\Psi)_0, \end{cases} \qquad (SW^\gamma)$$

considered as equations for pairs $(A, \Psi) \in \mathcal{A}(\det \Sigma^+) \times A^0(\Sigma^+)$. In the first equation $\slashed{D}_A$ stands for the Dirac operator $\slashed{D}_A^\gamma$ associated with the Clifford map γ, the Levi-Civita connection of g_γ, and the connection A. In the second equation F_A^+ denotes the selfdual component of the curvature F_A with respect to the metric g_γ, and γ stands for the identification $i\Lambda_+^2 \xrightarrow{\simeq} \mathrm{Herm}_0(\Sigma^+)$ induced by the Clifford map.

The gauge group $\mathcal{G} := C^\infty(X, S^1)$ acts on the configuration space $\mathcal{A}(\det \Sigma^+) \times A^0(\Sigma^+)$ by the formula

$$(A, \Psi) \cdot f = (A + 2f^{-1}df, f^{-1}\Psi).$$

The stabilizer of a pair (A, Ψ) is either trivial when $\Psi \ne 0$, or coincides with S^1 if $\Psi = 0$. In the latter case the pair is called *reducible*. The gauge group $\mathcal{G}$ leaves the space of solutions of (SW^γ) invariant. It is well known that the moduli space $\mathcal{M}^\gamma$ of solutions of (SW^γ) modulo $\mathcal{G}$ is finite dimensional and compact.

The interesting, gauge equivariant order 0-perturbations of (SW^γ), which preserve the compactness of the moduli space of solutions, are:

— (Witten's perturbation) Perturbing the term F_A^+ by a pure imaginary selfdual form $i\mu \in iA_+^2(X)$.

— Perturbing the Dirac operator $\slashed{D}_A$ by a term of the form $\gamma(\alpha)$, with $\alpha \in A^1(X, \mathbb{C})$.

Note that this class of perturbations of $\slashed{D}_A$ contains in particular the Dirac operators associated with Riemannian non-Levi-Civita connections in the bundle (Λ_X^1, g_γ). Note also that the additional perturbation obtained by multiplying the right hand term of the second equation by a positive function can be reduced to a corresponding rescaling of the Clifford map combined with a perturbation of the second type.

When $b_+(X) = 0$, the first class of perturbations can be reduced to the second. Indeed, in this case, the operator $d^+ : A^1(X) \to A_+^2(X)$ is surjective, so that $\gamma(F_A^+ + i\mu)$ can be written as $\gamma(F_{A'}^+)$, where $A' = A + a$, and $a \in iA^1(X)$ is chosen so that $d^+a = i\mu$. On the other hand one has $\slashed{D}_A = \slashed{D}_{A'} - \tfrac{1}{2}\gamma(a)$. This shows that the moduli spaces corresponding to the perturbations $(i\mu, \alpha)$ and $(0, \alpha - \tfrac{1}{2}a)$ can be naturally identified.

Now we fix a class $\mathfrak{c} \in \mathrm{Spin}^c(X)$ of Spin^c-structures with Chern class c, and we choose a pair $(\Sigma^\pm, \iota)$ as above with $c_1(\det \Sigma^\pm) = c$ and $\mathbb{R}SU(\Sigma^+, \Sigma^-) \simeq \Lambda_X^1$ as oriented bundles.

Define the set $\mathcal{P}_{X,\mathfrak{c}}$ of parameters associated with $\mathfrak{c}$ by

$$\mathcal{P}_{X,\mathfrak{c}} := \mathrm{Cliff}_{X,\mathfrak{c}} \times A^1(X, \mathbb{C}),$$

where $\text{Cliff}_{X,c}$ stands for the set of Clifford maps $\gamma : \Lambda_X^1 \to \mathbb{R}SU(\Sigma^+, \Sigma^-)$ representing c. We denote by (SW_α^γ) and $\mathcal{M}_\alpha^\gamma$ ($[\mathcal{M}_\alpha^\gamma]^*$) the Seiberg–Witten equations and the Seiberg–Witten moduli space of (irreducible) monopoles associated with the pair $(\gamma, \alpha) \in \mathcal{P}_{X,c}$.

Note that, if $b_+(X) = 0$, $\mathcal{M}_\alpha^\gamma$ always contains the moduli space $\mathcal{T}_{g_\gamma}$ of Hermitian connections in the line bundle $\det \Sigma^\pm$ with g_γ-harmonic, or equivalently anti-selfdual, curvature; this moduli space $\mathcal{T}_{g_\gamma}$ is a torus which is isomorphic to $i\mathbb{H}_{g_\gamma}^1/4\pi i H^1(X, \mathbb{Z})$. In particular this means, using the terminology of [OT1], that in the case $b_+ = 0$ every parameter (γ, α) is c-bad. Nevertheless, we will see that, under certain topological conditions, one can define a Seiberg–Witten-type invariant; however, the structure of this invariant is very complicated. The idea of the formalism comes from Floer theory in dimension 3, and can be simply formulated as follows: count only the irreducible solutions.

In order to be able to use the moduli space $[\mathcal{M}_\alpha^\gamma]^*$ of irreducible solutions to define a numerical invariant in the usual way, one must be sure that the following conditions are fulfilled:

1. The moduli space $[\mathcal{M}_\alpha^\gamma]^*$ is compact.

2. For parameters (γ', α') in a sufficiently small neighbourhood of (γ, α), the moduli spaces $[\mathcal{M}_{\alpha'}^{\gamma'}]^*$ remains compact and is cobordant to $[\mathcal{M}_\alpha^\gamma]^*$ inside

$$\mathcal{B}^* := [\mathcal{A}(\det \Sigma^+) \times A^0(\Sigma^+)]^*/\mathcal{G}.$$

On the other hand, the invariant can take non-trivial values only if

3. The expected dimension $w_c = \frac{1}{4}[c^2 - (3\sigma(X) + 2e(X))] = \frac{1}{4}(c^2 + b_2(X)) + b_1(X) - 1$ is non-negative.

A pair (γ, α) for which the conditions 1.–2. are not satisfied must have the following property:

P. There exists a sequence (γ_n, α_n) of parameters converging to (γ, α), and irreducible solutions (A_n, Ψ_n) of $(SW_{\alpha_n}^{\gamma_n})$, which converge to a reducible solution $(A, 0)$ of (SW_α^γ).

By elliptic semicontinuity, this implies $\ker(\slashed{D}_A^\gamma + \gamma(\alpha)) \neq \{0\}$. Therefore, the locus of bad parameters $\mathcal{B}_{X,c}$ is the projection of the differential geometric Brill–Noether locus

$$\mathcal{BN}_{X,c} := \{(\gamma, \alpha, [A]) \in \mathcal{P}_{X,c} \times \mathcal{T}_{g_\gamma} \mid \ker(\slashed{D}_A^\gamma + \gamma(\alpha)) \neq 0\}$$

on $\mathcal{P}_{X,c}$.

One can easily show that the space

$$\widetilde{\mathcal{BN}}_{X,c} := \{(\gamma, \alpha, [A], \Psi) \in \mathcal{P}_{X,c} \times \mathcal{T}_{g_\gamma} \times [A^0(\Sigma^+) \setminus \{0\}] \mid (\slashed{D}_A^\gamma + \gamma(\alpha))\Psi = 0\}/\mathbb{C}^*$$

is (after suitable Sobolev completions) a smooth manifold, and the projection on $\mathcal{P}_{X,c}$ is real analytic, proper, and Fredholm of index equal to $\text{index}_\mathbb{R}(\slashed{D}_c) + b_1(X) - 2$. The number $-\,\mathrm{index}_{\mathbb{R}}(\rlap{/}{D}_c) - b_1(X) + 2$ is the expected codimension of the bad locus. Therefore, if $\mathrm{index}_{\mathbb{R}}(\rlap{/}{D}_c) + b_1(X) - 2 < 0$, the bad locus is a (possibly singular) real analytic subspace of codimension at least $-\,\mathrm{index}_{\mathbb{R}}(\rlap{/}{D}_c) - b_1(X) + 2$ of $\mathcal{P}_{X,\mathfrak{c}}$.

Concluding, we can state that the conditions 1.–3. are satisfied for generic $(\gamma, \alpha) \in \mathcal{P}_{X,\mathfrak{c}}$ if

$$0 < -\,\mathrm{index}_{\mathbb{R}}(\rlap{/}{D}_c) - b_1(X) + 2, \quad \mathrm{index}_{\mathbb{R}}(\rlap{/}{D}_c) - 1 + b_1(X) \geq 0,$$

i.e. if $w_{\mathfrak{c}} = 0$. Note that this implies that $b_1(X)$ must be odd. From now on we suppose that the topological condition $w_{\mathfrak{c}} = 0$ is satisfied. Since in this case the expected codimension of the bad locus is 1, the parameter space is divided into (at most) countably many chambers.

The usual transversality technique shows that for a generic parameter (γ, α) within a fixed chamber, the moduli space $[\mathcal{M}_\alpha^\gamma]^*$ is a smooth 0-dimensional manifold. Furthermore, an orientation of $H^1(X, \mathbb{R})$ determines an orientation of the moduli space $[\mathcal{M}_\alpha^\gamma]^*$ in any regular point.

Concluding, we can define our Seiberg–Witten invariant

$$SW_{X,\mathfrak{c}} : \pi_0(\mathcal{P}_{X,\mathfrak{c}} \setminus \mathcal{B}_{X,\mathfrak{c}}) \longrightarrow \mathbb{Z}$$

by $SW_{\mathfrak{c}}(C) := \#[\mathcal{M}_\alpha^\gamma]^*$, where $(\gamma, \alpha) \in C$ is a generic parameter for which the moduli space $[\mathcal{M}_\alpha^\gamma]^*$ is regular.

It is easy to see that the bad locus and the chambers are invariant under the natural action of the group

$$\mathrm{Aut}(\Sigma^\pm, \iota) := \{(f_+, f_-) \in \mathrm{Aut}(\Sigma^+) \times \mathrm{Aut}(\Sigma^-) \mid \det f_+ = \det f_-\}$$

on the space $\mathrm{Cliff}_{X,\mathfrak{c}}$. This group is precisely the gauge group of the principal $\mathrm{Spin}^c(4)$-bundle associated with $(\Sigma^\pm, \iota)$, and the quotient $\mathrm{Cliff}_{X,\mathfrak{c}}/\mathrm{Aut}(\Sigma^\pm, \iota)$ can be identified with the space of metrics Met_X [OT1]. Therefore one can alternatively regard the Seiberg–Witten invariant as a map

$$SW_{X,\mathfrak{c}} : \pi_0([\mathrm{Met}_X \times A^1(X, \mathbb{C})] \setminus \bar{\mathcal{B}}_{X,\mathfrak{c}}) \longrightarrow \mathbb{Z},$$

where $\bar{\mathcal{B}}_{X,\mathfrak{c}}$ is the $\mathrm{Aut}(\Sigma^\pm, \iota)$-quotient of $\mathcal{B}_{X,\mathfrak{c}}$.

3. Wall crossing

As in the case $b_+(X) = 1$, one can prove a wall-crossing formula, which shows that the invariant is in principle determined by its value on a single chamber, assuming the chamber configuration known. The wall crossing formula below refers to a transversal crossing of the wall in a regular point.

Definition 3.1. A pair $(g, \alpha) \in \bar{\mathcal{B}}_{X,\mathfrak{c}}$ is called regular if, choosing a Clifford map $\gamma \in \mathrm{Cliff}_{X,\mathfrak{c}}$ compatible with g, the following conditions are satisfied:

1. There is only one point $(\gamma, \alpha, [A]) \in \mathcal{B}\mathcal{N}_{X,\mathfrak{c}}$ lifting (γ, α).
2. $\ker[\slashed{D}_A^\gamma + \gamma(\alpha)]$ is 1-dimensional.
3. $\mathcal{B}\mathcal{N}_{X,\mathfrak{c}}$ meets the torus $\{\gamma, \alpha\} \times \mathcal{T}_g$ transversally in the point $(\gamma, \alpha, [A])$.

Note that by 2. (after taking suitable Sobolev completions), $\mathcal{B}\mathcal{N}_{X,\mathfrak{c}}$ is a smooth manifold of codimension $-\text{index}_\mathbb{R}(\slashed{D}_\mathfrak{c}) + 2 = b_1(X) + 1$ in the point $(\gamma, \alpha, [A])$, so that condition 3. has sense. The normal space of $\mathcal{B}\mathcal{N}_{X,\mathfrak{c}}$ in a point $(\gamma, \alpha, [A])$ satisfying 2. can be identified with $\text{coker}[\slashed{D}_A^\gamma + \gamma(\alpha)]$, so it comes with a canonical complex orientation. Therefore, fixing an orientation of $H^1(X, \mathbb{R})$ orients the normal line of any regular point of the wall $\bar{\mathcal{B}}_{X,\mathfrak{c}}$.

If we choose an orientation of $H^1(X, \mathbb{R})$, any regular point $\mathfrak{r} = (g, \alpha) \in \bar{\mathcal{B}}_{X,\mathfrak{c}}$ of the wall determines uniquely two chambers $C_\mathfrak{r}^\pm$ defined by the condition that there exists a smooth path $\lambda : [-1, 1] \to \text{Met}_X \times A^1(X, \mathbb{C})$ with the properties:

i) $\lambda(\pm 1) \in C_\mathfrak{r}^\pm$, $\lambda(0) = \mathfrak{r}$.

ii) λ meets the wall only in 0.

iii) $\dot\lambda(0)$ is transversal to the wall and is compatible with the orientation of the normal line.

Theorem 3.2. *Let X be a closed, connected, oriented 4-manifold with $b_+(X) = 0$. Fix an orientation of $H^1(X, \mathbb{R})$, and let $\mathfrak{r} = (g, \alpha)$ be a regular point of the wall. Then the following wall-crossing formula holds*

$$SW_{X,\mathfrak{c}}(C_\mathfrak{r}^+) - SW_{X,\mathfrak{c}}(C_\mathfrak{r}^-) = 1.$$

Proof. The proof is based on the same method as in the case $b_+ = 1$ [OT1]. One shows that, for generic λ, the closure $\bar{\mathcal{M}}$ of the fibre product

$$\bigcup_{t\in[-1,1]} \{t\} \times \mathcal{M}_{\lambda(t)}^*$$

is a manifold with boundary, whose boundary consists of $\mathcal{M}_{\lambda(1)}^* - \mathcal{M}_{\lambda(-1)}^*$ and a single point corresponding to the reducible solution $([A], 0)$ defined by the unique lift of $\mathfrak{r}$ in the Brill–Noether locus. The cobordismus $\bar{\mathcal{M}}$ is obtained by performing a "real blow up" of the reducible locus in the parametrized moduli space. □

Remark 3.3. The set of all functions from $\pi_0([\text{Met}_X \times A^1(X, \mathbb{C})] \setminus \bar{\mathcal{B}}_{X,\mathfrak{c}})$ to $\mathbb{Z}$ which satisfy this wall crossing formula is a $\mathbb{Z}$-torsor. The Seiberg–Witten invariant associated with $\mathfrak{c}$ is a distinguished element of this torsor.

4. Comparison theorem

In [OT2] we introduced parameter dependent Seiberg–Witten equations on 3-manifolds and we proved a comparison theorem, which reduces the computation of the Seiberg–

Witten moduli spaces on a 3-manifold M to the computation of the Seiberg–Witten moduli spaces associated with the pull-back data on the four-dimensional product $S^1 \times M$. In the 3-dimensional case, the parameters of the equations vary in the product $\operatorname{Met}_M \times A^0(M) \times Z^2_{\operatorname{DR}}(M)$. Using the same substitution as above, one can see that in the case $b_1(M) = 0$, one can alternatively work with the parameter space $\operatorname{Met}_M \times A^0(M) \times iA^1(M)$. The invariant associated with an element $\mathfrak{c} \in \operatorname{Spin}^c(M)$ can then be regarded as a map

$$SW_{M,\mathfrak{c}} : \pi_0([\operatorname{Met}_M \times A^0(M) \times iA^1(M)] \setminus \bar{\mathcal{B}}_{M,\mathfrak{c}}) \longrightarrow \mathbb{Z},$$

where $\bar{\mathcal{B}}_{M,\mathfrak{c}}$ is the bad locus

$$\bar{\mathcal{B}}_{M,\mathfrak{c}} := \{(g, f, a) \in \operatorname{Met}_M \times A^0(M) \times iA^1(M) |\ \ker(\slashed{D}^\gamma + f\operatorname{id} + \gamma(a)) \neq \{0\}\}.$$

Here γ is chosen such that $g_\gamma = g$, and $\slashed{D}^\gamma$ is the Dirac operator associated with a flat connection in $\det \Sigma^\pm$, which is unique up to gauge transformation.

We define a pull-back map

$$\pi_0([\operatorname{Met}_M \times A^0(M) \times iA^1(M)] \setminus \bar{\mathcal{B}}_{M,\mathfrak{c}}) \longrightarrow \pi_0([\operatorname{Met}_X \times A^1(X, \mathbb{C})] \setminus \bar{\mathcal{B}}_{X,p_M^*(\mathfrak{c})})$$

by

$$p_M^*[g, f, a] = [\operatorname{can}_{S^1} \times g, -p_M^*(f)\operatorname{vol}_{S^1} + p_M^*(a)].$$

With this definition, our comparison theorem [OT2] takes the following simple form in the case $b_1 = 0$

Theorem 4.1. *Let M be a closed oriented 3-manifold with $b_1(M) = 0$. Then*

$$SW_{M,\mathfrak{c}}(C) = SW_{S^1 \times M, p_M^*(\mathfrak{c})}(p_M^*(C)).$$

Remark 4.2. Combining Theorem 3.2 with Theorem 4.1, one gets a universal wall crossing formula for the 3-dimensional Seiberg–Witten invariant in the case $b_1 = 0$.

5. Complex geometric interpretation

Suppose now that X is a complex surface. In [OT2][1] we have shown that the moduli spaces of irreducible monopoles which are associated with certain complex geometric parameters, can be identified with Douady moduli spaces of divisors with bounded volume.

This result is based on the non-Kählerian version of the Kobayashi–Hitchin correspondence, proved in [L], which we recall here for the convenience of the reader.

The canonical $\operatorname{Spin}^c(4)$-structure of a Hermitian surface (X, J, g) is the triple $(\Sigma_{\operatorname{can}}^\pm, \iota_{\operatorname{can}}, \gamma_{\operatorname{can}})$ given by

$$\Sigma_{\operatorname{can}}^+ = \Lambda^{00} \oplus \Lambda^{02},\ \ \Sigma_{\operatorname{can}}^- = \Lambda^{01},$$

---
[1] The sign of $\frac{\theta}{4}$ in Theorem 4.1 in [OT2] is wrong. The correct sign is given in Theorem 5.1 below.

where ι_{can} is the natural identification of $\det(\Sigma_{\text{can}}^+)$ with $\det(\Sigma_{\text{can}}^-)$, and

$$\gamma_{\text{can}}(u)(\cdot) = \sqrt{2}(u^{01} \wedge \cdot - i\Lambda_g(u^{10} \wedge \cdot)).$$

If N is an arbitrary Hermitian line bundle on X with $c_1(N) = n$, the $\text{Spin}^c(4)$-structure $(\Sigma_N^\pm, \iota_N, \gamma_N)$ is defined by

$$\Sigma_N^\pm := \Sigma_{\text{can}}^\pm \otimes N, \quad \iota_N := \iota_{\text{can}} \otimes \text{id}_{N^{\otimes 2}}, \quad \gamma_N := \gamma_{\text{can}} \otimes \text{id}_N.$$

In other words, $(\Sigma_N^\pm, \iota_N, \gamma_N)$ is the n-twist of the canonical structure, via the $H^2(X, \mathbb{Z})$-torsor structure of $\text{Spin}^c(X)$. Every Spin^c-structure on (X, g) has this form.

Denote by $\text{Dou}(n)$ the Douady space of effective divisors on X representing the homology class Poincaré dual to n. In the non-Kähler case, the spaces $\text{Dou}(n)$ can be non-compact, since fixing the homology class does not controle the volume. However, the subspace of $\text{Dou}(n)$ consisting of divisors with volume bounded by a constant is always relatively compact.

Define the operator $j : A^0(X) \longrightarrow \mathbb{R}$ in the following way: $j(f)$ is the unique constant function which is congruent to f modulo the image of the operator $i\Lambda\bar{\partial}\partial$.

Using the operator j, one can define a degree map

$$\deg_g : \text{Pic}(X) \longrightarrow \mathbb{R},$$

by

$$\deg(\mathcal{L}) := j(\Lambda c_1(\mathcal{L}, h)) \text{Vol}_g(X),$$

where $c_1(\mathcal{L}, h)$ is the Chern form of a Hermitian metric h in $\mathcal{L}$ [LT]. Let θ_g be the torsion form of g.

Theorem 5.1 ([L]). *Let (X, J, g) be a Hermitian surface, let β be a real $(1, 1)$-form on X, and let $\left[\mathcal{M}_{\beta, -\frac{1}{4}\theta_g}^{\gamma_N}\right]^*$ be the moduli space of irreducible monopoles corresponding to the parameters γ_N, $\mu = 2\pi\beta^+$, $\alpha = -\frac{1}{4}\theta_g$. There exists a canonical isomorphism of real analytic spaces*

$$\left[\mathcal{M}_{2\pi i \beta^+, -\frac{1}{4}\theta_g}^{\gamma_N}\right]^* \simeq \mathcal{V}(n, \beta) \coprod \mathcal{V}(k - n, -\beta),$$

where

$$\mathcal{V}(n, \beta) := \{D \in \text{Dou}(n) | \; 2\deg_g(\mathcal{O}(D)) < \deg_g(K_X) + j(\Lambda_g \beta) \text{Vol}_g(X)\}.$$

This theorem, together with the transformation explained in 2. gives

Corollary 5.2. *Let (X, J, g) be a Hermitian surface with $b_+(X) = 0$. Let $a \in iA^1(X)$ be an imaginary 1-form with da of type $(1, 1)$. There exists a natural isomorphism of moduli spaces*

$$\left[\mathcal{M}_{-\frac{1}{4}\theta_g - \frac{1}{2}a}^{\gamma_N}\right]^* \simeq \mathcal{V}\left(n, \frac{1}{2\pi i}d^+a\right) \coprod \mathcal{V}\left(k - n, -\frac{1}{2\pi i}d^+a\right).$$

This corollary yields a complex geometric interpretation of all moduli spaces associated with parameters $(g, \alpha) \in \text{Met}_X \times A^1(X, \mathbb{C})$, satisfying the conditions
1. g is Hermitian.
2. $\text{re}(\alpha) = -\frac{1}{4}\theta_g$, $d(\text{im}(\alpha)) \in i A^{1,1}(X)$.

Such parameters will be called *complex geometric*. As we have seen, interesting for differential topological purposes are only the moduli spaces of expected dimension 0. With our notations, this condition becomes $n(n - k) = 0$, where $k := c_1(K_X)$. In this case, a chamber $C \in \pi_0([\text{Met}_X \times A^1(X, \mathbb{C})] \setminus \mathcal{B}_{X,\mathfrak{c}})$ will be called complex geometric, if it contains a complex geometric pair. The corollary above reduces the computation of the Seiberg–Witten invariant in a complex geometric chamber to the "counting" of divisors in certain homology classes with a certain bound for the volume. The method can be used effectively if the corresponding parts of the Douady spaces are finite, i.e. no continuous families of divisors occur in the considered moduli space. The property of a chamber to be complex geometric depends of course on the complex structure J.

6. Explicit computations

1. Let (X, J, g) be an Inoue surface endowed with an arbitrary Hermitian metric. Such a surface has $b_2(X) = 0$, $b_1(X) = 1$, $c_1(K_X) = 0$, and $\deg_g(K_X) > 0$ [P]; it has no curves [BPV]. Let a_0 be a pure imaginary 1-form such that $d^+ a_0 = 2\pi i \omega_g$. Corollary 5.2 leads to the identification

$$\left[\mathcal{M}^{\gamma_{\text{can}}}_{-\frac{\theta_g}{4} - \frac{t}{2}a_0}\right]^* = \mathcal{V}(0, t\omega_g) \coprod \mathcal{V}(0, -t\omega_g),$$

where $\mathcal{V}(0, t\omega_g)$ is the Douady space of divisors which are homologically trivial with volume bounded by $\frac{1}{2}(\deg_g(K_X) + 2t \text{Vol}_g(X))$. Therefore $[\mathcal{M}^{\gamma_{\text{can}}}_{-\frac{\theta}{4} - t\frac{1}{2}a_0}]^*$ consists of two points (two copies of the empty divisor) if $|t| < \frac{\deg_g(K_X)}{2 \text{Vol}_g(X)}$, and of only one point if $|t| > \frac{\deg_g(K_X)}{2 \text{Vol}_g(X)}$ ([L]). All these points come with positive orientation in the moduli space, if one uses the natural complex orientation as explained in [OT2], Proposition 4.3. Thus the line of parameters $(g, -\frac{\theta}{4} - \frac{t}{2}a_0)_{t \in \mathbb{R}}$ meets the wall twice, and the Seiberg–Witten invariant $SW_{X, \mathfrak{c}_{\text{can}}}$ takes the values 1, 2, 1 in the three chambers through which this line passes.

2. Fix $(a_1, a_2) \in [\mathbb{R} \setminus \{0\}]^2$, and let $(X_{a_1, a_2}, J, g_{a_1, a_2})$ be the primary Hopf surface constructed in [OT2] p. 284. When $\frac{a_1}{a_2} \notin \mathbb{Q}$, this surface has only two holomorphic

curves F_1, F_2 with $\mathrm{Vol}(F_i) = \frac{4\pi^2}{|a_i|}$. Choosing a form a_0 as above, we find

$$\left[\mathcal{M}^{\gamma_{\mathrm{can}}}_{\theta g_{a_1,a_2} - \frac{t}{2} a_0} \right]^* = \mathcal{V}(0, t\omega_{g_{a_1,a_2}}) \coprod \mathcal{V}(0, -t\omega_{a_1,a_2}).$$

For $t \geq 0$, the second component is empty, and the first can be identified with

$$\{(n_1, n_2) \in \mathbb{Z}^2_{\geq 0} |\ 2[n_1 \mathrm{Vol}(F_1) + n_2 \mathrm{Vol}(F_2)] < \kappa_{a_1,a_2} + t\, \mathrm{Vol}(X_{a_1,a_2})\},$$

where $\kappa_{a_1,a_2} := \deg_{g_{a_1,a_2}}(K_{X_{a_1,a_2}})$ is negative [OT2]. One sees that the half-line $(g, -\frac{\theta}{4} - \frac{t}{2}a_0)_{t \geq 0}$ meats the wall in infinitely many points, and the Seiberg–Witten invariant $SW_{X_{a_1,a_2}, \mathfrak{c}_{\mathrm{can}}}$ takes on all non-negative integer values in the chambers through which this half-line passes. A similar result holds for a multiple blow-up of $(X_{a_1,a_2}, J, g_{a_1,a_2})$.

3. According to [N], p. 227, there exist class VII_0 surfaces X with $b_2(X) > 0$ admitting no effective divisors in the homology classes 0 or $PD(c_1(K_X))$. For such a surface X and any Hermitian metric g, one sees that the line $(g, -\frac{\theta}{4} - \frac{t}{2}a_0)_{t \in \mathbb{R}}$ meets the wall only once. The invariant $SW_{X, \mathfrak{c}_{\mathrm{can}}}$ takes the values 0, 1 in the two chambers through which this line passes. The non-vanishing value corresponds to a moduli space consisting only of the empty divisor.

References

[BPV] Barth, W., Peters, C., Van de Ven, A., Compact complex surfaces,, Springer-Verlag, 1984.

[B] Biquard, O., Les équations de Seiberg–Witten sur une surface complexe non kählerienne, Comm. Anal. Geom. 6 (1998), 173–197.

[D] Donaldson, S., The Seiberg–Witten equations and 4-manifold topology, Bull. Amer. Math. Soc. 33 (1996), 45–70.

[G1] Gauduchon, P., Sur la 1-forme de torsion d'une variété hermitienne compacte, Math. Ann. 267 (1984), 495–518.

[LT] Lübke, M., Teleman, A., The Kobayashi–Hitchin correspondence, World Scientific Publishing Co., 1995.

[L] Lupaşcu, P., Seiberg–Witten equations and complex surfaces, Ph. D. thesis, Zürich 1998.

[N] Nakamura, I., Towards classification of non-Kählerian complex surfaces, Sugaku Expositions 2, No 2, (1989), 209–229.

[OT1] Okonek, Ch., Teleman, A., Seiberg–Witten invariants for manifolds with $b_+ = 1$, and the universal wall crossing formula, Internat. J. Math. 7 (6) (1996), 811–832.

[OT2] Okonek, Ch., Teleman, A., 3-dimensional Seiberg–Witten Theory and non-Kählerian Geometry, Math. Ann. 312 (1998), 261–288.

[P]   Plantiko, R., A rigidity property of class VII$_0$ surface fundamental groups, J. Reine Angew. Math. 465 (1995), 145–163.

[W]   Witten, E., Monopoles and four-manifolds, Math. Res. Lett. 1 (1994), 769–796.

A geometric proof of Ax' theorem

Jeroen G. Spandaw

In this short note we give a geometric proof of the following theorem.

Theorem. *Every injective polynomial map $\mathbb{C}^n \to \mathbb{C}^n$ is surjective.*

This assertion follows by model theory[1] (see e.g. [2, theorem 1.14]) from the corresponding statement for $\overline{\mathbb{F}}_p$, which is trivial. (This magic proof is due to Ax.) An algebraic proof which does not use model theory can be found in [3]. The statement, however, is geometric and so one would like to have a *geometric* proof. In this short note we give such a proof for the case $n = 2$. I do not know of a geometric proof for $n > 2$. It is known that the statement is no longer true if one replaces "polynomial" with "holomorphic". Indeed, Bieberbach constructed for each $n \geq 2$ a holomorphic injective map $f : \mathbb{C}^n \to \mathbb{C}^n$ with the property that the complement of the image has a non-empty interior (see e.g. [1, p. 152])!

We now give the geometric proof for the case $n = 2$. We interpret f as a birational map from $\mathbb{P}^2$ to $\mathbb{P}^2$. By blowing up the points of indeterminacy of f and f^{-1} we obtain a smooth projective surface Z and birational morphisms $p, q : Z \to \mathbb{P}^2$ with $q = f \circ p$. Note that q is surjective. We say that a curve C in Z is q-exceptional if $q(C)$ is a point or contained in $L_\infty := \mathbb{P}^2 \setminus \mathbb{C}^2$. We denote by $\mathrm{Exc}(q)$ the union of such curves. $\mathrm{Exc}(p)$ is defined analogously. Since all points of indeterminacy of f are contained in L_∞, we have $\mathrm{Exc}(p) = p^{-1}(L_\infty)$. We write

$$\mathrm{Exc}(p) = A_1 \cup \cdots \cup A_r \cup B_1 \cup \cdots \cup B_s$$
$$\mathrm{Exc}(q) = C_1 \cup \cdots \cup C_t \cup B_1 \cup \cdots \cup B_s$$

with $A_1, \ldots, A_r, B_1, \ldots, B_s, C_1, \ldots, C_t$ irreducible, pairwise distinct curves. Since $r + s = e(Z) - 2 = t + s$, where $e(Z)$ is the Euler characteristic, we have $r = t$. If $t > 0$ then $D := p(C_1) \not\subseteq L_\infty$, since C_1 is not p-exceptional. Furthermore, $q(C_1) = f(D) \not\subseteq L_\infty$ since $f(\mathbb{C}^2) \subseteq \mathbb{C}^2$. Hence $f(D)$ is a point, contradicting the injectivity of $f|_{\mathbb{C}^2}$. We conclude that $r = t = 0$, i.e. $\mathrm{Exc}(q) = p^{-1}(L_\infty)$. Note that $p^{-1}(L_\infty)$ is connected, hence so is $q(\mathrm{Exc}(q))$. The set $S := q(\mathrm{Exc}(q)) \setminus L_\infty$ consists by definition of contracted curves. In particular, it is a finite set. Hence the connectedness of $q(\mathrm{Exc}(q))$ implies that $S = \emptyset$ and thus $qp^{-1}(L_\infty) = L_\infty$. In other words, $f(\mathbb{C}^2) = \mathbb{C}^2$.

---

[1] I am grateful to Prof. J. Reineke, who explained this model theory to me.

References

[1] H. Grauert, K. Fritsche, Einführung in die Funktionentheorie mehrerer Veränderlicher, Springer-Verlag, 1974.

[2] Ch.U. Jensen, H. Lenzing, Model theoretic algebra, Gordon and Breach Science Publishers, 1989.

[3] W. Rudin, Injective polynomial maps are automorphisms, Amer. Math. Monthly 102 (6) (1995), 540–543.

A note on the Cayley–Bacharach property for vector bundles

Sheng-Li Tan and Eckart Viehweg*

Abstract. We study the Cayley–Bacharach property on smooth complex projective varieties for zero-dimensional subschemes, defined as the zero set of a global section of a rank n vector bundle, and for codimension 2 subschemes, defined by global sections of rank 2 vector bundles.

1991 Mathematics Subject Classification: 14J60, 14F05, 14M06, 14N99.

The main purpose of this note is to present and to generalize results from [17] and to use them to study properties and the construction of vector bundles on smooth complex projective varieties X of dimension $n \geq 2$.

In [17], the first author proved that the Cayley–Bacharach property of a zero-dimensional complete intersection in X is equivalent to the k-very ampleness of some adjoint linear systems. In this paper, we show that the result remains true for the zero-dimensional subscheme defined by the zero set of a global section of a rank n vector bundle (Theorem 4), generalizing a theorem of Griffiths and Harris [8], p. 677. Due to the Bogomolov inequality for rank 2 semistable vector bundles [4], [12], we can establish the Cayley–Bacharach theorem for codimension 2 subschemes defined by global sections of rank 2 vector bundles (Theorem 5 and Corollary 6). This result can be used to reprove Paoletti's theorem [14], [13], a generalization of the classical theorem of Halphen. As an application, we give an explicit construction of rank 2 vector bundles from codimension 2 subschemes (Theorem 7).

Throughout this paper we use the notion "k points" for any zero-dimensional subscheme of length k, not requiring the points to be distinct. The degree of an object is defined with respect to an fixed ample divisor A on X, hence the degree of a codimension r subscheme Y of X is defined by $\deg Y = A^{n-r} Y$, although A is not mentioned in the statements.

---

*This work is supported by the DFG Forschergruppe "Arithmetik und Geometrie". The first author is also supported by the NSF for Outstanding Youths.

1. An exact sequence

Let X be a smooth projective variety over $\mathbb{C}$ of dimension $n \geq 2$, and let Z be a subscheme of X of pure codimension $r \geq 2$.

Given a subscheme $Z' \subset Z$, the "complement" Z'' of Z' in Z is the canonical closed subscheme $Z'' \subset Z$ with sheaf of ideals $\mathcal{I}_{Z''} = [\mathcal{I}_Z : \mathcal{I}_{Z'}]$, i.e., for any open set $U \subset X$, we define

$$\mathcal{I}_{Z''}(U) := \{g \in \mathcal{O}_X(U) \mid g\mathcal{I}_{Z'}(U) \subset \mathcal{I}_Z(U)\},$$

or equivalently,

$$\mathcal{I}_{Z''}/\mathcal{I}_Z = \mathcal{H}om_{\mathcal{O}_X}(\mathcal{O}_{Z'}, \mathcal{O}_Z).$$

The second description implies that $Z'' = Z$ if the support of Z' does not contain some of the irreducible components of Z. Moreover, if Z is reduced, then Z'' is the closure of $Z - Z'$. We call Z'' the *residual subscheme* of Z' in Z and denote it by

$$Z'' = Z - Z'.$$

Let E be a vector bundle on X of rank $r \geq 2$, let s be a global section of E and let $Z = Z(s) \subset X$ be its zero scheme. As above we will assume that Z is of pure codimension r, hence it is a local complete intersection. For a divisor L and a subscheme $\Delta \subset Z(s)$, we want to study hypersurfaces F in X satisfying the equations

$$\begin{cases} \Delta = Z(s) - Z(s)F, \\ L \equiv \det E - F. \end{cases} \quad (*)$$

Given Δ and L we will call (E, s, F) a solution of $(*)$ if $Z(s)$ is of pure codimension $r = \text{rank}(E)$ and if the equation $(*)$ holds true.

Here and throughout this note $Z(s)F$ denotes the intersection subscheme of $Z(s)$ and a hypersurface (or effective divisor) F in X. If a hypersurface F satisfies the first equation in $(*)$ we will say that F does not pass through Δ. In a similar way, if Z' is a subscheme of F, we will say that F passes through Z'.

If $Z(s)$ is a reduced subscheme of X then F satisfies the first equation in $(*)$, if Δ is the union of all irreducible components of $Z(s)$ which are not contained in F.

Theorem 1. *Let E be a vector bundle on X of rank $r \geq 2$, and let s be a section whose zero subscheme $Z = Z(s)$ is of pure codimension r. Let $Z'' \subset Z'$ are two codimension r subschemes of Z and let L be a divisor. Then there exists a complex of vector spaces*

$$0 \longrightarrow H^0(\mathcal{I}_{Z-Z''}(\det E - L)) \xrightarrow{\alpha} H^0(\mathcal{I}_{Z-Z'}(\det E - L)) \xrightarrow{\mu}$$
$$H^{n-r+1}(\mathcal{I}_{Z'}(K_X + L)) \xrightarrow{\beta} H^{n-r+1}(\mathcal{I}_{Z''}(K_X + L)) \longrightarrow 0,$$

exact except at $H^{n-r+1}(\mathcal{I}_{Z'}(K_X + L))$. If E is sufficiently ample, then the complex is exact everywhere.

Remark 1. The condition "E is sufficiently ample" we used in the theorem stands for the following vanishing conditions:

$$H^j(X, \wedge^i E^\vee(\det E - L)) = 0, \quad \text{for } i, j = 1, \ldots, r-1. \tag{1}$$

If $X = \mathbb{P}^n$ and E splits (the hypersurface case), then (1) is always true. In general (1) can be enforced by replacing E by $E \otimes H$, for a sufficiently ample line bundle H (cf. Lemma 1.3 and the end of the proof of Theorem 1).

The connecting map μ is not "natural", but there is natural map to the dual of $\ker \beta$.

Throughout the proof of Theorem 1, F will denote an effective divisor on X with $F \equiv \det E - L$. We consider the Koszul complex of (E, s):

$$0 \longrightarrow \mathcal{E}_{r-1} \longrightarrow \cdots \longrightarrow \mathcal{E}_0 \xrightarrow{s} \mathcal{I}_Z \longrightarrow 0,$$

where $\mathcal{E}_0 = E^\vee$, $\mathcal{E}_i = \wedge^{i+1}\mathcal{E}_0$, and where s is the dual map of $\mathcal{O} \to \mathcal{E}_0^\vee$ given by the global section s of $\mathcal{E}_0^\vee$. Because $Z = Z(s)$ is a local complete intersection, the Koszul complex is exact (see [9], p. 245). We split the Koszul complex as follows:

$$\begin{aligned}
0 &\longrightarrow \mathcal{F}_1 \longrightarrow \mathcal{E}_0 \xrightarrow{s} \mathcal{I}_Z \longrightarrow 0, \\
0 &\longrightarrow \mathcal{F}_2 \longrightarrow \mathcal{E}_1 \longrightarrow \mathcal{F}_1 \longrightarrow 0, \\
&\vdots \qquad\qquad \vdots \\
0 &\longrightarrow \mathcal{F}_{r-1} \xrightarrow{s} \mathcal{E}_{r-2} \longrightarrow \mathcal{F}_{r-2} \longrightarrow 0,
\end{aligned} \tag{2}$$

where $\mathcal{F}_{r-1} \cong \mathcal{E}_{r-1} \cong \det E^\vee$.

Lemma 1.1. *Assume that Z' is a subscheme of Z. If (1) holds, then*

$$H^{n-r+1}(\mathcal{I}_{Z'}(K_X + L))^\vee \cong \operatorname{Ext}^1(\mathcal{I}_{Z'}, \mathcal{F}_1(F)).$$

Proof. By Serre duality ([9], Theorem 7.6) one has an isomorphism

$$H^{n-r+1}(\mathcal{I}_{Z'}(K_X + L))^\vee \cong \operatorname{Ext}^{r-1}(\mathcal{I}_{Z'}, \mathcal{O}(-L)).$$

On the other hand, (1) implies that

$$\operatorname{Ext}^j(\mathcal{O}_X, \mathcal{E}_i(F)) \cong H^j(\mathcal{E}_i(F)) = 0, \quad \text{for } i \leq r-2, \ 1 \leq j \leq r-1. \tag{3}$$

From (3) and from the exact sequence

$$0 \longrightarrow \mathcal{I}_{Z'} \longrightarrow \mathcal{O}_X \longrightarrow \mathcal{O}_{Z'} \longrightarrow 0, \tag{4}$$

we obtain easily that

$$\begin{aligned}
\operatorname{Ext}^j(\mathcal{I}_{Z'}, \mathcal{E}_i(F)) &\cong \operatorname{Ext}^{j+1}(\mathcal{O}_{Z'}, \mathcal{E}_i(F)) \\
&\cong H^{n-j-1}(\mathcal{O}_{Z'}(\mathcal{E}_i^\vee(-F + K_X)))^\vee \\
&= 0,
\end{aligned}$$

for $i \leq r - 2$, $1 \leq j \leq r - 2$. Considering the long exact sequences obtained from the short exact sequences in (2), we thereby have isomorphisms

$$\mathrm{Ext}^1(\mathcal{I}_{Z'}, \mathcal{F}_1(F)) \cong \mathrm{Ext}^2(\mathcal{I}_{Z'}, \mathcal{F}_2(F)) \cong \cdots \cong \mathrm{Ext}^{r-2}(\mathcal{I}_{Z'}, \mathcal{F}_{r-2}(F))$$

and an exact sequence

$$0 \longrightarrow \mathrm{Ext}^{r-2}(\mathcal{I}_{Z'}, \mathcal{F}_{r-2}(F)) \longrightarrow \mathrm{Ext}^{r-1}(\mathcal{I}_{Z'}, \mathcal{F}_{r-1}(F)) \xrightarrow{\tau} \mathrm{Ext}^{r-1}(\mathcal{I}_{Z'}, \mathcal{E}_{r-2}(F)).$$

Since $\mathcal{F}_{r-1}(F) \cong \mathcal{O}(-L)$, it remains to prove that the morphism τ is zero. Indeed, $\mathcal{E}^{\vee}_{r-2} \cong \mathcal{E}_0 \otimes \det E$ and by Serre duality τ is the dual morphism of

$$H^{n-r+1}(\mathcal{I}_{Z'} \otimes \mathcal{E}_0(K_X + L)) \xrightarrow{s} H^{n-r+1}(\mathcal{I}_{Z'}(K_X + L)).$$

On the other hand, from (4), we obtain a commutative diagram

$$\begin{array}{ccccc}
H^{n-r}(\mathcal{O}_{Z'} \otimes \mathcal{E}_0(K_X + L)) & \longrightarrow & H^{n-r+1}(\mathcal{I}_{Z'} \otimes E_0(K_X + L)) & \longrightarrow & 0 \\
\downarrow {\scriptstyle s|_{Z'}} & & \downarrow {\scriptstyle s} & & \\
H^{n-r}(\mathcal{O}_{Z'}(K_X + L)) & \longrightarrow & H^{n-r+1}(\mathcal{I}_{Z'}(K_X + L)) & &
\end{array}$$

Since s is vanishing on Z', we find $s|_{Z'}$ to be zero, which implies that the morphism s is zero as well. $\square$

Lemma 1.2. *Under the assumptions made in Lemma 1.1, there is an exact sequence*

$$H^0(\mathcal{E}_0(F)) = \mathrm{Hom}(\mathcal{I}_{Z'}, \mathcal{E}_0(F)) \xrightarrow{s} H^0(\mathcal{I}_{Z-Z'}(F)) \longrightarrow \mathrm{Ext}^1(\mathcal{I}_{Z'}, \mathcal{F}_1(F)) \longrightarrow 0.$$

Proof. Applying the functor $\mathrm{Hom}(\mathcal{I}_{Z'}, \cdot)$ to

$$0 \longrightarrow \mathcal{F}_1(F) \longrightarrow \mathcal{E}_0(F) \xrightarrow{s} \mathcal{I}_Z(F) \longrightarrow 0,$$

we obtain the exact sequence

$$\mathrm{Hom}(\mathcal{I}_{Z'}, \mathcal{E}_0(F)) \xrightarrow{s} \mathrm{Hom}(\mathcal{I}_{Z'}, \mathcal{I}_Z(F)) \longrightarrow \mathrm{Ext}^1(\mathcal{I}_{Z'}, \mathcal{F}_1(F)) \longrightarrow 0.$$

Note that the 0 term on the right hand side comes from (3) if $r \geq 3$, and for $r = 2$ from the morphism

$$\tau : \mathrm{Ext}^{r-1}(\mathcal{I}_{Z'}, \mathcal{F}_{r-1}(F)) \longrightarrow \mathrm{Ext}^{r-1}(\mathcal{I}_{Z'}, \mathcal{E}_{r-2}(F))$$

which is zero as we have seen in the proof of Lemma 1.1.

Because $Z - Z'$ is the residual subscheme of Z' in Z, we have (cf. [17])

$$\mathrm{Hom}(\mathcal{I}_{Z'}, \mathcal{I}_Z(F)) \cong H^0(\mathcal{I}_{Z-Z'}(F)),$$

completing the proof of Lemma 1.2. $\square$

Proof. [Proof of Theorem 1 for E sufficiently ample.] By Lemma 1.1 and Lemma 1.2 for Z' and Z'', we obtain a commutative diagram

$$\begin{array}{ccccccccc} 0 & \longrightarrow & \operatorname{Im} s & \longrightarrow & H^0(\mathcal{I}_{Z-Z'}(F)) & \xrightarrow{\mu_{Z'}} & H^{n-r+1}(\mathcal{I}_{Z'}(K_X+L))^\vee & \longrightarrow & 0 \\ & & \| & & \uparrow & & \uparrow & & \\ 0 & \longrightarrow & \operatorname{Im} s & \longrightarrow & H^0(\mathcal{I}_{Z-Z''}(F)) & \xrightarrow[\mu_{Z''}]{} & H^{n-r+1}(\mathcal{I}_{Z''}(K_X+L))^\vee & \longrightarrow & 0. \end{array} \quad (5)$$

Note that the middle and right vertical morphisms are injective and by the Five Lemma we can see that they have the same cokernel Q, hence

$$0 \longrightarrow H^0(\mathcal{I}_{Z-Z''}(F)) \longrightarrow H^0(\mathcal{I}_{Z-Z'}(F)) \longrightarrow Q \longrightarrow 0,$$

and

$$0 \longrightarrow Q^\vee \longrightarrow H^{n-r+1}(\mathcal{I}_{Z'}(K_X+L)) \longrightarrow H^{n-r+1}(\mathcal{I}_{Z''}(K_X+L)) \longrightarrow 0.$$

Choosing any isomorphism $Q \cong Q^\vee$ one obtains Theorem 1 from the two exact sequences above. $\square$

For the general case we will replace the vector bundle E by $E \otimes H$, for some sufficiently ample line bundle H.

Lemma 1.3. *Assume that (E, s, F) is a solution of $(*)$ for fixed L and Δ. Let H be a sufficiently ample line bundle and $M \in H^0(E \otimes E^\vee \otimes H)$ a sufficiently general section, viewed as a morphism $M : E \to E \otimes H$. Let*

$$\widetilde{E} = E \otimes H, \quad \widetilde{s} = sM, \quad \widetilde{F} = F + Z(\det M).$$

Then $(\widetilde{E}, \widetilde{s}, \widetilde{F})$ is also a solution of $(*)$ for L and Δ.

Proof. We can assume that the divisor of $\det M$ does not contain any component of $Z(s)$. Let

$$\widetilde{\Delta} = Z(\widetilde{s}) - Z(\widetilde{s})\widetilde{F}$$

be the new residual subscheme. We only need to prove that $\widetilde{\Delta} = \Delta$, i.e., $\mathcal{I}_{\widetilde{\Delta}} = \mathcal{I}_\Delta$.

Indeed, by definition, it is clear that $\mathcal{I}_\Delta \subset \mathcal{I}_{\widetilde{\Delta}}$. Conversely, $\mathcal{I}_{\widetilde{\Delta}}$ consists of the local sections $\widetilde{g}$ such that $\widetilde{g} f \det M$ vanishes on $Z(\widetilde{s})$, where f is the local defining equation of F. Hence it also vanishes on $Z(s)$. Because $\det M$ does not vanish on any component of $Z(s)$, this implies that $\widetilde{g} f$ vanishes on $Z(s)$. Now we know that $\widetilde{g}$ is contained in $\mathcal{I}_\Delta$. So $\mathcal{I}_{\widetilde{\Delta}} \subset \mathcal{I}_\Delta$. $\square$

Proof. [Proof of Theorem 1 for arbitrary E.] Keeping the notations and assumptions of Lemma 1.3 we have a diagram

$$
\begin{array}{ccccccccc}
0 & \longrightarrow & H^0(\mathcal{I}_{\widetilde{Z}-Z''}(\widetilde{F})) & \longrightarrow & H^0(\mathcal{I}_{\widetilde{Z}-Z'}(\widetilde{F})) & \longrightarrow & \widetilde{Q} & \longrightarrow & 0 \\
& & \uparrow \phi_1 & & \uparrow \phi_2 & & \uparrow \psi & & \\
0 & \longrightarrow & H^0(\mathcal{I}_{Z-Z''}(F)) & \longrightarrow & H^0(\mathcal{I}_{Z-Z'}(F)) & \longrightarrow & Q & \longrightarrow & 0
\end{array}
$$

where $Q = \mathrm{coker}\,\alpha$, and where ϕ_1 and ϕ_2 are defined as the multiplication by $\det M$. In particular ϕ_1 and ϕ_2 are injective. For H sufficiently ample, Theorem 1 holds true for $\widetilde{E}$, and $\widetilde{Q} = \ker \beta$. Thus we only need to prove that ψ is injective. By the Five Lemma, it is enough to prove that the induced natural map

$$\mathrm{coker}\,\phi_1 \to \mathrm{coker}\,\phi_2$$

is injective.

Indeed, let $G \equiv \widetilde{F}$ represent an element of $\mathrm{coker}\,\phi_1$, then G passes through $\widetilde{Z} - Z''$. If its image in $\mathrm{coker}\,\phi_2$ is zero, i.e., if $G = G' + \det M$ and G' passes through $Z - Z'$, we need to prove that G is also zero in $\mathrm{coker}\,\phi_1$, i.e., G' passes through $Z - Z''$. This is obvious because $\det M$ does not pass through $Z - Z''$, but G does. □

2. Solutions of the equation (∗)

Theorem 2. *Let Δ be a subscheme of X of pure codimension r and let L be a divisor on X. Then the following conditions are equivalent.*

1) (∗) has a solution (E, s, F) for Δ and L, i.e., there are a hypersurface F, a rank r vector bundle E and a nonzero global section s of E whose zero set $Z = Z(s)$ is an $n - r$ dimensional subscheme such that (∗) holds, so Δ is the residual subscheme of $Z(s)F$ in $Z(s)$.

2) There is an element η in $H^{n-r+1}(\mathcal{I}_\Delta(K_X + L))^\vee$ such that for any proper codimension r closed subscheme $\Delta' \subsetneq \Delta$, η is not in the image of the following natural inclusion map:

$$H^{n-r+1}(\mathcal{I}_{\Delta'}(K_X + L))^\vee \longrightarrow H^{n-r+1}(\mathcal{I}_\Delta(K_X + L))^\vee.$$

Proof. If (∗) has a solution, then by Lemma 1.3, (∗) has a solution with E sufficiently ample. Hence we are allowed to use the diagram (5) from the previous section for $\Delta' \subset \Delta \subset Z$ instead of $Z'' \subset Z' \subset Z$. Recall that the vertical maps in (5) are natural inclusions.

We claim that $\eta = \mu_\Delta(f)$ is the desired element, where $f \in H^0(\mathcal{I}_{Z-\Delta}(F))$ denotes the section with $F = Z(f)$. Indeed, if for some $\Delta' \subsetneq \Delta$ one has $\eta = \mu_{\Delta'}(f')$ with $f' \in H^0(\mathcal{I}_{Z-\Delta'}(F))$, then $\mu_\Delta(f) = \mu_{\Delta'}(f') = \mu_\Delta(f')$, thus $\mu_\Delta(f - f') = 0$.

This implies that $f - f'$ as an element of the image of

$$s : \text{Hom}(\mathcal{I}_\Delta, \mathcal{E}_0(F)) \to H^0(\mathcal{I}_{Z-\Delta}(F))$$

vanishes on Z and hence f vanishes on $Z - \Delta'$. We find $\Delta = Z - ZF \subset \Delta'$, contradicting the assumptions made.

Conversely, from a class η as in 2), we have to construct a solution (E, s, F). Let $F_1, \ldots, F_r$ be sufficiently ample hypersurfaces containing Δ. Assume that $Z = F_1 \ldots F_r$ is a complete intersection. From Theorem 1, we can find an $f \in H^0(\mathcal{I}_{Z-\Delta}(F))$ such that $\eta = \mu_\Delta(f)$, let $F = Z(f)$. Then we have the above commutative diagram for $Z' = \Delta$ and $Z'' = Z - ZF \subset \Delta$. If $Z'' \neq \Delta$, then $\eta = \mu_{Z''}(f)$, which contradicts our assumption. So $\Delta = Z - ZF$, and $(*)$ has a solution with $E = \oplus_{i=1}^r \mathcal{O}_X(F_i)$. $\square$

Remark 2. *From the proof of this theorem, we see that if (E, s, F) is a solution of $(*)$, then $(*)$ has a solution with splitting E, i.e., we can find hypersurfaces $F_1, \ldots, F_{r+1}$ such that $F_1 \ldots F_r$ is a complete intersection and*

$$\begin{cases} \Delta = F_1 \ldots F_r - F_1 \ldots F_{r+1}, \\ L \equiv F_1 + \cdots + F_r - F_{r+1}. \end{cases} \quad (**)$$

Corollary 3. *Let Δ be a codimension r subscheme of X and L a divisor on X. Then the following conditions are equivalent.*

1)
$$h^{n-r+1}(\mathcal{I}_\Delta(K_X + L)) > h^{n-r+1}(K_X + L)$$

but for any subscheme $\Delta' \subsetneq \Delta$,

$$h^{n-r+1}(\mathcal{I}_{\Delta'}(K_X + L)) = h^{n-r+1}(K_X + L).$$

2) $(*)$ has a solution (E, s, F) for Δ and L, but for any subscheme $\Delta' \subsetneq \Delta$, $(*)$ has no solution for Δ' and L.

3. A generalization of Griffiths–Harris theorem

Let us first recall the definition of *k-very ampleness* (cf. [2], [3] or [11]). A linear system $|D|$ on X is called k-very ample if for any zero-dimensional subscheme Y of degree $k + 1$, the restriction map

$$\rho_Y : H^0(\mathcal{O}(D)) \longrightarrow H^0(\mathcal{O}_Y(D))$$

is surjective, which is equivalent to the injectivity of

$$\beta_Y : H^1(\mathcal{I}_Y(D)) \longrightarrow H^1(\mathcal{O}(D)) \longrightarrow 0.$$

Note that "0-very ample" is equivalent to "base point free", and "1-very ample" is equivalent to "very ample".

Theorem 4. *For a fixed divisor L on X and a positive integer k, the following conditions are equivalent.*

1) Let E be a rank n vector bundle with a nonzero global section s such that $Z = Z(s)$ is a zero-dimensional subscheme, and let $F \in |\det E - L|$. If F passes through a subscheme Z' of Z whose degree $\geq \deg Z - k$, then F passes through Z.

2) $|K_X + L|$ is $(k-1)$-very ample.

Proof. The first condition means that for any F in the linear system, the degree of $\Delta := Z(s) - Z(s)F$ is zero or bigger than k. From Corollary 3, this is equivalent to the condition that for any zero dimensional subscheme Y of degree $\leq k$,

$$\beta_Y : H^1(\mathcal{I}_Y(K_X + L)) \longrightarrow H^1(\mathcal{O}_X(K_X + L))$$

is injective. Now by definition, this is just saying that $|K_X + L|$ is $(k-1)$-very ample. □

If $L = -K_X$, then obviously $|K_X + L|$ is base point free. Hence the first part of Theorem 4 is true for $k = 1$. This is the Cayley–Bacharach Theorem due to Griffiths and Harris without the assumption that $Z(s)$ is reduced (cf. [8], p. 677).

Since $\mathcal{O}_{\mathbb{P}^n}(k)$ is k-very ample, one obtains a Cayley–Bacharach theorem on $\mathbb{P}^n$ [17]. In fact, this theorem is sharp.

4. Rank 2 vector bundle case

A divisor L is called numerically effective (nef) if the intersection number $L.C$ is non negative, for all irreducible curves C on X.

Theorem 5. *Let L be a nef divisor, let E be a rank 2 vector bundle on X, and let s be a global section such that $Z = Z(s)$ is of pure codimension 2. For some $F \in |\det E - L|$ let*

$$\Delta = Z(s) - Z(s)F$$

be the residual subscheme of $Z(s)F$ in $Z(s)$. If $\deg \Delta < \deg L^2/4$, then either Δ is empty or there exists an effective divisor D passing through Δ such that

$$\deg DL - \deg \Delta \leq \deg D^2 < \frac{1}{2} \deg DL \leq$$

$$\frac{1}{4}\left(\deg L^2 - \sqrt{\deg L^2}\sqrt{\deg L^2 - 4\deg \Delta}\right) < \deg \Delta.$$

Proof. By Lemma 1.3, we can assume that E is sufficiently ample. So we have a class $\eta = \mu_\Delta(f)$ satisfying the condition 2) of Theorem 2. As in the proof of

Theorem 2, we can find a new solution (E', s', F_3) of $(*)$ with $E' = \mathcal{O}(F_1) \oplus \mathcal{O}(F_2)$, $s' = (f_1, f_2)$, i.e.,

$$\begin{cases} \Delta = F_1 F_2 - F_1 F_2 F_3, \\ L \equiv F_1 + F_2 - F_3. \end{cases}$$

Now we use [17], Corollary 2.2, to complete the proof. □

If we take $\deg \Delta = 1$ or 2, then we can find a theorem of Reider's type (cf. [16]).
The following Corollary is a generalization of the classical Cayley–Bacharach Theorem [1], [6].

Corollary 6. *Keeping the notations introduced above, let H be an ample divisor on X, let ℓ be a positive integral, and let $F \in |\det E - \ell H|$. If F passes through an $(n-2)$-dimensional subscheme of $Z(s)$ whose degree is larger than or equal to $\deg Z(s) - \ell + 2$, then F passes through $Z(s)$.*

Remark 3. Theorem 5 can be used to study the codimension 2 subvarieties in projective space. We assume that Y is a codimension 2 projective subscheme of $\mathbb{P}^n$. We are interested in the following invariants.

$$d = \deg Y,$$
$$s = \min\{m \mid H^0(\mathcal{I}_Y(m)) \neq 0\},$$
$$e = \max\{m \mid H^{n-1}(\mathcal{I}_Y(m)) \neq 0\}.$$

Note that $H^{n-1}(\mathcal{I}_Y(m)) = H^{n-2}(\mathcal{O}_Y(m))$.

Let $\ell = e + n + 1$. For a reduced and irreducible subscheme Y Theorem 2 implies that $(**)$ has a solution. Hence there are 3 hypersurfaces F_1, F_2, F_3 of degree d_1, d_2 and d_3, respectively, such that

$$\begin{cases} Y = F_1 F_2 - F_1 F_2 F_3, \\ \ell = d_1 + d_2 - d_3. \end{cases}$$

In fact, we can choose F_1 such that $\deg F_1 = s$. If $\ell \geq 2\sqrt{d}$, then by Theorem 5,

$$s \leq \frac{1}{2}\ell - \frac{1}{2}\sqrt{\ell^2 - 4d}.$$

This reproves a theorem of Paoletti in [14].

5. An explicit construction of rank 2 vector bundles

As is well known, a codimension 2 subscheme Δ of X is the zero subscheme of a global section of a rank 2 vector bundle, provided it satisfies certain cohomological conditions. In this section, we are going to give an explicit construction of the corresponding vector bundle using the methods developed above.

We will prove that under those cohomological conditions on Δ, we can find three hypersurfaces F_1, F_2 and F_3 such that F_1 and F_2 have no common components,

$$\Delta = F_1 F_2 - F_1 F_2 F_3,$$

and $F_1 F_2 F_3$ is pure codimension 2 and Cohen–Macaulay. We denote by $\mathcal{F}$ the syzygy sheaf of F_1, F_2, F_3, i.e.,

$$0 \longrightarrow \mathcal{F} \longrightarrow \bigoplus_{i=1}^{3} \mathcal{O}_X(-F_i) \longrightarrow \mathcal{I}_{F_1 F_2 F_3} \longrightarrow 0.$$

Then we will construct a global section of the rank 2 vector bundle $\mathcal{E} = \mathcal{F}(F_1 + F_2)$ whose zero subscheme is Δ.

Theorem 7. *Let $\Delta \subset X$ be a subscheme of pure codimension 2. Then the following are equivalent:*

1) Δ *is the zero subscheme of a section of a rank 2 vector bundle $\mathcal{E}$.*

2) Δ *is a local complete intersection, ω_Δ can be extended to an invertible sheaf $\mathcal{W}$ on X, and there is an element $\eta \in H^{n-1}(\mathcal{I}_\Delta(\mathcal{W}))^\vee$ such that for any codimension 2 subscheme $\Delta' \subset \Delta$ with $\deg \Delta' < \deg \Delta$, η is not contained in the image of the following inclusion map*

$$H^{n-1}(\mathcal{I}_{\Delta'}(\mathcal{W}))^\vee \longrightarrow H^{n-1}(\mathcal{I}_\Delta(\mathcal{W}))^\vee.$$

3) *There are three hypersurfaces F_1, F_2 and F_3 such that F_1 and F_2 have no common components, $\Delta = F_1 F_2 - F_1 F_2 F_3$, and such that $F_1 F_2 F_3$ is of pure codimension 2 and Cohen–Macaulay.*

Furthermore, if 1), 2) and 3) hold true, then

$$c_1(\mathcal{E}) \equiv \mathcal{W} - K_X \equiv F_1 + F_2 - F_3.$$

Proof. 1) $\Longrightarrow$ 2): It is well known that Δ is a local complete intersection and

$$\omega_\Delta = (\det \mathcal{E} + K_X)|_\Delta,$$

so $\mathcal{W} = \det \mathcal{E} + K_X$. For $Z = \Delta$, $F = 0$ and for $L = \det \mathcal{E}$ the equation (∗) is satisfied. By Theorem 2, we obtain an element η satisfying the desired conditions.

2) $\Longrightarrow$ 3): Let F_1 and F_2 be two sufficiently ample hypersurfaces containing Δ. Assume that they have no common components. Due to Theorem 2, there is an $F_3 \equiv F_1 + F_2 - L$ (here $L = \mathcal{W} - K_X$) such that

$$\Delta = F_1 F_2 - F_1 F_2 F_3.$$

Now we only need to prove that $F_1 F_2 F_3$ has pure codimension 2. In fact, if S is the pure codimension 2 part of $F_1 F_2 F_3$, then Δ and S are linked, so S is also Cohen–Macaulay [15]. Because Δ is a local complete intersection, we can assume that Δ and

S have no common components. On the other hand, we claim that for $Z = F_1 F_2$ the sequence

$$0 \longrightarrow \omega_\Delta \longrightarrow \omega_Z \longrightarrow \mathcal{O}_S \otimes \omega_Z \longrightarrow 0 \tag{6}$$

is exact. Indeed, let Z be embedded in some $\mathbb{P}^N$, and let s be its codimension. Because Z is Cohen–Macaulay, $\Delta \cap S$ is a divisor in Δ (see [7], p. 454). So $\Delta \cap S$ has codimension $s + 1$, thus we have

$$\mathcal{E}xt^s_{\mathbb{P}^n}(\mathcal{O}_{\Delta \cap S}, \omega_{\mathbb{P}^N}) = 0.$$

Since Δ and S are Cohen–Macaulay, we have

$$\mathcal{E}xt^{s+1}_{\mathbb{P}^n}(\mathcal{O}_\Delta, \omega_{\mathbb{P}^N}) = 0, \quad \mathcal{E}xt^{s+1}_{\mathbb{P}^n}(\mathcal{O}_S, \omega_{\mathbb{P}^N}) = 0.$$

Applying $\mathcal{E}xt(., \omega_{\mathbb{P}^N})$ to the exact sequence

$$0 \longrightarrow \mathcal{O}_Z \longrightarrow \mathcal{O}_\Delta \oplus \mathcal{O}_S \longrightarrow \mathcal{O}_{\Delta \cap S} \longrightarrow 0,$$

one obtaines an exact sequence,

$$0 \longrightarrow \omega_\Delta \oplus \omega_S \longrightarrow \omega_Z \longrightarrow \omega_{\Delta \cap S} \longrightarrow 0. \tag{7}$$

Thus the sequences

$$0 \longrightarrow \omega_S \longrightarrow \omega_Z/\omega_\Delta \longrightarrow \omega_{\Delta \cap S} \longrightarrow 0 \tag{8}$$

and

$$0 \longrightarrow \omega_\Delta|_S \oplus \omega_S \longrightarrow \omega_Z|_S \longrightarrow \omega_{\Delta \cap S} \longrightarrow 0$$

are exact. Since $\omega_Z|_S$ is invertible and since $\omega_\Delta|_S$ is a torsion sheaf, $\omega_\Delta|_S = 0$. One obtains an exact sequence

$$0 \longrightarrow \omega_S \longrightarrow \omega_Z|_S \longrightarrow \omega_{\Delta \cap S} \longrightarrow 0. \tag{9}$$

Note that there is a natural surjective morphism $\phi : \omega_Z/\omega_S \to \omega_Z|_S$. Comparing (8) and (9), we find ϕ to be an isomorphism. This proves (6).

From $\omega_\Delta = W|_\Delta$ and $\omega_Z = (F_1 + F_2 + K_X)|_Z$, we obtain an exact sequence

$$0 \longrightarrow \mathcal{O}(-F_3)|_\Delta \longrightarrow \mathcal{O}_Z \longrightarrow \mathcal{O}_S \longrightarrow 0.$$

On the other hand,

$$0 \longrightarrow \mathcal{O}_\Delta(-S \cap \Delta) \longrightarrow \mathcal{O}_{F_1 F_2} \longrightarrow \mathcal{O}_S \longrightarrow 0,$$

so $F_3 \cap \Delta = S \cap \Delta$. This implies that $F_1 F_2 F_3$ has pure codimension 2.

3) $\Longrightarrow$ 1): Let $\mathcal{F}$ be the syzygy sheaf of F_1, F_2, F_3. Since $F_1 F_2 F_3$ is of pure codimension 2 and Cohen–Macaulay, we know that $\mathcal{F}$ is locally free (cf. [5], [10] or [18]). Considering the composition

$$\phi : \mathcal{F} \longrightarrow \bigoplus_{i=1}^{3} \mathcal{O}_X(-F_i) \longrightarrow \mathcal{O}(-F_3),$$

we can see that the image of ϕ in $\mathcal{O}_X(-F_3)$ is $\mathcal{I}_\Delta(-F_3)$ (cf. the definition of $\mathcal{I}_\Delta$). Thus ker ϕ is an invertible sheaf. By comparing the first Chern classes, we obtain

$$0 \longrightarrow \mathcal{O}_X(-F_1 - F_2) \longrightarrow \mathcal{F} \longrightarrow \mathcal{I}_\Delta(-F_3) \longrightarrow 0,$$

i.e., $$0 \longrightarrow \mathcal{O}_X \longrightarrow \mathcal{F}(F_1 + F_2) \longrightarrow \mathcal{I}_\Delta(F_1 + F_2 - F_3) \longrightarrow 0.$$

Thus the rank 2 vector bundle $\mathcal{E} = \mathcal{F}(F_1 + F_2)$ has a section s whose zero subscheme is Δ, and det $\mathcal{E} = F_1 + F_2 - F_3$. □

References

[1] Bacharach, I., Über den Cayley'schen Schnittpunktsatz, Math. Ann. 26 (1886), 275–299.

[2] Beltrametti, M. C., Sommese, A. J., Zero cycles and k-th order embeddings of smooth projective surfaces (with an appendix by L. Göttsche), in: Problems in the Theory of Surfaces and Their Classification (F. Catanese, C. Ciliberto, M. Cornalba, eds.), Sympos. Math. 32 (1991), Academic Press, London, 33–48.

[3] Beltrametti, M. C., Sommese, A. J., The Adjunction Theory of Complex Projective Varieties, de Gruyter Exp. Math. 16, Walter de Gruyter, Berlin–New York 1995.

[4] Bogomolov, F. A., Holomorphic tensors and vector bundles on projective varieties, Math. USSR Izv. 13 (1978), 499–555.

[5] Burch, L., On ideals of finite homological dimension in local rings, Proc. Cambridge Philos. Soc. 64 (1968), 941–948.

[6] Cayley, A., On the intersection of curves, Cambridge Math. J. 3 (1843), 211–213; Collected Math. Papers I, Cambridge University Press, 1889, 25–27.

[7] Eisenbud, D., Commutative algebra, Grad. Texts in Math. 150, Springer-Verlag, New York 1995.

[8] Griffiths, P., Harris, J., Principles of Algebraic Geometry, Wiley-Interscience, 1978.

[9] Hartshorne, R., Algebraic Geometry, Grad. Texts in Math. 52, Springer-Verlag, New York 1977.

[10] Kaplansky, I., Commutative Algebra, W. A. Benjamin Inc., New York 1970.

[11] Lazarsfeld, R., Lectures on Linear Series, in: Complex Algebraic Geometry (J. Kollar, ed.), IAS/Park-City Math. Ser. 3, Amer. Math. Soc., Providence 1997, 163–219.

[12] Miyaoka, Y., The Chern classes and Kodaira dimension of a minimal variety, in: Algebraic geometry, Sendai 1985, Adv. Stud. Pure Math. 10, Kinokuniya, Tokio 1987, 449–476.

[13] Paoletti, R., Seshadri constants, gonality of space curves, and restriction of stable bundles, J. Differential Geom. 40 (1994), 475–504.

[14] Paoletti, R., On Halphen's speciality theorem, in: Higher dimensional complex varieties (M. Andreatta, T. Peternell, eds.), Trento 1994, Walter de Gruyter, Berlin–New York 1996, 341–355.

[15] Peskine, C., Szpiro, L., Liaison des variétés algébriques, Invent. Math. 26 (1974), 271–302.

[16] Reider, I., Vector bundles of rank 2 and linear systems on algebraic surfaces, Ann. of Math. 127 (1988), 309–316.

[17] Tan, S.-L., Cayley–Bacharach property of an algebraic variety and Fujita's conjecture, J. Algebraic Geom., to appear.

[18] Tan, S.-L., On the integral closure of a cubic extension and applications, preprint 1998.

The scientific work of Michael Schneider

Thomas Peternell

1. The scientific career of Michael Schneider

Michael Schneider studied mathematics and physics with O. Forster and K. Stein at the University of Munich, interrupted by one semester in Geneva. He obtained the diploma in Munich in 1966 and the doctoral degree in 1969 with a dissertation on complete intersections in Stein manifolds (see no. [1] in the list of papers). From 1969–1974 he hold an assistant position in Regensburg where he received the habilitation in 1974.

In 1975 he got a professorship at the University of Göttingen. Five years later he obtained a chair for mathematics at the University of Bayreuth, rejecting an offer from Münster in 1982. He hold his chair in Bayreuth until his tragic death in the French Alps near Nice on august 29, 1997.

Michael Schneider began his mathematical studies under the strong influence of the German school in complex analysis (Forster, Grauert, Remmert, Stein) and began to move in the mid 70's to algebraic geometry. His main interest in the Göttingen period doubtlessly were vector bundles on projective spaces, one of the active areas at that time. One of its culminations is the standard text book by himself, C. Okonek and H. Spindler.

In the 80's and 90's his interests shifted more and more to geometric problems: subvarieties of projective spaces, global geometry of projective and more general compact complex manifolds, and automorphisms. But occasionally he turned back to his analytic roots studying old problems with new technology.

Michael Schneider was very active in the organisation of science, too. He was editor of the Journal für die reine und angewandte Mathematik from 1984–1995. He organised several conferences in Bayreuth, Oberwolfach, Alghero and Warsaw and was involved in the organisation of the AMS Summer Institute 1985 in Santa Cruz and of the special year in complex analysis 1995/96 at the Mathematical Sciences Research Institute at Berkeley (with Y. T. Siu). In 1996 he was elected as Fachgutachter of the Deutsche Forschungsgemeinschaft in pure mathematics, an important and influential position. Last but not least he was very much involved in the business of the Mathematische Forschungsinstitut Oberwolfach.

Michael Schneider was one of the main initiators of a Forschungsschwerpunkt Komplexe Mannigfaltigkeiten of the Deutsche Forschungsgemeinschaft, 1987-1993. From 1991 on he was running with H. Kerner, T. Peternell and F. O. Schreyer the Graduiertenkolleg Komplexe Mannigfaltigkeiten at Bayreuth and organised several summer schools. He also participated in the Euro networks "Algebraic Geometry in Europe" (AGE) and "Europroj".

2. Complex analytic roots

The 1969 dissertation of Michael Schneider (see [1]) treats closed complex submanifolds Y of Stein manifolds X. Forster and Ramspott had shown under the assumption $\dim Y < \frac{1}{2} \dim X$, that $Y \subset X$ is a complete intersection if and only if the normal bundle is trivial. Schneider considered the limit case $\dim Y = \frac{1}{2} \dim X$ and proved that Y is a complete intersection if and only if

(1) the normal bundle N_Y is trivial and

(2) the fundamental class $[Y] \in H^n(X, \mathbb{Z})$ vanishes ($n = \dim X$).

In 1981 Schneider came back to these type of questions, established in [19] the singular version of the above theorem and proved a series of results stating that locally complete intersections $Y \subset X$ are set-theoretically complete intersections under certain conditions. There has been a lot of activities both on affine and Stein manifolds under which conditions subvarieties are complete intersections but it would lead much too far to discuss details here. Even more attention has been paid to the projective situation, and Michael Schneider was also very much involved in this area of research; see Sections 2 and 3.

After his dissertation Schneider worked on various problems in complex analysis. One was dealing with Grauert's fundamental results in bimeromorphic geometry (criteria for blowing down, geometry of exceptional sets). The papers [2] with Knorr and [12] settle many of Grauert's results in the relative case which is important for applications.

The papers [3] and [4] arose from the joint efforts of O. Forster, K. Knorr and M. Schneider to find an easier proof of Grauert's direct image theorem. They treat very basic questions on the behaviour of cohomology in families of compact complex spaces which are nowadays subsumed in standard theorems most of them first proved by Grauert. We recommend the book of Banica and Stanasila on algebraic methods in the global theory of complex spaces and skip all details. However the influence of Grauert on the work of Schneider becomes already apparent.

We find this influence again in the paper [5]. In 1962 Andreotti and Grauert had introduced the notion of q-*convex* and q-*complete* complex space (as well as q-concave spaces) and proved their famous finiteness and vanishing theorems. In that context

Schneider found — solving a conjecture of Hartshorne in the smooth case — the following

Theorem. *Let X be a compact complex manifold and $Y \subset X$ a submanifold of codimension k with Griffiths-positive normal bundle. Then $X \setminus Y$ is k-convex.*

It seems still unknown whether it is sufficient that the normal bundle is ample. Notice also that K. Fritzsche has proved singular versions which get technically rather complicated.

In the paper [9] Schneider solved another interesting analytic problem: he showed that given a complex space X and $Y \subset X$ a closed Stein subspace, then for every compact set $K \subset Y$ one can find arbitrary small open Stein neighborhoods of K in X. Of course the restriction to K was expected to be unnecessary, indeed Y. T. Siu showed around the same time that every Stein subspace admits arbitrary small Stein neighborhoods.

Recently Schneider came back to classical analytic questions on extending meromorphic functions [53]. Together with Badescu he proved that a submanifold Y of a projective manifold X whose normal bundle has a weak positivity property (N_Y is $(d-1)$-ample in the sense of Sommese, $d = \dim Y$), has the property $G2$, i.e. the field of formal meromorphic functions along Y is a finite field extension of the field of meromorphic functions on X. Furthermore Badescu–Schneider gave various applications, e.g. they found again the following result of Faltings: the diagonal $\Delta \subset Z \times Z$, Z rational homogeneous, is $G3$, i.e. every meromorphic function on the formal completion extends uniquely to $Z \times Z$.

3. Vector bundles on projective spaces

Vector bundles on projective space were one of the most active areas in algebraic geometry in the 70's and early 80's. The most influential and important contribution of Schneider is certainly the monography [59] with C. Okonek and H. Spindler which is still *the* standard reference on the subject. A very good introduction is also provided by Schneider's Bourbaki article [13]. We shall now describe his further work on bundles on projective spaces.

The paper [8] gives a numerical criterion for a stable rank 2 bundle E on $\mathbb{P}_n$ to be ample, the condition on the Chern classes reads

$$c_1 \geq 2c_2 - \frac{c_1^2}{2}.$$

Moreover a criterion for global generatedness is given.

In [11] unstable rank 2 bundles on projective 4-space are investigated; a final result (on non-existence of those bundles) is still very much open.

The paper [14] with Elencwajg and Hirschowitz deals with uniform bundles E on $\mathbb{P}_n$. Recall that E is uniform if E has the same splitting type on every line. The structure theorem is as follows.

Theorem. *Let E be a holomorphic vector bundle of rank r on $\mathbb{P}_n$. Suppose that E is uniform and $r \leq n$. Then E is splits or $E \simeq \Omega^1_{\mathbb{P}_n}(a)$ or $E \simeq T_{\mathbb{P}_n}(b)$.*

The subsequent paper [15] with Forster and Hirschowitz treats more generally semistable bundles on projective manifolds (fix a polarisation). The result generalises the theorem of Grauert–Mülich–Spindler on the splitting type of semistable bundles on lines in projective space. The lines have to be substituted by curves and the splitting by the Harder–Narasimhan filtration. As a consequence, semistable bundles with fixed Hilbert polynomial form bounded families, a step towards construction of moduli spaces.

In [16] and [17] Schneider investigates 3-bundles E on $\mathbb{P}_n$. For semistable vector bundles of rank r one has the Bogomolov–Miyaoka inequality

$$c_1^2(E) \leq \left(\frac{2r}{r-1}\right) c_2(E).$$

In case of rank 2 this is completely satisfactory. A rank 3 bundle however has a third Chern class $c_3(E)$. So the problem arises whether there are inequalities also invoking c_3. Schneider gives such inequalities for 3-bundles which depend a little bit on the normalisation. If say $c_1(E) = 0$, e.g. the inequality reads

$$|c_3| \leq c_2^2 + c_2.$$

Whether there is a general inequality for normalised bundles on an arbitrary manifold is still open and potentially very interesting. However there are no Chern class inequalities in degree 3 which are invariant under twists. Problems like this have also been treated in the dissertation of Schneider's student H.Kratz. In both papers [16] and [17] restrictions to hyperplanes are studied intensively, the general results of Mehta–Ramanathan and Flenner being only concerned with hypersurfaces of higher degree. The result can be summarised as follows

Theorem. *Let E be a semistable rank 3 bundle on $\mathbb{P}_n, n \geq 3$. Let H be a general hyperplane. Then $E|H$ is semistable unless $n = 3$ and E is a twist of the tangent or cotangent bundle.*

Much later Schneider came back to the study of vector bundles on projective manifolds. In the paper [40] he proved in collaboration with F. Catanese polynomial estimates for the degree of the zero set of sections of vector bundles on a polarised manifold (X, H) with H very ample and applied this to Weierstrass schemes:

Theorem. *Let X be a projective manifold of dimension n with ample canonical bundle K_X. Then there are constants C and N such that the degrees of the Weierstrass schemes W_k^m, taken with respect to K_X, can be bounded by $C(K_X^n)^N$.*

As an application Catanese and Schneider are able to give estimates of the dimension of the moduli spaces of stable vector bundles and furthermore they can bound the higher Chern classes of a semistable vector bundle on a fixed projective manifold by the first two Chern classes.

4. Subvarieties of projective spaces

A main guide to Schneider's reserarch interest in projective geometry has been the famous (still unsolved) conjecture of Hartshorne

Conjecture. *Let $X \subset \mathbb{P}_n$ be a smooth subvariety of dimension $m > \frac{2}{3}n$. Then X is a complete intersection.*

Barth and Van de Ven showed that if the degree d of X is small compared to n (i.e. $n > N(d)$), then X is a complete intersection. Schneider improved the bounds in case X has semistable normal bundle [21]. The paper [25] with Holme treats the same problem without any stability assumption, however only in case X has codimension 2. The bound they obtain reads

$$d < (n-1)(n+5).$$

Note that for $n \geq 6$ the submanifold X defines a rank 2 vector bundle on $\mathbb{P}_n$ so that this section is closely related to the previous one.

One way to approach Hartshorne's conjecture is via k-normality. Recall that $X \subset \mathbb{P}_n$ is k-normal if the restriction map

$$H^0(\mathbb{P}_n, \mathcal{O}(k)) \longrightarrow H^0(X, \mathcal{O}(k))$$

is surjective. In this context Hartshorne's conjecture is equivalent in case of codimension 2 and $n \geq 6$ to k-normality for all k; in any codimension complete intersections are k-normal for all k. The first step in this direction was taken by Zak who proved linear normality ($k = 1$) in the Hartshorne range. The paper [28] with Le Potier and Peternell proves new vanishing theorems for symmetric powers of vector bundles and applies this to prove quadratic normality (in codimension 2) for $n \geq 12$. This bound was later improved by Alzati and Ottaviani to $n \geq 7$.

A special role play surfaces in $\mathbb{P}_4$ which is still an active area of research. This topic is particularly difficult since here we are completely outside the range where Lefschetz theorems apply. In particular there is a priori no control on the irregularity $q(X)$ (= number of 1-forms) of a surface $X \subset \mathbb{P}_4$. However there is no surface known with $q(X) \geq 3$ and the only surface with $q(X) = 2$ is the famous Horrocks–Mumford torus. There are only a few examples with $q(X) = 1$. Making things a little easier one requires that the canonical bundle K_X is the restriction of a bundle from $\mathbb{P}_4$, i.e. $K_X = \mathcal{O}_X(e)$. This is to say that X is induced by a rank 2 bundle on $\mathbb{P}_4$ by the Serre correspondence. If $e = 1$ it is known by Ballico and Chiantini that X is a complete intersection. The paper [32] deals with the case $e = 2$ (half-canonical surfaces). Excluding complete intersections, it is proved that either X has degree 16 in which case X comes from the Horrocks–Mumford bundle, or X has degree 18 or 22 in which cases the irregularity fulfilles $1 \leq q \leq 4$ resp. $5 \leq q \leq 9$. More can be said in these potential cases whose existence is still unsolved.

A main topic of Schneider's research starting in the late 80's were threefolds X in $\mathbb{P}_5$. Up to degree 8 these were classified by Okonek and Ionescu. Together with Beltrametti and Sommese, Schneider classified all threefolds of degree 9, 10 and 11 making essentially use of adjunction theory [34, 38]. Schneider's student G.Edelmann studied the case of degree 12, although it seems still open whether every combination in his list of invariants can be realised. See [37] for a survey. The paper [41] with Braun, Ottaviani and Schreyer in contrast proves a theoretical result:

Theorem. *There are only finitely many families of threefolds in $\mathbb{P}_5$ not of general type.*

So in principle all threefolds in $\mathbb{P}_5$ not of general type can be classified in a finite list (up to deformation). For surfaces in $\mathbb{P}_4$ the corresponding has been known by Ellingsrud and Peskine. The paper [41] gives also an explicit bound for the degree of a "special" threefold in $\mathbb{P}_5$, the best possible bound still being unknown. Consequently, the same authors [54] completely classified conic bundles in $\mathbb{P}_5$ and then are able to give a complete list of log-special threefolds $X \subset \mathbb{P}_5$, this is to say that $K_X + H$ is not at the same time big and nef where H is a hyperplane section. In turns out that the degree of those threefolds is at most 12. A refined study of threefolds in $\mathbb{P}_5$ from the point of view of adjunction theory is presented in [48].

The above finiteness theorem should hold actually in a much more general context. Experts believe that there are only finitely many families of submanifolds $X_m \subset \mathbb{P}_n$ not of general type provided $m \geq n - m$. Schneider verified this in case $m \geq \frac{n+2}{2}$ [39].

We finish these type of questions by mentioning a much broader problem. Since every projective manifold of dimension n can be embedded into $\mathbb{P}_{2n+1}$, it is interesting to ask which manifold can be embedded into $\mathbb{P}_k$ for $k \leq 2n$. In other words we ask for embedding obstructions. Via vanishing theorems for twisted symmetric powers of the cotangent bundle, Schneider was e.g. able to prove [42] that if the cotangent bundle of X is ample, then X cannot be embedded into $\mathbb{P}_{2n-1}$, $n = \dim X$.

The last result in projective geometry is concerned with projections from positive dimensional linear subspace in projective space. In the paper [56] with Beltrametti, Howard and Sommese, published in this volume, the situation is investigated when a subvariety X has a lower dimensional image by the projection mapping from the above specified linear subspace.

5. Global theory of projective varieties and compact complex manifolds

The first global result we want to discuss here is Schneider's proof of the Le Potier vanishing theorem:

Let X be a projective manifold, E an ample rank r vector bundle on X. Then

$$H^q(X, E \otimes \Omega_X^p) = 0, \ p+q \geq n+r.$$

For $r = 1$ this is nothing but the Kodaira–Akizuki–Nakano vanishing theorem. Schneider's method is very simple: he considers the projectivised bundle $\mathbb{P}(E)$ and applies there the Kodaira–Akizuki–Nakano vanishing theorem to the ample line bundle $\mathcal{O}_{\mathbb{P}(E)}(1)$.

We would like to mention here that there is a differential geometric analogue of ampleness, called Griffiths positivity. For line bundles these both notions coincide but for higher rank it is still unknown whether they are quivalent. Interestingly, if we consider Griffiths semipositivity and *nefness* for vector bundles (i.e. E is nef if $\mathcal{O}_{\mathbb{P}(E)}(1)$ is nef), then both notions are different (see [46, 58]).

Le Potier's vanishing theorem has a lot of refinements, e.g for symmetric powers etc. These are mainly due to Sommese, Demailly and Manivel. The vanishing theorem and its refinements have also interesting geometric consequences, for example to quadratic normality [28] and to the Barth–Lefschetz theorem [44].

Often it is hard to check whether a given vector bundle E is ample. In [24] Schneider and Tancredi derived a criterion for a *semi-stable* rank 2 bundle E (with respect to the determinant of E, which is assumed to be ample) on a surface to be ample. The necessary conditions are

$$c_1(E)^2 > 0, \ c_2(E) > 0$$

and the ampleness $E|C$ on every curve C. In [26] this criterion is applied to check that Kodaira surfaces have ample cotangent bundles. The paper [30] treats a generalisation of ampleness, called "ampleness almost everywhere." This is an important notion since Miyaoka has shown that surfaces of positive index ($c_1^2 > c_2$) have almost everywhere ample cotangent bundles.

In [46] Schneider investigates in collaboration with Demailly and Peternell compact Kähler manifolds whose tangent bundle are nef. This is the algebro-geometric analogon of Kähler manifolds with semipositive bisectional curvature in differential geometry which were classified by Mok. However the algebraic class is much larger; the main result can be summarised as follows

Theorem. *Let X be a compact Kähler manifold whose tangent bundle T_X is nef. Then*
(a) *The Albanese map $\alpha : X \longrightarrow A = \mathrm{Alb}(X)$ is a surjective submersion;*
(b) *after possibly substituting X by a finite étale cover, the fibers F of α are Fano manifolds with nef tangent bundles T_F and $\pi_1(X) = \pi_1(A)$.*

This theorem essentially reduces the study of Kähler manifolds with nef tangent bundles to the study of *Fano* manifolds with nef tangent bundles. Recall that a manifold F is Fano, if the anticanonical bundle is ample or, equivalently, if F admits a Kähler metric of positive Ricci curvature. The structure of Fano manifolds X with nef tangent bundles is still not understood. In dimension at most 3, X is rational homogeneous (Campana–Peternell). It is expected that this holds in every dimension.

The subsequent papers [43] and [52] investigate a larger class of "positively curved" manifolds, namely manifolds whose anticanonical class $-K_X = \det T_X$ is nef. We should recall the definition of a line bundle L to be nef: it is required that for any positive ϵ there is a metric on L whose curvature satisfies the inequality

$$\Theta \geq -\epsilon \omega$$

where ω is a fixed positive $(1, 1)$-form. If X is projective, L is nef if and only if $c_1(L) \cdot C \geq 0$ for every irreducible curve $C \subset X$. The differential geometric analogues of manifolds with nef anticanonical bundles are Kähler manifolds with semipositive Ricci curvature. It is shown that the class of manifolds with nef anticanonical bundle is strictly larger than the class of manifolds with semipositive Ricci curvature. The general structure now is as follows.

Theorem. *Let X be a compact Kähler manifold with nef anticanonical bundle.*
(a) *If the line bundle $-K_X$ even admits a metric with semipositive curvature, then the universal cover $\tilde{X}$ is of the form*

$$\tilde{X} \simeq \mathbb{C}^m \times X_1 \times \cdots \times X_r,$$

where the X_i are Calabi-Yau or symplectic manifolds or have the property that $H^0(X_i, (\Omega^1_{X_i})^{\otimes k}) = 0$ for all positive k. In particular $\pi_1(X)$ has an abelian subgroup of finite index.
(b) *If $-K_X$ is merely nef, then $\pi_1(X)$ has subexponential growth, in particular X does not admit a map onto a curve of genus at least 2.*

Also the Albanese map of manifolds with nef anticanonical bundles was studied and several partial results led to the conjecture that it should be a surjective submersion as in the metric case. Meanwhile surjectivity is proved by Zhang in case X is projective and in the threedimensional non-algebraic Kähler case in [58]. Smoothness has been proved in the threedimensional algebraic case by Peternell–Serrano. Of course one might even expect part (1) of the last theorem to be true in the general situation ($-K_X$ nef).

The paper [50] with Demailly–Peternell studies line bundles with some positivity, signatures of curvature and applications to curves in higherdimensional varieties (questions on various cones, movability and the normal bundle). We are not going into details. This study is continued and extended in [58], especially generically nef and pseudo-effective line bundles are considered there.

Schneider's interest focused often on the general question how the global geometry of X is influenced by a submanifold $Y \subset X$. Of course Y has to have a sufficiently positive normal bundle. We already met this type of problems in the preceeding sections. One of the most fascinating questions in this context is the following conjecture of Hartshorne:

Conjecture. Let X be a projective manifold, $Y, Z \subset X$ submanifolds with ample (or Griffiths positive) normal bundles. Suppose $\dim Y + \dim Z \geq \dim X$. Is then $Y \cap Z \neq \emptyset$?

Schneider's student Lübke solved the case where X is homogeneous, Barlet gave a positive anwer when X is a hypersurface in a homogeneous manifold and in [35] the case is settled where X is a $\mathbb{P}_2$-bundle over a surface. Here concavity of $X \setminus Y$ and convexity of small neighborhhoods of Z play an important, concepts which were familiar to Schneider since the early 70's (see sect.1). Hartshorne's conjecture is still wide open (if say Y has codimension 1, then the solution is easy) and is much related to movability questions for cycles.

A rather different problem, although of the same flavour, was treated in [57], namely to relate the Kodaira dimension $\kappa(X)$ with the Kodaira dimension $\kappa(Y)$ under certain assumptions on the normal bundle $N_{Y|X}$. It is proved that

$$\kappa(X) \leq \kappa(Y) + \operatorname{codim} Y$$

if some symmetric power of N is generically generated by global sections (or generically nef) or if N is nef and Y admits a good minimal model in the sense of Mori theory.

In the paper [31] Schneider took up the problem of compactifactions of $\mathbb{C}^n$. This problem was first posed by Hirzebruch in his famous problem list from 1953. In the above paper projective compactifactions of $\mathbb{C}^3$ with normal boundary at infinity were studied, i.e. smooth projective threefolds X with $b_2(X) = 1$ and an irreducible divisor $Y \subset X$ such that $X \setminus Y \simeq \mathbb{C}^3$. Then X is automatically a Fano manifold and [31] classifies the case of index 2. Recall that the index is by definition is the maximal number r such that $-K_X = rL$ for some line bundle L. If (for threefolds) $r = 4$ then X is projective space (which is a compactification) and if $r = 3$, then X is a quadric (which is a compactification, too). The case $r = 2$ has also been treated by Furushima, in particular he constructed examples. The case $r = 1$ finally has been solved by Furushima and Peternell afterwards, also non-algebraic situations have been considered; we are not going into details. See also the survey paper [36].

Turning to a completely different subject, Catanese and Schneider studied rather recently automorphisms of manifolds of general type. Here is the problem is to find good extimates for the number of automorphisms in terms of invariants of the underlying variety. Using effectivity results of Demailly and Kollár, they were able to obtain polynomial estimates for abelian groups of automorphisms of varieties of general type. More precisely:

Theorem. *Let X be a projective manifold of general type, $n = \dim X$. Let G be an abelian group of birational automorphisms of X. Then*
(1) *If K_X is nef, then $\text{card}(G) \leq C_n(K_X^n)^{2^n}$, where C_n is a constant only depending on $\dim X$.*
(2) *If $n = 3$, and if m is a positive integer admitting a G-eigenspace in $H^0(mK_X)$ of dimension at least 2, then $\text{card}(G) \leq \max(6P_2(X), P_{3m+2}(X))$, where $P_i(X) = \dim H^0(iK_X)$.*

Although there was a lot of work on surfaces (Xiao, Corti, Huckleberry–Sauer etc.), this is the first result in higher dimensions apart from earlier work of Howard–Sommese in 1982.

6. Publications of Michael Schneider

[1] Vollständige Durchschnitte in Steinschen Mannigfaltigkeiten, Math. Ann. 186 (1970), 191–200.

[2] (with K. Knorr) Relativexzeptionelle analytische Mengen, Math. Ann. 193 (1971), 238–254.

[3] Halbstetigkeitssätze für relativanalytische Räume, Invent. Math. 16 (1972), 161–176.

[4] Bildgarben und Fasercohomologie für relativ-analytische Räume, Manuscripta Math. 7 (1972), 67–82.

[5] Über eine Vermutung von Hartshorne, Math. Ann. 201 (1973), 221–229.

[6] Ein einfacher Beweis des Verschwindungssatzes für positive holomorphe Vektorbündel, Manuscripta Math. 11 (1974), 95–101.

[7] Lefschetzsätze und Hyperkonvexität, Invent. Math. 31(1975), 183–192.

[8] Stabile Vektorraumbündel vom Rang 2 auf der projektiven Ebene, Nachrichten Akad. Wiss. Göttingen, II. Mathem.-Physikal. Klasse, 6 (1976).

[9] Tubenumgebungen Steinscher Räume, Manuscripta Math. 18 (1976), 391–397.

[10] Lefschetz theorems and a vanishing theorem of Grauert–Riemenschneider, Proc. Sympos. Pure Math. 30 (1977), 35–39,

[11] (with H. Grauert) Komplexe Unterräume und holomorphe Vektorraumbündel vom Rang 2, Math. Ann. 230 (1978), 75–90.

[12] Familien negativer Vektorraumbündel und 1-konvexe Abbildungen, Abh. Math. Sem. Hamburg 47 (1978), 150–170.

[13] Holomorphic Vector Bundles on $\mathbb{P}$, Séminaire Bourbaki 530 (1978–1979).

[14] (with G. Elencwajg und A. Hirschowitz) Les fibrés uniformes de rang au plus n sur $\mathbb{P}$ sont ceux qu'on croit, Progr. Math. 7, Birkhäuser, 1980, 37–63.

[15] (with O. Forster und A. Hirschowitz) Type de scindage généralisé pour les fibrés stables, Progr. Math. 7, Birkhäuser, 1980, 65–81.

[16] Chernklassen semi-stabiler Vektorraumbündel vom Rang 3 auf dem komplex-projektiven Raum, J. Reine Angew. Math. 315 (1980), 211–220.

[17] Einschränkung stabiler Vektorraumbündel vom Rang 3 auf Hyperebenen des projektiven Raumes, J. Reine Angew. Math. 323 (1981), 177–192.

[18] Stable bundles of rank 3 on $\mathbb{P}_3$, Inst. Elie Cartan, Nancy (1981), 1–25.

[19] Vollständige, fast-vollständige und mengentheoretisch-vollständige Durchschnitte in Steinschen Mannigfaltigkeiten, Math. Ann. 260 (1982), 151–174.

[20] Moishezon spaces and almost positive coherent sheaves. Math. Ann. 264 (1983), 517–524.

[21] Submanifolds of projective space with semi-stable normal bundle, in: Several Complex Variables, Proc. of the 1981 Hangzhou Conference, Birkhäuser, 1984, 151–160.

[22] On the number of equations needed to describe a variety in projective or affine space, Proc. Sympos. Pure Math. 41 (1984), 163–180.

[23] Some remarks on vanishing theorems for holomorphic vector bundles, Math. Z. 186 (1984), 135–142.

[24] (with A. Tancredi) Positive vector bundles on complex surfaces, Manuscripta Math. 50 (1985), 133–144.

[25] (with A. Holme) A computer aided approach to codimension 2 subvarieties of $\mathbb{P}_n$, $n \geq 6$, J. Reine Angew. Math. 357 (1985), 205–220.

[26] Complex surfaces with negative tangent bundle, in: Complex Analysis and Algebraic Geometry, Lecture Notes in Math. 1194, Springer-Verlag, 1986, 150–157.

[27] Vector bundles and submanifolds of projective space: 9 open problems, Proc. Sympos. Pure Math. 46 (1987), 101–107.

[28] (with T. Peternell und J. le Potier) Vanishing theorems, linear and quadratic normality, Invent. Math. 87 (1987), 573–586.

[29] (with T. Peternell und J. le Potier) Direct images of sheaves of differentials and the Atiyah class, Math. Z. 196 (1987), 75–85.

[30] (with A. Tancredi) Almost-positive vector-bundles on projective surfaces, Math. Ann. 280 (1988), 537–547.

[31] (with T. Peternell) Compactifications of $\mathbb{C}^3$, I, Math. Ann. 280 (1988), 129–146.

[32] (with W. Decker, T. Peternell und J. le Potier) Half-canonical surfaces, in: Algebraic Geometry, L'Aquila 1988, Lecture Notes in Math. 1417, Springer-Verlag, 1990, 91–110.

[33] Vector bundles and low-codimensional submanifolds of projective space: a problem list, in: Topics in Algebra, Banach Center Publ. 26, Part 2, PWN-Polish Scientific Publishers, Warsaw 1990, 209–222.

[34] (with M. Beltrametti und A. Sommese) Threefolds of degree 9 and 10 in $\mathbb{P}^5$, Math.Ann. 288 (1990), 413–444.

[35] (with D. Barlet und T. Peternell) On two conjectures of Hartshorne's, Math. Ann. 286(1990), 13–25.

[36] (with T. Peternell) Compactifications of $\mathbb{C}^n$: A survey, Proc. Sympos. Pure Math. 52 (1991), Part 2, 455–466.

[37] 3-folds in $\mathbb{P}_5$ — classification in low degree and finiteness results, in: Geometry of Complex Projective Varieties (A. Lanteri, M. Palleschi, D. C. Struppa, eds.), Cetraro (Italy), June 1990, Mediterranean Press, Rende 1993, 275–288.

[38] (with M. Beltrametti und A. Sommese) Threefolds of degree 11 in $\mathbb{P}^5$, in: Complex Projective Geometry, London Math. Soc. Lecture Notes Ser. 179, Cambridge Iniversity Press, 1992, 59–80.

[39] Boundedness of low-codimensional submanifolds of projective space, Internat. J. Math. 3 (1992), 397–399.

[40] (with F. Catanese) Bounds for stable bundles and degrees of Weierstraß schemes, Math. Ann. 293 (1992), 579–594.

[41] (with R. Braun, G. Ottaviani und F. O. Schreyer) Boundedness of 3-folds in $\mathbb{P}_5$, In Complex Analysis and Geometry (V. Ancona and A. Silva, eds.), Plenum Press, New York 1992, 311–338..

[42] Symmetric differential forms as embedding obstructions and vanishing theorems, J. Algebraic Geom. 1 (1992), 175–181.

[43] (with J. P. Demailly und T. Peternell) Kähler manifolds with semipositive Ricci curvature, Compositio Math. 89 (1993), 217–240.

[44] (with J. Zintl) The theorem of Barth–Lefschetz as a consequence of Le Potier's vanishing theorem, Manuscripta Math. 80 (1993), 259–263.

[45] (with M. Beltrametti und A. Sommese) Applications of the Ein–Lazarsfeld criterion for spannedness of adjoint bundles, Math. Z. 214 (1993), 593–599.

[46] (with J. P. Demailly und T. Peternell) Kähler manifolds whose tangent bundle is nef, J. Algebraic Geom. 3 (1994), 295–345.

[47] (with T. Peternell) Neuere Entwicklungen in der komplexen Geometrie, in: Duration and Change: fifty years at Oberwolfach, Springer-Verlag, 1994, 275–306.

[48] (with M. Beltrametti und A. Sommese) Special properties of the adjunction theory for 3-folds in $\mathbb{P}_5$, Mem. Amer. Math. Soc. 116, n° 554, Amer. Math. Soc., Providence 1995, 1–62.

[49] (with F. Catanese) Polynomial bounds for abelian groups of automorphisms, Composition Math. 97 (1995), 1–15.

[50] (with J. P. Demailly und T. Peternell) Holomorphic line bundles with partially vanishing cohomology, Israel Math. Conf. Proc. 9, Bar-Ilan, 1993 (M. Teicher, ed.), 1996 165–198.

[51] (with M. Beltrametti und A. Sommese) Inequalities for Chern classes of ample and spanned vector bundles, Israel Math. Conf. Proc. 9, Bar-Ilan, 1993 (M. Teicher, ed.), 1996, 97–107.

[52] (with J. P. Demailly und T. Peternell) Compact Kähler manifolds with hermitian semipositive anticanonical bundle, Compositio Math. 101 (1996), 217–224.

[53] (with L. Badescu) A criterion for extending meromorphic functions, Math. Ann. 305 (1996), 393–402.

[54] (with R. Braun, G. Ottaviani und F. O. Schreyer) Classification of log-special 3-folds in $\mathbb{P}_5$, Ann. Scuola Norm. Sup. . Pisa, Serie IV, 23 (1996), 69–97.

[55] (with L. Badescu) Formal rational functions in homogeneous manifolds, in preparation.

[56] (with M. Beltrametti, A. Howard und A. Sommese) Projections from subvarieties, this volume.

[57] (with T. Peternell und A. Sommese) Kodaira dimension of subvarieties, to appear in Internat. J. Math.

[58] (with J. P. Demailly und T. Peternell), in preparation.

[59] **Monography** (with C. Okonek und H. Spindler) Vector Bundles on Complex Projective Spaces, Progr. Math. 3, Birkhäuser 1980.

Michael Schneider – an alpine vita

Ulf Persson

A few miles inland from Nice, one encounters the ochra-colored cliffs of Le Baou. Those are the exposed rocks, dropping some 300 feet, of an otherwise gently sloping hill. At the foot of the hill nestles the small town of Saint-Jeanette, with its steep and narrow winding streets, so typical of the hillside town, of which the mountainous backcountry of the Cote d'Azur abounds. Above the small town, one may follow a narrow path, leading through an undergrowth of thorny shrubs, amidst straggling oaks giving way further up for pines, finally ending up at the rocky foot of the cliffs. The cliffs in fact serve as a popular place for local climbers to practise their technical skills, engaging in what is known as 'sports-climbing' a supposedly safe form of climbing, as far as it is exercised in a controlled environment, free from extraneous hazards. Turning around, one has a nice view of green undulating hills meeting the dark blue sea, with the sprawling city of Nice hidden from view in the intercept. It was here that Michael Schneider, on August 29, 1997, so tragically fell some 80 feet to his death, while descending. The exact circumstances will never to be known, as Schneider was climbing alone, and definitely will be of no interest to his family, friends and colleagues. And there is no reason to dwell on the moment and circumstances of death, whose main significance is that it comes last, not that it in anyway constitutes a summing-up of a man.

Schneider was a climber, a passionate climber since some forty years, a fact that not all of his colleagues may have been aware of. He was never once to brag about his exploits, the like of which, would certainly have tempted lesser men to be insufferable; but of course it was inevitable that he now and then would drop a tantalizing hint. He would sit in a restaurant somewhere in Italy, and his vivacious conversation would suddenly come to a halt, and he would lose himself in thought looking at a poster of a mountain face. His companions would ask him what he was looking at, and he would answer, - the North Face of Eiger, and then admit that he once climbed it. And that would be it. At some other time, he may casually mention that he had been bivouacking on a climb, and his mathematical colleagues would be flabbergasted, and hardly believe their ears, and the matter would be tactfully dropped. Or he would admit to his discomfort once, of being marooned at an altitude in excess of twenty thousand feet, lending support to a fellow-climber, worrying about the effects of high-altitude. The mountains of Pamir, the Andes, Kilimanjaro, you name it, inevitably they would come up fleetingly in conversation.

The outward life of a mathematician is seldom very exciting to talk or write about, but there are of course exceptions. With Schneider you realised that mathematics was just one facet of his personality, and that there were others, equally important to him. Although as most mathematicians, Schneider showed an early aptitude, and he was definitely foremost among his schoolmates, he did initially waver between different academic options as he was about to enter University. In fact there were so many things he could have done, and done well. German literature in particular tempted him. He sought professional counseling, was given a test, in which he of course did well. The counselor, however, advised him against mathematics, giving as his reason, that he had not stopped writing when he was told to do so. Schneider liked to explain that the stupidity of the man provoked him to take up mathematics. This detachment gave to his endeavours a perspective, and saved him, as a mathematician, of falling into the trap of being too 'hung-up' on his mathematics, as the sources of his self-esteem were not confined to his mathematical prowess, but were multi-varied, and hence unlikely to run dry. This allowed him a relaxed attitude, which many of his colleagues would have been well advised to envy.

His multi-varied interests and talents, gave to him an urbanity of manners, seldom recognised in mathematicians. (Some may argue that this social obtuseness among mathematicians is charming, others may point out that this is their only charm). He was equally at ease in almost any context, be it at a formal occasion with the President of a University, or at a dinner table with his colleagues, or sleeping attached to a mountain slope. He was well-rounded, an excellent cook, no doubt a connoiseure of wines, and to the delight of his friends and colleagues, a most entertaining table companion and conversationalist.

A man is certainly more than the sum of his parts, and while this account will narrowly focus on his climbing adventures, just as the more professional account by Peternell will limit itself to mathematics; neither together, nor let alone by themselves, will do justice to his personality. The reason to interject this extra-curricular account is to introduce the mathematical reader to another side of his life, a side of which they would certainly have no coherent picture, probably no suspicion of its true scope, maybe not even being aware of at all. Although the connection of climbing and mathematics is not unusual, some even sarcastically attributing the dual passion to the same source, - that of the irresistable challenge of goals, as pure and difficult as they are ultimately futile: one should beware of drawing facile parallels or let alone morals, as the two occupations are basically disjoint, but of course not incompatible.

Schneider was born in Munich in May 1942. One certainly could have picked a more propitious time to be born. Hence he grew up during the materially austere immediate post-war years of Germany, fatherless to boot, as his father had died in the War, when Schneider was only two. But this is a fate which he shared with many of his contemporaries, and one should perhaps not make too much of it. His accidental position in time set him apart from his younger colleagues, who grew up in more indulgent times, and gave to his life-span a certain depth, deprived others. Thus he was already a teen-ager in the mid fifties, and found himself roaming around Italy on a bicycle. This was in retrospect an idyllic time, with so few cars on the roads. He remembered fondly that the Italians were so hospitable, and treated him so well. It is hard not to see this as an early instance of exercising his great natural charm. Anyway more than anything else it showed his initiative and love of adventure, not necessarily of the reckless kind, and presaged his later world-wide travels, which due to the accident of the timing of his birth, could take place, before such excursions became common-place.

His passion for the mountains was kindled as a student at an Internat at Berchtesgaden, up in the Alps, just at the Austrian border. Here he quickly displayed a natural attitude for the sport, as well as an obsession almost to the detriment of his school-work. Thus concomitantly with graduating academically (doing his Abitur) he had matriculated as an expert climber as well. (And incidentally, on his mothers advise, concluded an apprenticeship as an electrician)

During his initial ten years as a climber he had explored all the classical challenges of the Alps, a list of which would be too long to include here. Among the most spectacular climbs being the ascents of the north faces of Eiger and Matterhorn. Northfaces (at least on the northern hemisphere) being singled out by their added inclemency of weather. Such ascents are too high and steep to normally be executed in a single day, and thus involve one or several nights of bivouacking, i.e. spending the night sleeping attached to the face. But a serious Alpinist needs further challenges, and the natural places to look for such elsewhere are in the Andes and the Himalayas, where the mountains are far less well-trodden, and far higher and wilder to boot.

In May 1966 we find Schneider aboard the ship 'Verdi', which had left Genua in the beginning of the month, crossing the Atlantic, squeezing through the Panama Canal, and about to dock in Callao, the port of Lima, at the end of the month. A trip which normally would by itself be a most memorable experience. Predictably the Peruvian custom officials caused troubles at the unloading of 2.5 tonnes of equipment, but by early June the matters had been resolved, and the climbing gear could be loaded on a truck heading for the mountains of Cordillera Vilcanota in southern Peru. What now followed was a Bonanza for the young expedition of alpine enthusiasts, who during the course of some six weeks 'bagged' some 24 peaks, 13 of which were actual first climbs. Pictures of the expedition show young men in sunglasses and widebrimmed sombreros against a backdrop of rugged mountain faces and towering snow mounds.

A typical climb involved the three neighbouring peaks of Jatunhuma (6142 m), two of which had never been previously ascended. Schneider and his companions decided in youthful exuberance to climb all three in a single go, dispensing with bivouacking. They set out in darkness early in the morning on June 28, and found themselves soon ahead of their schedule. But complications eventually arose. A freezing wind started to blow, it was too cold to stop and rest. In the words of Schneider himself:

Der Abstiegweg ist nicht klar zu erkennen, riesige Spalten und Schründe zwingen zu Zickzackwegen, immer wieder kleine steile Blankeiswände, die uns zu Sicherung und Aufmerksamkeit zwingen. Unser Zeitplan gerät ins Wanken, die Sonne verschwindet hinter dem Auzangate [another peak]. Bei beginnender Dämmerung sind wir auf einem grossen Plateau, das durch ein Spaltengewirr von Gletscher getrennt wird. Nach einigen erfolglosen Versuchen, einen geeigneten Durchschlupf zu finden, scheint es so, als ob wir doch biwakieren müssen. Wir sind müde, aber Hans [-Albert Mayer] gelingt es doch noch überraschend einen Weg durch gewaltige Spalten zu finden. Wir sind fast froh, daß es dunkel ist, so erscheint uns alles weniger gefährlich.

[The descent is not easy to find. Giant crevices and precipees force us to zig-zag. Once and again steep walls of exposed ice force us to close ranks and pay close attention. Our time-schedule goes out of the window, the sun disappears behind the peak of Auzangate. As dusk settles, we find ourselves on a large plateau, which is cut through by glacial crevices. After a few unsuccessful attempts to find a way through, it seems as that we have to bivouack after all. We are tired, but then surprisingly Hans-Albert Meyer is able nevertheless to find a way through the giant crevices. We are almost happy that it is dark, as then everything appears to us less terrifying.]

And so they find themselves in the end happy and contended lying with their backs on the glacier, gazing at the stars in the pitchdark subtropical night, soon to descend to their camp.

Three years later Schneider is sweltering in the summer heat of Moscow, on his way to the mountains of Pamir. A week later he writes from the base camp (3500 m) below Pik Lenin. Getting there had been an interesting trip just by itself. Flying from Moscow in a propeller plane, touching down in Osh, and then continuing by bus with an overnight stay at a Spa, known for its sulphuric waters. The final leg had been quite an ordeal, holding on for dear life on a truck bouncing up and down a dusty road littered with potholes, at the end of which he admits to total exhaustion, covered from head to foot with dust. Yet, a week later on, he is impatient to get climbing, feeling quite well acclamatised, in spite of digestional problems caused by the unfamiliar food. He admits though that the scenery is fantastically beautiful, and writes about flowery meadows interspersed with lakes, and of Kirgisian Nomads camping in the neighbourhood. He describes a visit to one of their round 'tents', with its floor covered with beautiful rugs; enjoying their hospitality, drinking tea out of exquisite porcelain cups, and being treated to tender lamb.

But the continuation was not to be as idyllic. The climbing of Pik Lenin (7134 m) would turn out to be the most extreme of all his adventures. After having reached the summit along a daring and unprecedented northern route, his climbing companion, the well-known alpinist Toni Hiebeler, was stricken with altitude sickness. Rather than to descend to his own safety, Schneider decided to put his own life at very serious risk, in order to lend assistance to his friend. It meant bivouacking for five days, just below the peak, without the benefit of supplemental oxygen, holding out until help finally arrived. It is safe to assume that had he not gallantly sacrificed himself, Hiebeler would have perished. Needless to say the experience was a very harrowing one.

The incident was an eye-opener to Schneider, who realised that he would have to choose between a career as a high altitude mountaineer, and that was a serious option, as he was in his hey-day considered one of the foremost alpinists in Germany; and his burgeoning career as a mathematician. (He had at that time just finished his Ph.D. and obtained a position as an assistant at Regensburg.) Not being a dare-devil, or rather not a mono-maniac, he chose the latter, and did from then on totally forego further high-altitude climbs, and in particular never went to the high altitudes of Himalayas, which otherwise would have been a logical step. It did basically mean that he altogether stopped embarking on ambitious climbing expeditions with one notable exception. A few years later in 1974, he went on another expedition, this time to the Caucasus, more specifically Uschba, climbing Mt Elbrus.

In 1965 he had met Angelika von Johnson, and they married in 1970. She joined him on many hiking and climbing expeditions, including one to Kilimanjaro. The most notably though of those, took place in Nepal in the spring of 1973, after an extended stay at the Tata Institute. They hired porters in Jumla, and covered some 300 km in toto, walking ten hours at day, sleeping in their own tent at nights. Part of the trek went through Tibet, which at the time was off-limits to tourists. However she eventually had to reluctantly forego her participations in her husbands climbs, as in 1974 their first son Florian was born, and a year and a half later their second son Martin. Soon thereafter Schneider also finished his apprenticeship at Regensburg and become a professor at Göttingen. But it did not mean that he was giving up climbing. From then on he pursued his climbing as a passionate hobby, not to set records, nor to subject himself to extraneous risks, but just simply to enjoy. And there were plenty of opportunities, abbetted by the mobility accorded a succesful mathematician. His favourite climbing involved, smooth, firm and reliable rock. He was naturally drawn to the pursuit of so called sportsclimbing, which fitted him well, both temperamentally and physically. With his intellectual attitude to climbing, combined with his natural agility, he was able to excel and up to the very end succeed in mastering difficulties up to the eighth grade, as he did just four weeks prior to his fall, climbing by the 'Wilden Kaiser'. He kept records of his climbs, and in a notebook of his from 1991, which probably would be typical of his later years, one may see on almost every week, one or two days, the records of a half a dozen of climbs or so, rated by $7, 7^+, 8^-$ etc refering to a standard classification, well-known to all climbers.

One characteristic of his climbing was the emphasis he put on planning, strategically thinking things through, and also of always putting the highest priority on looking out for others, and as a consequence his climbing expeditions were singled out by their unusual margin of safety. No climber associated with him ever reported that in their success, the element of luck and good fortune, had ever intervened. His concern for climbing safety was present from the start, and already in 1969 he was one of the founders of the '*Sicherheitskreis*' of DAV (Deutsche Alpen Verein). It did not mean that he was entirely spared mishaps. In 1987 he suffered an accident at Gorges du Verdun, breaking a leg; and although he always made light of it, claiming that it had been too trivial to be properly medically managed, it did hamper him, and in particular prevent him from taking long hikes. (Although it did not seem to have prevented him from taking part in cross-country skiing races, an activity he picked up when he moved to Bayreuth; the tangible results of which were stacked away in his basement) From then on, he had to rely increasingly on the bicycle to get the work-outs he craved.

But climbing, no matter how conscientiously performed, involves serious risks, even fatal ones, and it would certainly be hypocrisy to claim otherwise. Admittedly death is always a component of life, but in climbing its presence is particularly palpable. In fact death gives to climbing its ultimate syntax, rendering mistakes absolutely unpardonable. This is particularly true for the kind of adventourous climbing that Schneider sought out in his youth; but the possibility of disaster can never be dismissed in any type of technical climbing, no matter how favourable the external circumstances may be.

To Schneider the dangers of climbing were not a generalized abstract awareness. Two of his best friends had succumbed to fatal accidents. Once during his youth when Harald Seibold, with whom he had started out to climb in Berchtesgaden, died, only twenty four years old. And once again in his middle age when Gerhard Rebitzer, his best friend of his Bayreuth years, perished during a climb in Peru, incidentally about the same time Schneider himself fell and broke a leg. It is not easy to know exactly how this affected him, and his continued attitude to climbing. It is reported that he hoped that death would be swift and preferably in the mountains, and he was once quoted as saying that *Wenn es morgen zu Ende sein sollte, dann hat es sich bis heute jeder Tag gelohnt (If the end has to be tomorrow, then every day up to now has been worthwhile)*. This might be interpreted that he was granted to die, doing what he most of all enjoyed. However one should be wary of drawing such sentimental conclusions. Most of us profess to prefer a swift death; but once we are confronted with examples thereof, we recoil in horror from their pitiless brutality, just as we shudder at the contemplation of the final moments of Schneiders life.

And although we may nevertheless grant that Schneider fatalistically had come to terms with the inherent dangers of climbing, this certainly never was the case with his wife and sons. And why should it have been? It was certainly easier for him, who in each particular instance was feeling confident and in control, than for his family, who did not have this option of active engagement. To them, his continued persistence of repeatedly putting his life at risk, was a source of unremitting anguish, concluded only by its tragic vindication.

Yet in spite of its tragic, but not inevitable, conclusion: we should not repress his at times spectacular climbing-career. It certainly meant a lot to him, it furnishes us with fascinating stories, and it expresses his dynamic, vivacious personality.

Acknowledgment. The writing of this text would not have been possible but for the assistance of two people. Foremost that of his widow Angelika Schneider, who not only gave me permission to undertake this initiative, painful as the connotations must have been, but also most graciously lent me material, including notebooks and personal letters of Michael's, to which this short note has not been able to do full justice. I also would like to thank his climbing friend Herwig Sedlmayer, whose climbing obituary, published through the DAV, was of crucial importance, and out of which I have taken the liberty to crib entire portions.

Lectures of the Symposium at Bayreuth, June 11–13, 1998

André Hirschowitz: The scientific work of Michael Schneider

Yum-Tong Siu: Deformational invariance of plurigenera

Christian Okonek: Seiberg–Witten theory and complex geometry

Otto Forster: Complete intersections

Jean-Pierre Demailly: Almost complex projective embeddings of compact symplectic real manifolds

Fabrizio Catanese: A topological characterization of constant moduli fibrations

Andrew Sommese: Reducible hyperplane sections

Ciro Ciliberto: New results on linear systems of plane curves

List of Authors and Participants

Alberto ALZATI, Dipartimento die Matematica, Università di Milano, Via C. Saldini 50, 20133 Milano, Italy.
E-mail address: `alzati@umimat.mat.unimi.it`

Vincenzo ANCONA, Dipartimento die Matematica, Università di Firenze, Viale Morgagni 67/A, 50134 Firenze, Italy.
E-mail address: `ancona@udini.math.unifi.it`

Lucian BADESCU, Department of Mathematics, University of Bucharest, and Institute of Mathematics, Romanian Academy, P.O. Box 1-764, 70700 Bucharest, Romania.
E-mail address: `lbadescu@stoilow.imar.ro`

Daniel BARLET, Institut Elie Cartan, Université Henri Poincaré Nancy I, UMR 7502 CNRS - INRIA - UHP, BP 239, 54506 Vandoeuvre-lès-Nancy Cédex, France.
E-mail address: `laboratoire@iecn.u-nancy.fr`

Wolf P. BARTH, Mathematisches Institut der Universität Erlangen, Bismarckstr. 1 1/2, 91054 Erlangen, Germany.
E-mail address: `barth@mi.uni-erlangen.de`

Gottfried BARTHEL, Fakultät für Mathematik und Informatik, Universität Konstanz, Postfach 5560, 78434 Konstanz, Germany.
E-mail address: `Gottfried.Barthel@uni-konstanz.de`

Thomas BAUER, Mathematisches Institut, Universität Bayreuth, 95440 Bayreuth, Germany.
E-mail address: `thomas.bauer@uni-bayreuth.de`

Arnaud BEAUVILLE, DMA-École Normale Supérieure, 45 rue d'Ulm, 75230 Paris Cédex C5, France.
E-mail address: `beauville@dmi.ens.fr`

Mauro C. BELTRAMETTI, Dipartimento di Matematica, Università di Genova, Via Dodecaneso 35, 16146 Genova, Italy.
E-mail address: `beltrame@dima.unige.it`

Idranil BISWAS, School of Mathematics, Tata Institute of Fundamental Research, Homi Bhabha Road, Bombay 400005, India.
E-mail address: `idranil@math.tifr.res.in`

Hans-Christian von BOTHMER, Mathematisches Institut, Universität Bayreuth, 95440 Bayreuth, Germany.
E-mail address: `bothmer@btm8x5.mat.uni-bayreuth.de`

Robert BRAUN, Mathematisches Institut, Universität Bayreuth, 95440 Bayreuth, Germany.
E-mail address: `braun@btm8x5.mat.uni-bayreuth.de`

Mark de CATALDO, Department of Mathematics, Harvard University, Cambridge, Massachusetts 02138-2901, U.S.A.
E-mail address: `mde@math.harvard.edu`

Fabrizio CATANESE, Mathematisches Institut, Universität Göttingen, Bunsenstr. 3–5, 37073 Göttingen, Germany.
E-mail address: `catanese@uni-math.gwdg.de`

Ciro CILIBERTO, Dipartimento di Matematica, Università di Roma "Tor Vergata", Viale della Ricerca Scientifica, 00133 Roma, Italy.
E-mail address: `ciliberto@axp.mat.uniroma2.it`

Jean-Pierre DEMAILLY, Institut Fourier, Laboratoire de Mathématiques, Université de Grenoble I, URA 188 du CNRS, BP 74, 38402 Saint-Martin d' Hères, France.
E-mail address: `demailly@fourier.ujf-grenoble.fr`

Klas DIEDERICH, Fachbereich Mathematik, Universität Wuppertal, Gaußstr. 20, 42097 Wuppertal, Germany.
E-mail address: `klas.diederich@math.uni-wuppertal.de`

Thomas ECKL, Mathematisches Institut, Universität Bayreuth, 95440 Bayreuth, Germany.
E-mail address: `eckl@btm8x5.mat.uni-bayreuth.de`

Lawrence EIN, Department of Mathematics, University of Illinois at Chicago, 841 South Morgan St., M/C 249, Chicago, Illinois 60607-7045, U.S.A.
E-mail address: `ein@uic.edu`

Geir ELLINGSRUD, Matematisk Institutt, Postboks 1053, Blindern, 0316 Oslo, Norway.
E-mail address: `ellingsrud@math.uio.no`

Detlev ELSNER, Mathematisches Institut, Universität Bayreuth, 95440 Bayreuth, Germany.
E-mail address: `elsner@btm8x5.mat.uni-bayreuth.de`

John-Erik FORNAESS, Department of Mathematics, University of Michigan, Ann Arbor, Michigan 48109, U.S.A.
E-mail address: `fornaess@math.isa.umich.edu`

Otto FORSTER, Mathematisches Institut, Universität München, Theresienstr. 39, 80333 München, Germany.
E-mail address: `forster@rz.uni-muenchen.de`

David GARBER, Department of Mathematics and Computer Sciences, Bar-Ilan University, 52900 Ramat Gan, Israel.
E-mail address: `garber@macs.biu.ac.il`

Helmut HAMM, Mathematisches Institut, Universität Münster, Einsteinstr. 62, 48149 Münster, Germany.
E-mail address: `hamm@math.uni-muenster.de`

Peter HEINZNER, Fakultät für Mathematik, Ruhr-Universität Bochum, Universitätsstr. 150, 44780 Bochum, Germany.
E-mail address: `heinzner@cplx.ruhr-uni-bochum.de`

André HIRSCHOWITZ, Laboratoire J.-A. Diendonne, Université de Nice, Sophia Antipolis, Parc Valrose, 06108 Nice Cédex 02, France.
E-mail address: `ah@math.unice.fr`

Alan HOWARD, Department of Mathematics, Notre Dame, Indiana 46556, U.S.A.
E-mail address: `howard.1@nd.edu`

Alan T. HUCKLEBERRY, Fakultät und Institut für Mathematik, Gebäude NA 4/73, Universität Bochum, 44780 Bochum, Germany.
E-mail address: `ahuck@cplx.ruhr-uni-bochum.de`

Klaus HULEK, Institut für Mathematik, Universität Hannover, 30060 Hannover, Germany.
E-mail address: `hulek@math.uni-hannover.de`

Bo ILIC, Department of Mathematics, U.C.L.A., Los Angeles, California 900095, U.S.A.
E-mail address: `ilic@math.ucla.edu`

Paltin IONESCU, Department of Mathematics, University of Bucharest, and Instiute of Mathematics, Romanian Academy, P.O. Box 1-764, 70700 Bucharest, Romania.
E-mail address: `pionescu@pompeiu.imar.ro`

Priska JAHNKE, Mathematisches Institut, Universität Bayreuth, 95440 Bayreuth, Germany.
E-mail address: `priska.jahnke@uni-bayreuth.de`

Ludger KAUP, Fakultät für Mathematik und Informatik, Universität Konstanz, Postfach 5560, 78434 Konstanz, Germany.
E-mail address: `ludger.kaup@uni-konstanz.de`

Stefan KEBEKUS, Mathematisches Institut, Universität Bayreuth, 95440 Bayreuth, Germany.
E-mail address: `kebekus@btm8x5.mat.uni-bayreuth.de`

Hans KERNER, Mathematisches Institut, Universität Bayreuth, 95440 Bayreuth, Germany.

János KOLLÁR, Department of Mathematics, University of Utah, Salt Lake City, Utah 84112, U.S.A.
E-mail address: `kollar@math.utah.edu`

Matthias KRECK, Mathematisches Forschungsinstitut Oberwolfach, Lorenzhof, 77709 Oberwolfach-Walke, Germany.
E-mail address: `kreck@mfo.de`

Herbert KURKE, Mathematisches Institut, Humboldt-Universität Berlin, PSF 1297, 10117 Berlin, Germany.
E-mail address: `kurke@mathematik.hu-berlin.de`

Herbert LANGE, Mathematisches Institut, Universität Erlangen, Bismarkstr. 1 1/2, 91054 Erlangen, Germany.
E-mail address: `lange@mi.uni-erlangen.de`

Antonio LANTERI, Dipartimento die Matematica, Università di Milano, Via C. Saldini 50, 20133 Milano, Italy.
E-mail address: `lanteri@umimat.mat.unimi.it`

Robert LAZARSFELD, Department of Mathematics, University of Michigan, Ann Arbor, Michigan 48109, U.S.A.
E-mail address: `lazarsfeld@math.isa.umich.edu`

Angelo Felice LOPEZ, Dipartimento di Matematica, Università di Roma Tre, Largo San Leonardo Murialdo 1, 00146 Roma, Italy.
E-mail address: `lopez@matrm3.mat.uniroma3.it`

Martin LÜBKE, Mathematisch Instituut, Rijksuniversiteit Leiden, P.O. Box 9512, 2300 RA Leiden, The Netherlands.
E-mail address: `lubken@wi.leidenuniv.nl`

Jón MAGNUSSON, Science Institude, University of Iceland, Raunvisindastofnun Háskólans, Dunhaga 3, 107 Reykjavik, Iceland.
E-mail address: `jim@raunvis.hi.is`

Gerriet MARTENS, Mathematisches Institut, Universität Erlangen, Bismarckstr. 1 1/2, 91054 Erlangen, Germany.
E-mail address: `martens@mi.uni-erlangen.de`

Rick MIRANDA, Department of Mathematics, Colorado State University, Ft. Collins, Colorado 80523, U.S.A.
E-mail address: `miranda@math.colostate.edu`

Michael MÜDSAM, Mathematisches Institut (Mathematik I), Universität Bayreuth, 95440 Bayreuth, Germany.
Phone: +49 (0) 911 711067

Gerhard MÜLICH, Mathematisches Seminar, Universität Hamburg, Bundesstr. 55, 20146 Hamburg, Germany.
E-mail address: `muelich@math.uni-hamburg.de`

Uwe NAGEL, FB 17 - Mathematik - Informatik, Universität Paderborn, Warburger Str. 100, 33098 Paderborn, Germany.
E-mail address: `uwen@uni-paderborn.de`

Hans-Joachim NASTOLD, Mathematisches Institut, Universität Münster, Einsteinstr. 62, 48149 Münster, Germany.
E-mail address: `ernstin@math.uni-muenster.de`

Karl OELJEKLAUS, U.F.R. M.I.M. de l'Université de Provence Aix-Marseille I, Centre Mathématique et Informatique, 39, rue Julio-Curie (Ch. Gombert), 13453 Marseille Cédex 13, France.
E-mail address: `karloelj@gyptis.univ-mrs.fr`

Christian OKONEK, Institut für Mathematik, Universität Zürich, Winterthurerstr. 190, 8057 Zürich, Switzerland.
E-mail address: `okonek@math.unizh.ch`

Takeo OHSAWA, Graduate School of Mathematics, Nagoya University, Chijusa-ku, Nagoya 464-8602, Japan.
E-mail address:`ohsawa@math.nagoya-u.ac.jp`

Giorgio OTTAVIANNI, Dipartimento di Matematica, Università di Firenze, Viale Morgagni 67/A, 50134 Firenze, Italy.
E-mail address: `ottavian@udini.math.unifi.it`

Ulf PERSSON, Department of Mathematics, Chalmers University of Technology, 41296 Goteborg, Sweden.
E-mail address: `ulfp@math.chalmers.se`

Thomas PETERNELL, Mathematisches Institut, Universität Bayreuth, 95440 Bayreuth, Germany.
E-mail address: `thomas.peternell@uni-bayreuth.de`

Roberto PIGNATELLI, Dipartimento di Matematica, Università di Pisa, Via F. Buonarroti 2, 56127 Pisa, Italy.
E-mail address: `pignatel@dm.unipi.it`

Herbert POPP, Fakultät für Mathematik und Informatik, Universität Mannheim, 68131 Mannheim, Germany.
E-mail address: `popp@math.uni-mannheim.de`

Ivo RADLOFF, Mathematisches Institut, Universität Bayreuth, 95440 Bayreuth, Germany.
E-mail address: `ivo.radloff@uni-bayreuth.de`

Kristian RANESTAD, Matematisk Institutt, Postboks 1053, Blindern, 0316 Oslo, Norway.
E-mail address: `ranestad@math.uio.no`

Reinhold REMMERT, Mathematisches Institut, Universität Münster, Einsteinstr. 62, 48149 Münster, Germany.

Oswald RIEMENSCHNEIDER, Mathematisches Seminar, Universität Hamburg, Bundesstr. 55, 20146 Hamburg, Germany.
E-mail address: `riemenschneider@math.uni-hamburg.de`

Alessandra SARTI, Mathematisches Institut, Universität Erlangen, Bismarkstr. 1 1/2, 91054 Erlangen, Germany.
E-mail address: `sarti@mi.uni-erlangen.de`

Frank-Olaf SCHREYER, Mathematisches Institut, Universität Bayreuth, 95440 Bayreuth, Germany.
E-mail address: `schreyer@btm8x5.mat.uni-bayreuth.de`

Georg SCHUMACHER, Fachbereich Mathematik, Universität Marburg, 35032 Marburg, Germany.
E-mail address: schumac@mathematik.uni-marburg.de

Wolfgang K. SEILER, Fakultät für Mathematik und Informatik, Universität Mannheim, 68131 Mannheim, Germany.
E-mail address: seiler@math.uni-mannheim.de

Edoardo SERNESI, Dipartimento di Matematica, III Università di Roma, Via C. Segre 2, 00146 Roma, Italy.
E-mail address: sernesi@matrm3.mat.uniroma3.it

Yum-Tong SIU, Department of Mathematics, Harvard University, Cambridge, Massachusetts 02138-2901, U.S.A.
E-mail address: siu@math.harvard.edu

Andrew J. SOMMESE, Department of Mathematics, Notre Dame, Indiana 46556, U.S.A.
E-mail address: sommese@nd.edu

Jeroen SPANDAW, Institut für Mathematik, Universität Hannover, Postfach 6009, 30060 Hannover, Germany.
E-mail address: spandaw@math.uni-hannover.de

Robert SWITZER, Mathematisches Institut, Universität Göttingen, Bunsenstr. 3–5, 37073 Göttingen, Germany.
E-mail address: switzer@namu01.gwdg.de

Shen-Li TAN, Department of Mathematics, East China Normal University, Shanghai 200062, P.R. OF CHINA.
E-mail address: sltan@math.ecnu.edu.cn

Andrei TELEMAN, Institut für Mathematik, Universität Zürich, Winterthurerstr. 190, 8057 Zürich, Switzerland.
E-mail address: teleman@math.unizh.ch

Mina TEICHER, Department of Mathematics and Computer Sciences, Bar-Ilan University, 52900 Ramat Gan, Israel.
E-mail address: teicher@macs.biu.ac.il

Peter ULLRICH, Mathematisches Institut, Universität Münster, Einsteinstr. 62, 48149 Münster, Germany.
E-mail address: ullricP@math.uni-muenster.de

A. J. H. M. VAN DE VEN, Mathematisch Instituut, Rijksuniversiteit Leiden, P.O. Box 9512, 2300 RA Leiden, The Netherlands.
E-mail address: `ven@rulwinw.leidenuniv.nl`

Eckart VIEHWEG, FB 6 Mathematik, Universität Essen, 45117 Essen, Germany.
E-mail address: `viehweg@uni-essen.de`